Klaus Grasshoff
Klaus Kremling
Manfred Ehrhardt

Methods of
Seawater Analysis

 WILEY-VCH

Klaus Grasshoff/Klaus Kremling/
Manfred Ehrhardt

Methods of Seawater Analysis

Third, Completely Revised and Extended Edition

With Contribution by:

Leif G. Anderson, Meinrat O. Andreae, Brigitte Behrends,
Constant van den Berg, Lutz Brügmann, Kathryn A. Burns,
Gustave Cauwet, Jan C. Duinker, David Dyrssen, Manfred Ehrhardt,
Elisabet Fogelqvist, Stig Fonselius, Hans Peter Hansen,
Arne Körtzinger, Wolfgang Koeve, Folke Koroleff,
Klaus Kremling, Joachim Kuss, Gerd Liebezeit, Willard S. Moore,
Thomas J. Müller, Andreas Prange, Michiel Rutgers van der Loeff,
Martina Schirmacher, Detlef Schulz-Bull, Peter J. Statham,
David R. Turner, Günther Uher, Petra Wallerstein,
Margareta Wedborg, Peter J. le B. Williams, Bengt Yhlen

 WILEY-VCH

Weinheim · New York · Chichester · Brisbane · Singapore · Toronto

Prof. Dr. Klaus Grasshoff †
Institut für Meereskunde an der Universität Kiel
Düsternbrooker Weg 20
D-24105 Kiel

Dr. Klaus Kremling
Dr. Manfred Ehrhardt
Institut für Meereskunde an der Universität Kiel
Abt. Meereschemie
Düsternbrooker Weg 20
D-24105 Kiel

Cover Illustration: Dr. Arne Körtzinger

Library of Congress Card No.:

British Library Cataloguing-in-Publication Data:

Die Deutsche Bibliothek Cataloguing-in-Publication Data:
Methods of seawater analysis / ed. by Klaus Graßhoff ... With contributions by Leif Anderson; Constant van den Berg. – 3., completely rev. and extended ed. – Weinheim; New York; Chiester; Brisbane; Singapore; Toronto: Wiley-VCH, 1999
ISBN 3-527-29589-5

Composition: Kühn & Weyh, D-79111 Freiburg
Printing: Strauss Offsetdruck, D-69509 Mörlenbach
Bookbinding: Wilh. Osswald, D-67433 Neustadt

Printed in the Federal Republic of Germany.

Dedication

(.

In memoriam
Prof. Dr. Klaus Grasshoff
09. 06. 1932 – 11. 03. 1981

    With this third edition of Methods of Seawater Analysis we (K.K. and M.E.) should like to commemorate the late Klaus Grasshoff, our friend and colleague of many and yet too few years. It was he who, in the foundation years of the 1970s, saw the need to accumulate the analytical expertise of the then small number of marine chemists and to form a common experimental basis for generating reliable data on chemical variables in the sea. In this spirit as guiding principle we strive to cultivate the original seed.

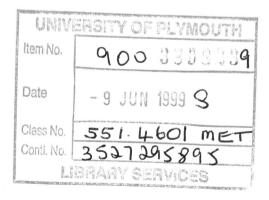

Preface to the third edition

Fifteen years have passed since the second edition of this book was published, a twinkling on the time scale of many oceanic processes, but an eaon if compared with the rapid evolution of analytical techniques needed to describe and measure them. The pace of development varied; it was most rapid among techniques still immature a decade and a half ago and more sedate among methods such as batchwise nutrient analyses and their automated versions that rightfully bear the epithet classical. Methods to determine trace elements as well as natural radioactive tracers and to analyse the complex assemblages of organic trace constituents of seawater, however, have grown in scope and considerably advanced in refinement. Thus, separate and sometimes new chapters are devoted to electrochemical methods, analysis by X-ray fluorescence, and on sampling and analysis of suspended particles. Clean, large scale sample collection at almost arbitrary depths is described providing sufficient material for organic trace analyses at the single compound level, in extreme cases made possible by multi-dimensional gas chromatography. Chapters on pH and total dissolved inorganic carbon determinations were thoroughly revised and expanded. Those interested in material exchange processes between ocean and atmosphere will find detailed instructions for measuring the partial pressure of dissolved carbon dioxide, concentrations of dimethyl sulphide, and how to analyse volatile halocarbons. HPLC analysis of photosynthetic pigments is another topic not dealt with in the previous editions. The book does not attempt to include every conceivable aspect of analytical marine chemistry; rather, its authors describe methods they themselves have introduced or refined and that consistently yielded reproducible results in everyday work.

Expanding the scope of the book was made possible thanks to the dedicated efforts of colleagues, who authored ground-breaking publications before yielding to our suggestions to make their abundant expertise available to those looking for advice. Substantial contributions were also made by the technical staff of the Department of Marine Chemistry, Institute for Marine Research at the University of Kiel, Germany. We are specially indebted to – in alphabetical order – Hergen Johannsen, Gert Petrick, and Peter Streu for inspired work and creative ideas.

Compiling a multi-authored book puts considerable strain on the typists. We would have been in dire straits without Ute Weidingers vast skills and unfailing enthusiasm, ably assisted by Annelore Paulsen. Ilona Oelrichs of the Graphic Arts Department spent uncounted hours reproducing the figures many of which needed the helping hands of Maike Heinitz and Reinhold Hellwig.

Kiel, November 1998

Klaus Kremling
Manfred Ehrhardt

Contents

8   Determination of total alkalinity and total dissolved inorganic carbon   127
           L.G. Anderson, D.R. Turner, M. Wedborg and D. Dyrssen

9   **Determination of carbon dioxide partial pressure ($p(CO_2)$)**   **149**
          A. Körtzinger

10   **Determination of nutrients**   **159**
          H.P. Hansen and F. Koroleff

22   Determination of selected organochlorine compounds in seawater   479
            J. C. Duinker and D. E. Schulz-Bull

24   Determination of dimethyl sulphide in seawater   521
            G. Uher

List of Contributors

Meinrat O. Andreae
Max-Planck-Institut für Chemie
Postfach 3060
55122 Mainz
Germany

Leif G. Anderson
Department of Analytical and Marine Chemistry
Göteborg University
41296 Göteborg
Sweden

Brigitte Behrends
Zentrum für Flachmeer-, Küsten- und
Meeresumweltforschung
Forschungszentrum Terramare
Schleusenstraße 1
26382 Wilhelmshaven
Germany

Constant M. G. van den Berg
Oceanographic Laboratory
University of Liverpool
Liverpool L69 3BX
United Kingdom

Lutz Brügmann
Hartkrögen 30
22559 Hamburg
Germany

Detlef E. Schulz-Bull
Institut für Meereskunde
an der Universität Kiel
Abteilung Meereschemie
Düsternbrooker Weg 20
24105 Kiel
Germany

Kathryn A. Burns
Australian Institute for Marine Science
P. O. Box 3
Townsville, Qld 4810
Australia

Gustave Cauwet
Observatoir Océanologique
Labortoire d'Océanographie Biologique
(CNRS-URA 2071)
BP 44
66651 Banyuls sur Mer
France

Jan C. Duinker
Institut für Meereskunde
an der Universität Kiel
Abteilung Meereschemie
Düsternbrooker Weg 20
24105 Kiel
Germany

David Dyrssen
Department of Analytical and Marine Chemistry
Göteborg University
41296 Göteborg
Sweden

Manfred G. Ehrhardt
Institut für Meereskunde
an der Universität Kiel
Abteilung Meereschemie
Düsternbrooker Weg 20
24105 Kiel
Germany

Elisabet Fogelqvist
Department of Chemistry
Analytical and Marine Chemistry
Göteborg University
S-41296 Göteborg
Sweden

Stig Fonselius
SMHI Oceanographical Laboratory
Nya Varvet 31
S-42671 Västra Frölunda
Sweden

Hans Peter Hansen
Institut für Meereskunde
an der Universität Kiel
Abteilung Meereschemie
Düsternbrooker Weg 20
24105 Kiel
Germany

Wolfgang Koeve
Institut für Meereskunde
an der Universität Kiel
Abteilung Planktologie
Düsternbrooker Weg 20
24105 Kiel
Germany

Folke Koroleff
Institute of Marine Research
P. O. Box 33
SF-00931 Helsinki 93
Finland

Arne Körtzinger
Institut für Meereskunde
an der Universität Kiel
Abteilung Meereschemie
Düsternbrooker Weg 20
24105 Kiel
Germany

Klaus Kremling
Institut für Meereskunde
an der Universität Kiel
Abteilung Meereschemie
Düsternbrooker Weg 20
24105 Kiel
Germany

Joachim Kuss
Institut für Meereskunde
an der Universität Kiel
Abteilung Meereschemie
Düsternbrooker Weg 20
24105 Kiel
Germany

Gerd Liebezeit
Zentrum für Flachmeer-, Küsten- und
Meeresumweltforschung
Forschungszentrum Terramare
Schleusenstraße 1
26382 Wilhelmshaven
Germany

Michiel Ruttgers van der Loeff
Alfred-Wegener-Institut für Polar- und
Meeresforschung
Postfach 120161
27568 Bremerhaven
Germany

Willard S. Moore
Geology Department
University of South Carolina
Columbia SC 29208
U.S.A.

Thomas J. Müller
Institut für Meereskunde
an der Universität Kiel
Abteilung Meeresphysik
Düsternbrooker Weg 20
24105 Kiel
Germany

Andreas Prange
GKSS-Forschungszentrum
Postfach 1160
21494 Geesthach
Germany

Martina Schirmacher
GKSS-Forschungszentrum
Postfach 1160
21494 Geesthach
Germany

Peter J. Statham
University of Southampton
Department of Oceanography
Waterfront Campus, European Way
Southampton SO14 3ZH
United Kingdom

David R. Turner
Department of Analytical and
Marine Chemistry
Göteborg University
41296 Göteborg
Sweden

Günther Uher
University of Newcastle upon Tyne
Department of Marine Sciences and
Coastal Management
Ridley Building (Claremont Road)
Newcastle upon Tyne NE1 7RU
United Kingdom

Bengt Yhlen
SMHI Oceanographical Laboratory
Nya Vervet 31
42671 Västra Frölunda
Sweden

Petra Wallerstein
Zentrum für Flachmeer-, Küsten- und
Meeresumweltforschung
Forschungszentrum Terramare
Schleusenstraße 1
26382 Wilhelmshaven
Germany

Margareta Wedborg
Department of Analytical and
Marine Chemistry
Göteborg University
41296 Göteborg
Sweden

Peter J. le B. Williams
School of Ocean Sciences
Menai Bridge
Gwynedd LL59 5EY
United Kingdom

1 Sampling

L. Brügmann and K. Kremling

1.1 Introduction

An analytical chemist should be able to produce correct results. However, no chain is stronger than its weakest link, and this axiom also applies to the quantitative determination of seawater components. The quality of data obtained from seawater studies is related to three principal factors: (a) the design and performance of a representative sampling programme; (b) the adoption of analytical protocols which enable measurements of appropriate accuracy and precision (see Chapters 3-27); and (c) if necessary, the selection and use of suitable storage procedures for samples to minimize changes in analyte concentrations and speciation prior to further treatment in the laboratory (see Chapter 2). Unless these factors are given adequate consideration before and during a study, its aims will not be achieved and valuable efforts and facilities will be wasted. Sampling is not only the first but also often the most critical step in seawater analysis.

The dissolved constituents of seawater are grouped into two categories, the major or conservative and the trace components. By definition, the conservative constituents are not influenced significantly by biological processes, and the time required for concentration changes due to chemical and geochemical processes is very long. Therefore, their distribution is controlled mainly by the physical processes of advection and convection, turbulence diffusion, *etc.*

The trace constituents, in addition to being affected by physical processes, are also influenced by the biological processes of uptake, excretion and biodegradation. Furthermore, physico-chemical exchange reactions take place at surfaces of lithogenic and organic particles and at boundaries, such as air-sea or water-sediment interfaces. Rapid chemical reactions also frequently occur in transition layers between oxic and anoxic environments. Other types of chemical reactions are the pressure and temperature dependent readjustments of equilibria. Therefore, the optimal approach to a particular sampling problem requires an understanding of the environment and the wide variety of physical, chemical and biological processes that influence the distribution of the analyte in question; the marine analytical chemist must, therefore, have a reasonable knowledge of oceanography.

The topic of sampling will be treated in four sections, the first one discussing adequate sampling strategies. This is followed by the main section in which a selection of samplers serving different purposes is described. Analytical errors to be considered especially during sampling are then critically discussed. Finally, a brief overview is presented of the quality control measures necessary within the analytical laboratory.

1.2   Sampling strategy

The sampling strategy is primarily controlled by the objectives of the investigation and by the expected or known spatial and temporal variability of the analyte concentrations in the study area. Based on this information, an adequate sampling scheme is developed outlining the station grid, vertical resolution and frequency of sampling. Very often, outlined programmes have to be reduced due to constraints in available resources (*e.g.*, shiptime, cost of laboratory analyses) or by the fact that the available sampling equipment will not adequately serve the purpose of the intended studies. However, even a curtailed programme must be optimized to guarantee acquisition of valid data.

During field work, the marine analytical chemist must be aware that the constituents dissolved or dispersed in the sea vary from place to place, with depth and with time (season) because of physical and biogeochemical processes. On the other hand, samples taken from the ship should, as far as possible, represent the conditions in a water body or at a given geographical location. The 'true' description of the momentary pattern of three-dimensional distribution can, however, merely be approximated by narrow-spaced sampling under quasi-synoptic conditions. With respect to chemical trend monitoring programmes, for example, sampling should be concentrated in winter times when the analyte concentrations are less affected by biogeochemical processes.

For many applications the so-called standard depths are sampled (Table 1-1), as recommended in 1936 by the International Association for the Physical Sciences of the Ocean (IAPSO) and recently confirmed by *UNESCO* (1991). In practical work, however, these depths can only be approximated because of the wire angle, yawing of the vessel and distortion of the curvature of the wire by the ship's drift and ocean currents.

Table 1-1. IAPSO standard depth levels (in decibars; *Sverdrup et al.,* 1942). In addition, the values listed in parentheses were added in keeping with *Levitus* (1982).

1	(125)	600	(1300)	3500	7000
10	150	(700)	(1400)	4000	7500
20	(200)	800	1500	4500	8000
30	(250)	(900)	(1750)	5000	8500
50	300	1000	2000	5500	9000
75	400	(1100)	2500	6000	9500
100	500	1200	3000	6500	10000

In practiced field work, selection of the sampling sites and depths often depends not only on the demands of the chemists, but also on those of hydrographers, biologists and/or scientists from other disciplines. So, the chemists often have to share water samples with other groups. Obviously, sampling at standard depths is unsuitable if specific problems are to be studied, such as near bottom gradients, the chemistry of the oxygen minimum layer or the conditions in discontinuity layers.

When planning the sampling network, one also has to take into account that the concentrations and the variability of the analytes may increase considerably when approaching continental shelf edge areas or input sources such as river estuaries or coastal urban and industrial centres. Then the programmes have to compensate for this higher variability by denser networks and increased sampling frequencies. Other sampling problems may arise due to enrichment of materials at the air-sea interface. Under conditions of low turbulent mixing, high organic phytoplankton production and/or high atmospheric deposition, fairly stable surface microlayers are formed at the sea-atmosphere interface. These microlayers exhibit particular biochemical environments characterized *inter alia* by enrichment of trace inorganic and organic constituents including microorganisms by orders of magnitude as compared with the composition of subsurface waters (*Liss and Duce,* 1997). Specially designed samplers have been constructed to sample the microlayer (see Section 1.3.1). For subsurface sampling, on the other hand, these layers may create problems. Many substances accumulated in the surface microlayer are either adhesive themselves or are associated with surface active compounds. Samplers passing through the surface of the sea open may adsorb such substances and later release them into the enclosed samples; much higher analyte concentrations may then be found than really exist at the investigated depth. Samplers passing through the surface of the sea closed will minimize this error risk.

When concentrations of the compounds under study differ greatly between samples taken at different depths from the same station or at a certain area, prevention of memory effects due to adsorption or insufficient flushing of samplers used may become all important for producing accurate data. Simple approaches might help, such as starting sampling at depths where the lowest concentrations are expected or intensive washing and re-conditioning of samplers between repeated uses.

During multi-disciplinary cruises, standard hydrographic or hydrochemical variables such as salinity, temperature, oxygen and nutrient concentrations are commonly recorded directly by CTD sensors (see Chapter 3) and/or determined in aliquots of the same sample collected with multi-purpose rosette samplers. The determination of trace constituents, however, requires samplers specifically designed for the collection of single constituents or groups of substances (see next section). Using aliquots of these specific samples is recommended for the determination of some standard parameters (*e.g.*, salinity, nutrients), to prove the identity of selected sample depths or water layers and to aid in the interpretation of data of the specific constituents. This is less important in deep-ocean waters, but is of great relevance in continental shelf areas with significant variations of water masses in space and time.

1.3 Sampling techniques

In the following paragraphs different sampling techniques are discussed. They include the description of some typical equipment for the collection of discrete seawater samples at various depths as well as for continuous sampling using pumping systems. It must be stressed, however, that a comprehensive description of the large number of commercially available water samplers or sampling systems is far beyond the scope of this book. The selection of the techniques outlined here focuses on effective and robust sampling methods commonly applied within the marine scientific community, or on specific equipment and procedures developed or used successfully by us. Furthermore, reference is made to a number of specific sampling methods applied by other workers and outlined in detail in those chapters of this book where the determination of the respective constituents is described. We have also included a section on procedures for the collection of marine particles, recognizing the increasing interest of chemical oceanographers in this type of sample (see Section 1.3.5, Chapter 2 and Section 12.6).

1.3.1   Surface water sampling

Collection of undisturbed discrete surface seawater samples from research vessels is not possible. To leave the contaminated water plume of the main vessel and to approach closer to the sea surface, suitable small tenders have to be used. Sampling is performed from the lee-side of the tender, > 200 m up-wind from the ship in an area not previously passed by any vessel. During sampling, the tender may be moved slowly at right angles to the direction of the prevailing surface currents.

Surface samples (≥ 0.2 m) are taken with bottles attached to the top of 2–4 m long telescopic bars of non-contaminating material. Even with the use of arm-long gloves, hand sampling involves certain contamination risks. From the tender, water samples from depths of between 0.5 and 10 m, depending on the draught of the main vessel, may be taken manually with suitable samplers attached to appropriate ropes. This is especially important for analytes where the ship may constitute a significant source of contamination. Figure 1-1 shows an example of a surface skimmer type oil sampler. It consists of a 250 mL Pyrex bottle attached at right angles to a cylindrical float. The sampler is pulled with a line along the water surface and turns through 90° into the vertical position once the sampler is filled.

For surface sampling with pumping systems at fixed positions, a buoy placed at some distance from the main ship may be used. From the buoy, water is drawn *via* polytetrafluoroethylene (PTFE) tubes to a pump on the vessel. Sampling is often combined with in-line filtration. A successfully tested system which delivers about 2 L/min of sample to a working height of 2 m above the sea surface with an air supply pressure of 4 bar has been described by *Harper* (1987); see also *Tokar et al.* (1981).

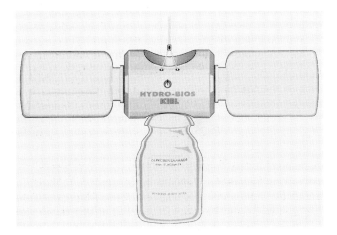

Fig. 1-1. Surf bottle in sampling position (*Hydrobios*, 1997; oil sampler according to Schomaker).

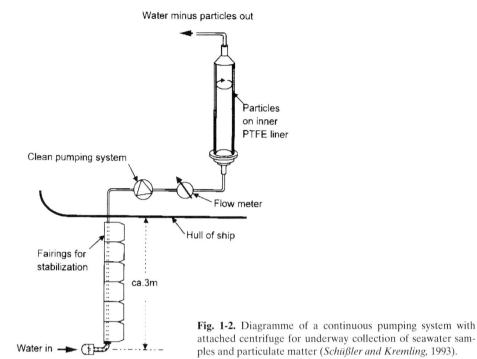

Fig. 1-2. Diagramme of a continuous pumping system with attached centrifuge for underway collection of seawater samples and particulate matter (*Schüßler and Kremling*, 1993).

A pumping system for sampling when underway, which has been used extensively for simultaneous collection of inorganic and organic trace constituents in open-ocean and continental shelf surface waters, has been described by *Schüßler and Kremling* (1993). In this system (Fig. 1-2) a chemically inert air-driven membrane pump moves water from below the ship through a polyethylene (PE) tube mounted inside a stainless-steel tube extending around 3 m beneath the keel through the ship's 'moon pool'. The seawater is in contact only with PE, polypropylene, Teflon and titanium. Seawater flow rates range up to $1.3 \, \text{m}^3/\text{h}$ providing high resolution for horizontally integrated signals. During operation, the speed of the vessel should not be less than 6 knots to minimize the effects of the ship's hull on sample composition. Owing to the high flow rates, several subsampling facilities for different purposes can be fed simultaneously, *i.e.,* the collection of large amounts of particles by application of a continuous-flow centrifuge (see Section 1.3.5 and Chapter 12, Section 12.6).

The pumping system can be used on almost all research ships, provided the vessel is equipped with a 'moon pool' and a specific mounting plate. It combines the advantages of relatively low ship costs per sample (when passing from station to station) with the possibility of interdisciplinary studies.

Chemical oceanographers often are interested in the composition of the 'sea-surface microlayer' and related processes. This interfacial boundary (in the µm to mm range) between the bulk ocean water and the atmosphere not only acts to mediate the air-sea exchange processes, both globally and regionally, but also accumulates a large number of identified and unknown organic substances many of which are surface active (*e.g., Liss and Duce,* 1997). However, taking measurements in the sea-surface microlayer is not an easy

Fig. 1-3. Stainless-steel net (right side) and nylon net in a frame made of acrylesters for sampling of trace organic compounds and trace elements, respectively, in the surface microlayer (*Brügmann et al.,* 1985).

task, since the immersion of the measuring or sampling device may alter the conditions in an indeterminate way.

Therefore, in practice the term 'sea-surface microlayer' has been defined as that thin layer of water that adheres to sampling devices, such as wiremesh screens (Fig. 1-3), glass plates, Teflon disks and rotating drums. For a comparison and discussion of the large variety of different surface film samplers with regard to their sampling efficiencies reference is made to *Van Vleet and Williams* (1980), *Hühnerfuss* (1981) or *Hardy et al.* (1988).

1.3.2   Water samplers for major hydrochemical variables

For recording vertical profiles of basic hydrochemical pararmeters in the sea (such as salinity, pH, oxygen and nutrient concentrations), water samples from different depths should be taken quasi-synchronously. On a number of ships, the famous *Nansen* bottle (Fig. 1-4) introduced more than 70 years ago (*Knudsen,* 1923, 1929) is still part of standard oceanographic equipment. Even on modern research vessels it is often carried for comparison purposes and as a back-up.

The original *Nansen* bottle was made of brass, suitable for use at all water depths, robust and safe to handle even under rough weather conditions. A large number of these samplers of about 1–3 L volume can be fastened to hydrowires 4–8 mm in diameter at a specified distance from each other. Reversing thermometers (2–4 in number), protected as well as unprotected against the water pressure, are attached to the samplers. The samplers are supplied with messengers and lowered open, by a winch, to the desired depth. After about 10 min, which is necessary for conditioning, *i.e.,* proper flushing and temperature equilibration, the hydrocast is released by a messenger sent down the wire. An upper spring clip is activated which permits the sampler to turn around its lower fastening screw by a maximum of 180°. As it swings round, the sampler is closed and the next messenger released. By overturning the thermometers, the mercury in the capillary is fixed. Based on repeated careful calibrations in pressure tanks, the difference between temperatures

Fig. 1-4. Transparent plastic *Nansen* (*TPN*) water sampler (*Hydrobios*, 1997). The body of the 1.7 L *TPN* sampler is a volume-graduated PC tube.

recorded with protected and unprotected thermometers is used as a relatively precise depth indicator.

The use of *Nansen* bottles has some disadvantages, one being based on their construction. The closing mechanism impairs flushing of the sampler due to 'dead volumes' and turbulent mixing. Modern samplers of the *Nansen* type are usually made of plastic materials such as polycarbonate (PC) and poly (vinyl chloride) (PVC) instead of brass (Table 1-2). Such samplers are less robust, but lighter, they are non-corrosive, greasing of the valves becomes unnecessary and interference with sample composition is minimized.

Table 1-2 Selected seawater samplers.

Name	Characteristics	Volume L	Mass kg[a]	Range m	Analyte[b]	Firm
Shallow-water samplers	*simple construction, limited depth of operation, manual operation from small vessels possible*					
MICROS	stainless-steel frame, glass bottle fixed in PTFE, sealed by silicone rubber tubes during lowering, not usable in sampling series, sterilizable	0.5		≤ 100		HB
LIMNOS	PVC frame, glass bottle, sealed by silicone rubber tubes during lowering, not usable in sampling series	2 × 1		≤ 30		HB
MERCOS	Ti frame, PTFE bottles, sealed by silicone rubber tubes during lowering, usable in sampling series, sterilizable	2 × 0.5		≤ 100	TM	HB
Serial samplers	*as part of a sampling system (plus winch, hydrowire, bottom weight) used from ships, optionally supplied with reversing thermometers, triggered by messengers*					
PWS	plastic water sampler, PVC tube, internal rubber spring, suitable for use with *MWS* (see below)	1.0–30		ul	Std.	HB
COC	COC water sampler, PVC tube, external metal spring and pressure cylinder which open sampler at about 10 m depth, good flushing characteristics	1.7–10		ul	Std.	HB
UWS	universal water sampler, tube made of piacryl or opaque fiber glass, good flushing characteristics	3.5, 5	8.5, 10	ul	Bio, (Std.)	HB
TPN	transparent plastic *Nansen* water sampler, PC tube, Teflon O-rings, ball valve, reduced flushing	1.7	3	ul	Std., (TM)	HB
GO-FLO	PVC tube, Teflon coating optional, external rubber spring, for single use and in rosette-type samplers, COC, reduced flushing	1.7–100	3.9–47	ul	Std., TM	GO
Niskin 1010	PVC tube, Teflon coating optional, internal rubber spring, for single use and in rosette-type samplers, reduced flushing	1.2–30	2–12.7	ul	Std., (TM)	GO

Table 1-2 Selected seawater samplers *(continued)*.

Name	Characteristics	Volume L	Mass kg[a]	Range m	Analyte[b]	Firm
Niskin 1011	PVC tube, Teflon coating optional, external spring for closing lids, for single use and in rosette-type samplers, good flushing characteristics	1.7–12	3.3 –7.8	ul	Std., (TM)	GO
Rosette samplers	*Rosette of standard ('serial') samplers on top of CTD profilers, maximum operation depths governed by properties of sensors and length of wire*					
Model 1016[c]	12–36 *Niskin* or *GO-FLO* samplers (see above)	12 × 1.2 to 36 × 12	127 – 1203	ul	Std.	GO
SVS[d]	12 or 24 PVC samplers with a volume of either 30, 60 or 500 mL, external closing by spring, for studies of analyte gradients	12 × 0.03 to 24 × 0.5	≈ 250	ul	Std.	GO
FSI[e]	12–36 samplers, *inter alia* of the *Niskin* or *GO-FLO* type	12 × 1.7 to 36 × 9	up to 590	ul	Std.	FSI
MWS	multi-water sampler, for PWS samplers (see above), in four specifications, including 3 *PWS* (30 L), 24 *PWS* (1.7–10 L) or 6 *PWS* (1.0 L).	6 × 1.0 to 12 × 30		≤ 3000 (6000)	Std.	HB
Special samplers	*For specific use only*					
LVWS	large volume water sampler, stainless steel, internal reversing thermometers, partition wall in the middle part, one opening with deflector plate, poor flushing characteristics	270, 400	130, 220	ul	(TO)	HB
Model 1010C[f]	*Niskin*-type sampler, special holding device allows single use in vertical and horizontal direction, designed for surface sampling	1.7, 2.5	2.9, 3.6	ul	Std.	GO
Model 1050	a syringe is used as sampler, materials: stainless steel, Teflon, PC and glass, for specific sampling of thin layers (≥ 5 cm)	0.01, 0.06		ul	Std.	GO
HWS	horizontal water sampler, materials and construction as *UWS* but fixed differently to wire, with stabilizing rudder, for near-bottom sampling	2	10	ul	Bio, (Std.)	HB

Table 1-2 Selected seawater samplers *(continued).*

Name	Characteristics	Volume L	Mass kg[a]	Range m	Analyte[b]	Firm
Oil sampler	according to *Schomaker*, glass bottle, lateral buoyancy floating bodies, sampling of surface layers of a few mm thickness	0.15		surface	TO	HB
CIT	California Inst. Technol., USA, stainless-steel frame, PE bag, Teflon, COC, *Schaule and Patterson* (1978)	5–10	ul	TM		CIT

[a] Deadweight in air (kg); [b] in parantheses: limited suitability; [c] intelligent rosette multi-bottle water sampling system; [d] small volume sampler; [e] FSI SURE-FIRE submersible array; [f] *Niskin* non-metallic convertible surface water sampling bottles.
Bio – biological parameters; COC – close-open-close principle; FSI – Falmouth Scientific, Inc.; GO – General Oceanics, Inc., Miami, USA; HB – Hydrobios Apparatebau GmbH, Kiel, Germany; PVC – poly (vinyl chloride); Std. – hydrographic and hydrochemical standard parameters (*e.g.,* temperature, salinity, oxygen, nutrients); TM – trace metals; TO – trace organic compounds; ul – unlimited.

Today, CTD profilers are standard tools in oceanography. In addition to transducers for conductivity (salinity), temperature and hydrostatic pressure (depth), they may be equipped with sensors for other determinants such as oxygen, hydrogen sulphide, pH, light scattering or fluorescence (see Chapter 14). The lowering velocity of the CTD must be adjusted to the response time of the slowest sensor. Investigations on nutrients and other trace constituents as well as the calibration of the sensors demand the sampling of seawater. Therefore, it was obvious to combine the CTD with a rosette of remote-controlled samplers. Sample collection with a rosette is much more practical than a conventional hydrocast and allows precise selection and identification of the actual sampling depth.

Comparison of commercially available rosette samplers (Table 1-2, Fig. 1-5) shows that different constructions follow almost the same principle. A circular protective frame accommodates 3–36 samplers which commonly can be used as serial samplers in hydrocasts. In addition to their usage in basic hydrochemical studies, standard rosette samplers (*e.g.,* 12 L *Niskin* samplers; see Table 1-2) are also suitable for the collection of oceanographic tracers such as fluorocarbon compounds, helium-3, tritium or even carbon-14 (except for ^{14}C studies in the deep waters of the Pacific and Indian Oceans, where large special samplers of ≥ 250 L volume are needed; see Table 1-2 and Chapter 13).

The sensors of the CTD are located in a central position below the rosette. Therefore, CTD profiles are recorded mostly during lowering whereas sampling is preferably performed during lifting. Rosette systems are usually hung on a wire which also serves for transmission of energy, triggering signals and records. This intimate mechanical coupling between rosette and ship may lead to disturbances under rough weather conditions. With a strongly yawing ship, the signals may be disturbed due to changes in lowering and lifting velocities. Free-falling and telemetric-controlled CTD systems have been developed which are lowered with constant speed independent of the ship's movement. This approach results in improved CTD profiles but is not considered to be an advantage for the sampling operation (for further details on CTD profilers see Chapter 3).

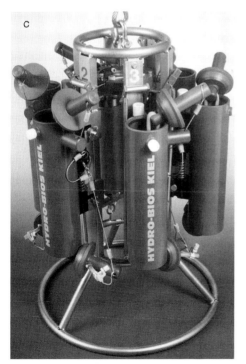

Fig. 1-5. Examples of rosette ring water samplers

(a) 36-fold rosette with 10 L *Niskin* samplers (Model 10163610; *General Oceanics*, 1997).

(b) 24-fold rosette with 12 L *Niskin* samplers from General Oceanics. (Photograph by W. Zenk, IfM Kiel).

(c) 6-fold rosette *MWS* (multi-water sampler; Model 'Slimline') with 1 L plastic (PVC) water samplers. Diameter and height are 0.5 and 0.7 m, respectively. Total weight of the rosette is around 30 kg. Maximum sampling depth is 3000 m (*Hydrobios*, 1997).

1.3.3   Water samplers for trace constituents

1.3.3.1   Trace elements

To avoid contamination of seawater with trace elements during sampling, samplers should meet the following conditions:

(a) Those parts of the sampler in contact with the seawater should be manufactured of acid-resistant plastic material only (*e.g.*, Teflon, PTFE, polypropylene or PE; see also Chapter 12). Acryl esters and polycarbonates are of limited use, mainly because they cannot be cleaned thoroughly with concentrated acids. Uncoated PVC is not suitable at all. External metal parts, which are inevitable, should be coated with plastic materials. O-rings except for those made from PTFE, may leak trace metals. (b) Ideally, the seawater should not be in contact with the sampling system before being sampled. This requirement is in contradiction, however, with the wish to record trace metal profiles based on quasi-synchronous sampling from different water layers. (c) The sampler must pass the 'dirty' surface closed, *i.e.*, function according to the COC (close-open-close) principle.

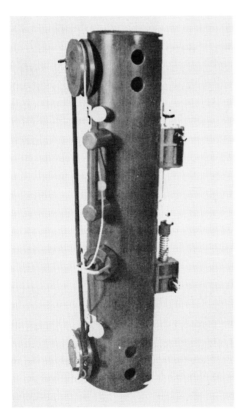

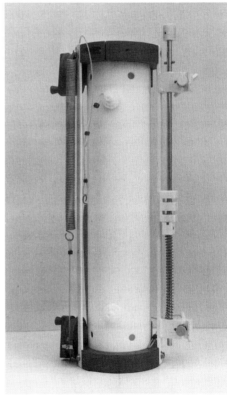

Fig. 1-6. *GO-FLO* sampler made of PVC with optional internal Teflon-coating (Model 1080, General Oceanics). The sampler passes through the water surface closed and opens at about 10 m depth *via* a pressure-activated 'releaser'.

Fig. 1-7. *WATES* sampler made of PTFE (*Brügmann et al.,* 1987). The sampler passes the water surface closed and opens at a depth of about 5–10 m following manual activation with a plastic line.

Table 1-2 lists some samplers used at present and in the past for trace metal studies in seawater (TM). Among them, the Teflon-coated *GO-FLO* PVC samplers (Fig. 1-6) are in fairly widespread use (*Berman and Yeats*, 1985). Figure 1-7 shows the *WATES* sampler made entirely of PTFE thus avoiding contamination risks due to damaged or insufficient coating. The *CIT* sampler (*Schaule and Patterson*, 1978) was designed to take single samples from deep oceans with very low risk of contamination. This sampler hangs at the end of a plastic wire, with a retractable cup to protect the entry port; upon triggering with a messenger, withdrawing of a piston fills a PE bag. This sampler was considered to provide the first trustworthy data on dissolved and particulate lead in deep-ocean water. The *MERCOS* sampler was constructed for the determination of mercury but also of other trace elements in shallow seas (Fig. 1-8; *Hydrobios*, 1997). The two removable 500 mL PTFE sampling bottles can also be used for storage and further treatment of the samples.

In addition to appropriate samplers, components of an optimum sampling system for trace element studies involve plastic (coated) 'wires', such as made of *Kevlar*, or plastic coated hydrowires, serial winches with plastic drums and plastic-coated bottom weights and messengers.

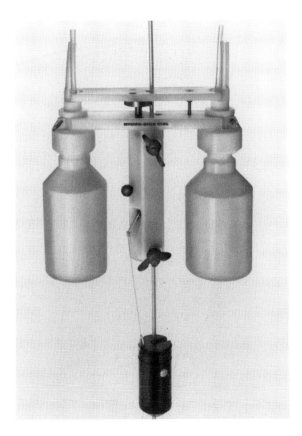

Fig. 1-8. *MERCOS* water sampler made of PTFE (*Hydrobios*, 1997). A messenger activates the filling of two 500 mL PTFE screw-cap bottles *via* silicone tubes down to a depth of 100 m.

1.3.3.2 Trace organic compounds

Most organic compounds occur at extremely low concentrations in seawater. Their determination involves an extraction/sorption step to increase their concentration levels to the sensitivity range of instrumental analytical methods. Compared with other groups of analytes, the risk of adsorptive losses is most important for trace organic compounds. Therefore, techniques are preferable which involve direct extraction within the sampler or *in situ* enrichment onto solid adsorbents. Using very large samplers with a comparably high ratio between volume and internal sampler surface, adsorption losses can be minimized but never excluded entirely.

For the monitoring of relatively high concentrations of organic compounds, however, comparatively small seawater volumes from depths down to 50 m might be sufficient. A fairly simple and robust technique has been developed for this purpose which can also be performed from small vessels in the absence of winches. Commercially available standard ground-glass bottles of maximum 2.5 L volume are fixed in a metal frame and lowered to the desired depth with a Nylon line. The messenger hits a sharpened metal bar which ruptures a PTFE disk that seals the bottle. The filled bottle is heaved up and may be used directly for extraction of trace organic substances, *e.g.*, oil equivalents using ultraviolet-fluorescence (UVF) spectroscopy (*Brügmann et al.*, 1985). A similar sampler is described in Chapter 21.

Fig. 1-9. Glass-sphere water sampler (*Stadler and Schomaker,* 1977; *Theobald et al.,* 1990). Shown is a 23-L glass sampler including the stainless-steel frame with clamps to fasten it to the hydrowire. Left: in sampling position; two inlets closed by inserted glass vials; spring-loaded cutter fixed to the messenger-operated releaser. Right: in upright position with glass funnel extruder for separation of solvent following extraction.

For more than 20 years, spherical flasks for industrial use with volumes of 5–100 L have found application as samplers to study trace organic components in seawater (*e.g., Stadler and Schomaker*, 1977; *Theobald et al.*, 1990). The glass flasks have stainless-steel lids and are fixed in a frame made of the same material to protect them against damage and to fasten the sampler to the hydrographic wire (Fig. 1-9). The empty samplers are lowered to the desired depth. A messenger-activated spring cuts two glass vials which close the entrance ports. The filled sampler is retrieved. Following exchange of the lid and addition of solvents, extraction is performed directly in the sampler. Using turbo-stirrers (*Brügmann et al.*, 1985), extraction of a 20 L sample volume is complete in < 15 min. Using solvents with densities lower than seawater (*e.g.*, hexane), extracts are simply separated using displacers. The operating depth of the glass samplers is limited by their implosion threshold which depends on the quality of the glass and the volume of the sampler. Depths of about 500 m are possible with ordinary 25 L glass flasks. Specially manufactured (tempered) flasks allow operating depths of down to 1000 m (*Brügmann et al.*, 1985).

In situ sample enrichment, especially for the measurements of persistent anthropogenic organics such as organochlorines, is described in detail in Chapter 22. Using an air-pressure operated membrane pump such as described in Section 1.3.1 (*Schüßler and Kremling*, 1993), sampling for accurate PCB determinations in seawater could even be performed from a sailing ship (*Schulz-Bull et al.*, 1995). More recently, a new *in situ* enrichment technique based on semi-permeable membrane devices (SPMDs) appears to be potentially useful for passive sampling of trace organics in seawater (*Huckins et al.*, 1993; *Lebo et al.*, 1995; *Prest et al.*, 1995).

1.3.4 Specific samplers

A number of samplers have been designed specifically for rarely investigated variables. Microbiological studies, for example, often require previously sterilized samplers which are lowered tightly closed to the desired depth. Several types of such samplers have been developed, the most common systems by *Zobell* (1946) and *Niskin* (1962). The *Zobell* sampler (Fig. 1-10) consists of a strong glass bottle (*e.g.*, a champagne bottle with a volume of about 700 mL), sealed with a rubber bung through which passes a glass tube. A short length of pressure tubing connects it with a flame-sealed small-bore glass tube. Following initial autoclaving the bung is pushed firmly into the bottle thus creating a partial vacuum as it cools. In operation, the outer tube is broken with a lever activated by the hydrographic messenger, and the glass bottle fills according to the partial vacuum and the hydrostatic pressure.

When sampling near the surface is required, the glass bottle must be evacuated. A disadvantage of the *Zobell* system is the limitation to an operating depth of around 200 m. In contrast, the *Niskin* sampler (Fig. 1-11) allows collection of about 3 L of seawater from all depths within a sterile plastic bag. When lowered to the desired depth, the sampler is triggered by a messenger which in turn activates a knife blade to sever a sealed tube leading to the bag. Simultaneously, the spring-loaded hinged frame is opened and water flows into the bag. Before raising the filled bag, it is tightly closed again. Compared with the *Zobell* system, the *Niskin* sampler is more difficult to operate in rough weather and the plastic bag sometimes leaks during raising.

A large range of home-made equipment is used to sample at interfaces to the sediment and the atmosphere or across internal density discontinuity layers or redox (chemo-)clines (*e.g., Bale and Barrett*, 1995; *Eversberg*, 1990; *Hallberg et al.*, 1977; *Schwedhelm et al.*, 1988;

Fig. 1-10. *Zobell* sampler for collection of microbiological samples down to water depths of *ca.* 200 m.

Fig. 1-11. *Niskin* sampler for the collection of microbiological samples from deep ocean waters (General Oceanics).

Sternberg et al., 1991). For different vertical resolutions, the water is collected in multi-chamber samplers. The transport is performed across filter and dialysis membranes, either actively by suction *via* pumps and vacuums or passively driven by concentration gradients and diffusion.

Sampling techniques for subsequent analysis of radioactive tracers in seawater are described in detail in Chapter 13.

1.3.5   Collection of marine particles

Observations of particle concentration and flux in the ocean have shown that most of the particulate matter is in a slowly sinking or non-sinking reservoir (defined as suspended particulate matter or SPM), while the particulate matter flux is carried mainly by a relatively small number of larger, rapidly sinking particles (*e.g., McCave*, 1984, *Clegg and Whitfield,* 1990). Therefore, to understand the processes controlling the distribution of the chemical components as well as their biogeochemical cycles, the relative masses and composition of the two particle types must be known, implying the collection and analysis of both SPM and sinking particles.

1.3.5.1   Collection of suspended particulate matter (SPM)

SPM may be collected in various ways and is described in more detail in Chapter 2 (see also Section 12.6). In general, the most popular technique is vacuum filtration which has, however, a number of disadvantages with regard to contamination risks when collecting samples for studies on trace constituents. These problems are minimized by pressure filtration. With this method, the water sampler itself is pressurized to force the sample *via* an inline filter into the receiving bottle.

In waters with extremely low particulate concentrations, *i.e.*, in offshore regions or in the deep ocean, the analysis of most particulate constituents requires the filtration of large volumes of seawater to collect enough material to match the sensitivity range of instrumental analytical methods. Here, *in situ* filtration systems (*via* pumps) offer many advantages over bottle samples, mainly with respect to: (a) the large sample volumes that can be filtered; (b) the wider range of determinations possible on a single particulate sample; and (c) reducing blank problems. For a more detailed review on *in situ* filtration techniques see Chapters 2 and 13.

Another method for processing large volumes of near-surface seawaters is continuous-flow centrifugation. Under ideal circumstances, this method might combine sufficiently close spatial resolution with the capacity of effectively collecting large amounts of particulate matter of low contamination. This approach was used for the simultaneous collection of samples for analyses of both particulate and dissolved trace constituents (*Schüßler and Kremling,* 1993) and has already been described in Section 1.3.1. Further details on the centrifuge and its capacity are provided in Chapter 2. Handling, digestion and analysis of collected particles are outlined in Chapters 12, 17 and 22.

1.3.5.2   Collection of sinking particulates

Sediment traps for measuring the downward flux of particles in the water column have been used for nearly 100 years. Since the pioneering investigation by *Heim* (1900), the literature on sediment-trap techniques in marine and limnic environments has grown steadily (for broad reviews see *Reynolds et al.*, 1980; *Blomquvist and Håkanson*, 1981). In recent decades, the use of sediment traps has expanded, especially in open-sea areas, promoted by advances in microprocessor technologies and new plastic materials (*e.g.*, *Honjo*, 1982; *Deuser*, 1986; *Jickells et al.*, 1990; *Wefer and Fischer*, 1990; *Kremling and Streu*, 1993). Such sediment traps are capable of measuring particle fluxes at deep-water stations, over long periods, either continuously or in time-fractionated sequences. Unfortunately, there is no standard design for sediment traps, nor has there been any general agreement concerning the handling of the entrapped material, although a serious effort was launched in this direction during the 'Joint Global Ocean Flux Study' (*Knauer and Asper*, 1989).

In this section we offer some practical guidelines for the operation of sediment traps, with emphasis on problems associated with construction of apparatus and the collection of trace inorganic and organic substances. A detailed description of handling and analysis of particulates with emphasis to trace elements is given in Section 12.6.

General design objectives of ocean sediment traps are as follows (as modified from *Honjo and Doherty*, 1988):

(1) The trap aperture should be large enough to collect a sufficient mass of sample, given that each fractionated deployment period is relatively short (days to weeks).

(2) Deployment of the trap should be possible at any, except extreme, depths from any ordinary research vessel by a few skilled technicians.

(3) The time-series (microprocessor) control unit should possess a large degree of programming flexibility (with intervals from 1 min to 1 year). The control unit should be able to communicate with any standard PC.

(4) After recovery of the trap, a comprehensive sampling protocol should be readable from the microprocessor control unit.

(5) Each time-fractionated sample should be stored individually *in situ* and sufficiently sealed from the surrounding water, and should withstand substantial physical shock until being recovered on deck.

(6) Samples must be protected against contamination caused by the trap, ship and surface water.

Two of the time-series sediment traps which satisfy most of these conditions have been described by *Honjo and Doherty* (1988) and by *Kremling et al.* (1996). Hence, these models are in widespread use and will therefore be outlined here, but briefly. Both traps consist of three major components: (1) a multi-celled baffel-funnel combination; (2) a rotary sampler assembly with sampling bottles, drive motor and controlling electronics; and (3) a supporting protective frame with bridles to connect to the host mooring array. They both utilize a number of high-quality synthetic materials and inert metal components thus ensuring long-lasting reliable performance in the marine environment. The traps of *Honjo and Doherty* (models 'Mark 5' and 'Mark 6') have apertures of 0.5 and $1.15\,m^2$ (with cone-angles of 42 and 36°, respectively) and are capable of collecting 12–25 samples.

The 'Kiel Sediment Trap' (Fig. 1-12) by *Kremling et al.* (1996) has a $0.5\,m^2$ aperture, a funnel slope of 34° and is capable of collecting 20 samples at programmed intervals. This trap has been designed especially to sample for reliable analyses of both trace inorganic and

Fig. 1-12. The 'Kiel Sediment Trap' with 1.90 m overall height, 1.08 m width, 0.5 m^2 aperture and a funnel slope of 34°. The trap is capable of collecting 20 samples at programmed intervals from 1 min to 1 year (*Kremling et al.*, 1996).

organic compounds (such as trace elements, n-alkanes, polychlorinated biphenyls (PCBs), polycyclic aromatic hydrocarbons (PAHs), amino/fatty acids), in addition to the standard biogeochemical variables in the collected particles. Blank values determined in the open ocean were as low as 1 % of the amounts present in trapped material, even at low particle loads. Tests with dissolved tracers during field studies proved that their loss from supernatants in the sampling bottles to the surrounding seawater during an one year deployment was as low as 10 %. This is important, since it has been shown by several studies that poisoning of the sampling cups (*e.g.*, by addition of formaldehyde or sodium azide) cannot entirely prevent the degradation of biogenic material by microbial and chemical processes. Thus, large portions of certain constituents (such as cadmium, phosphorus or fatty acids) may be released from collected particles into solution. Both particles and supernatants, therefore, have to be analysed for reliable flux data.

Independent of trap design, there are other problems in determining accurate flux data. For example, resuspension or advective transport of bottom sediments may cause anomalously high results. Therefore, it is recommended for the deep ocean that traps should remain at least 1000 m above the sea floor. Another significant problem may be caused by zooplankton 'swimmers' in the trapped sample, potentially disturbing the results of purely sedimenting material. This was observed especially in upper-ocean traps (*e.g.*, *Michaelis et al.*, 1990).

Recent tests by other workers (*Gust et al.,* 1994) seem to indicate that the flow around and inside sediment traps may have considerable effects on the trap collection efficiency at shallow depths, whereas the deep-ocean traps appear to collect the actual flux. We recommend, therefore, to equip the instruments with current meters, and depth and tilt sensors. This allows a description of the dynamics of the hydrographical field and its effects on mooring line motions as well as on the excursions of traps from the vertical position, and may even offer the basis for corrections of hydrodynamic biases in particle flux measurements.

On the other hand, individual measurements of natural decay-series of radionuclides (*e.g.,* ^{230}Th and ^{231}Pa) in the trap material can provide useful information on the trap collection efficiency during the experiment (*Bacon,* 1996). It is also important to note that under- or overtrapping will neither affect the composition data nor the relationship between flux variables as long as the material collected is representative of the vertical transport of the constituents to the deep waters.

1.4   Sampling errors

The analytical chemist has to deal with two types of errors, systematic (bias) and random errors. Systematic errors tend to influence the measurement and to produce results that consistently are either too low or too high. They may occur in the sampling procedure, caused for example by an incorrect determination of the sampling depth, by inefficient flushing of the sampler, or, especially in the determination of trace constituents of seawater, by a sink or a contamination source for these compounds in the sampling device itself. Another commonly introduced bias (*e.g.,* in the determination of oxygen) is contamination of the sample during subsampling; *i.e.,* the subdivision of the water sample from the sampler into different bottles or flasks for individual determinations.

The ship as a sampling platform is a potential source of contamination for either the seawater to be investigated, the sampling equipment to be deployed or the sample brought aboard for analysis. Besides physical mixing of the water column on the way to and during the stay at the station, there are multiple risks of permanent and often unavoidable contamination of the surface water by the ship's operation. They include continuous release of cooling waters, corrosion and dissolution of unprotected metal parts (*e.g.,* propellers, anode plates) and coatings, and seeping of compounds from intential and unintential leaks, *e.g., via* the propeller shaft. In addition, exhaust gases from main or auxiliary engines may contain contaminating particles which deposit onto the ship and the surrounding water area. Therefore, cleaning and de-rusting operations should not be undertaken immediately before or during sampling. The same applies to the discharge of any wastes. The working environment within the ship's laboratories, which have often been constructed for multipurpose use, may deviate significantly from the conditions in land laboratories. This may constitute an additional source of cross-contamination for samplers and seawater samples, *e.g.,* due to unsuitable construction materials or *via* air-conditioning systems. By using clean benches or, more preferably, purpose-designed container laboratories these problems can be minimized and controlled (see also Chapter 12).

Commonly, the sampling equipment involves hydrographic winches, wires, bottom weights, samplers and (mostly) messengers. Except for the winch, all these parts are deployed into the sea and as a rule will release material when in contact with seawater. For example, the use of bottom weights made of lead, of zinc-coated hydrowires or messengers made of brass will not affect basic hydrochemical and nutrient data. For trace metal studies, however, these parts of the sampling equipment may be responsible for severe contamination of the seawater to be sampled. In addition, following closure of the sampler, the samples might stay therein for hours before being retrieved. Contamination or (adsorptive) losses of the analyte during enclosure may cause significant systematic sampling errors. Effective control and quantification of this type of error is still one of the most challenging analytical problems in chemical oceanography, especially with regard to the determination of trace components. Storage tests using artificial seawater to determine sampling blanks in a classical analytical sense would complicate the problem further. The required salts and water must be of controlled very high purity, which is difficult to achieve and would also be very expensive.

Very often it is difficult to detect a systematic deviation from the true value, because in field work the true value is usually unknown. Sometimes the application of different sampling methods or sampling devices made of different materials may uncover systematic errors. However, the use of different types of samplers might be also misleading, because repeated sampling does not automatically ensure that the same body of water is hit each time. The ship may drift, waves influence the depth at which the sampler closes and currents move the water. Comparative studies, therefore, are only useful if one can be sure that the body of water is large enough and is in itself homogeneous with respect to the constituent in question. One way of testing equipment for bias in sampling is to attach different types of samplers to the hydrowire at the same depth or as close together as possible, but even the microstructure in the environment may result in different samples being enclosed in different samplers. Since the true value is unknown, systematic errors during the sampling procedure are difficult to detect and, therefore, troublesome. In some cases, bias in the sampling procedure may be detected by means of an obvious break in a known correlation of two constituents or properties in ocean waters (*e.g.*, the cadmium-phosphate relationship).

The second category is random errors. They are such that the results of repeated analyses of the same sample under identical conditions will vary. These errors are caused by the type of instrumentation used, personnel errors and errors inherent in the method itself. In most cases it is possible to recognize these random errors and obtain a reasonable estimate of their magnitude by, for example, comparison of independent analyses performed by different analysts, and, if possible, by applying different methods or by comparing independent analyses from the same sample using spiking techniques. The magnitude of random errors can be reduced by repetition of the analysis. Random errors are always present and independent of the existence of systematic errors.

Even when most of the severe sampling error sources are excluded, certain risks remain which could be reduced further only with very great effort. Whether or not to spend this effort on these risks has to be considered critically. For instance, sampling errors will be higher if the circulation pattern of the water is complicated, or if different water masses occur in close proximity and mixing processes take place. Here the variance of the sampling procedure may be much larger than that of the analytical procedure, so that time-consuming and elaborate chemical methods will by no means improve the quality of the total analysis. On the other hand, environmental conditions in the deep sea call for extreme accuracy both in sampling and in the chemical procedures in order to detect often very small but meaningful differences in the characteristics of deep-sea water masses.

1.5   Quality control

Because sampling and handling prior to instrumental analysis are part of the overall data generation, it is appropriate to define in this first chapter of the book some terms that are commonly used in conjunction with measurements. The focus here is on a brief overview of quality control measures with regard to sampling which can be performed by analytical laboratories. This subject has been covered comprehensively by *Keith et al.* (1983) and *Keith* (1991).

1.5.1   Precision

Precision or reproducibility of a method is a multi-level concept, involving all steps from sampling of the water to the final data. Precision usually is expressed as relative standard deviation (RSD) or coefficient of variation (c.v.). It can be obtained by analysing repeatedly the same sample to identify the 'within-laboratory precision', or over a longer time period, thereby including the sampling process. A common procedure is to fill all samplers of a rosette (*e.g.*, 12 or 24 bottles) with water from the same depth representing a homogeneous water mass. Following subsampling and analysis, RSD values are estimated. In such examples, however, systematic errors, *e.g.*, due to the time lapse between sampling and subsampling, can distort the normal (Gaussian) error distribution. For example, particulate material initially distributed throughout the sample volume will tend to settle. If not filtered off, or if the sampler is not shaken before subsampling, the uneven distribution of the particulate material among different subsamples from the same sampler may influence the results of the analysis. Similar problems are observed in the analysis of dissolved gases. When, for instance, the oxygen partial pressure in the samplers increases because of an increase in temperature of the water samples, then a bias by outgassing is usually unavoidable. A simple but effective method for detecting the source of such 'floating systematic errors' is to note the numbers of the successive subsamples together with the histogramme of analytical data.

RSDs tend to increase with lower concentration levels. Most of the RSDs available from the literature are related to the respective laboratory only, *i.e.*, they are measures of the *intra*-laboratory reproducibility. The *inter*-laboratory reproducibility, however, indispensible for comparison of data sets from different working groups, is more difficult to obtain, unless intercomparison exercises are arranged. Usually, the application of several analytical methods by different analysts introduces an additional uncertainty in determining a certain analyte, and RSDs increase in comparison with *intra*-laboratory RSDs.

1.5.2   Accuracy

Accuracy refers to the correctness of data, *i.e.*, it is related to the deviation of the determined value from the true value. Inaccuracy results from imprecision (random error) and bias (systematic error) in the measurement process. Whereas precision is easier to determine, accuracy is more informative and important but the most difficult quality criterion to quantify. Certified reference materials are most useful in evaluating the laboratory analysis (see subsequent chapters). In addition, the recovery of spikes of the analyte from a sample can also give an indication of accuracy. When interpreting the recovery, however, it should be remembered that the behaviour of such spikes might differ from the analyte in the sample.

Intercomparisons between laboratories, including the exchange of reference materials and application of different methods, is another way to check what the 'true' concentration might be. It should be stressed here, however, that accuracy is not necessarily proven by good agreement among different laboratories because of the possibility that their result might be similarly biased.

Another approach to establish accuracy is to prove the plausibility of the data. In the marine field, this means looking for changes in concentrations and relationships of the analyte with other well known variables in the water column, if they exist at all ('oceanographic consistency').

1.5.3 Limit of detection

The limit of detection (LD) or sensivity is commonly defined as the lowest concentration level that is statistically different from a blank at a specified level of confidence. In practice, the LD for a given analyte is seldom limited by the sensivity of the analytical technique, but governed by the level and variability of the blank value; *i.e.*, impurities introduced with reagents, procedural steps, apparatus, air and instrumental variations. It is generally accepted by analysts (*e.g.*, *Kaiser*, 1970; *Tölg*, 1972; *Keith*, 1991) and will be adopted here that an analytical result (X_A) is considered as 'real' and different from the blank if it is at least as great as the mean blank value (X_{bl}) plus 3 standard deviations of the blank value (s_{bl}):

$$X_A \geq (X_{bl} + 3s_{bl}) \text{ or}$$
$$X_A - X_{bl} \geq 3s_{bl}$$

Signals below $3s_{bl}$ (= LD) should be reported as 'not detected' (ND) and the limit of detection should be given in parentheses. This statistical definition of the LD implies that the precision of the overall blank value is at least as important as its size. Nevertheless, as a rule the experimental approach should always try to keep the blank level at about $\leq 10\,\%$ of the measured quantity. Effective ways to guarantee high sample-to-blank ratios are either to start with relatively high quantities of sample or to minimize the blank by rigorous control procedures similar to those described in the following chapters. It is essential that blank determinations should be done in the same way and at the same time as sample processing.

For trace analyses in marine chemistry, the determination of proper s_{bl} values is very complicated and might even be misleading, as outlined in the discussion of sampling errors. Therefore, an often used approach for estimating LDs is the determination of the standard deviations (s) from several measurements of a low-concentration sample. This can be done during an expedition by repeatedly taking duplicate samples of the same type at different stations, *e.g.*, from deep-water layers with similar concentration levels. The standard deviation (s) can be estimated (*Kaiser*, 1970) by

$$\sqrt{\frac{\sum d^2}{2n}}$$

where d is the difference between a pair of results, and n is the number of pairs. For the limit of detection, a $3s$ value is a reasonable estimate.

References to Chapter 1

Bacon, M.P. (1996), in: *Particle Flux in the Ocean*: Ittekot, V., Schäfer, P., Honjo, S., Depetris, P.J. (Eds.). Chichester: John Wiley & Sons, 1996; pp. 85–90.

Bale, A.J., Barrett, C.D. (1995), *Netherlands J. Sea Res.*, 34, 259.

Berman, S.S., Yeats, P.A. (1985), *CRC Crit. Rev. Anal. Chem.*, 16, 1.

Blomquvist, S., Häkanson, L. (1981), *Arch. Hydrobiol.*, 91, 101.

Brügmann, L., Franz, P., Fröhlich, K., Gellermann, R., Hebert, D., Lange, D., Mohnke, M., Rohde, K.-H., Thiele, J., Weiß, D. (1985), *Geod. Geoph. Veröff.*, Series IV, 40, 110 pp.

Brügmann, L., Geyer, E., Kay, R. (1987), *Mar. Chem.*, 21, 91.

Clegg, S.L., Whitfield, M. (1990), *Deep-Sea Res. I*, 37, 809.

Deuser, W.G. (1986), *Deep-Sea Res.*, 33, 225.

Eversberg, U. (1990*), Helgoländer Meeresunters.*, 44, 329.

General Oceanics, Inc. (1997), *Product information sheets*, Miami.

Gust, G., Michaels, A.F., Johnson, R., Deuser, W.G, Bowles, W. (1994), *Deep-Sea Res. I*, 41, 831.

Hallberg, R.O., Bågander, L.E., Engvall, A.-G., Lindström, M., Schippel, F.A. (1977), *Baltica*, 6, 117.

Hardy, J.T., Coley, J.A., Antrim, L.D., Kiesser, S.L. (1988), *Can. J. Fish. Aquatic Sci.*, 45, 822.

Harper, D.J. (1987), *Mar. Chem.*, 21, 183.

Heim, A. (1900), *Vierteljahresschrift d. Naturforschenden Gesellschaft in Zürich*, 45, 164.

Honjo, S. (1982*), Science*, 218, 883.

Honjo, S., Doherty, K.W. (1988*), Deep-Sea Res.*, 35, 133.

Huckins, J.N., Manuweera, G.K., Petty, J.D., Mackay, D., Lebo, J.A. (1993), *Environ. Sci. Technol.*, 27, 2489.

Hühnerfuss, H. (1981), *Meerestech. Mar. Technol.*, 12, 1170.

Hydrobios (1997), *Product information sheets*, Kiel.

Jickells, T.D., Deniser, W.G., Fleer, A, Hemleben, C. (1990), *Oceaonologica Acta*, 13, 291.

Kaiser, H. (1970), *Anal. Chem.*, 42, 26A.

Keith, L.H., Crummett, W., Deegan, J.Jr., Libby, R.A., Taylor, J.K., Wentler, G. (1983), *Anal. Chem.*, 55, 2210.

Keith, L.H. (1991), *Environmental Sampling and Analysis: A Practical Guide*. Chelsea: Lewis Publishers.

Knauer, G., Asper, V. (Eds.) (1989), *Sediment Trap Technology and Sampling*, U.S. JGOFS Report 10.

Knudsen, M. (1923), *Publ. Circ. Cons. Expl. Mer*, 77, 1.

Knudsen, M. (1929), *J. Cons. Int. Explor. Mer*, 4, 192.

Kremling, K., Streu. P. (1993), *Deep-Sea Res. I*, 40, 1155.

Kremling, K., Lentz, U., Zeitzschel, B., Schulz-Bull, D.E., Duinker, J.C. (1996), *Rev. Sci. Instrum.*, 67, 4360.

Lebo, J.A., Gale, R.W., Petty, J.D., Tillitt, D.E., Huckins, J.N., Meadows, J.C., Orazio, C.E., Echols, K.R., Schroeder, D.J., Inmon, L.E. (1995), *Environ. Sci. Technol.*, 29, 2886.

Levitus, S. (1982), *NOAA Professional Paper*, no. 13, Washington.

Liss, P. S., Duce, R. A. (1997), *The Sea Surface and Global Change*. Oxford: University Press.

McCave, I.N. (1984), *Deep-Sea Res.*, 31, 329.

Michaelis, A.F., Silver, M.W., Gowing, M.M., Knauer, G.A. (1990), *Deep-Sea Res. I*, 37, 1285.

Niskin, S.J. (1962), *Deep-Sea Res.*, 9, 501.

Prest, H.F., Huckins, J.N., Petty, J.D., Herve, S., Paasivirta, J., Heinonen, P. (1995), *Mar. Poll. Bull.*, 31, 306.

Reynolds, C.S., Wiseman, S.W., Gardner, W.D. (1980), *Occasional Publication*, no. 11. Ambleside Freshwater Biological Association.

Schaule, B., Patterson, C.C. (1978), in: *Occurrence, Fate and Pollution of Lead in the Marine Environment:* Branica, M. (Ed.). Oxford: Pergamon Press, 1978.

Schüßler, U., Kremling, K. (1993), *Deep-Sea Res. I,* 40, 257.

Schulz-Bull, D.E., Petrick, G., Kannan, N., Duinker, J.C. (1995), *Mar. Chem.,* 48, 245.

Schwedhelm, E.,Vollmer, M., Kersten, M. (1988), *Fresenius J. Anal. Chem.,* 332, 756.

Stadler, D., Schomaker, K. (1977), *Dt. Hydrogr. Z.,* 30, 20.

Sternberg, R.W., Kineke, G.C., Johnson, R. (1991), *Cont. Shelf Res.,* 11, 109.

Sverdrup, H.U., Johnson, M.W., Fleming, R.W. (1942), *The Oceans: Their Physics, Chemistry, and General Biology.* New York: Prentice-Hall, Inc.

Theobald, N., Lange, W., Rave, A., Pohle, U., Koennecke, P. (1990), *Dt. Hydrogr. Z.,* 43, 311.

Tölg, G. (1972), *Talanta,* 19, 1489.

Tokar, J.M., Harvey, G.R., Chesal, L.A. (1981), *Deep-Sea Res.,* 28, 1395.

UNESCO (1991), *Processing of Oceanographic Station Data. JPOTS Editorial Panel,* UNESCO.

Van Dorn, W.G. (1956), *Trans. Am. Geophys.,* 37, 682.

Van Vleet, E.S., Williams, P.M. (1980), *Limnol. Oceanogr.,* 25, 764.

Wefer, G., Fischer, G. (1990), *Deep-Sea Res. I,* 40, 1613.

Zobell, C.E. (1946), *Marine Microbiology.* Waltham: Chronica Botanica Co.

2 Filtration and storage

K. Kremling and L. Brügmann

2.1 Filtration

2.1.1 General remarks

It has generally been accepted in marine geochemistry, that the term 'dissolved' refers to that fraction of seawater and its constituents which passes through a 0.45 μm (or 0.4 μm) filter. Clearly this classification is defined operationally rather than based on objective criteria. It has been known for a long time, however, that standard filtration fails to distinguish between true solutes and colloidal phases in seawater as well as between suspended and the larger sinking particles. Only in the early 1990s did evidence appear that a large fraction of the marine dissolved organic carbon (DOC) pool is in the colloidal size range (*Koike et al.*, 1990; *Wells and Goldberg*, 1991). Given their tremendous specific surface area it is postulated that marine colloids play an important role in the cycling of many reactive trace elements and hydrophobic organic compounds (*e.g.*, *Honeyman et al.*, 1988; *Dai et al.*, 1995).

The operational definition of the colloidal fraction is somewhat ambiguous. In recent years, the upper limit has been defined as material passing through a 0.20–0.45 μm filter; the lower size limit for the colloidal phase usually is considered to be $\approx 10^3$–10^4 Da (*e.g.*, *Reitmeier et al.*, 1996). A promising technique to separate sub-micrometer particles and colloids from dissolved solutes in natural water samples of 10–100 L is tangential cross-flow (ultra) filtration (CFF). For a thorough overview see the special volume of *Marine Chemistry*, 55, Nos. 1–2 (1996). However, major problems still arise from blanks, changes and definitions of effective cut-offs and physical and chemical interactions between the sample and CFF equipment during processing. At the time of writing no one standardized CFF system has been identified. Therefore, this filtration technique will not be outlined here in detail.

Nevertheless, differentiation by filtration (and centrifugation) between the dissolved phase, as defined, and the suspended particulate material (SPM) seems to be both reasonable and practical. There are several reasons why sample solutions are often filtered prior to analysis. Valuable information may be obtained from the composition of the particles; high concentrations of solids can cause analytical interference (*e.g.*, scattering of light in spectrophotometry). In addition, if the total sample is to be acidified (*e.g.*, for storage), then labile fractions of the particles might be released into solution and will be wrongly considered as part of the dissolved phase. On the other hand, the concentration of the dissolved phase may be reduced through adsorption of trace constituents onto reactive particles (such as bacteria).

Many other filtration-related factors exist that may influence the concentration of constituents in the 'dissolved' fraction. This was demonstrated by *Horowitz et al.* (1992) in a systematic investigation of membrane filtration artefacts when determining Fe and Al in synthetic and natural waters. They found that the dissolved concentration of these elements was significantly influenced by the filter type, filter diameter, method of filtration, concentration of SPM, the size of the particles or the volume of sample processed. Furthermore, mechanical treatment of particles, especially those of biological origin (*e.g.,* plant cells) during the filtration procedure may lead to destruction of cell membranes and thus release intercellular fluid into the filtrate. In an extended study, *Hall et al.* (1996) investigated the effects of four different 0.45 μm pore size filters on the 'dissolved' concentration of 28 elements in natural water samples. They found a mean standard deviation in elemental concentration across the four filter types of between 9 and 21 %, and that the nature and degree of filtration artefacts was matrix-dependent. Thus the potential presence of such filtration artefacts must be kept in mind, when, for example, data from various sources are being combined.

2.1.2 Filters

It must be emphasized that none of the existing filters and filter materials meet all requirements (from the analytical point of view) for universal application when seawater samples are to be filtered. As modified from *Riley et al.* (1975) and *Howard and Statham* (1993), an ideal filter should:

(1) have a uniform and reproducible pore size;
(2) have a high filtration rate and not clog easily;
(3) readily equilibrate with the surrounding atmosphere during gravimetric determinations;
(4) retain particles at the filter surface to aid analysis by microscopy and spectroscopic techniques;
(5) not adsorb dissolved constituents to be determined;
(6) not contain significant amounts of constituents to be determined;
(7) have reasonable mechanical strength;
(8) not shed fibres.

In practice all of these requirements cannot be met, not only as already discussed with respect to the separation of truly dissolved matter from the particulates, but also with respect to the extent of undesired but unavoidable side effects. A description of filters commonly used in oceanographic work is given in Table 2-1. Figure 2-1 shows the structures of some selected filter media.

Filters may be grouped into two main types (*Hurd and Spencer,* 1991). The pore sizes of 'depth filters' are not well defined, and the separation relies on physical trapping and surface contact. Generally, the thicker the filter, the smaller the effective size of particle which passes through. The filter materials include cellulose, metal oxides or glass fibres (see Table 2-1 and Fig. 2-1.). The pore sizes of 'sieve filters' are much better defined. They are made principally from plastic films; *e.g.,* polycarbonate (see Table 2-1 and Fig. 2-1). Some types (*e.g.,* Nuclepore filters) have very well defined holes, but the amount of material which can be collected on these filters is normally lower than that recovered on the 'depth filters'. It is also important to note that the effective pore size will change too as more and more particles accumulate on the filter.

The final choice of filter is usually a compromise between the various requirements of the analysis. Glass fibre filters are most commonly used for the analysis of biogenic core parameters (such as particulate organic carbon and particulate organic nitrogen (POC/PON) or chlorophyll; see Chapter 17) and trace organic substances (Table 2-1). Owing to their usually low metal concentrations, polycarbonate filters (*e.g.,* the Nuclepore type) are widely applied

Table 2-1. Commonly used filter materials and filter brands in seawater analysis.

Name	Material	Pore diameter, manuf. specific./μm	Thickness, manuf. specific./μm
Syringe filter			
Gelman Steril Acrodisc	Versapor	5	
Millipore Millex-SV	Durapore	5	
Nuclepore Syrfil-NP	Nylon with glass fibre	5	
Sartorius Minisart	cellulose acetate	5	
Schleicher & Schüll FP 030/10	cellulose nitrate	5	
Filter			
Gelman GN-6 Metricel	mixed cellulose esters	0.45	
Gelman Typ A/E	glass fibre		
Millipore HTTP	polycarbonate	0.4	
Millipore AP 15	glass fibre		
Millipore AP 20	glass fibre		
Millipore AP 25	glass fibre		
Nuclepore PC	polycarbonate	0.4	10
Nuclepore ME	mixed cellulose esters	0.45	125
Sartorius Type 111	cellulose acetate	0.45	
Sartorius Type 113	cellulose nitrate	0.45	100
Sartorius Type 13400	glass fibre		
Schleicher & Schüll ME 25	mixed cellulose esters	0.45	135
Schleicher & Schüll OE 67	cellulose acetate	0.45	115
Schleicher & Schüll BA 85	cellulose nitrate	0.45	
Schleicher & Schüll GF 92	glass fibre		350
Whatman WCN	cellulose nitrate	0.45	145
Whatman WME	mixed cellulose esters	0.45	150
Whatman WCA	cellulose acetate	0.45	125
Whatman GF/A	glass fibre		260
Whatman GF/C	glass fibre		260
Whatman GF/F	glass fibre		260

for trace element analyses. These membranes, however, are unsuitable for investigations of Hg because of adsorption and contamination problems (silica fibre filters may be used instead). Some Nuclepore filters also appear to suffer from Cr contamination. If seawater samples have to filtered for the determination of nutrients, then glass fibre (GF/C) filter disks of around 50 mm diameter with a nominal pore size of $\approx 1\ \mu$m are widely used (*Kirkwood*, 1996). Polycarbonate filters are used for the separation of particulate silicate as well as for the microscopic identification of particles collected on the filter.

All filters must be cleaned thoroughly before use. The treatment of filters depends on the analyte and the filter material. For detailed procedures see the sections on the determination of individual compounds. As discussed already, filters can affect samples by adsorption

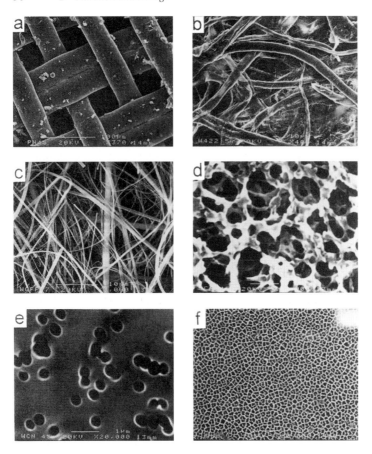

Fig. 2-1. Scanning electron micrographs of commonly used filtration materials. The nominal diameter of particles retained on these materials is given in parentheses, together with the magnification used: (a) phytoplankton netting (45 μm, × 185); (b) paper filter (Whatman 42; 2.5 μm, × 200); (c) glass fibre filter (Whatman GF/F; 0.7 μm, × 200); (d) cellulose nitrate membrane (Sartorius, 0.45 μm, × 10 000); (e) polycarbonate membrane (Nuclepore; 0.4 μm, × 10 000); (f) aluminium oxide filter (Whatman Anopore; 0.1 μm, × 10 000) (after *Howard and Statham*, 1993; with permission of authors and *John Wiley & Sons*).

or release of constituents or by disturbing the natural particulate composition. Therefore, it is essential to examine filtration procedures carefully, *i.e.,* to test the blanks of filters and to assess losses due to sorption onto the filter for each constituent of interest.

2.1.3   Filtration techniques

Generally, filtration serves two purposes, the separation of the SPM in order to receive a 'clean' water sample and/or separation (and enrichment) of the SPM for a subsequent analysis of its composition. In open-ocean seawater sampling, filtration is usually avoided, firstly, because SPM concentrations in these waters are relatively low (this does not hold, however, for all substances; see, for example, the survey on particulate Fe concentrations in the Atlantic Ocean by *Helmers et al.,* 1991), and secondly, because the filtration step itself is a potential source of serious contamination for many constituents. In spectrophotometric measurements of unfiltered samples it may be necessary to compensate for possible non-specific absorbance. It should be borne in mind that several reagents contain acid which may dissolve part of the suspended matter. Turbidity corrections should therefore be made with samples to which the corresponding amount of acids contained in the reagents has been added. This procedure is described under the sections dealing with the methodology for the determination of individual compounds.

If, for example, for sampling in marginal seas and the coastal (shelf) area filtration is necessary, it can be performed in different ways, briefly described in the following sections.

2.1.3.1 Vacuum filtration

Filtration is usually carried out by suction under aspirator vacuum. The filtration units have a filter support made of sintered glass or ceramic material, although all-plastic filtration apparatus are commercially available (*e.g.,* from Sartorius). Vacuum filtration has several disadvantages, mainly because of the potential contamination risks through the nature and number of transfer steps whereby the samples are in contact with different materials and the laboratory air.

2.1.3.2 Pressure filtration

Pressure filtration avoids problems observed with vacuum filtration. When the water sampler itself can be pressurized to force the sample through an in-line filter into the receiving bottle, the number of transfer steps is minimized, thus reducing the potential risk of contamination. If the redox state of the water sample must be retained to prevent the precipitation of, for example, iron and manganese hydroxides in higher valence states, an inert gas such as nitrogen must be used for pressurization.

Filter holders made of different materials are commercially available (*e.g.,* polycarbonate system from Sartorius; 'Swinnex' polypropylene apparatus from Millipore, or TFE–Teflon cylinders from various companies). An intercomparison study of the different filtration systems, used for the preparation of coastal seawater samples for the determination of dissolved trace elements, showed good agreement between the various systems. The filtrates from off-line systems did not exhibit marked differences when compared with filtrates from the in-line systems (*Bewers et al.,* 1985).

2.1.3.3 *In situ* filtration

In offshore waters most of the particulate trace inorganic and organic constituents occur at extremely low particle concentrations. Their determination involves a filtration step to raise the concentration levels to the sensitivity range of instrumental analytical methods. This is specially evident for sampling of deep ocean waters where SPM contents are in the range of about 10–50 $\mu g/L$ (*e.g., Sherrell and Boyle,* 1992). For example, to allow the detection of organochlorine compounds, such as polychlorinated biphenyls (PCBs) (see Chapter 22) using gas chromatography with electron capture detection (GC-ECD) with a detection limit for CB congeners of 0.1 pg, the processed water volume must be at least 100–500 L for particulate CB concentrations of between 0.01 and 0.05 pg/L *(Schulz-Bull et al.,* 1998). Large volume samplers of up to 400 L are available for the deep ocean (*e.g., Bodman et al.,* 1961; *Roether,* 1971; *Schulz et al.,* 1988). However, their use implies a number of problems; *e.g.,* they are often difficult to handle, cannot be controlled reliably with respect to contamination and have a serious risk of sample water exchange during recovery due to leakage. Finally, the most serious problem with bottle sampling is sedimentation of particulate material to the bottom of the sampler and incomplete recovery of these particles during filtration. Therefore, if samplers are applied it is recommended that the entire contents of the bottle are filtered. Swirling or shaking of the sampler to overcome the settling bias can alter partitioning between dissolved and particulate size fractions and therefore is not recommended.

Instead, *in situ* filtration systems offer many advantages over bottle sampling (see also Chapter 13). The clear advantages of pump systems are the large volume of water that can be filtered, the greatly expanded suite of determinations possible on a single sample and the minimization of blank problems. Pump systems can be easily deployed from moorings and are capable of providing a temporally integrated sample. Although a full depth profile using pumps requires about 12 h extra shiptime (when compared to a deep ocean CTD bottle cast; see Chapter 3), the return of orders of magnitude more material per unit shiptime can offset the additional shiptime required for pumping.

A number of systems have proven their usefulness under particular conditions (*e.g., Lisitzin,* 1972; *Krishnaswami et al.,* 1976; *Simpson et al.,* 1987; *Sherrell,* 1991; *Buesseler et al.,* 1992; *Baskaran et al.,* 1993). They were mainly designed for the specific task of collecting large amounts of particles for the determination of trace elements and radioisotopes (see also Chapter 13). Here, a system which has been applied successfully for the sampling of dissolved and particulate organic and particulate inorganic trace constituents in the deep ocean down to 6000 m is shown in Fig. 2-2 (see also Chapter 22; Fig. 22-1) and described in some more detail (*Petrick et al.,* 1996). It allows the *in situ* separation of suspended particulate matter and the extraction of 'dissolved' forms of individual organics by means of different resins (*e.g.,* XAD-2; see Chapter 18). The pumping rate can be selected and kept constant between 1 and 200 L/h. Glass fibre or polycarbonate filters (diameter of 140 or 290 mm) are used in an all-Teflon filter holder (see Fig. 2-2). If desired, for example for trace element analysis, the water does not contact materials other than Teflon and polyethylene. If the filter is downstream of the pump, for example when sampling the suspended and dissolved fractions of organic substances, a careful check should be made whether or not the pump leaves the particles intact. The unit is PC controlled, and all essential data (*e.g.,* pumping rate, time of start and end of pumping, volume sampled) are stored on disk.

2.1.3.4   Centrifugation

To avoid the potential problems of filtration, centrifugation immediately after sampling might be advantageous. *Kérouel and Aminot* (1987), for example, proposed a device whereby the subsample for nutrient analyses from the rosette sampler is collected in a polypropylene bottle and, without further handling, is centrifuged and injected directly into the

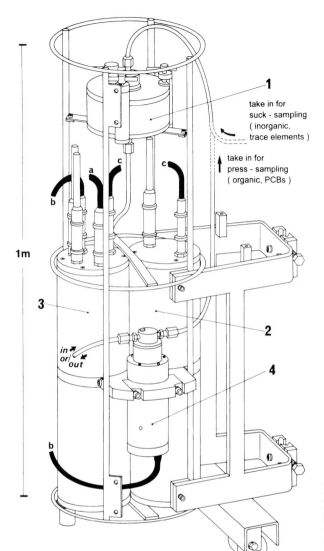

1

take in for
suck - sampling
(inorganic,
trace elements)

take in for
press - sampling
(organic, PCBs)

1m

3

2

in
or
out

4

Fig. 2-2. Line drawing of the *in situ* pump: 1 = filter holder; 2 = standard power pack housing; 3 = housing for electronics; 4 = pump; a, b, c = cable connections, (modified after *Petrick et al.*, 1996). See also Fig. 22-1.

autoanalyser system. This batch centrifugation also offers a tool for collecting larger amounts of particles. However, exchanging the supernatant with unprocessed water inevitably involves the risks of losses and contamination.

Where large samples of seawater are to be processed, *e.g.*, for studies of SPM in near-surface waters, continuous flow centrifugation procedures are appropriate. Ideally, they combine a sufficiently small spatial resolution of sampling with the capacity of efficiently collecting large amounts of particulate matter under conditions of low contamination. Such an approach is described in detail by *Schüßler and Kremling* (1993) and has been outlined in Chapter 1, Fig.1-2. The volume of seawater processed during a single open-ocean transect typically ranges from 5 to 12 m^3 corresponding to an SPM sample of about 1 g dry mass (assuming SPM concentrations in remote surface waters to be of the order of 100–200 mg/m^3). The seawater is in contact with polyethylene, polypropylene, Teflon and titanium only and this method has proved suitable for the collection of trace organic constituents as well (*e.g., IOC*, 1993). Thus flow centrifugation is a powerful tool for interdisciplinary work in near-surface waters. However, the possible damage of plankton cells is a disadvantage of the procedure. This should be checked carefully, *e.g.*, by microscopy.

For further information on filtration/separation systems see Chapters 12 (Section 12.6) and 13.

2.2   Storage

2.2.1   General remarks

The biological activity in seawater does not stop with sample collection, since bacteria and micro- and nano-plankton continue to digest and excrete material. In many cases the walls of the sampler and subsampling bottles are excellent substrates for bacteria, often enhancing bacterial growth rates by several orders of magnitude. In addition, the material of the sampler and the subsampling vessels may interact with dissolved micro-constituents. Even the chemically most inert materials may adsorb certain constituents (see also Chapter 1). Furthermore, photolytic breakdown of organic compounds or changes in the speciation of inorganic constituents (due to changes of pH or redox potential) may also alter the composition of samples.

The literature on the preservation of seawater samples is voluminous and often contradictory. It contains only a few systematic studies where samples have been stored for several months or even years. In addition, each analyte (or at least class of constituents) has its own reaction chemistry and consequently different requirements for storage in solution. Therefore, no general procedure can be recommended for the storage of seawater samples. It is always preferable to carry out the analysis as soon as possible after the sample has been retrieved. However, the determination of certain constituents requires the use of sophisticated equipment, and thus these analyses can often not be performed on board ship. The size of the ship or weather conditions may also be factors in the decision whether to store or analyse a sample. Very often the risk of contaminating the sample in a shipboard laboratory is high, and transport of the sample to a 'clean' laboratory is recommended. It is often advantageous to carry out some steps of the analysis immediately after sampling and continue later in the shore laboratory, thus avoiding the complications arising during storage and transport. (For details see chapters of the individual constituents.)

2.2.2   Storage for the determination of major compounds

Despite the fact that the major ions are present in seawater at relatively high concentrations (see Chapter 11), difficulties in storage have been observed by several workers. For example, most types of glass contain alkali and alkaline earth metals capable of ion-exchange reactions with the dissolved ions of seawater and thus are unsuitable materials for

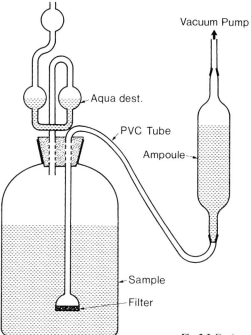

Vacuum Pump

Aqua dest.

PVC Tube

Ampoule

Sample

Filter

Fig. 2-3. Equipment for filtration and storage of the main components.

long-term storage containers. Certain types of plastics, *e.g.,* low-density polyethylene, tend to let water vapour and gases pass through slowly. Thereby, they may be a serious source of error in the determination of ion concentrations. High-density polyethylene proved to be satisfactory. Losses are assumed to be of minor importance if large volumes are utilized (about 5 L) and analyses are performed within a few days.

For long-term storage of seawater, the special glass ampoules used by the 'Standard Sea Water Service' should be employed. Seawater sealed in this type of resistant glass has been found to remain unchanged with respect to sample composition, electrical conductivity and density over a period of several years (see also Chapters 3 and 11).

In our experience, the following procedure is to be recommended for long-term storage of seawater:

The samples (about 5 L) taken from an ordinary hydrocast are dispatched to the laboratory in purified high-density polyethylene bottles. Each sample is filtered through glass fibre filters of low porosity and sealed in the special ampoules of approximately 300 mL volume as illustrated in Fig. 2-3. The filtration rate is regulated *via* the vacuum pump so that 80 % of the ampoule is filled in about 30 s. Residual drops of seawater in the neck of the ampoule are removed with filter paper before flame-sealing. The ampoules should be stored in the dark.

2.2.3   Storage for the determination of nutrients

2.2.3.1   General remarks

The concentrations of nutrients and other bioactive elements are liable to change due to the activity of microorganisms naturally present in seawater. Therefore, as a general rule, samples should not be exposed unnecessarily to light and analysed within a few hours after collection. Relatively simple and fast analytical procedures, and automated techniques in particular, make nutrient determinations possible even on board smaller vessels and under unfavourable weather conditions (see Chapter 10).

Nevertheless, it is sometimes necessary to postpone the analysis for some hours or days because of rough weather or shortage of personnel and laboratory space. There is ample literature on this subject (for overviews see, *e.g., Kirkwood, 1992, 1996; Dore et al., 1996*) indicating that no single universal preservation regime will satisfy all requirements. For example, glass containers are not suitable if silicate is to be determined; in different seasons samples from the same location may contain microorganisms of different species and concentrations, so that a given preservation regime could be effective in spring but not in fall. With this in mind the following two approaches to preservation are outlined; refrigeration and poisoning.

2.2.3.2   Refrigeration

Freezing

Freezing (to $-20\,°C$) is the method of choice of many workers if nutrient samples have to be stored for several weeks or even months (*e.g., Macdonald and McLaughlin, 1982; Macdonald et al., 1986; Kremling and Wenck, 1986; Chapman and Mostert, 1990; Kirkwood, 1996*). If the samples are visibly turbid, they should be filtered rapidly. Subsamples should be placed in carefully cleaned bottles and frozen, stored and thawed in an upright position. For storage, hard-glass bottles with Teflon-lined screw caps should be used or, preferably, high-density polyethylene, polycarbonate or polypropylene bottles. For silicate samples only plastic bottles are recommended. They should only be filled to $\approx 2/3$ of their volume to prevent squeezing of the liquid through the screw caps during the freezing process.

If possible, 'quick-freezing' in liquid nitrogen or in a dry ice-methane slurry (to $-20\,°C$ within about 20 min) is recommended. This procedure may offer considerable advantages especially for samples to be analysed for ammonia and nitrite. Otherwise these nutrient compounds might rapidly undergo biological transformations (*i.e.,* oxidation of ammonia to nitrite and further to nitrate or fixation as organically bound nitrogen in organisms).

In low-salinity or high-concentration samples ($> 120\,\mu mol/L$ of Si) dissolved silicate is likely to polymerize during the freezing process. As a consequence, when samples are thawed prior to analysis, sufficient time (preferably > 24 h) must be allowed for depolymerization. For such samples, it is recommended to store a separate aliquot of the silicate subsample under acidic conditions (see next section).

Non-frozen refrigeration

When coastal seawater is collected and a delay of no more than 2 h is expected between sampling and analysis, the nutrient samples should be stored in the dark in a refrigerator (at <8 °C). Samples with lower bacterial acitivity (such as deep-ocean waters) may be stored much longer. We recommend, however, to restrict the storage time of nutrient samples under these conditions to a maximum of 10 h. Silicate is best preserved by acidifying the seawater sample with sulphuric acid to a pH of about 2.5. With respect to nitrate, the same amount of ammonium chloride buffer should be added immediately after subsampling, as required for the subsequent nitrate analysis (see Chapter 10), and the solution should be stored in the dark at <8 °C. This buffer solution prevents further bacterial activities, *i.e.*, oxidation of ammonia and nitrite to nitrate. With respect to nitrite, in most cases, the storage of refrigerated samples in glass bottles for up to 3 h causes no significant changes in nitrite concentrations, provided the original ammonia level is <1 μmol/L. With regard to short-term storage of dissolved phosphate samples, neither the addition of acid (causing hydrolysis of polyphosphates and release of phosphate from plankton and bacteria) nor the addition of analytical reagents can be recommended since arsenate and silicate will also react slowly and could cause erroneous phosphate readings (see Chapter 10).

2.2.3.3   Poisoning

When refrigeration of samples is not possible, addition of poisoning chemicals is another option. The intention of this procedure is to poison those species which are responsible for consumption of nutrients. Of the various agents investigated, only three have found widespread application: acidification (mostly with sulphuric acid), chloroform and mercury(*II*) chloride. For a thorough review see *Kirkwood* (1992).

Sulphuric acid and chloroform

Neither of these two preservatives can be recommended here. With the exception of silicate (see Section 2.2.3.2), adding sulphuric acid may be even worse than no treatment. This is particularly true for the preservation of nitrogen compounds. At low pH, nitrite can be lost rather rapidly as a result of its oxidation to nitrate or because of its reaction with ammonia or α-amino acids. The unsuitability of this preservative for the storage of dissolved phosphate has already been outlined in Section 2.2.3.2.

There is general agreement in the literature that chloroform is unsatisfactory for various reasons. The major disadvantage is its volatility, which may cause losses by penetrating ill-fitting seals or even through the container materials in the case of some types of plastic bottles. In addition, interferences with analyser tubing material may occur. There are also reports on nutrient depletion apparently induced by the addition of chloroform (*Kirkwood*, 1992).

Mercury(*II*) chloride

The poisoning effect of mercury(*II*) ions is probably due to its binding with sulphydryl groups on proteins inhibiting enzyme activity. Therefore, the effective concentration of added mercury is highly dependent on the concentration of organic material and on the number of grazing species in the sample. Estuarine and coastal waters must be filtered before poisoning, otherwise added mercury could be scavenged by suspended particulate material (see also Chapter 12).

The use of mercury(II) chloride ($HgCl_2$) as preservative for seawater samples has a long history (for a review see *Kirkwood,* 1992). Over the years, however, it has lost some favour, most probably due to concurrent measurements of mercury in seawater at sub-nanogram per litre levels, which suffered contamination problems when any form of mercury was used on research vessels (see Chapter 12). The second objection to its use is that higher mercury concentrations may lead to erratic nitrate measurements if copperized cadmium reductors are used (see Chapter 10). There are, however, some contradictory results on the critical level at which mercury(II) chloride causes these analytical problems. *Kremling and Wenck* (1986) found that concentrations $> 10\,mg/L$ of $HgCl_2$ may cause problems in the determination of nitrate. On the other hand, these concentrations were inadequate to preserve unfiltered Atlantic Ocean water samples at $4\,°C$ containing around $10\,\mu mol/L$ and $0.5\,\mu mol/L$ of nitrate and phosphate, respectively, for more than two weeks. *Kirkwood* (1992) has shown that an $HgCl_2$ concentration of $20\,mg/L$ in filtered North Sea samples (with nitrate and phosphate levels of about $5.5\,\mu mol/L$ and $0.5\,\mu mol/L$, respectively) causes no problems in the determination of nitrate, and in addition, that this $HgCl_2$ concentration was an effective preservative at room temperature. Therefore, some further investigations are clearly required to establish a well-founded recommendable dosage concentration with regard to the analytical requirements.

2.2.4   Storage for the determination of trace elements

In contrast to the major components and nutrients, trace elements in seawater are extremely dilute analytes. Even minor absolute losses or sample contamination in the ultratrace range may significantly influence the results. In fact, the surface area of a container can be large enough to remove a trace element from solution completely by adsorption. Therefore, if possible, determination of trace elements should be performed on board ship within a few hours after collection. Relatively simple and fast electrochemical methods are now available for reliable on-board determinations of a number of trace elements if special conditions are met (see Chapter 12).

In most cases, however, long-term storage will be necessary. In the authors' experience and based on evaluation of published and unpublished data, the safest procedure to ensure minimum alteration of concentrations of the more often determined elements such as Cd, Co, Cu, Fe, Mn, Ni, Pb or Zn for a period of up to 2 years (or possibly longer) is to filter the samples (except open-ocean water), to acidify them to pH 1.5–2.0 with ultraclean acid (approximately $1\,\mu L$ of HNO_3 or HCl per mL of seawater sample; see Chapter 12) and to store them at about $4\,°C$ in properly cleaned containers made of Teflon, high-density polyethylene or quartz (see Chapter 12). Mercury samples are best stored in Pyrex or quartz bottles and are pre-acidified to give a final concentration of 2 % HNO_3. In the past, it was common practice to add oxidants such as permanganate or dichromate so as to keep the element in its Hg(II) ionic form and thus to prevent diffusive loss of elemental Hg through the container wall. This caused, however, severe contamination and blank problems. It is now acknowledged that the Hg concentrations of strongly acidified seawater samples in tightly closed bottles (under cool und dark conditions) will remain unchanged over weeks to months.

If the speciation of trace elements is to be studied, measurements must be performed immediately after sampling. If for some elements minor changes in original speciation are acceptable, then the samples should be stored unacidified but frozen (see also the storage of As, Sb and Ge outlined in Chapter 12) to preserve as much as possible the original distribution of species. A few studies indicate, however, that some metal associations with humic substances might be stable even down to pH 2.3 (*e.g., Helmers*, 1994).

An alternative 'storage' procedure for a number of elements is their separation from seawater by ion-exchange resins (*e.g.,* Chelex 100) on-board, with subsequent elution in onshore laboratories (see Chapter 12). This procedure has the advantage of small sample quantities to be stored and can be carried out on board smaller vessels if a clean bench is available to reduce contamination risks.

References to Chapter 2

Baskaran, M., Murphy, S.J., Santschi, P.M., Orr, J.C., Schink, D.R. (1993), *Deep-Sea Res.,* 40, 849.

Bewers, J.M., Yeats, P.A., Westerlund, S., Magnusson, B., Schmidt, D., Zehle, H. Berman, S.S., Mykytiuk, A., Duinker, J.C., Nolting, R.F. (1985), *Mar. Poll. Bull.,* 16, 277.

Bodman, R.H., Slabaugh, L.V., Bowen, V.T. (1961), *J. Mar. Res.,* 19, 141.

Buesseler, K.O., Cochran, J.K., Bacon, M.P., Livingston, H.D., Casso, S.A., Hirschberg, D., Hartman, M.C., Fleer, A.P. (1992), *Deep-Sea Res.,* 39, 1103.

Chapman, P., Mostert, S. A. (1990), *S. Afr. J. Mar. Sci.,* 9, 239.

Dai, M., Martin, J.-M., Gustave, C. (1995), *Mar. Chem.,* 51, 159.

Dore, J.E., Houlihan, T., Hebel, D.V., Tien, G., Tupas, L., Karl, D.M. (1996), *Mar. Chem.,* 53, 173.

Hall, G.E.M., Bonham-Carter, G.F., Horowitz, A.J., Lum, K., Lemieux, C., Quemerais, B. (1996), *Appl. Geochem.,* 11, 243.

Helmers, E., Mart, L., Schrems, O. (1991), *Fresenius J. Anal. Chem.,* 340, 580.

Helmers, E. (1994), *Fresenius J. Anal. Chem.,* 350, 62.

Honeyman, B.D., Ballistrieri, L.S., Murray, J.W. (1988*), Deep-Sea Res.,* 35, 227.

Horowitz, A.J., Elrick, K.A., Colberg, M.R. (1992), *Water Res.,* 26, 753.

Howard, A.G., Statham, P.J. (1993), *Inorganic Trace Analysis.* Chichester: John Wiley & Sons, 1993.

Hurd, D.C., Spencer, D.W., (Eds.) (1991), *Marine particles: Analysis and characterization.* Washington: American Geophysical Union, Geophysical Monograph 63.

IOC (1993), *Chlorinated biphenyls in open-ocean waters: sampling, extraction, clean-up and instrumental determination.* Paris: *UNESCO Tech. Rep.* Ser. 25.

Kérouel, R., Aminot, A. (1987*), Mar. Environ. Res.,* 22, 19.

Kirkwood, D.S. (1992), *Mar. Chem.,* 38, 151.

Kirkwood, D.S. (1996), Nutrients: Practical notes on their determination in seawater. Copenhagen: ICES Tech. *Mar. Environ. Sci.,* 17, 25 pp.

Koike, I. S., Hara, S., Terauchi, K., Kogne, K. (1990), *Nature,* 345, 242.

Kremling, K., Wenck, A. (1986), *Meeresforschung,* 31, 69.

Krishnaswami, S., Lal, D., Somayalulu, B.L.K., Weiss, R.F., Craig, H. (1976), *Earth Planet Sci. Lett.,* 32, 420.

Lisitzin, A.P. (1972), *Sedimentation in the world ocean.* Tulsa: *S.E.P.M. Spec. Publ.,* 17.

Macdonald, R.W., McLaughlin, F.A. (1982), *Water Res.,* 16, 95.

Macdonald, R.W., McLaughlin, F.A., Wong, C.S. (1986), *Limnol. Oceanogr.,* 31, 1139.

Petrick, G., Schulz-Bull, D.E., Martens, V., Scholz, K., Duinker, J.C. (1996), *Mar. Chem.*, 54, 97.

Reitmeier, R., Powell, R.T., Landing, W.M., Measures, C.J. (1996*), Mar. Chem.*, 55, 75.

Riley, J.P., Robertson, D.E., Dutton, J.W.R., Mitchell, N.T., Williams, P.J. leB. (1975), in: *Chemical Oceanography,* 2nd ed.: Riley, J.P., Skirrow, G. (Eds.). London: Academic Press, 1975; vol. 3, pp. 193–514.

Roether, W. (1971), *J. Geophys. Res.,* 76, 5910.

Schulz, D.E., Petrick, G., Duinker, J.C. (1988), *Mar. Poll. Bull.,* 19, 526.

Schulz-Bull, D.E., Petrick, G., Bruhn, R., Duinker, J.C. (1998), *Mar. Chem.*, 61, 101.

Schüßler, U., Kremling, K. (1993), *Deep-Sea Res.*, 40, 257.

Sherrell, R.M. (1991), in: *Marine particles; Analysis and characterization:* Hurd, D.C., Spencer, D.W. (Eds.). Washington: American Geophysical Union, Geophysical Monograph 63, pp. 285–294.

Sherrell, R.M., Boyle, E.A. (1992), *Earth Planet. Sci. Lett.*, 111, 155.

Simpson, W.R., Gwilliam, T.J.P., Lawford, V.A., Fasham, M.J.R., Lewis, A.R. (1987), *Deep-Sea Res.,* 34, 1477.

Wells, M.L., Goldberg, E.D. (1991*), Nature*, 353, 342.

3 Determination of salinity

T. J. Müller

3.1 Introduction

In oceanography, there is interest in the composition of seawater. Firstly, chemical oceanographers want to know which elements in what concentration can be found in seawater, and what their regional and temporal distributions are. Secondly, (mainly) physical oceanographers want to derive from its composition the density, mostly because its variability in time and space is important for understanding the ocean circulation. The first group can achieve its goal by chemically analysing seawater samples using various methods as described in this book.

Since the end of the 19th century it has been known that the composition of seawater is almost constant in space and time (the concept of 'conservatism', see Chapter 11). Therefore, to a good first approximation, oceanographers assume that seawater consists of just two components, the first one being pure water and the second one representing all dissolved ions that contribute to the mass of seawater, namely salinity. By this two-component assumption, three thermodynamic parameters are needed to derive the state of seawater. Besides salinity, it is convenient to choose temperature (T) and pressure (P) since they are relatively easy to measure at the required accuracies and, together with salinity, are also valuable for water mass analysis.

The methods to determine salinity have changed with time. From the beginning of the 20th century until 1978, the constituent selected to represent and estimate the salinity was chlorinity (see Sections 11.1 and 11.2.4 for definition). Salinity then, together with the other two basic parameters, served to calculate density.

With the introduction of electrical measurement methods and shipboard computers into oceanography in the 1960s, electrical conductivity replaced chlorinity as the third basic parameter. Two reasons were responsible for this change. Firstly, conductivity can be measured electrically *in situ* along with pressure and temperature, and salinity and density can be derived on-line once the fundamental determinations needed for such calculations have been made. Secondly, it turned out that based on electrical conductivity, the accuracy in density that could be achieved is an order of magnitude better than that based on chlorinity.

In this chapter, a brief overview of the early definitions of salinity based on the chlorinity concept is given. It is followed by a description of the definition of the Practical Salinity Scale 1978 (PSS78), which is based on the measurement of electrical conductivity. Finally, methods are described that are used to derive salinity (and thus density) with modern instrumentation, both from bottle samples on a bench and *in situ*.

3.2   Symbols and abbreviations

C	=	Conductivity as a function of S, T and P
CTD	=	Conductivity Temperature Depth profiler
EOS80	=	Equation of State of Seawater on the 1980 scale
IAPSO	=	International Association for the Physical Sciences of the Ocean
ICES	=	International Council for the Exploration of the Sea
IPTS68	=	International Practical Temperature Scale of 1968
ITS90	=	International Temperature Scale of 1990
JPOTS	=	UNESCO/ICES/SCOR/IAPSO Joint Panel on Oceanographic Tables and Standards
K_{15}	=	Conductivity ratio of seawater to a KCl solution as defined in the PSS78 scale, both at one standard atmosphere pressure and at T=15°C
P	=	Pressure
PSS78	=	International Practical Salinity Scale of 1978
R	=	Conductivity ratio of seawater to SSW at one standard atmosphere pressure, T=15°C, S=35
R_{15}	=	Conductivity ratio of seawater to SSW, both at 15 °C
r_T	=	Conductivity ratio of seawater to SSW at one standard atmosphere
R_T	=	Conductivity ratio of seawater to SSW at one standard atmosphere and temperature T
R_P	=	Conductivity ratio of seawater to SSW: pressure term
R_1, R_2, R_p, R_w	=	Variable electrical resistances
SCOR	=	Scientific Committee on Oceanic Research
S	=	Practical salinity (PSS78 scale)
S_A	=	Absolute salinity (mass of salts in 1 kg of seawater)
SSW	=	Standard seawater
t	=	Time
T	=	Temperature
τ	=	Time constant
UNESCO	=	United Nations Educational, Scientific and Cultural Organization
WOCE	=	World Ocean Circulation Experiment

3.3   Definition of salinity

3.3.1   Early concepts

The problem of how to determine the mass fraction of salts (S_A) in seawater quickly and with sufficient accuracy (10^{-5} kg/m^3 for derived density) has occupied oceanographers over the last century and led to three definitions of salinity (S) that approximate S_A. These have been reviewed thoroughly by *Lewis* (1980) and by *Fofonoff* (1985), and will be briefly outlined in this and the next section.

Based on the relatively constant composition of the bulk of seasalts, Knudsen and co-workers (*Forch et al.*, 1902) suggested salinity, S‰, to be determined from the titration of chlorinity with silver nitrate using Mohr's (1856) method as

$$S‰ = 0.03 + 1.8050 \; Cl‰. \tag{3-1}$$

To derive Eq. (3-1), Knudsen and co-workers mainly used samples from the Baltic and from the Mediterranean and the Red Sea. The offset of 0.03 in Eq. (3-1) reflects that the salt composition especially in the Baltic Sea is not exactly constant (*Millero and Kremling*, 1976) and thus contradicts the basic assumption of constancy that led to Eq. (3-1). However, the proposed titration method also had advantages: it could be performed in reasonable time onboard a ship, and for salinity measurements in open ocean areas where S is close to 35‰, the error induced by the non-constancy is less than that from titration (0.02‰ in salinity). Therefore, the so-called 'Mohr-Knudsen' titration method (*Mohr*, 1856; see Chapter 11) and the 'Knudsen' formula (3-1) served oceanographers for more than 60 years to determine salinity from chlorinity.

To achieve the accuracy of 0.02‰ in salinity for a single determination with the above method, and also to make salinity determinations from different institutions comparable, Knudsen required frequent comparison with an internationally accepted standard of known chlorinity. On behalf of the International Council for the Exploration of the Sea (ICES), standard seawater (SSW), later often called Copenhagen Water as it was prepared in Copenhagen for a long time, served this purpose. Each batch was numbered and distributed to oceanographers in sealed ampoules with the chlorinity indicated. For a detailed description of the history of standard seawater see *Culkin and Smed* (1979).

In the early 1960s, bench salinometers were developed that allowed measurement of the electrical conductivity of a seawater sample relative to that of a standard with high precision. *Cox et al.* (1967) had related chlorinity and conductivity ratios of seawater to standard seawater at temperatures higher than 10 °C and tabulated their results (*UNESCO*, 1966). Following their work, the responsible international oceanographic organizations adopted a redefinition of salinity (*Wooster et al.*, 1969). Firstly, it was assumed that salinity was proportional to chlorinity, to be consistent with the assumed constancy of the ionic composition. The constant was chosen so that for S=35‰, both the Knudsen formula (3-1) and the new relationship

$$S‰ = 1.806\,55 \; Cl‰ \tag{3-2}$$

yielded the same chlorinity. Also, within the oceanic range of salinities the difference between both methods of calculation is much less than the titration error. Next, Eq. (3-2) was used to convert chlorinity into salinity for the results from *Cox et al.* (1967). This gave the defining relationship of salinity as a polynomial of 5th order in R_{15}, where R_{15} was the measured ratio of the specific electrical conductivity of a seawater sample to that of standard seawater that had S=35‰, both samples being at 15 °C and at a pressure of one standard atmosphere.

The 1969 definition and the available corrections for temperatures higher than 10 °C were suitable for use with bench salinometers. However, for use with *in situ* measuring systems, it was insufficient because it did not cover the whole oceanic range of parameters. It was this severe lack of validity and the demand to define salinity solely from physical parameters that led to the definition of the Practical Salinity Scale of 1978.

3.3.2   The practical salinity scale of 1978 (PSS78)

The required new scale was to (i) be independent of the exact knowledge of the composition of seawater, (ii) be solely based on physical measurements and (iii) maintain continuity with the previous definitions. Following the suggestions of *Lewis and Perkin* (1978), the major international oceanographic organizations as represented in the Joint Panel of Oceanographic Tables and Standards (JPOTS) adopted the definition of a new scale which is in use today (*UNESCO*, 1981a). Firstly, it relates a quantity termed practical salinity S to the mass fraction S_A (absolute salinity) of salts in seawater by the linear relationship

$$S_A = a + b\, S \tag{3-3}$$

Next, it defines the practical salinity S in terms of the ratio K_{15} of the specific electrical conductivity (hereafter termed conductivity) of seawater to that of a reference potassium chloride (KCl) solution, both at a temperature of $15\,°C$ and under a pressure of one standard atmosphere:

$$S = a_0 + a_1 K_{15}^{1/2} + a_2 K_{15} + a_3 K_{15}^{3/2} + a_4 K_{15}^2 + a_5 K_{15}^{5/2} \tag{3-4}$$

where

$$
\begin{aligned}
a_0 &= +\ 0.0080 \\
a_1 &= -\ 0.1692 \\
a_2 &= +25.3851 \\
a_3 &= +14.0941 \\
a_4 &= -\ 7.0261 \\
a_5 &= +\ 2.7081 \\
\Sigma\, a_i &= 35
\end{aligned}
$$

To preserve continuity with the previous 1969 definition, the reference KCl solution has a mass fraction of 32.4356×10^{-3} of KCl, to have the same conductivity as a standard seawater with S=35 (Cl=19.3739 ‰ in the 1969 definition, see Eq. (3-2)). Equation (3-4) is valid for a practical salinity S from 2 to 42. For $K_{15}=1$, the practical salinity is S=35. As the coefficients in Eq. (3-4) were determined by comparison of diluted and evaporated fractions of standard seawater with the KCl solution, a standard sea water with known K_{15} value may replace the KCl solution as a reference. This is the general practice for preparing standard seawater.

On the 1978 scale, practical salinity S no longer depends on the ionic composition of seawater. Any improvement in the knowledge of the ionic composition of seawater will change the coefficients a and b in Eq. (3-3), but not S. For standard seawater and its dilutions, $a = 0$ and b has a value close to 1 depending on the ionic composition of the batch.

Note that the practical salinity S has no unit assigned. In the background papers that were published prior to the final adoption of its definition and that were reprinted later by *UNESCO* (1981a), the symbol ‰ is still used with practical salinity S. However, the symbol ‰ was omitted in the final adoption. In the literature, sometimes the term psu appears to denote practical salinity values although its usage is not recommended.

Instruments, termed CTD, are available for the *in situ* measurements of conductivity, temperature and pressure (and derived depth). However, as temperature and pressure values change *in situ*, Eq. (3-4) cannot be used directly to calculate S. Additional relationships are needed to link S to the entire oceanic temperature and pressure ranges. Let

$$R = C(S,T,P)/C(35,15,0) \tag{3-5}$$

denote the measured ratio of the conductivity of seawater $C(S,T,P)$ at salinity S, temperature T and pressure P to that of the standard KCl solution or equivalent standard seawater, $C(35,15,0)$. From the definition of S, no value for $C(35,15,0)$ needs to be fixed. *Culkin and Smith* (1980) have determined a value that quite frequently is used to convert the ratio R to conductivity: 4.2914 S/m^2 (42.914 mS/cm^2), where S here is the unit of conductance ($1\,S = 1\,\Omega^{-1}$).

The ratio R can be factorized with sufficient accuracy (see *Lewis* (1980) and *Fofonoff* (1985) for more detailed reviews):

$$R = R_P\,R_T\,r_T \tag{3-6}$$

where

$$
\begin{aligned}
R_P &= C(S,T,P)/C(S,T,0) \\
R_T &= C(S,T,0)/C(35,T,0) \\
r_T &= C(35,T,0)/C(35,15,0)
\end{aligned}
$$

Given the measured quantities R, $T/°C$ and $P/dbar$, the following steps (Eqs. (3-7) to (3-0)) are performed to derive practical salinity S in the form and with coefficients as approved by JPOTS (*UNESCO*, 1981a).

The pressure term R_P solely depends on the measured quantities R, T and P:

$$R_P = 1 + (e_1\,P + e_2\,P^2 + e_3\,P^3)/(1 + d_1\,T + d_2\,T^2 + (d_3 + d_4\,T)\,R) \tag{3-7}$$

with

$$
\begin{aligned}
e_1 &= +2.070 \times 10^{-5} & d_1 &= +3.426 \times 10^{-2} \\
e_2 &= -6.370 \times 10^{-10} & d_2 &= +4.464 \times 10^{-4} \\
e_3 &= +3.989 \times 10^{-15} & d_3 &= +4.215 \times 10^{-1} \\
& & d_4 &= -3.107 \times 10^{-3}
\end{aligned}
$$

The temperature term of standard seawater is given by a polynomial:

$$r_T = \Sigma\,(c_i\,T^i) \tag{3-8}$$

with

$$
\begin{aligned}
c_0 &= +6.766\,097 \times 10^{-1} \\
c_1 &= +2.005\,64 \times 10^{-2} \\
c_2 &= +1.104\,259 \times 10^{-4} \\
c_3 &= -6.9698 \times 10^{-7} \\
c_4 &= +1.0031 \times 10^{-9}
\end{aligned}
$$

Now, Eq. (3-6) can be solved for R_T:

$$R_T = R/(R_p\,r_T) \tag{3-9}$$

Salinity S then is linked to R_T and temperature T:

$$S = \Sigma\,(a_i\,R_T^{i/2} + f\,b_i\,R_T^{i/2}) \tag{3-10}$$

with

$$
\begin{aligned}
f &= (T-15)/(1+ k\,(T-15)) \\
k &= 0.0162
\end{aligned}
$$

coefficients a_i as in Eq. (3-4)
and

$$b_0 = +0.0005$$
$$b_1 = -0.0056$$
$$b_2 = -0.0066$$
$$b_3 = -0.0375$$
$$b_4 = +0.0636$$
$$b_5 = -0.0144$$
$$\Sigma\, b_i = 0$$

Note that for T=15 °C the second term in Eq. (3-10) vanishes and (3-10) reduces to the defining Eq. (3-4) for S.

The above set of formulae is valid for S from –2 to 42, for T from 2 to 35 °C, and for P from 0 to 10 000 dbar. Figure 3-1a shows that the conductivity at atmospheric pressure $C(S,T,0)$ increases with both salinity and temperature. The dependency of $C(35,T,P)$ on pressure and temperature is shown in Figure 3-1b. The conductivity increases with both temperature and pressure. A typical oceanic profile will first show the conductivity decreasing with temperature as the temperature effect dominates. Then, as temperature changes become weak in the deep ocean conductivity will increase with pressure.

The precision of the fits to the basic data is better than 0.001 in salinity. At S=35 and at atmospheric pressure, a 0.01 apparent salinity change is caused by changes in temperature or in conductivity ratio of 0.01 K or 0.0002, respectively. At 1000 dbar pressure, a 20 dbar change causes a 0.001 change in apparent salinity. From these numbers it is clear that the accuracy that can be achieved for salinity critically depends on the quality of calibration and performance of the basic measurements, *i.e.*, on well calibrated bench and of *in situ* instruments, so-called salinometers. However, achievable accuracy also depends on slight deviations of standard seawater batch salinity from its labelled salinity value. Using the KCl standard as reference, deviations of up to 0.003 have been documented by *Mantyla* (1987) for standards that were prepared up to the 1980s. From then on (batch P93), deviations decreased to less 0.001 between measured and labelled practical salinity (*Takatsuki et al.*, 1991).

Algorithms in FORTRAN computer language to calculate S and other fundamental properties of seawater have been published by *Fofonoff and Millard* (1984). Given S, T and P, the conductivity ratio R can be calculated by numerical back iteration. The accompanying test values for correct programming should be used when algorithms are newly developed or are copied. *Siedler and Peters* (1986) have compiled a set of formulae for the calculation of some physical properties of seawater taking into account the PSS78 and the equation of state of seawater on the 1980 scale EOS80 (see *UNESCO*, 1981b).

Note that in the definition of the salinity scale, temperature is measured on the International Practical Temperature Scale 1968 (IPTS68). After definition of the PSS78, the International Temperature Scale 1990 (ITS90) has been defined, and reporting temperature in this scale has been required since then. To be consistent when using the PSS78, temperatures that are reported in the ITS90 scale have to be converted into the IPTS68 scale prior to using the PSS78 (and prior to the calculation of other derived oceanographic quantities). In the oceanic range of temperature, the difference in both scales is up to 7 mK at 30 °C. A relationship for conversion that is valid from –10 to 40 °C with an error of less 0.4 mK has been proposed by *Saunders* (1990) and is now widely accepted:

$$T_{68} = 1.000\ 24\ T_{90} \tag{3-11}$$

a)

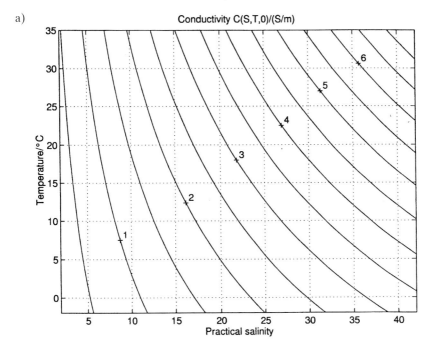

b)

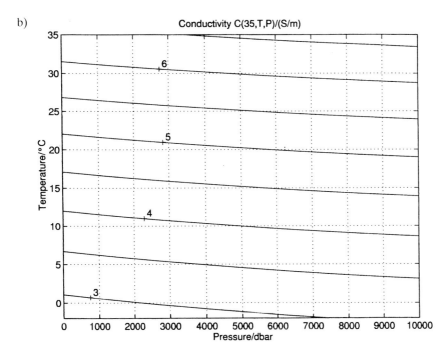

Fig. 3-1. Specific electrical conductivity $C/$(S/m) (a) in T–S space at atmospheric pressure and (b) in T–P space for S=35. Relative values have been converted into absolute values by multiplying with the factor 4.2914 S/m, a value which often is used to represent $C(35,15,0)$.

Practical salinity is not compatible with salinity values derived from other conductivity based algorithms. *Lewis and Perkin* (1981) have published corrections for the most frequently used formulae that may be used to match those salinity values to the PSS78. Some of these corrections can amount to 0.02 in salinity S.

3.4   Measurement of the conductivity ratio

The principles of the two methods still in use in oceanography to measure the specific electrical conductivity of seawater relative to that of a reference, have been described by *Dietrich et al.* (1975). The galvanic method uses electrodes in Wheatstone type bridges. Alternating current is needed to avoid polarization effects at the electrodes. In the simplest case (Fig. 3-2a), the resistor R_w of a seawater volume is measured between two electrodes and compared in a Wheatstone bridge with a potentiometer R_p and two high precision resistors R_1 and R_2. If R_p is changed such that the bridge is balanced, *i.e.*, the meter G measures no current, the relationship $R_w/R_1 = R_p/R_2$ holds, from which R_w can be determined. Further, if both R_1 and R_2 are compared to standard seawater by standardization of the instrument, the desired conductivity ratio is measured directly (incorporating a factor known from the standardization). The electrodes are mainly metallic rings inside a glass tube, the cell. A two-electrode cell has an outer electrical field. To minimize effects from irregularities of the outer field, high precision cells with more than two electrodes are in use. In these constructions, the outer electrodes are kept at the same electrical potential thus avoiding outer fields, and the measurements are made between the inner electrodes.

The inductive method uses seawater as the coupling loop between two transformers (T_1 and T_2 in Fig. 3-2b). Through this loop, the transformer T_1 induces a current in T_2 which is measured with a galvanometer (G in Fig. 3-2b). The induced current depends on the coupling in the compensating loop with resistor R_p. This loop induces a counter current in T_2. If R_p is changed such that the current through G vanishes, $R_w = R_p$.

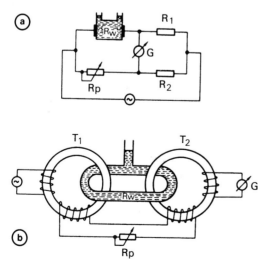

Fig. 3-2. Principle of conductivity measurement with (a) electrode and (b) inductive cells.

3.5 Salinity from bench salinometers

3.5.1 Purpose

Although diverse electronic systems are available on the market that measure salinity *in situ*, the determination of salinity from bottle samples is still important. The main reason is that despite the expectation of early constructors, experience has shown that high accuracy (better than 0.002) of *in situ* derived salinity can only be achieved with an almost station-by-station calibration using salinity values from bottle samples.

To determine salinity from bottle samples, bench salinometers have now replaced the chlorinity titration method almost completely, initially for practical reasons. Subsequently, as the 1978 Practical Salinity Scale has been adopted, consistent application of this scale requires the use of salinometers.

Nevertheless, there are only a few bench salinometers commercially available that fulfill the requirements for accurate and simple operation. The one that has become a standard for high precision measurements in temperature-stable laboratories during recent international programmes, such as the World Ocean Circulation Experiment (WOCE), is the AUTOSAL Model 8400 B made by Guildline (Section 3.5.4). The less expensive Beckman Model RS10 (Section 3.5.5) is also frequently used. It is portable and particularly valuable for coastal and estuarine research, and for waters of high biological production as biological contamination has little effect on its operation.

High accuracy salinity measurements (0.002) require knowledge of the interpretation of standard seawater measurements (Section 3.5.2), and careful sampling, storage and logging (Sections 3.5.3 and 3.5.6). Along with Section 3.5.4 on the operation of the AUTOSAL, these sections describe procedures as recommended for the WOCE (see *Stalcup*, 1991) with some supplementary instructions and remarks added.

To reduce costs, users sometimes replace standard seawater with substandards to calibrate their salinometer. The use of substandards will inevitably result in reduced accuracy, but under certain circumstances may be acceptable. Their usage is described in Section 3.5.7.

3.5.2 Standard seawater

Only IAPSO recognized SSW provides a reliable standard for conductivity ratio measurements. It should be used to standardize each bench salinometer before being used for sample salinity measurements and to detect and eventually trace any drift. The use of so-called substandards for standardization is not recommended, as it will significantly decrease accuracy (see Section 3.5.7 for exceptions if a decrease in accuracy is acceptable).

IAPSO recognised SSW is sold in vials that are labelled as such. Its filling date along with its K_{15} and salinity values are printed on the label. Vials older than 4–5 years should be compared with younger batches before being used to detect any changes in salinity due to ageing. SSW is usually from a batch with salinity slightly below 35 to enable highest accuracies for salinities from the deep ocean to be obtained. For non-open-ocean applications, SSW from batches with salinities of 10, 30 and 40 are available.

Experience shows that about 5 % of SSW vials from the same batch may deviate in salinity from their printed value by more than the claimed accuracy (0.001). After an instrument has been standardized if a vial shows differences larger than twice the instrument's precision (0.001 for both, the AUTOSAL and the RS10), it should immediately be checked with a follow-up vial to identify a possibly 'bad' vial.

After a salinometer has been standardized, SSW measurements are required on a regular basis to detect any drift or malfunction. Small offsets of SSW measurements from its printed salinity value can be used for linear trend corrections. Large offsets usually indicate non-stable room temperature or a malfunction of the salinometer.

3.5.3 Sampling

In situ water sampling

Station salinity samples for analysis with a salinometer are taken from water samplers that are closed at desired depths or pressures. In many cases, a rosette sampler is used that carries Niskin or comparable water sampling bottles (see Chapter 1). The bottles are mounted in one or more rings on a frame. Closing the bottles at desired pressure levels is triggered from the deck unit in the ship's laboratory. New systems allow selection of bottles individually and not just in the order of their position on the frame. Most frequently, and necessary for WOCE standards, the rosette sampler is operated together with a CTD that is mounted on the rosette's frame. The CTD's on-line pressure information is then used for controlled closing of the water samplers.

Other systems allow stand-alone operation of a rosette sampler. Some have implemented a pressure sensor. Its signal triggers the closing of the bottles at prescribed levels that are set before-cast. In a simpler application, a rosette sampler may be closed acoustically at certain wire lengths. For special applications, hydrocasts with water samplers mounted on a hydrographic wire are still used.

Samples for salinity measurements are transferred from water samplers to salinity sample bottles and stored in the laboratory before they are analysed with a salinometer.

Water sampling not only serves the purpose of salinity determinations. For purposes other than salinity sampling, the design of the water bottles may be adjusted; *e.g.,* for some purposes the springs that keep the caps tightly closed are made of stainless-steel while for others they are made of silicon. It has not been so far reported whether these changes or the specific material from which the bottles are made influences the salinity content of water samples significantly during the relatively short time (3 hours) that they are in contact with seawater between closing and sampling.

Salinity sample bottles

The minimum capacity of salinity sample bottles should be 120 mL to ensure sufficient volume of sample for sufficient rinsing of the salinometer cell. Larger volumes will decrease the relative error due to contaminations during sampling. Bottles made of flint glass with Poly-Seal screw caps prevent leakage by evaporation and chemical interaction of seawater with the glass. With these bottles, storage times have been reported of up to 6 months with salinity shifts of less than 0.001 (*Stalcup*, 1991). The screw caps of the sample bottles should be replaced after 2–3 years and bottles after 8–10 years.

Other sample bottles in common use include glass beer bottles of 330 mL volume with rubber sealed porcelain caps and a stainless steel-clipping mechanism. After a typical maximum storage time before measurement at sea, *i.e.,* a few days to 2 weeks, salinity samples show less than 0.001 variability. After storage times of 6 months, variability may increase slightly up to 0.002, which is probably due to interaction of the glass with the seawater (*Sy and Hinrichsen,* 1986). The rubber seals need to be replaced after 1–2 years.

Salinity sampling

From the water samplers, salinity sample bottles are a quarter filled while avoiding direct contact between the bottle neck and the outlet of the water sampler. The salinity sample bottle is shaken well for at least 10 s to clean the inside and then emptied. This procedure is repeated three times before filling the bottle. If the sample bottle has caps fixed to the bottle neck, the bottle is shaken upside down with the cap loosely held on the outlet to rinse the cap; otherwise, the cap must be rinsed separately.

For storing a sample, the bottle is filled up to the shoulder which leaves space for thermal expansion of the sample. Screw caps must be dried on the inside before closing the bottle to avoid the formation of salt crystals in the threads during storage.

While rinsing and closing the bottle, the operator must take care never to touch the cap's sealing mechanism or the bottle outlet with bare hands. Also, the operator should be aware of surface spray, rain and other sources of contamination.

Storage

Sample bottles are kept in cases large enough to store all bottles from a deep-ocean station. The cases with the samples are stored in the temperature controlled laboratory where the salinometer is operated. Depending on the size of the sample bottles, the samples will have reached laboratory temperature between a few hours and one day. Good ventilation may speed up the equilibration.

Logging

To minimize errors while logging, it is recommended that sample bottles have the same numbers as the water samplers from which the samples are taken. However, other methods are acceptable as long as they allow clear identification.

Each cast has one logsheet for salinity sampling. It must contain the cruise identification, the station and cast numbers, the numbers of the water samplers from which the samples are taken, the storage case identification and the salinity sample bottle numbers. It is not recommended solely to log nominal depth or pressure instead of the number of the water sampler. Table 3-1 shows an example of a salinity sampling logsheet where salinity samples were taken from samplers of a CTD–rosette system. In this example, the CTD data acquiring programme created the form while the water bottles were closed at nominal pressures. Averages and standard deviations over a 20 s recording interval (the typical swell period) of CTD pressure, temperature and salinity values are written to file for later use while the averages are included on the printout of the form. The column for the storage case and the bottle numbers of salinity samples are filled in later when the salinity samples are taken from the water samplers.

Table 3-1. Salinity sampling logsheet with number of water sampler BNO, time of bottle closure; CTD values for pressure PCTD, temperature TCTD and salinity SCTD at bottle closings; salinity sample bottle number SBNO.

Cruise: POSEIDON 233/1   Station: 671   Cast: 2
Filename: HIEVP233.DAT   Date: 04.10.1997
Remark: ESTOC October 1997

BNO	Time UTC	PCTD	TCTD	SCTD	Box no. 39 SBNO
00	12:35	2	22.535	0.007	before-cast deck value
01	13:39	3500	2.570	34.902	617
02	13:51	3000	2.818	34.930	1367
03	14:06	2500	3.351	34.989	1348
04	14:18	1996	4.398	35.115	1344
05	14:30	1498	6.144	35.318	not closed
06	14:36	1300	7.873	35.560	855
07	14:39	1196	8.518	35.634	1261
08	14:41	1100	8.619	35.566	850
09	14:45	1000	9.067	35.595	621
10	14:49	800	9.940	35.528	614
11	14:54	598	11.211	35.565	609
12	14:57	399	13.301	35.805	611
13	14:59	300	14.467	36.017	601
14	15:02	196	16.236	36.309	612A
15	15:04	150	17.314	36.508	600
16	15:05	125	17.867	36.600	598
17	15:06	101	18.592	36.683	604
18	15:07	74	20.730	36.844	606
19	15:08	49	22.536	36.964	619
20	15:10	24	23.918	37.062	587
21	15:11	14	24.013	37.060	603
99	15:13	1	22.662	0.445	post-cast deck value

3.5.4   The Guildline AUTOSAL Model 8400 B

Principle of operation

The AUTOSAL 8400 B (Fig. 3-3; *Guildline*, 1997) uses the galvanic method with a 4-electrode cell that is immersed in a constant temperature bath. Using a potentiometer, the circuit is standardized with SSW as reference. Twice the value of R_T is displayed. The measurement procedure assumes that the sample is at the prescribed bath temperature. A digital BCD output is optional.

Accuracy

It is stated in the manual, that the AUTOSAL has a resolution of 0.0002 and an accuracy of 0.003 in salinity. However, an accuracy of 0.001 in salinity can be achieved if some of the operating conditions that are also described in the manual are carefully observed. WOCE and some earlier cruises of leading institutes have proven this (*Stalcup*, 1991).

Three general conditions must be fulfilled to achieve an accuracy close to 0.001. Firstly, the instrument must be operated in a constant temperature laboratory with variability < 1 K at a temperature 1–2 K below the bath temperature. Only then is the standardization almost free from temperature drift, and only then can some of the potentially occurring irregularities be detected, which are described below. Secondly, the instrument must be maintained and checked before the cruise strictly along the manual's guidelines. Some of the checks can be repeated easily during the cruise. Thirdly, special care must be taken in standardization of the instrument and in sample handling.

Users of the AUTOSAL may decide or be forced not to fulfil some of the requirements. On older or smaller vessels, for example., a constant temperature laboratory may not be available. In these cases, salinities may still be measured with better accuracy than with any other instrument, but they will certainly not achieve the highest possible accuracy of 0.001 in salinity.

Before-cruise maintenance

Before a major cruise, each salinometer needs all the checks and the maintenance recommended in the manual to be carried out including checks of the electric circuits. As an accurately controlled bath temperature is essential for high accuracy in salinity measurement,

Fig. 3-3. The AUTOSAL Model 8400 B salinometer.

the calibration of the bath temperature regulating thermistors must be checked. In many cases, a slow drift in standardization may be due to a drift in the calibration of one (or both) of these thermistors. In particular, those older than *ca.* 7 years may have to be replaced.

Laboratory temperature

Stalcup (1991) reported that changes in laboratory temperatures of more than 1 K can be read from the standardization history of the salinometer during a cruise. A 3 K amplitude in laboratory temperature could lead to a correlated 0.003 standardization amplitude, i.e., well above the required accuracy. It is virtually hopeless to trace these changes by repeated standardization on a routine basis and at reasonable costs during a long cruise. During another cruise, however, an amplitude of < 1 K temperature changes showed only 0.0005 correlated changes in standardization, which are well within the required accuracy.

An experience to be kept in mind is that well maintained AUTOSAL salinometers are extremely stable under stable temperature conditions. Any salinity drift larger than 0.001 during a cruise is probably due either to non-stable laboratory conditions or to a malfunction of one of the salinometer's component. Under no circumstances should it simply be assumed that that 'the salinometer has drifted'.

Installation

The instrument is best installed amidships in a for/aft facing position to minimize effects of pitch and roll (which also affects the operator's concentration). The overflow tubing must be electrically isolated from the ground, *i.e.,* especially from the (rinsing) waste water. Therefore, the pipe should not be immersed into the waste water reservoir. Also, a hole in the tube just beyond the outlet will allow all water to drain from the tube when the cell is rinsed, thus breaking any possible electrical contact.

The instrument needs about 24 hours to warm up. It should not be switched off during the cruise. During major cruises a second instrument should be available as back-up.

Electromagnetic fields

Radio and other operations that create strong electromagnetic fields may disturb operation of the salinometer. Jumps in readings of a single sample of more than 10 digits have been observed during telex traffic on a WOCE cruise. Whenever such disturbances occur in short time scales of less than one minute, electromagnetic fields are reasonable candidates. As radio operators are usually on duty on a fixed schedule, it may be wise to avoid their times-on-duty for salinometer operations.

The power cable and the connecting interface cable from the salinometer's digital output to a personal computer may act as antennae. Shielded power and data cables should therefore be used. Computers should be de-coupled electrically from the salinometer and therefore communicate with the salinometer through opto-coupling interfaces only.

In-cruise checks

A daily drift in spite of stable temperature conditions indicates malfunction of part of the instrument. In some of these cases, one of the bath controlling thermistors may drift slowly. A large jump in standardization usually indicates sudden changes of several degrees in the bath temperature due to a failure of one of these thermistors. Guildline recommends a bi-weekly check of the thermistors which would require (stable) thermometers of better than 0.01 K accuracy on-board. As this is a cumbersome operation, a crude check can be done by switching the temperature sensors from 'NORM' to 'CK1' and 'CK2.' The heater lamps should cycle within a few minutes. If this is not the case it is highly likely that one or both thermistors are no longer calibrated.

Zero readings also check the calibration. They can be performed easily and logged before a set of samples, *e.g.,* from one station, is measured. If it exceeds 5 digits, the instrument needs checking and recalibrating.

Stand-by readings monitor stability of the electronics. Small changes of up to 3 digits typically occur along with small changes in laboratory temperature. Jumps larger than 5 digits since the last standardization (under stable laboratory temperature conditions) indicate a malfunction which needs repair and recalibration.

Using the suppression dial must produce continuous readings. A simple check is to measure seawater with salinity at the upper limit of a change in suppression, *e.g.,* 5 digits below the change limit. Then switch the suppression. If the reading in the higher suppression mode differs by more than 1–2 digits (the usual uncertainty), the contacts of the suppression switch may be corroded or a recalibration of the instrument may be necessary.

Both heating lamps should cycle while heating. Control by the operator, however, may be annoying due to bad 'observation' conditions. This check can be made easier by installing two small (red) control lamps on the front panel that flash together with the two heater lamps.

Cell cleaning

If the cell does not fill completely in all side arms repeatedly or if air bubbles stay in the cell after filling, the cell might be dirty and require cleaning. If repeated soaking with seawater or distilled water does not help, the best way is to dismount the cell according to the instructions in the manual and to clean it with a bottle brush.

Good results can be obtained with a cleaning solution such as Mukasol while the cell is still in operation. However, thorough flushing with distilled water is needed afterwards to remove any residuals solution, and the instrument's standardization must be checked.

Never use acid solutions for cleaning as it will polarize the electrodes and seriously change the calibration.

Preparing for a measurement

Clean both the filling tube and the rubber stopper with lint-free tissue to dry them and to remove salt crystals from earlier samples. Do not touch the tube with bare hands. Shake each bottle before opening to eliminate any gradients.

Standardization

The cell is rinsed with SSW from one vial until the reading repeats within the required precision (2–3 digits in readings). If starting from distilled water, at least 10 rinses are necessary to reach an accuracy of 0.001. In these cases, the first 5 rinses may be performed with seawater close to $S=35$. If the reading no longer changes within the precision, the potentiometer is changed such that the display reads twice the vials printed K_{15} value. Then the potentiometer is fixed. To control the fixing, the cell is again flushed with SSW from the same vial. The reading should be the same. The instrument is now standardized.

To check the quality of the vial that has been used for standardization, a second vial from the same batch is measured immediately after. If the reading is not the same within the precision, a third vial is measured to identify the possibly 'bad' vial and possibly to repeat the standardization procedure.

SSW should be measured before each station and after finishing the daily work. If the reading differs by more than the accuracy from the standardization, another vial should be measured to identify a possible 'bad' vial. Usually, an AUTOSAL is stable within the acceptable precision for several weeks. Therefore, a 'drift' in the standardization in most cases indicates either a bad vial or instrumental malfunction which needs repairing. Some simple checks can be done during the cruise (see *In-cruise checks* in this section).

Switching to the back up AUTOSAL should help in these situations. If no spare is available, the drift must be traced with SSW. It is probably better not to re-standardize the instrument as long as the drift does not differ by more than 0.005 in salinity from the SSW value. Drift may then be corrected for linearly in time. An accuracy of 0.001, however, cannot be claimed in such cases.

Only vials from a single batch should be used throughout one cruise. If under certain circumstances a change to another batch is necessary, a cross check between the two batches will log any batch-to-batch changes in labelled salinity. Three vials of the new batch are measured as samples with consistent congruent salinity before the instrument is re-standardized with the new batch.

Measurement of samples

To minimize rinsing errors, samples are measured station-wise from the surface to the bottom, and from the bottom to the surface for the following station. Rinsing is required until the reading stays stable within the precision (2–3 digits between 2 consecutive reading). Usually, this takes 4–5 rinses.

3.5.5   The Beckman Model RS10

Principles of operation

The RS10 salinometer (Fig. 3-4; *Beckman*, 1986) operates with an inductive cell (see Section 3.4) with operator-controlled balancing of the circuit. As the instrument is portable, it has no temperature stabilizing bath. Instead, the temperature of the sample is measured directly in the cell during operation, and the temperature effect on the standardization is compensated for within a given tolerance, *e.g.*, ± 3 K, around the temperature at which the RS10 was standardized. This range of temperature compensation is different for each instrument and is given in the accompanying manual.

Accuracy

The manufacturer gives a resolution of 0.0004 in salinity. Thus, the precision is 0.0008 at best. Accuracy is claimed to be within 0.003 and may be achieved with careful operation and in a laboratory which is stabilized to within 1 K of the standardization temperature.

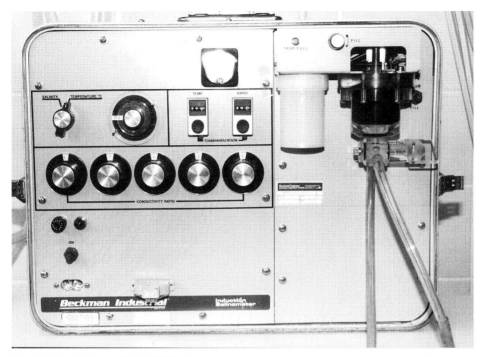

Fig. 3-4. The Beckmann Model RS10 salinometer.

Before-cruise maintenance

All the tests recommended by the manufacturer should be performed before the cruise. In particular, the temperature compensation circuit requires attention.

Effect of laboratory temperature

The instrument works with a temperature compensating circuit that corrects for changes in the cell temperature relative to the temperature at which the RS10 was standardized. This tolerance is given in the manual, and is of the order of ± 3 K. Keeping the laboratory temperature stable within these limits to the standardization temperature is essential for the accuracy of the results. If the laboratory temperature is beyond the limits, the drift of the RS10 needs to be checked with SSW, and a new standardization made.

Installation

See Section 3.5.4.

Electromagnetic fields

Such influence has not been reported.

Cell cleaning

See Section 3.5.4.

Preparing for a measurement

See Section 3.5.4.

Standardization

SSW is used for standardization at room temperature. The cell is rinsed until 3 successive readings for each of 2 successive vials yield the same readings within the required precision (0.0008 in salinity). During operation, the instrumental drift should be traced with further SSW measurements, *e.g.,* at the beginning of a batch from a new station.

The RS10 needs re-standardization with SSW if: (i) the laboratory temperature is outside the limits, *e.g.,* $\pm 3\,K$ of the last standardization temperature because then the temperature compensation is no longer valid; or (ii) a SSW reading differs by more than the achievable accuracy (0.003 in salinity) from the last standardization.

Measurement of samples

To minimize errors from rinsing, the samples are grouped to neighbouring salinities (see Section 3.5.4). For a sample measurement, the cell is rinsed twice while the stirrer is switched on for a short time. The cell should be free of bubbles from the second rinse on. After the third rinse, a reading can be performed until it is stable within the precision required. The reading is valid if the temperature (with the stirrer switched on) is within the allowed tolerance from the standardization temperature.

3.5.6   Data logging

Principles

The log must be complete and consecutive. It must contain the general information of the cruise, the identification of the salinometer and the operator's name. Each measurement is logged with date, time, station, cast, number of water sampler, case identification of salinity sample, salinity sample bottle number, the salinometer reading, its uncertainty (quality), the bath or sample bath temperature and the calculated salinity. Readings and uncertainties are estimated by eye (which is a reasonable method) or through the computer interface and a data acquiring computer programme.

Each SSW measurement is logged too, with date and time, batch number (instead of the station number), the number of SSW measurement series (instead of the cast number) and the consecutive vial number (instead of the number of the water sampler).

Each salinity is recorded to at least 4 decimal places to assure accuracy of further calculations. Table 3-2 shows an example of a logsheet.

Special instrumental tests, such as STAND-BY and ZERO readings with the AUTOSAL are also recorded.

Table 3-2. Start of a salinometer protocol. Some corrections were added by the operator and included with reference to the documenting file at the beginning of the data file. Computer data logging during R/V POSEIDON cruise 233/1. Standard seawater (SSW) batch p129, K15 value is input from operator; 2xK15 and batch SSW salinity is output from programme; STAT is station or SSW batch number; CAST is cast number or standardization sequence; BNO is bottle number of water sampler or SSW vial number; SBNO is salinity sample bottle number; 2 Rt is salinometer reading; 2*STDDV denotes twice the standard deviation of ca 100 readings. Standardization (STAT=p129; CAST=0001) with 3 different vials; substandard SUB1 with 3 different bottles has salinity SSUB1 = 34.8613. Stability control 3 hours after standardization with 4th bottle of SUB1 to better 0.002 in salinity.

Corrections in file made for stations: 573, 576, 629, 605
see cor_p233.doc   M. Knoll, 20 Nov 1997

Operator	: M. Knoll
EXPEDITION	: P233/1
DATE (MM-DD-YY)	: 9-17-1997
SSW batch no.	: P129
Batch K15 value	: 0.99996 => 2 x K15 : 1.99992
Batch salinity	: 34.9984
Bath temperature	: 24.000
Guildline 8400A, IfM Kiel	: AS3
Laboratory temperature	: 22.000

STAT	CAST	BNO	SBNO	Time	2xRt	2xSTDDV	Salinity
p129	0001	0001	0001	8:39:23	1.999924	0.000012	34.9985
p129	0001	0001	0001	9: 8:35	1.999897	0.000021	34.9980
p129	0001	0001	0001	9:10:29	1.999910	0.000000	34.9982
p129	0001	0002	0002	9:12:13	1.999913	0.000014	34.9983
p129	0001	0002	0002	9:13:38	1.999932	0.000010	34.9987
p129	0001	0002	0002	9:14:56	1.999902	0.000025	34.9981
p129	0001	0003	0003	9:16:36	1.999920	0.000008	34.9984
p129	0001	0003	0003	9:17:50	1.999922	0.000008	34.9985
p129	0001	0003	0003	9:19: 7	1.999930	0.000007	34.9986
p129	0001	0003	0003	9:20:16	1.999916	0.000014	34.9983
sub1	0001	0001	0001	9:28: 3	1.992947	0.000009	34.8612
sub1	0001	0001	0001	9:30:19	1.992921	0.000008	34.8607
sub1	0001	0001	0001	9:31:21	1.992983	0.000009	34.8619
sub1	0001	0002	0002	9:32:25	1.993011	0.000010	34.8625
sub1	0001	0002	0002	9:33:25	1.993007	0.000013	34.8624
sub1	0001	0002	0002	9:34:38	1.992969	0.000007	34.8617
sub1	0001	0003	0003	9:35:49	1.992931	0.000006	34.8609
sub1	0001	0003	0003	9:36:55	1.992902	0.000008	34.8604
sub1	0001	0003	0003	9:37:57	1.992909	0.000011	34.8605
sub1	0001	0003	0003	9:39: 0	1.992941	0.000007	34.8611
0566	0006	0020	1355	10: 2:21	2.092192	0.000021	36.8219
0566	0006	0020	1355	10: 3:32	2.092175	0.000012	36.8215
0566	0006	0020	1355	10: 6: 6	2.092354	0.000012	36.8251
0566	0006	0020	1355	10: 7:26	2.092275	0.000012	36.8235
0566	0006	0020	1355	10: 8:36	2.092186	0.000010	36.8218
0566	0006	0020	1355	10: 9:41	2.092158	0.000009	36.8212
0566	0006	0020	1355	10:11: 7	2.092111	0.000011	36.8203

Table 3-2. *(Continued)*

STAT	CAST	BNO	SBNO	Time	2xRt	2xSTDDV	Salinity
0566	0006	0020	1355	10:12:46	2.092089	0.000023	36.8198
0566	0006	0020	1355	10:14: 6	2.092096	0.000023	36.8200
0566	0006	0020	1355	10:15:16	2.092104	0.000023	36.8201
0566	0006	0020	1355	10:16:18	2.092116	0.000012	36.8204
0567	0007	0020	1484	10:26:41	2.094046	0.000010	36.8587

– other stations –

sub1	0002	0004	0599	11:40:19	1.992961	0.000006	34.8615
sub1	0002	0004	0599	11:41:13	1.992961	0.000012	34.8615
sub1	0002	0004	0599	11:42: 9	1.992953	0.000010	34.8614

– more stations –

Computerized logging

If the salinometer has a digital output, a computer can be used for direct logging (see Section 3.5.4 for connecting). Suitable computer programmes are available commercially or have been developed in research institutes. Observing the following principles in using such programmes has shown the highest accuracy in salinity results with an AUTOSAL. They may be adapted to any programmes and salinometers.

Computer programmes may not be able to decide at which step of the process the cell has been thoroughly rinsed and then to measure a sample. It therefore must be the operator's and not the programme's decision at which point of the process data for a salinity value are to be read from the interface. On the AUTOSAL, this is executed by pressing the 'Data Log' button. Having read a set of data for a single salinity measurement, the programme then can easily deliver the statistics from which the operator will finally accept the measurement or not.

For a single salinity determination, at least 19 readings are taken to assure statistical reliability. On the AUTOSAL, this will last less than 10 s. Of these, all values are removed that lie outside an interval of 3 times the precision around the median. This step will remove outliers and will leave the rest of the data with a distribution close to normal around the mean. It is emphasized to use the median and not the average in this step to minimize the influence of spikes. From the remaining values, the average is offered as a best estimate to the operator with twice the standard deviation as the approximate 95 % confidence limit. The salinity is calculated from the average. Finally, the operator decides to accept and store the result along with the station and the sample information or to repeat the measurement of the sample. Table 3-2 shows an example of a computerized logsheet.

3.5.7   Substandards

Principles

All bench salinometers must be standardized with SSW. While strongly discouraged for achieving WOCE standards in salinity, substandards may be used together with SSW to control the stability of the salinometer if less than the best achievable accuracy is acceptable in order to save costs. In these cases, the following procedure has proven to provide sample salinities with prescribed expected accuracies.

Sampling

At the beginning of a cruise, probably during a test station, seawater is sampled at a single station from a single depth that is more than 1000 m deep to minimize biological activity, and that has a weak salinity gradient. Water from the mixed layer should not be sampled unless the vessel is operating solely in shallow seas. The sample seawater is poured into a closed 30–50 L reservoir with enough space left for thermal expansion. When the temperature of the reservoir has adapted to the temperature of the salinometer laboratory, it is thoroughly shaken to eliminate any salinity gradients that may have developed. Then, samples are quickly poured into salinity bottles observing the filling rules given in Section 3.5.3.

Salinity

Immediately after the salinometer has been standardized with SSW (see Sections 3.5.3 and 3.5.4 for the AUTOSAL and the RS10, respectively), the salinities of three substandard bottles are measured successively. If each salinity differs from their average by less than the accuracy that is required for individual samples during the cruise, the substandard is accepted, with the average as the salinity of the substandard and the maximum difference between the three individual substandard sample salinities as its uncertainty.

Usage

Individual substandard samples then serve to identify any drift in the salinometer calibration. If a substandard measurement differs by more than the substandard's uncertainty from its estimated salinity, a second substandard sample is taken for control. If the second value is outside the uncertainty interval as well, a SSW sample is measured for control of any instrumental drift within the required accuracy.

Accuracy

If an AUTOSAL is operated under otherwise WOCE conditions (see Section 3.5.4), the accuracy for individual salinity samples can still be of the order of 0.002 when substandard samples instead of SSW are used to identify instrumental drifts. For the RS10, it can be better than 0.01 depending on the operating conditions.

3.6   Salinity from *in situ* measurements: CTD profilers

3.6.1   Principles

Oceanographers were always keen to have visible and available *in situ* salinity on-line. The most effective method to achieve this goal was to construct an instrument that measures three basic thermodynamic parameters simultaneously with high precision from a research vessel, transfer the data *via* a one-conductor cable to the deck, calculate salinity from the three basic parameters, make them visible and store them. The parameters easiest to measure electrically, are pressure P, temperature T and the specific electrical conductivity of sea water C relative to a fixed reference. With the introduction of the PSS78 (see Section 3.3.2), a unique procedure was prescribed to calculate salinity S from these three basic measurements.

Among the first instruments to measure almost continuously the three basic parameters in the water column (*Hinkelmann*, 1957; *Brown*, 1974), the one developed by Brown was termed Conductivity Temperature Depth profiler (CTD). This name now is in general use for such instruments. Additional sensors are often attached, e.g., sensors to measure dissolved oxygen and pH values (see Chapter 14), fluorescence, light attenuation, sound velocity and others.

CTDs are designed for many different applications and requirements of accuracy. Attainable accuracies are 0.05 % for pressure, 2 mK for temperature and 0.002 in derived salinity on down profiles. Up-profiles have lower data quality since the sensors then are in the turbulent wake of the wire, the underwater unit and possibly a rosette water sampler.

Highest quality in CTD salinity can be obtained only if the CTD is operated together with a rosette sampler (Fig. 3-5) for *in situ* calibration (see Section 3.6.3 for the calibration procedure and Section 3.6.4 for data processing). In these cases, a one-conductor cable serves to transfer CTD data from the underwater unit to the deck unit and a personal computer, and to control bottle closures by the operator. Recently, some manufacturers started to offer combined CTD-rosette sampler systems that have internal CTD data storage and that close bottles at prescribed pressure levels. Closing the bottles is controlled using the CTD's pressure sensor signal.

If a CTD is towed by a moving vessel it is not possible to take salinity samples for calibration. Calibration then solely depends on laboratory calibrations or on nonsynchronous comparisons with other data.

Less accurate measurements are often made from vessels that have no one-conductor cable. In these cases, the CTD records data internally. Data are read from the storage unit after the CTD has returned to the deck using an interface to a personal computer. Accuracies of 0.02 in salinity may be achieved depending on the quality of the CTD and its laboratory calibration.

Inexpensive instruments are available that have the temperature sensor and the conductivity cell attached directly to a three-conductor cable. The cables which carry the underwater unit are up to 100 m long, and they are integrated into the measuring circuit. They have no pressure sensor, and salinity is derived assuming constant (zero) pressure. Such instruments can be accurate to 0.2 in salinity and are valuable for coastal and estuarine measurements.

Fig. 3-5. A CTD-rosette system is lowered from the R/V METEOR.

Other applications require internal data storage for at least some time as they are not vessel based: CTDs that are moored at a fixed depth to obtain a time series at a fixed position; CTDs that glide along a mooring line up and down at prescribed times to obtain time series of vertical profiles at fixed positions; and CTDs that are incorporated into freely drifting subsurface floats to obtain vertical profiles when the float surfaces to transmit its data *via* satellite. In all these cases, it is impossible to have bottle samples for calibration, and consequently accuracy in salinity critically depends on the laboratory calibration of the conductivity sensor and its stability.

The following sections describe methods of handling, calibration and data processing that are necessary to obtain high quality salinity results from CTD data. They summarize procedures that are described in more detail by *UNESCO* (1988), the *World Ocean Circulation Experiment* (1994), and *Millard and Yang* (1993).

3.6.2 Operation of CTD-rosette sampler systems

Installation

The CTD-rosette system is best operated from amidships to minimize effects of pitch and roll on the data. When all electrical connections work properly, the CTD should not be switched off throughout the cruise.

To make near-bottom measurements one must know the distance to the bottom when the CTD-rosette approaches the bottom, to avoid damage. Most frequently, the distance is measured independently with an acoustic pinger that is attached to the rosette frame. In some CTDs a small high frequency echo sounder can be implemented. Others detect the bottom mechanically by a weight that is suspended from the CTD on a rope 10–20 m long.

Preparation of a cast

The logsheet is prepared. It contains a check list for preparation of the CTD-rosette that follows the manufacturer's recommendations and the operator's experience with individual systems. Make sure by intensive rinsing with fresh water that all mechanical parts of the rosette work properly. Data acquisition starts before the CTD is lowered into the sea as in-air data are needed for later processing.

Deployment

CTD sensors respond best when lowered at continuous speed with no upward movement in between that may be induced by the ship's pitch and roll. This requires high lowering speeds. However, if the speed of the winch exceeds the free-fall speed of the CTD-rosette in the water, it may tumble. The optimal lowering speed changes with vessel, sea state, winch, size and weight of the CTD-rosette and the length of the cable already out. To start with 0.5 m/s for the first 200 m and then to increase to 1 m/s seems to be a good approximation to the optimal lowering speed in many cases.

Sampling

In order not to avoid damage by compression, bottles are closed on the way up at the desired pressure levels. On the way up, the CTD sensors are in the wake of the system which leads to noisier data. It is therefore best to stop the winch before closing a bottle.

Samples for calibration of the CTD's conductivity sensor are taken from regions with weak vertical gradients, *i.e.,* mostly from the deep sea and from the mixed layer. To check the precision of the entire procedure it is recommended to close two bottles at the same depth from time to time.

After the cast

Rinse both the CTD and the rosette intensively with fresh water to avoid the formation of salt crystals that may falsify the first measurements on the next cast or block mechanical parts of the rosette.

3.6.3   Calibration

General

Accuracies achievable for measurements in the ocean are 0.002 mK for T, 0.05 % for P and 0.002 for S. Temperature and pressure sensors are calibrated in the laboratory. *In situ* comparison of the laboratory calibrations of these sensors only makes sense if additional sensors with digital output can be attached to the CTD. The accuracy and stability of reversing thermometers and pressure sensors sometimes attached to rosette bottles is not sufficient for *in situ* checks of possible small drifts in the CTD calibration. To account for these, pre- and post-cruise calibrations are recommended.

The situation is different for conductivity cells as laboratory calibration is not sufficient to achieve the highest accuracy for salinity, 0.002, and sensor drift can be significant and non-linear with time. Bottle salinities and the knowledge of calibrated CTD values at bottle closing pressures is required for high accuracy *in situ* calibration of the conductivity cell. Direct calibration of salinity is not recommended in these cases because of its strongly non-linear relationship to the three basic parameters, but may be performed if less accuracy is acceptable.

The three sensors are calibrated separately. The electronics of the underwater unit may be sensitive to changes in temperature during the calibration procedure. Therefore, preferably the whole CTD has to be immersed in a water bath, at least for temperature and conductivity calibrations as these sensors are most sensitive to temperature changes. In modern designs, CTDs have the electronics for digital output integrated into the sensor heads. In these cases and in cases where the electronics are known to be insensitive to, or are compensated for, temperature changes, the sensors may be detached for calibration for easier handling.

Usually, the manufacturer provides a basic calibration for each sensor. In many cases it is linear. Careful calibration with high resolution may detect any small non-linearity in this basic calibrations (see *Müller et al.*, 1995, for examples). However, it is recommended not to change the basic calibration coefficients, but rather add corrections to these.

Temperature

The temperature sensor is usually a platinum resistance thermometer or a thermistor. The sensor is calibrated according to the International Temperature Scale of 1990, ITS90 (*Rusby et al.*, 1991). On this scale, two primary fixed points are defined that are within the oceanic temperature range: the triple point of water at 273.16 K (0.01 °C) and the melting point of gallium at 302.9146 K (29.7646 °C). A platinum resistance thermometer serves to interpolate between these and other fixed points outside the oceanic range. The interpolation procedure has been encorporated in the definition of the ITS90. Such Platinum resistance thermometers are certified individually by national bureaus of standards. Used within a high standard temperature bridge, they serve as reference for the calibration of CTD temperature sensors.

Platinum resistance thermometers drift slightly. This requires that a reference thermometer is checked before each major calibration against primary fixed points, at least at the two points within the oceanic range. A water triple point cell is immersed together with the platinum reference thermometer in a bath that is temperature controlled to reach the fixed-point temperature. Any change in the offset of the reference thermometer is corrected. The procedure is repeated at the gallium melting point to control the slope. The use of two cells at both fixed points is recommended to detect any slight damage in these.

Calibration of the CTD temperature sensor to better than 2 mK accuracy thus requires three units: (i) a reference temperature bridge that is accurate to 1 mK at a resolution of at least 0.5 mK, (ii) a thermostat bath to obtain the two oceanic primary fixed-point temperatures with a water triple point cell and a gallium melting point cell to control the drift of the reference sensor and (iii) a temperature controlled bath that is stable to better than 0.5 mK while taking a calibration point with the reference thermometer and the immersed CTD.

To calibrate a CTD temperature sensor, first the reference sensor is checked against the two fixed points. Next, the CTD sensor is calibrated in the water bath over the whole range against the reference starting at $-1.5\,°C$ and going up to $28\,°C$. Note that calibration at a negative temperature requires the uses of salty water in the bath. To detect small non-linearities that may be present in the CTD sensor's response (see *Müller et al.,* 1995 for an example), it is recommended to take calibration points at small intervals, *e.g.,* 2 K. The corrections T_C are added to the basic sensor calibration T_{CTD} to obtain the calibrated temperature T:

$$T = T_{CTD} + T_C \tag{3-12}$$

Usually, the corrections T_C are smooth over the whole range. They may then be fitted as a function of T_{CTD} by polynomial regression. Otherwise, T_C is linearly interpolated from the calibration table.

For commercial CTDs, the precision that can be achieved following the above guidelines is better than 1 mK over the entire temperature range. For the duration of a cruise, however, accuracy is only better than 2 mK because of small drifts of the sensor. This drift may amount to 2 mK/a depending on the sensor type and the CTD. Therefore, to achieve highest accuracies, pre- and post-cruise calibrations are recommended.

Pressure

Pressure sensors are calibrated on dead-weight testers. On these instruments, a known force that acts on a known area is balanced hydraulically through an oil-filled tube by the surface of the pressure sensor. The acting force is given by a calibrated and certified mass under the local gravitation and the area by the cross section of the piston. For accurate calibration one has to observe that the earth's gravitation not only varies with latitude and height above sea level but also locally. In addition, the piston's cross section varies slightly with temperature. The manufacturers of dead-weight testers give instructions on how to deal with both of these variations.

Some sensors show a mechanical hysteresis that has to be corrected for. The hysteresis depends on the maximum pressure achieved during a profile. Some time after a cast, *e.g.,* one hour, the sensor will have readjusted to its before-cast value.

As electrical resistors, CTD pressure sensors respond not only to pressure changes but also to changes in temperature. This has to be taken into account for high accuracy measurements. Firstly, sensors respond statically as the resistance changes with temperature. Secondly, pressure sensors respond dynamically to fast temperature changes due to the different time scales for temperature equilibration at the outer and the inner parts of the sensor. Manufacturers adjust the sensor output to account for both effects. In some CTDs, the temperature inside the sensor cap is measured directly, and its signal is either used in an internal compensating circuit or for compensation of the digital output with a model of the sensor response. For accurate pressure measurements, the user must check if the corrections provided by the manufacturer still apply, because they may not have been adequately performed or may have drifted. Errors due to poor compensation of the static response may be as high as 2 dbar for the oceanic temperature range. For a 20 K fast change in temperature which may occur while profiling in the tropical ocean, the dynamic response could be as high as 3 dbar with a time constant of 2 hours.

Following the above, the CTD pressure sensor is calibrated on the dead-weight tester with increasing pressure (loading curve). To check for a hysteresis, unloading curves are recorded from 3 different maximum loads, *e.g.,* 6000, 4000 and 2000 dbar. The procedure is repeated at 3 different temperatures covering the oceanic range, *e.g.,* 1, 10 and 25 °C. For this, the CTD pressure sensor is immersed in a water bath that varies by not more than 1 K during the calibration. The corrections needed to bring the pressure sensor output to the reference pressure are tabulated in a matrix. The upper left element contains the laboratory temperature, and the remainder of the first row the sensor temperatures in increasing order at which the calibration was performed. The rest of the first column contains the reference pressure. The remainder of the matrix is filled with the pressure sensor output for loading and unloading at the lowest temperature followed by those pressures for the next highest temperature. Dummy values are inserted where no calibration values exist. Static correction P_{CS} to the basic sensor calibration P_{CTD} is derived from linear interpolation within this matrix. Values outside the calibration range are extrapolated constantly. For pressure sensors that have negligible hysteresis and temperature response, P_{CS} is fitted as a polynomial in P_{CTD}.

To estimate the dynamic response to fast temperature changes, the pressure sensor is first immersed in a warm water bath. After it has adjusted, the sensor is quickly placed in a cold water bath at the same pressure. The sensor response together with the temperature is recorded. If the response is still significant after all the manufacturer's compensations have been applied, the method of *Müller et al.* (1995) may be used to reduce the error. It assumes that the dynamic pressure correction P_{CD} is proportional to the temperature difference between the outer and the inner parts of the pressure sensor. If not measured, the temperatures at the outer and the inner parts of the pressure sensor are estimated from the measured main temperature of the CTD using a recursive exponential filter.

Assuming that both, static and dynamic corrections are independent, the calibrated pressure P is

$$P = P_{CTD} + P_{CS} + P_{CD} \tag{3-13}$$

Conductivity and salinity

All reference conductivities are derived from reference salinity values and calibrated CTD temperature and pressure by numerical back iteration of Eqs. (3-7) to (3-10). Traditionally, the measured conductivity ratio R is converted to conductivity assuming

$$C(0,15,35) = 4.2914 \, \text{S/m} \, (42.914 \, \text{mS/cm}).$$

The basic laboratory calibration of a conductivity cell is obtained by immersing the cell with the CTD in tanks filled with seawater of different salinities. To avoid effects of thermal expansion of the cell on the calibration, all bath temperatures are kept the same to within 2 K. The basic calibration C_{CTD} of the conductivity cell is referred to the conductivity of the baths and is valid for the overall bath temperature and at atmospheric pressure.

For highly accurate calibrations one has to consider that the cell constant of a conductivity cell changes as it deforms elastically under pressure and by thermal expansion. The combined effect may reach 0.0005 S/m (0.005 in salinity). One method to compensate for these effects has been proposed by *Millard and Yang* (1993). It assumes linearity and fixed constants prior to the *in situ* calibration:

$$C_{CC} = (1 + \alpha\,(T\text{-}T_0) + \beta\,(P\text{-}P_0))\,C_{CTD} \tag{3-14}$$

The constants α and β depend on the material the cell is made of. For cells of the classic CTD of *Brown* (1974), the values are $\alpha = -6.5 \times 10^{-6}$ and $\beta = 1.5 \times 10^{-8}$ (see *Millard and Yang*, 1993). Usually, $T_0 = 15°C$ and $P_0 = 0$ dbar are chosen.

In situ calibration of the conductivity sensor starts with calibration of the temperature and pressure values in the up profile at bottle closing pressures. Next, *in situ* reference conductivity C_{REF} is derived from bottle salinity S_{REF}, T and P. If necessary, C_{CTD} is compensated for pressure and temperature effects according to Eq. (3-14). The final correction $C_C = C_{REF} - C_{CC}$ depends on the sensor and the type of the CTD and on C_{CC}; the drift in terms of a group of cast numbers may be included and further pressure correction may be needed. As an example, the corrected conductivity C may be written as:

$$C = C_{CC} + C_C \tag{3-15}$$

with

$$C_C = a_0 + (a_2\,C_{CC} + a_1)\,C_{CC} + a_4\,Cast + \text{polynom}(P)$$

where *Cast* is a cast number.

The coefficients in Eq. (3-15) are derived from linear regression. The calibration is applied to the lowering profiles and salinity is calculated. An example of a conductivity cell calibration according to Eq. (3-15) is shown in Fig. 3-6. The coefficients were calculated for a group of 75 casts. The overall accuracy is estimated to better than 0.005 in calibrated CTD salinity. It may be improved by splitting into smaller cast groups.

Direct calibration of salinity from bottle salinity and salinity values that are derived from T, P and C_{CTD} will usually be less accurate because of the high non-linearity in the salinity formula.

3.6.4   Data processing

General

Processing of CTD data includes removal of spikes due to errors in data transmission, adjustment of different sensor response times, smoothing, calibration and interpolation to prescribed pressure intervals. Much of the background literature and general guidelines on how to process CTD data, were reviewed 10 years ago by *UNESCO* (1988) and more recently by *Millard and Yang* (1993). They are applicable for vertical profiling in combination with rosette samplers. The steps in processing CTD data as described below follow these guidelines with small modifications. If performed in the order given below, final data quality will meet WOCE Hydrographic Programme standards for vertical profiles. Steps may be skipped if they do not apply for a certain CTD type, for higher resolution applications or if lower quality data are acceptable.

CTD data processing requires a computer. In some computer languages, graphical editors are available or easy to write which assist in manual editing and determining of the editing parameters.

Time basis

To adjust sensor responses to each other in later steps of processing high resolution data, a common time basis is needed. If not incorporated by the data acquisition programme, it is created by combining cycle numbering and sampling rate, starting with the first record. It is kept throughout the processing procedure.

Basic calibration

Pressure, temperature and conductivity measurements are converted into physical units in their basic calibrations, P_{CTD}, T_{CTD} and C_{CTD}. This includes all necessary special corrections needed for certain CTD types. Preliminary salinity S_{CTD} is calculated.

Remove bad data cycles

Owing to poor data transmission, outliers may exist that are recognizable in the graphics. A median criterion (Sy, 1985) can be applied to remove them. It starts with P_{CTD} at the beginning of a profile. The centre value of a prescribed number of records (window length) is identified as outlier if it differs by more than a prescribed tolerance from the median. Records with outliers are deleted. Since a time basis has been created, it is not necessary to interpolate deleted values. The procedure is repeated for each record over the whole profile, then for T and S.

The median criterion is able to identify successively occurring outliers and outliers not equally distributed, if there are not too many in one window length. The window length for the median criterion should correspond to a vertical interval of about 1 dbar. Conservative tolerances on these scales are 0.5 dbar, 0.5 K and 0.5 in salinity. If necessary, the procedure may be repeated with smaller tolerances.

Dynamic correction of pressure

If the pressure sensor of the CTD responds to fast changes in temperature, it is corrected for according to the results of its dynamic calibration (Section 3.6.3). The correction starts with the pre-cast deck values thus assuming that the sensor is close to thermal equilibrium.

Pressure sensor offset and calibration

The pressure sensor may have changed its offset correction value compared with the laboratory calibration. The negative of the pressure sensor's pre-cast deck value replaces the laboratory offset correction. Calibration may be performed for all sensors at this stage (see Section 3.6.3). However, to save computing time it may be delayed until after the data have been reduced, but before smoothing and interpolation starts.

First in-water record

As a CTD cast starts, the conductivity cell is the sensor to identify best the transition from air to water because of the large gradient from zero measurement at deck to ocean surface values. However, waves and swell will not allow the CTD to be immersed into the water

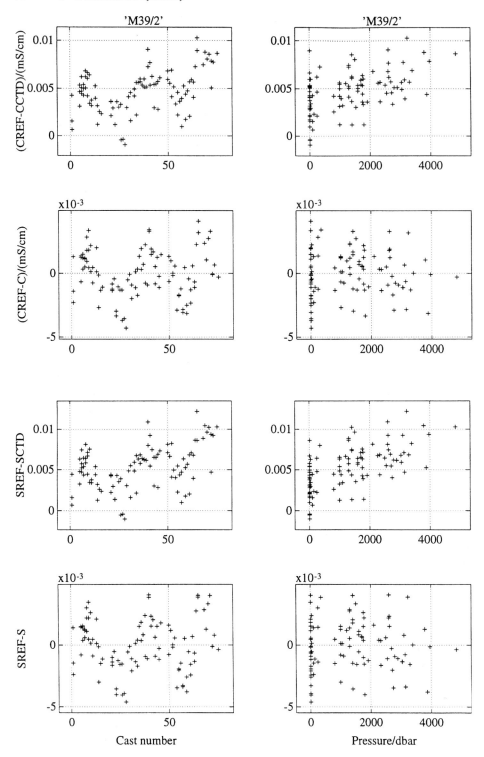

smoothly. Records are kept for which both, S is larger than an allowed minimum and offset-corrected pressure is larger than a pre-set minimum value, *e.g.,* 0.5 dbar.

Monotonizing and minimum downward velocity

Optimal sensor response requires a minimum and smooth downward speed of the CTD. Because of pitch and roll, this cannot be achieved on a ship. One particular approach is that records sampled at less than a pre-set minimum speed should be discarded. However, at least five records should be kept on the chosen vertical interpolation scale to avoid data with too many gaps if larger parts of the profile are at less than the minimum lowering speed for some reason.

In the first step, only records are kept that are strictly monotone in P, starting with the first in-water record. This criterion rejects all records sampled during an upward movement and those records sampled in the wake of the preceding upward movement. It also adjusts the vertical scale that results from the sampling rate and the actual lowering speed to that of the pressure sensor resolution in the case of over-sampling. For example, consider a CTD cast that is sampled at 32 Hz rate, has a pressure sensor resolution of 0.1 dbar and is lowered at a nominal speed of 1 m/s (1 dbar/s). Such a profile on average is over-sampled. Monotonizing will reduce the number of records in the lowering part of the cast by at least one third.

Next, the required minimum downward velocity is pre-set, *e.g.,* to half of the lowest nominal lowering speed during the cast, e.g., 0.5 m/s at the beginning of the cast. For each record, the CTD's actual downward velocity is estimated from the difference of the two adjacent pressure values and the corresponding time interval. If it is less the pre-set minimum velocity criterion and if the pressure difference to the last record accepted as good does not exceed one fifth of the interpolation interval, it is flagged. Finally, all flagged records are removed.

Response adjustments

Sensors have different response times which leads to spiky and systematically wrong salinity values if not corrected for. The strongest effect stems from the mismatch between slow temperature and fast conductivity sensors. A manufacturer may compensate for these effects by specific arrangement of the sensors and controlled flow past these sensors (conducted flow). Following the manufacturer's instructions will lead to good adjustments in these cases.

◄ **Fig. 3-6.** Conductivity/salinity calibration for a CTD (METEOR cruise M39/2). Upper row shows the corrections (CREFC-CTD) which is needed to match the conductivity measurements CCTD to the *in situ* reference value CREF as a function of the cast number (left) and pressure (right), the second row the residuals (CREF-C) after calibration of conductivity C. The third and fourth rows are corrections in salinity (SREFS-CTD) and residuals (SREF-S), respectively. Accuracy in C (95% confidence level) for all 74 casts is twice the root mean square error of the linear least square approximation, 0.0004 S/m (Siemens/m), corresponding to 0.004 in salinity. The result would be improved if the obvious cast-dependancies (see lower left part of panel) could be removed.

In other cases, a digital filter has to be applied. The method, which has shown acceptable results, uses recursive filtering to increase artificially the response time of the conductivity cell to match that of the temperature signal:

$$C_j := C_j + (C_{j-1} - C_j) \exp(-(t_j - t_{j-1})/\tau) \tag{3-16}$$

Here the index j refers to the jth record, t is the associated time, τ is the filter constant and the notation $:=$ indicates that the jth value of conductivity C is recursively replaced by the right hand side of Eq. (3-16), starting with $j=2$. For homogeneous conditions, there is no change in C.

The filter constant τ has to be determined so that salinity spikes in layers with strong gradients are minimized after Eq. (3-16) has been applied to the profile. Choosing a typical profile as an example and iterating τ with visual inspection of the resulting salinity is a simple way to do this. Using a more objective way, τ is iterated in such a way that the phase of the cross spectrum between C and T is brought closer to zero at high frequencies. Once τ has been determined, all profiles obtained with the same CTD during the cruise are filtered according to Eq. (3-16).

Calibration

If not carried out earlier, all data are calibrated according to the results from Section 3.6.3.

Smoothing

The data are smoothed by applying a symmetrical low pass filter on a vertical scale slightly less than the final interpolation depth. Neither moving averaging nor median filtering is recommended because of the poor response and possible disarrangement of data within a record, respectively. A cosine filter is preferred. The number of weights is chosen to cover approximately 1.5 of the interpolation interval on average over a profile. Considering the cost of computing time, the number of weights may be adjusted locally to the number of records actually available in that interval.

Interpolation

Data are interpolated to fixed pressure intervals, e.g., to 2 dbar for WOCE resolution of the final data. One method is to average all values around the centre value within one interpolation interval (basketing). Values are weighted by distance from the centre value.

The other method uses interpolation. With simple linear interpolation, just the two points next to the centre value will contribute. To better keep the shape of the profile, a 5-point Lagrangian interpolation is preferred.

Salinity

Because of the strong non-linearity in Eqs. (3-7) to (3-10), salinity is recalculated in the final step after all processing is finished. The above equations are applied.

References to Chapter 3

Beckman Industrial Corporation (1986), Model RS10 Portable Induction Salinometer. Cedar Grove, NJ, USA.

Brown, N.L. (1974), *IEE Conference on Engineering in the Ocean Environment*, vol. 2, 270.

Cox, R.A., Culkin, F., Riley, J.P. (1967), *Deep-Sea Res.*, 14, 203.

Culkin, F., Smed, J. (1979), *Oceanol. Acta*, 2 (3), 355.

Culkin, F., Smith, N.D. (1980), *IEEE J. Oceanic Eng.*, OE-5, No. 1, 22.

Dietrich, G., Kalle, K., Krauss, W., Siedler, G. (1975), *Allgemeine Meereskunde.* 3rd edition, Berlin: Gebrüder Borntraeger, 1975.

Fofonoff, N.P. (1985), *J. Geophys. Res.*, 98 (C2), 3332.

Fofonoff, N.P., R.C. Millard, R.C. (1984), *UNESCO* Tech. Pap. Mar. Sci., No. 44, Paris.

Forch, C., Knudsen, M., Sorensen, S.P. (1902), *D. Kgl. Danske Vidensk, Selsk. Skrifter, 6 Raekke, naturvidensk, og mathem. Afd*, XII 1.

Guildline Instruments Ltd. (1997), Technical Manual of the Model 8400B. Smith Falls, Ontario, Canada.

Hinkelmann, H. (1957), *Z. Angew. Phys.*, 9, 500.

Lewis, L.E. (1980), *IEEE J. Oceanic Eng.*, OE-5, 3–8. Reprinted by UNESCO (1981a).

Lewis, E.L., Perkin, R.G. (1978), *J. Geophys. Res.*, 83 (C1), 466.

Lewis, E.L., Perkin, R.G. (1981), *Deep-Sea Res.*, 28A, 307.

Mantyla, A. W. (1987), *J. Phys. Oceanogr.*, 17, 543.

Millard, R.C., Yang, K. (1993), *Technical Report WHOI-93-44*, Woods Hole, MA, USA.

Millero, F.J., Kremling, K. (1976), *Deep-Sea Res.*, 23, 1129.

Mohr, C.F. (1856), *Ann. Chem. Pharm.*, 97, 335.

Müller, T.J., Holfort, J., Delahoyde, F., Williams, R. (1995), *Deep-Sea Res.*, I, 42, 2113.

Rusby, R.L., Hudson, R.P., Durieux, M., Schooley, J.F., Steur, P.P.M., Swenson, C.A. (1991), *Metrologia*, 28, 9.

Saunders, P.M. (1990), *WOCE Newsl.* No. 10.

Siedler, G, Peters, H. (1986), *Properties of sea water.* Landolt-Börnstein, New Series, V/3a, 233.

Stalcup, M.C. (1991), in: *WOCE Report* No. 68/91, Revision 1. November 1994, Woods Hole, MA, USA.

Sy, A. (1985), *Deep-Sea Res.*, 32, 1591.

Sy, A., Hinrichsen, H.-H. (1986), *Dtsch. Hydrograph. Z.*, 39, 35.

Takatsuki, Y., Aoyama, M., Nakano, T., Miyagi, H., Ishihara, T., Tsutsumida, T. (1991), *J. Atmos. Ocean Technol.*, 8, 895.

UNESCO (1966), Int. Oceanogr. Tabl., Paris.

UNESCO (1981a), Tech. Pap. Mar. Sci., vol. 37, Paris.

UNESCO (1981b), Tech. Pap. Mar. Sci., vol. 38, Paris.

UNESCO (1988), Tech. Pap. Mar. Sci., vol. 54, Paris.

Wooster, W.S., Lee, A.J., Dietrich, G. (1969), *Deep-Sea Res.*, 16, 321.

World Ocean Circulation Experiment (1994), WOCE Report No. 68/91, Revision 1, November 1994, Woods Hole, MA, USA.

4   Determination of oxygen

H. P. Hansen

4.1   Introduction

Water, like all other solvents, has the ability to dissolve atmospheric gases such as nitrogen, oxygen, carbon dioxide and the noble gases. Other than, *e.g.*, carbon dioxide, only a very small amount of oxygen reacts with the ions of water according to *Breck* (1974)

$$\tfrac{1}{2}\,O_2 + H^+ + e^- = \tfrac{1}{2}\,H_2O_2$$

The amount of oxygen transformed by this reaction is negligible with respect to the analytical determination of the dissolved oxygen concentration. However, owing to the absence of significant concentrations of other redox couples (*Sillén*, 1965 a,b), it is responsible for the oxidation potential of seawater. The redox potential according to the above reaction was calculated to be 560–631 mV for a pH range of 8.2–7.0 and is in good agreement with redox potentials of 590 ± 30 mV observed in natural seawater for oxygen concentrations of 80–100 % of the theoretical saturation.

The oxygen concentration C (μmol/L) of water in equilibrium with the atmosphere is governed by Henry's Law

$$C = \alpha\,p$$

where α is the Bunsen coefficient and p is the partial pressure of oxygen in the gas phase.

The temperature dependence of the Bunsen coefficient is given by the van't Hoff equation

$$\log \alpha = K \cdot T^{-1}$$

where K is a constant and T is the absolute temperature.

The salt content of seawater reduces the solubility of oxygen. *Setschenow* (1875, 1889) formulated the salinity dependence of oxygen solubility as

$$\log (\alpha_{SW} / \alpha_{PW}) = -K_T$$

where α_{SW} and α_{PW} are the solubility coefficients of seawater and pure water, respectively.

Weiss (1970) fitted data of precision measurements by *Douglas* (1965), *Green and Carrit* (1967a, b), *Murray and Riley* (1969) and *Carpenter* (1966) by the least-squares method to thermodynamically consistent equations. The final oxygen solubility equation derived from the integrated van't Hoff equation and the Setschenow equation has the form

$$\ln C = A_1 + A_2(100/T) + A_3 \ln (T/100) + A_4 (T/100)$$
$$+ S [B_1 + B_2 (T/100) + B_3 (T/100)^2]$$

where T is the absolute temperatur, S is salinity (units) and $A_n....B_n$ are the best fit coefficients. The concentration C is either in mL/L or mL/kg depending on the set of coefficients applied.

The tables of oxygen saturation in Volume II of the International Oceanographic Tables (*UNESCO*, 1973) have been calculated using the above equation. The required vapor pressure of pure water and the vapor pressure depression caused by seasalt have been calculated using formulae of *Goff and Gratch* (1946) and *Witting* (1908).

Most of the experimental data used for the evaluation of the coefficients agree with the fit to within < 0.02 mL/L.

Oxygen equilibrium concentrations at standard temperature and pressure (0 °C, 1013.25 hPa), 100 % humidity and an atmospheric oxygen content of 20.95 % are calculated as either mL/L or mL/kg using the coefficients below

	C=mL/L	C=mL/kg
A_1	−173.4292	−177.7888
A_2	249.6339	255.5907
A_3	143.3483	146.4813
A_4	−21.8492	−22.2040
B_1	0.033 096	−0.037 362
B_2	0.014 259	0.016 504
B_3	−0.001 7000	−0.002 0564

Equilibrium saturation concentrations range from 4.24 to 10.22 for temperatures of 0 to 30 °C and salinities from 0 to 40 (Appendix, Table 3).

Oxygen concentrations in seawater are controlled by fluxes through the atmosphere-water interface and by biological assimilation and dissimilation. The concentration of dissolved oxygen, therefore, is considered to be a non-conservative parameter of a water body; next to salinity and temperature it is the most commonly determined constituent of seawater. Dissolved oxygen is a valuable tracer for water masses (*Millero and Sohn*, 1992) and is a sensitive indicator for biological and chemical processes occurring in the sea. The difference between the theoretical 100 % saturation value and the amount of oxygen actually found is called the apparent oxygen utilization (AOU). The AOU concept assumes that seawater downwelling from the surface is 100 % saturated with respect to the atmosphere. This concept can only be accepted with some reservations since physical and biological processes may occur in the surface layers which may lead to supersaturation or incomplete equilibration before the water breaks contact with the atmosphere.

Dissolved oxygen concentrations in water isolated from the atmosphere commonly range from 0 to 120 % of the saturation concentration. Determinations of oxygen in samples collected in shallow waters with very high phytoplankton productivity occasionally result in oxygen concentrations equivalent to more than 300 % oxygen saturation (*e.g., Krom et al.,*

1985). In these samples a considerable portion of the oxygen is probably gaseous oxygen contained in tiny assimilation generated bubbles which were adsorbed to particles and locally caused high concentrations of dissolved and gaseous oxygen. Oxygen saturation, however, refers to the concentration of dissolved oxygen only and not the gaseous phase.

Oxygen concentrations are traditionally reported as mL/L or μmol/L (1 mL/L = 44.615 μmol/L, based on the molar volume of ideal gases of 22.413 L/mol at standard conditions). Volume changes due to temperature changes of samples between sampling and oxygen fixation introduce a small error. Precision oxygen measurements, therefore, require monitoring of temperatures during the sampling and analytical procedure and recalculation of the determined oxygen values to mL/kg or μmol/kg (see Section 4.6.6).

4.2 Principle of the determination

Oxygen dissolved in seawater is almost exclusively determined by the chemical method first proposed by *Winkler* (1888) and modifications thereof. Other methods, *e.g.*, the microgasometric determination according to *Scholander et al.* (1955), the mass spectrometric method used by *Benson and Parker* (1961), the gas chromatographic procedure according to *Swinnerton et al.* (1962, 1964) and Weiss *and Craig* (1973) are used for special purposes only. Electrochemical sensors (see Chapter 14) are mainly used for *in situ* registrations or for a continuous record of oxygen consumption or oxygen profiles.

Modifications of the original Winkler method (*e.g.*, *Strickland and Parsons*, 1960; *Carpenter*, 1965; *Carrit and Carpenter*, 1966) have mainly improved the technical details of the analytical procedure. It is noteworthy that the precision of 'historical' Winkler determinations of oxygen is close to that of today's routine analyses.

The Winkler method is an iodometric titration. As dissolved oxygen in seawater does not directly oxidize the iodide ion to iodine, a multi-step oxidation is performed using manganese as a 'transfer' medium.

Manganese(*II*)chloride and an alkaline potassium iodide solution are added to a measured volume of water. Manganese(*II*) is precipitated as hydroxide

$$Mn^{2+} + 2OH^- \rightarrow Mn(OH)_2 \tag{4-1}$$

and oxidized to manganese(*III*) hydroxide in a heterogeneous reaction

$$2Mn(OH)_2 + \tfrac{1}{2}O_2 + H_2O \rightarrow 2Mn(OH)_3 \tag{4-2}$$

The large surplus of maganese(*II*) hydroxide forces the formation of manganese(*III*) instead of manganese(*IV*). Owing to the instability of manganese(*II*) in an alkaline medium the fixation of oxygen is rapid and quantitative.

Acidification of the sample to a pH between 1 and 2.5 dissolves the hydroxide precipitates, and the iodide ions added with the fixation reagents are oxidized to iodine by the manganese(*III*) ions which are reduced to manganese(*II*) ions

$$2Mn(OH)_3 + 2I^- + 6H^+ \rightarrow 2Mn^{2+} + I_2 + 6H_2O \tag{4-3}$$

The pH must not exceed 2.5, because at higher pH the oxidation of iodine by manganese(*III*) may be incomplete. On the other hand, a surplus of acid should also be avoided, because thiosulphate used to titrate the iodine (see below) is a weak anion and combines with hydrogen ions to form the unstable thiosulphuric acid, which in turn disproportionates into sulphur and sulphurous acid or sulphur dioxide (*Grasshoff,* 1962).

The iodine and the surplus iodide ions combine to give a complex of three iodine atoms with one negative charge, which has an iodine vapor pressure much lower than molecular iodine. The complex decomposes readily if iodine is removed from the system

$$I_2 + I^- \leftrightarrow I_3^- \qquad (4\text{-}4)$$

In the final step of the analysis, the iodine is titrated with thiosulphate. The iodine is reduced to iodide, and the thiosulphate in turn is oxidized to the tetrathionate ion. The concentration of the thiosulphate solution used for the titration must be known precisely. The endpoint of the redox titration is commonly indicated by a starch indicator or by photometric or amperometric endpoint detection. The starch indicator forms an enclosure compound with iodine. The large electron cloud of the iodine interacts with the hydroxo dipoles in the starch helix resulting in an intensely blue colour of the iodine starch complex. Nevertheless, the iodine molecules can leave the starch helix easily and thus can be reduced by thiosulphate. The endpoint of the titration is clearly marked by the change from blue to colourless.

The stoichiometric equation for the reaction of iodine with thiosulphate is

$$I_3^- + 2S_2O_3^{2-} \rightarrow 3I^- + S_4O_6^{2-} \qquad (4\text{-}5)$$

Modern versions of the Winkler method improve the sensitivity and accuracy of the method by computer control of the titration procedure and the endpoint detection. Instead of visual observation of the decolouration of the blue starch-iodine complex, either the starch-iodine complex colour or the iodine colour itself is measured photometrically in the visible to ultraviolet (UV) spectral range. The spectral absorbance of an I_3^- solution (oxygen sample before titration) is depicted in Fig. 4-1. *Grasshoff* (1981) described a dead-stop titration of iodine with thiosulphate using amperometric endpoint detection. *Bradburg and Hambly* (1952) have compared various endpoint detections for iodine-thiosulphate titrations in low concentration ranges and stated relative sensitivities for visual-starch, colourimetric-starch, amperometric, UV absorption as 1 : 0.2 : 0.002 : 0.0015.

The detected equivalence point (endpoint) of the titration depends on the detection method and differs considerably between the titration of iodine with thiosulphate and back titration of surplus thiosulphate with iodine-iodide solution (*Grasshoff,* 1981).

One mole (two atoms) of oxygen is equivalent to 4 moles of thiosulphate. Sodium thiosulphate, however, is not a primary standard, *i.e.*, a compound that does not change its composition, crystallizes with a constant number of water molecules and can be purified readily and dried to a constant weight. In addition, thiosulphate working solutions slowly deteriorate. Therefore, any solution of sodium thiosulphate must be standardized within 24 h before or after use. Generally accepted standards for calibrating a thiosulphate solution are potassium iodate (KIO_3) or potassium hydrogen biiodate ($KH(IO_3)_2$). Both compounds meet the requirements for a primary standard. The standardization is based on a co-proportionating reaction of iodide with iodate resulting in the formation of iodine, which in turn is bound by

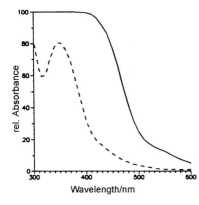

Fig. 4-1. Relative spectral absorbance of an I_3^- solution in a Winkler sample before titration (solid line) and of a 1:20 dilution thereof (dashed line).

the formation of the iodide-iodine complex by surplus iodide and thus protected from partial evaporation. The co-proportionating requires two hydrogen ions per mole of iodine formed. Hence, the reaction solution must be acidified. The co-proportionating reaction is stoichiometric and quantitative and follows the equation

$$IO_3^- + 5I^- + 6H^+ \rightarrow 3I_2 + 3H_2O \qquad (4\text{-}6)$$

The known amount of liberated iodine then is titrated with thiosulphate solution. The standardization of the thiosulphate solution is exactly the same as the last step in the determination of oxygen and thus any subjective error in recognizing the endpoint of the titration or any other bias affecting the indication of the endpoint is compensated for.

The above equations indicate that 1 mole of iodate produces 3 moles of iodine (I_2), an amount consumed by exactly 6 moles of thiosulphate. In order to ensure quantitative co-proportionating and complex formation, iodide and hydrogen ions are added in excess.

4.3   Error sources and interferences

The method is suitable to determine oxygen in seawater of all salinities as long as no H_2S is present.

Systematic and random errors in the quantitative determination of dissolved oxygen may have previously been introduced during the sampling and subsampling procedures, which will be described in Section 4.6.2 (see also Chapter 1). Therefore, this section will be confined to sources of bias in the chemical part of the analysis. Any systematic error introduced in the standardization of the thiosulphate solution will directly influence the determination of oxygen concentrations; therefore, only primary standards of the very best quality (analytical grade) should be used. Utmost care must be taken not to contaminate the stock reagent. The iodate must be dried carefully to constant weight at 115 °C and weighed with a high quality analytical balance. Calibrated glassware should be used to prepare the stock standard solution.

After acidification of the fixed samples, the iodide solution is sensitive to photochemical oxidation (*Carpenter*, 1965; *Carritt and Carpenter*, 1966), especially if the solution already contains iodine. Therefore, any exposure of the acidified iodide solution to sunlight or other UV light sources should be avoided. The liberated iodine should be titrated without delay.

Iodine, even bound to iodide as I_3^-, is barely soluble in water and has a significant vapor pressure. Again, speed of titration is essential.

Alkaline iodide solutions may be oxidized by atmospheric oxygen, especially in the presence of dust particles, or by contact with reagent bottles made of unsuitable materials that may enhance the oxidation

$$2I^- + 3O_2 \rightarrow 2IO_3^-$$

This reagent contamination is indicated in the reagent blank determination (Section 4.6.5).

4.4 Reagents

Pure water for the preparation of reagents and standards may be either distilled or deionized water.

1. *Manganese(II) chloride*: 60 g of $MnCl_2 \cdot 4H_2O$ are dissolved and made up to 100 mL with pure water.

2. *Alkaline iodide*: 60 g of KI and 30 g of KOH are dissolved separately in a minimum amount of water and combined. The solution is made up to 100 mL with pure water.

If the solution displays a yellowish-brown colour, discard and prepare again with fresh reagents. Store in a non-transparent polyethylene or polypropylene bottle.

Sodium iodide and hydroxide may be used instead of potassium iodide and hydroxide. However, if the reagents are stored for a longer period, sodium carbonate may precipitate because of its low solubility product and may block the dispenser tip.

3. *Sulphuric acid*: 50 mL of concentrated sulphuric acid are added carefully to 50 mL of pure water (the mixture must be cooled while mixing).

4a. *Sodium thiosulphate*, 0.2 mol/L: 49.5 g of $Na_2S_2O_3 \cdot 5H_2O$ are dissolved and made up to 1 L with pure water. 2.5 g of sodium borate ($Na_2B_4O_7$, analytical grade) may be added as a preservative. This stock solution is stored in a refrigerator.

4b. *Sodium thiosulphate*, 0.02 mol/L: The 0.02 mol/L working solution is a 1 : 10 dilution of 4a.

5. *Starch solution*: 1 g of soluble starch is dispersed in 100 mL of pure water. The solution is quickly heated to boiling point. (The starch solution should not be stored longer than one week and may be stabilized with 1 mL of phenol.) Instead of soluble starch, a commercial zinc starch compound may be used, which dissolves readily in water.

6. *Iodate standard*: Exactly 325 mg (0.833 mmol) of $KH(IO_3)_2$ or 356.7 mg (1.667 mmol) of KIO_3 are dissolved carefully and made up to 1 000 mL with pure water. The solution has an oxidation concentration of 0.0100 mol/L of electrons.

4.5   Instruments

Oxygen samples are subsampled with 100 mL (50 mL) bottles. A recommended bottle type is shown in Fig. 4-2. The stopper displaces a volume of about 18 mL which is sufficient to compensate for the volumes of the sulphuric acid (1 mL), the thiosulphate addition (*ca.* 8 mL of 0.02 mol/L or 0.8 mL of 0.2 mol/L), the magnetic stirring bar (*ca.* 1 mL) and the starch solution (1 mL). The titration may thus be performed in the subsample bottle.

The traditional method of oxygen determination of either 50 or 100 mL subsamples is still in use; *i.e.*, transfer of the acidified sample into a glass beaker and titration using visual (starch) endpoint detection. The achievable precision of ±0.03 mL/L is adequate for many applications, particularly in estuaries. Reagents, instruments and the analytical procedure are valid for all modifications of the method. Laboratory tests (*Hansen*, 1997 unpublished) indicate that the precision of visual endpoint detection is slightly better when a 0.02 mol/L thiosulphate solution (reagent 4b) is used for titration instead of the 0.2 mol/L solution (reagent 4a). The method described below refers to a 100 mL sample titration in the sample bottle with either visual or UV endpoint detection, but may easily be modified (a 10 mL burette instead of a 1 mL if a 0.02 mol/L thiosulphate solution is used, and 0.5 mL of each fixing reagent for a 50 mL sample instead of 1 mL for a 100 mL sample).

The bottle volumes must be determined by weighing the externally dry bottles (including stopper) empty and filled with pure water to an accuracy of ±0.05 g. The bottles, the balance and the water used for filling should be stored overnight in a controlled temperature room before performing the volume determination. The temperature should not change significantly during the volume determination, and must be recorded with the bottle volumes.

The fixing reagents are added with twin-automatic or semi-automatic reagent dispensers set to 1±0.01 mL dispensing volumes. Preferably, dispensers should be used that do not require filling *via* the dispenser nozzle. A suitable double-reagent dispenser is shown in Fig. 4-3.

A calibrated 1 mL (10 mL) precision piston burette with at least ±0.1 % precision and reading ability (display) to 0.1 % of the piston volume should be used for the titration. The piston drive should permit addition of at least 5 000 increments per burette volume.

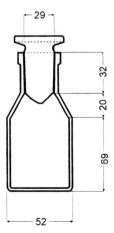

Fig. 4-2. Subsampling bottle with stopper volume displacement of about 18 mL (Figures given are in mm).

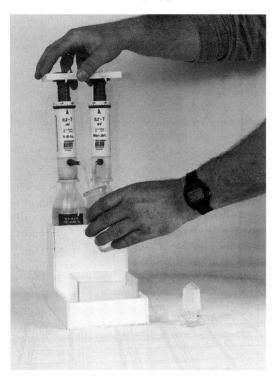

Fig. 4-3. Double reagent dispenser made from two Hirschmann (Germany) dispensers.

A computer-controlled automated titration unit with a remote controlled burette of the above specifications and photometric (UV) endpoint detection may be used (*Williams* and *Jenkinson*, 1982). Amperometric endpoint detection is also possible (*Grasshoff*, 1981).

Automatic oxygen titration units are commercially available, *e.g.*, the Metrohm 665 Dosimat Oxygen Titrator, which consists of a remote controlled burette and a UV detector for customer designed bottles. An analog-to-digital interface is required to connect the unit to a computer.

Figure 4-4 shows a plot of UV absorbances *versus* thiosulphate addition. The endpoint is calculated as the intercept of two linear regression lines just before and after the endpoint.

The principle of the automated UV titration is illustrated in Fig. 4-4. Suitable wavelengths are 450–470 nm for the I_3^- colour and about 660 nm for the colour of the starch indicator. Initially, the absorption of the I_3^- solution is beyond the validity of Beer-Lambert's law depending on the type and adjustment of the photometer. When the thiosuphate addition approaches the endpoint, the absorbance-thiosulphate addition relationship becomes linear. The piston increments of the burette are reduced (indicated by an expanded scale in Fig. 4-4) and data pairs of mL of thiosulphate added and relative absorbance are recorded. Regression lines are fitted (by computer) to the linear sections of the titration curve before and after the equivalence point. The equivalence thiosulphate addition is indicated by the calculated intersection of the regression lines.

In should be noted that automated oxygen titration does not accelerate the speed of determination but improves the precision.

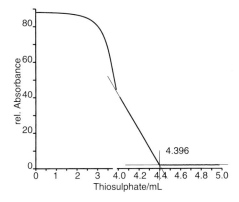

Fig. 4-4. UV absorbance (470 nm) of an oxygen sample during titration *versus* thiosulphate solution added.

A typical setup of a working system for manual determinations of oxygen as given in Fig. 4-5 is convenient for standard oxygen titrations applying visual (starch) endpoint detection and may be converted into an automated titration unit by adding a detector and a computer plus interface.

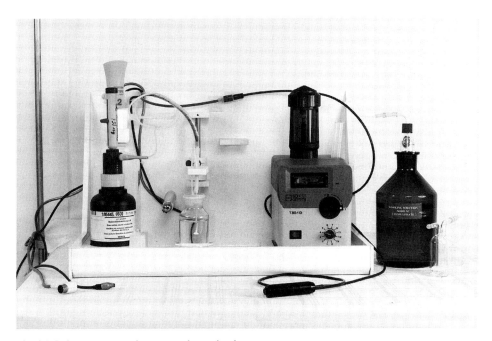

Fig. 4-5. Laboratory setup for oxygen determinations.

4.6   Procedure

4.6.1   Standardization of the thiosulphate solution

Pure water with a volume about half that of the sample is placed in a sample bottle and 1 mL of the 50 % sulphuric acid solution, 1 mL alkaline iodide solution and 1 mL manganese(II) chloride reagent are added separately. After each reagent addition the solution is thoroughly mixed to avoid any precipitation of manganese hydroxides.

Then 10.00 mL of the iodate standard solution are added with a calibrated pipette or a precision piston burette. The bottle is filled to about the sample volume (to the neck) with pure water.

Visual endpoint detection

The liberated iodine is titrated to a light yellow colour. A volume of 1 mL starch indicator solution is added, and the titration is continued until the blue colour then disappears. Near the endpoint the solution becomes 'cloudy' directly after addition of the thiosulphate solution. The endpoint is reached when this 'cloudy' effect can no longer be seen. Diffuse illumination from below and moderate ambient light facilitate detection of the endpoint.

UV endpoint detection

The volumes of the increments of the thiosulphate solution added are reduced near the equivalence point. The titration is continued after the endpoint until a group of consecutive data readings display no change. Linear regression lines are fitted to the linear segments of the absorbance curve immediately before and after the end point. The intersection of the two regression lines marks the endpoint.

4.6.2   Subsampling and fixation of dissolved oxygen

As a rule, subsampling for the determination of dissolved oxygen should be carried out as soon as possible after the sampler has been recovered. The bottles are unstoppered and after flushing the nozzle of the hydrocast sampler with sample water, the end of the nozzle tube is inserted into the sample bottle almost to the bottom. The nozzle should be transparent and sufficiently narrow to ensure that the sample stream carries with it any air bubbles and the air segment in the tube but wide enough to fill the bottle rapidly.

In order to flush the bottles, about twice the volume of the bottle should be allowed to flow through the oxygen bottle until it is finally filled. The sample stream should not generate too much turbulence in the bottle to avoid intrusion of atmospheric oxygen.

When the final stage of filling is reached, the nozzle is slowly withdrawn. If an air bubble has nevertheless been trapped on the walls of the bottle, it should be driven off by tapping the bottle gently. The bottle is now filled to the brim. Without intermediate stoppering, the reagents (1 mL each of manganese(II) chloride, reagent 1, and alkaline iodide, reagent 2) are added, preferably with a twin dispenser. The pipette tips are inserted almost to the bottom of the flask and slowly withdrawn while the reagents are added. Because of the high density of the reagents, they sink to the bottom of the bottle and do not mix with the sample. The stopper is then inserted, which thus replaces the top of the water sample which might have been contaminated by atmospheric oxygen.

The bottles are shaken vigorously for about 1 min to bring each molecule of dissolved oxygen into contact with manganese(*II*) hydroxide. After fixation of the oxygen, the precipitate is allowed to settle (which takes 10–20 min, depending upon the salinity of the sample).

Immediately after subsampling for oxygen determinations, the temperature of the water in the water sampler is measured with an accuracy of about $\pm 0.5\,°C$ and recorded in the protocol.

4.6.3 Storage

The oxygen samples may be stored for a few hours after addition of the reagents and after complete fixation. The bottles should be kept in the dark, and any change of temperature should be avoided because of the risk of 'breathing', *i.e.*, the volume of the sample may expand and contract and thus aspirate atmospheric oxygen. The necks of the bottles should be sealed with seawater or, for extended storage (maximum 12 h), the stoppers are fixed with clamps and the bottles submerged in seawater. If the necessary precautions are taken, the error will not exceed 0.5 %.

4.6.4 Titration

Before titration, the precipitated hydroxides are dissolved with sulphuric acid. For this, 1 mL of sulphuric acid (reagent 3) is pipetted into the sample bottle immediately after unstoppering. The tip of the pipette is inserted almost to the level of the precipitate and then slowly withdrawn without disturbing the precipitated hydroxides. A small magnetic stirring bar is carefully deployed in the bottle. The rotation of the stirrer is slowly accelerated. No hydroxide precipitate is to be whirled up beyond about half of the bottle height, *i.e.*, it should be kept well away from the surface.

In case of sample transfer and titration in a separate beaker, the bottle is stoppered again, care being taken that no air bubbles are trapped. The hydroxides are dissolved by shaking, the stopper is removed, the film of sample on the stopper now containing liberated iodine is washed down carefully into a titration beaker of about twice the size of the sample volume and the contents of the bottle are quantitatively transferred into the beaker. The bottle is rinsed twice with a few mL of pure water, and the wash water is combined with the bulk of the sample.

The titration is carried out immediately as described in Section 4.6.1.

4.6.5 Determination of the reagent blank

The reagents added to the seawater sample contain oxygen. This amount of oxygen is minimized by using reagents that are almost saturated with respect to their salt content. According to *Murray et al.* (1968) 1 mL of the reagents contains approximately 0.0017 mL

(0.0759 μmol) oxygen. If no total reagent blank is determined (see below), the oxygen concentration of the sample may be corrected by subtraction of a blank concentration DO_R

$$DO_R = 1\,000 \cdot V_R \cdot F_R / V_S$$

V_S and V_R are the sample and the reagent volumes (mL) respectively, and F_R is the oxygen content of 1 mL of reagent.

As the ratio of reagent addition and sample volume (V_R / V_S) is constant ($2:100$ in nearly all modifications of the Winkler method), DO_R is 0.034 mL/L or 1.52 μmol/L.

In addition to small amounts of oxygen, the reagents may contain impurities, *e.g.*, higher oxidation states of manganese or traces of iodine which result in elevated blank values. Before the start of the oxygen determinations the reagent blank has to be checked. Ignoring the type of impurity and the chemical reaction, this total reagent blank may be expressed as a thiosulphate solution volume consumed by the reagents. The proposed method to determine the reagent blank is similar to the standardization procedure of the thiosulphate solution (Section 4.6.1).

About 15 mL of pure water are placed in a sample bottle. A 1 mL volume of the 50 % sulphuric acid solution and the amounts of alkaline iodide solution and manganese(II) chloride used for fixation are added separately. After each reagent addition the solution is thoroughly mixed to avoid any precipitation of manganese hydroxides.

Then add 1.00 mL of the iodate standard solution, fill up to just below the bottle neck (about 5 mL less) and titrate with the thiosulphate solution until the exact equivalence point.

Add another 1.00 ml of the iodate standard and titrate again until the equivalence point. The reagent blank is the difference between the first and the second thiosulphate titration volume, *i.e.*,

$$a_R = (a_1 - a_2)$$

If the reagent blank exceeds the equivalence of 0.1 mL/L oxygen the reagents should be discarded.

4.6.6   Calculation of the result

The oxygen equivalent (E_{Ox}), *i.e.*, the amount of thiosulphate corresponding to 250.0 μmol (5.6035 mL) oxygen at standard temperature (0 °C) and pressure (1013.25 hPa) is 1 mmol.

Including the standardization of the thiosulphate solution the sample concentration of oxygen is calculated as:

$$C_{Ox} = \frac{(a - a_R) \cdot V_{STD} \cdot C_{STD} \cdot E_{Ox}}{(a_{STD} - a_R)\,(V_b - V_R)} - DO_R$$

where

a, a_R, a_{STD} are thiosulphate titration volumes in mL of the sample, the reagent blank and the iodate standard, respectively;

V_{STD}, V_b, V_R are volumes in mL of the iodate standard, sample (bottle) and fixing reagents (manganese(II) chloride plus alkaline iodide), respectively;

C_{STD} is the molar concentration of the iodate standard;

C_{Ox} is the sample concentration of oxygen in mL/L or μmol/L depending on the dimension of the oxygen equivalent E_{Ox};

and DO_R is the correction term for traces of oxygen contained in the fixation reagents (see Section 4.6.5).

For routine analyses it may be advantageous to simplify the calculation by combining the predetermined variables for a specific analytical procedure as a constant term, *i.e.*, nominal thiosulphate concentration, oxygen equivalent and bottle and reagent volumes.

The remaining input variables for the calculation are then:

(i) the thiosulphate volume of the titration (mL) corrected for a reagent blank if required;

(ii) the deviation of the thiosulphate concentration from the nominal concentration (0.02 or 0.2 μmol/L).

The deviation of the thiosulphate concentration from the nominal value is expressed by a factor f ($f = 1$ for the exact nominal concentration). Multiplication of a real concentration thiosulphate volume with f converts it into the respective volume of the nominal concentration.

The concentration (C) of the thiosulphate is determined from the titration with iodate standard (Section 4.6.1) and calculated as

$$C = 6 \cdot V_{STD} \cdot C_{STD} / V$$

where V is the thiosulphate volume (mL) used for standardization. The factor is calculated as

$$f = 6 \cdot V_{STD} \cdot C_{STD} / (C_{nom} \cdot V)$$

where C_{nom} is the nominal concentration of the thiosulphate solution.

The simplified calculation of the oxygen concentration of the sample is

$$C_{Ox} = \frac{C_{nom} \cdot E \cdot 10^3 (a - a_R) \cdot f}{(V_b - V_R)} - DO_R$$

where $V_b - V_R = 49$ for 50 mL samples plus 1 mL of reagents and 98 for 100 mL samples plus 2 mL of reagents.

$$E = 5.6035 \text{ for } C_{Ox} \text{ as mL/L and 250 for } C_{Ox} \text{ as } \mu\text{mol/L};$$

The nominal thiosulphate concentration C_{nom} is either 0.2 or 0.02;

The blank volume correction ($- a_R$) of the thiosulphate addition usually corresponds to less than 0.01 mL/L oxygen and may be ignored.

Without blank correction other than the oxygen content of the reagents (*Murray et al.*, 1968), the calculation of the titrated oxygen concentration of a nominally 100 mL sample with 2 mL reagent addition using a thiosulphate solution of nominally 0.2 mol/L is

$$C_{Ox} = \frac{1.1206 \cdot 10^3 \cdot a \cdot f}{(100-2)} - 0.034 \text{ mL/L}$$

$$= 11.435 \cdot a \cdot f - 0.034 \text{ mL/L}$$

(0.034 mL/L or 1.52 μmol/L is the term DO_R which corrects for the oxygen content of the fixing reagents; see Section 4.6.5).

For high-precision determinations of *in situ* oxygen concentrations the results have to be converted into μmol/kg because the amount of oxygen defined by a volume concentration depends on the sample temperature, which changes from *in situ* to subsampling conditions. The conversion of μmol/L concentrations into *in situ* μmol/kg requires the application of the equations for seawater density (*UNESCO*, 1983; see Appendix Table 4). The sample volume at subsampling temperature (fixation) is corrected to the *in situ* volume and mass. Temperature or volume change after fixation do not affect the analysis.

Thermal volume changes of the thiosulphate concentration between standardization and sample titration may also require a correction. A coefficient of expansion of 0.00025 K^{-1} may be used between 10 and 40 °C. Below 10 °C no correction is required; most probably standardization and oxygen determination will be performed at higher temperatures anyway. In the formula for the calculation of oxygen concentrations (C_{Ox}) the thiosulphate factor has to be divided by $(1 + 0.00025\,(t_D - t_S))$. The terms t_D and t_S are the temperatures of the thiosulphate solution during the oxygen determination and thiosulphate standardization. As can be seen, the required correction is rather small. A temperature difference of < 4 K between standardization and sample determination causes an error of < 0.1 %, which may be neglected. Greater temperature differences should be avoided or require correction.

Another, though small, error is introduced by temperature differences between subsampling and the volume determination of the sample bottles. The resulting volume error depends very much on the geometry and material of the sample bottles. For high-precision oxygen measurements the determination of the volume-temperature dependence of the type of sample bottle and resulting correction is recommended.

The software delivered with commercial autotitration units commonly includes routines for the conversion of mL/L oxygen into μmol/kg, the temperature correction of the thiosulphate solution and calculations of oxygen saturation and AOU if subsampling and *in situ* temperatures and salinity are provided.

4.6.7   Accuracy and precision

It is extremely difficult to estimate exactly the accuracy of the determination since the major contribution to the systematic error probably has its source in the sampling procedure itself (see Chapter 1). When attention is paid to all the sources of systematic errors (see Section 4.3), most of which result in an increased oxygen content, a field precision of ± 0.005 mL/L can be achieved using 100 mL samples and photometric endpoint detection and ± 0.03 mL with 50 mL samples and visual (starch) endpoint detection. The precision is about 25 % less for oxygen contents below 2 mL/L. If a good quality iodate standard is used for calibration the analytical accuracy is equal to the precision.

References to Chapter 4

Benson, B.B., Parker, P.D. M. (1961), *J. Phys. Chem.*, 65, 1489.

Bradburg, J.H, Hambly, A.N. (1952), *Aust. J .Sci. Res.*, A5, 541.

Breck, W.G. (1974), in: *The Sea*: Goldberg, E.D. (Ed.). New York: John Wiley & Sons, 1974; vol.5, pp. 153–179.

Carpenter, J.H. (1965), *Limnol. Oceanogr.*, 10, 135.

Carpenter, J.H. (1966), *Limnol. Oceanogr.*, 11, 264.

Carritt, D.E., Carpenter, J.H. (1966), *J. Mar. Res.*, 24, 286.

Douglas, E. (1965), *J. Phys. Chem.*, 69, 2608.

Goff, J.A., Gratch, S. (1946), *Trans. Am. Soc. Heat. Vent. Eng.*, 52, 95.

Grasshoff, K. (1962), *Kieler Meeresforschungen*, 18, 42.

Grasshoff, K. (1981), in: *Marine Electrochemistry:* Whitfield, M., Jagner, D., (Eds.). Chichester, John Wiley & Sons, 1981; pp. 327–420.

Green, E.J., Carritt, D.E. (1967a), *J. Mar. Res.*, 25, 140.

Green, E.J., Carritt, D.E. (1967b), *Science*, 157, 191.

Krom, M.D., Porter, C., Gordin, H. (1985), *Aquaculture,* 49, 159.

Millero, F.J., Sohn, M.L. (1992), *Chemical Oceanography.* Boca Raton: CRC Press.

Murray, C.N., Riley, J.P., Wilson, T.R.S. (1968), *Deep- Sea Res.*, 15, 237.

Murray, C.N., Riley, J.P. (1969), *Deep-Sea Res.*, 16, 311.

Scholander, P.F., Van Dam. L., Claff, C.L., Kauwisher, J.W. (1955), *Biol. Bull.,* 109, 328.

Setschenow, I. (1875), *Mem. Acad. Imp. Sci.*, St. Petersburg, 22, 6.

Setschenow, I. (1889), *Z. Phys. Chem.*, 4, 177.

Sillén, L.G. (1965a), *Arkiv Kemi*, 24, 431.

Sillén, L.G. (1965b), *Arkiv Kemi*, 25, 159.

Strickland, J.D.H., Parsons, T.R. (1960), *Fish. Res. Bd.Can., Bull.*, 125.

Swinnerton, J.W., Linnenbom, V.J., Cheek, C.H. (1962), *Anal. Chem.*, 34 (4), 483.

Swinnerton, J.W., Linnenbom, V.J., Cheek, C.H. (1964), *Anal. Chem.*, 36 (8), 1669.

UNESCO (1973), *Int. Oceanogr. Tab.*, Vol. II.

UNESCO (1983), *Techn. Pap. Mar. Sci.*, 44.

Williams, P.J.leB., Jenkinson, N.W. (1982), *Limnol. Oceanogr.*, 27, 576.

Weiss, R.F. (1970), *Deep-Sea Res.*, 17, 721.

Weiss, R.E., Craig, H. (1973), *Deep-Sea Res.*, 20, 291.

Winkler. L.W. (1888), *Ber. Dtsch. Chem .Ges.*, 21, 2843.

Witting, R. (1908), I. Finnland. Hydrog. Biol. Unters., 2, 173.

5 Determination of hydrogen sulphide

S. Fonselius, D. Dyrssen and B. Yhlen

5.1 Introduction

Many bacteria use the oxygen dissolved in seawater to oxidize organic matter to carbon dioxide, water and inorganic ions. In the deep water of stagnant basins, in many coastal lagoons and in sea areas with a very slow water exchange or high load of organic matter, *e.g.,* the Black Sea, the Baltic Sea, many fiords in Norway, Greenland and Canada and the Cariaco Trench off the Venezuelan coast, this mechanism may consume all the dissolved oxygen. In such anoxic waters sulphate-reducing bacteria use the oxygen bound in sulphate ions as an electron acceptor while reducing the sulphate ions to sulphide (*Skopintsev et al.,* 1959). The oxygen minimum layer in the Pacific off the Central American coast seems to be very close to the level for sulphide formation, but no sulphide has yet been detected in this region (*Cline and Richards,* 1972).

$$2CH_2O(org) + H_2SO_4 \rightarrow 2CO_2 + 2H_2O + H_2S \tag{5-1}$$

According to *Skopintsev et al.* (1959), only a few percent of the total sulphide content originates in organic sulphur compounds.

Hydrogen sulphide is a poisonous gas that readily dissolves in water. No higher life forms can survive in water containing hydrogen sulphide, and such areas are therefore transformed into oceanic deserts. Hydrogen sulphide in a water sample is detected easily by its characteristic smell, even at extremely low concentrations.

In water hydrogen sulphide is ionized; in seawater (pH 7–8) it occurs mainly as HS^- ions:

$$H_2O + H_2S \rightarrow H_3O^+ + HS^- \tag{5-2}$$

In an acidic medium, dissolved oxygen will rapidly oxidize hydrogen sulphide to elemental sulphur, but in neutral solutions, including seawater, it is slowly oxidized to sulphate (see Section 6.1):

$$H_2S + 0.5O_2 \rightarrow S(colloidal) + H_2O \tag{5-3}$$
$$HS^- + 2O_2 + H_2O \rightarrow SO_4^{2-} + H_3O^+ \tag{5-4}$$

Water containing hydrogen sulphide has a negative redox potential and therefore is a reducing medium.

5.2   Units

Hydrogen sulphide concentrations are usually expressed as μmol/L S^{2-} or sometimes as mL/L H_2S.

$$Y\ \mu\text{mol/L}\ S^{2-} = Y \cdot 22.41 \cdot 10^{-3}\ \text{mL/L}\ H_2S \tag{5-5}$$

or

$$Z\ \text{mL/L}\ H_2S\ \ = Z \cdot 10^3 / 22.41\ \mu\text{mol/L}\ S^{2-}$$

In some cases it may be convenient to express hydrogen sulphide as 'negative oxygen' *(Fonselius*, 1969). This unit does not represent the amount of hydrogen sulphide present in the water and should therefore only be used as an oxidation equivalent for comparison purposes. 'Negative oxygen' is the amount of oxygen equivalent to the amount of hydrogen sulphide produced through reduction of sulphate. Sulphate ions contain four atoms of oxygen, which are used for the bacterial oxidation of organic matter, and one sulphur atom which is reduced from S^{6+} to S^{2-}. The hydrogen sulphide concentration expressed in mL/L or μmol/L multiplied by 2 is the concentration of 'negative oxygen'.

5.3   Analytical methods

We recommend spectrophotometric determination of hydrogen sulphide, but other methods will be mentioned as well (see Sections 5.3.3 and 5.3.4).

The determination of hydrogen sulphide as methylene blue (3,7-bis(dimethylamino)phenothiazine-5-onium chloride) was introduced by *Fischer* (1883) and has been used for seawater in many investigations. *Cline* (1969) has closely reviewed and tested the methylene blue method and has suggested several improvements.

Methylene blue is formed from dimethyl-*p*-phenylenediamine, with an indammonium salt (Bindschedler's green) as an intermediate; this condenses with hydrogen sulphide giving the thiazine dye (Fig. 5-1). Iron(*III*) chloride is the usual oxidant for the condensation and cyclization reactions, but iron(*III*) sulphate or oxalate may be used instead. The colour intensity of the dye is pH dependent. According to *Cline* (1969) the optimum pH is 0.35.

Two methods are described below. The method by *Fonselius* (1969) is recommended for low concentrations of hydrogen sulphide and the second by *Cline* (1969) for high concentrations.

5.3.1   Method by Fonselius

The method is particularly suitable for measuring very low concentrations of hydrogen sulphide, *e.g.*, in the Baltic Sea. The blank values are negligible, and Beer's law is applicable up to around 60 μmol/L, but determinations of up to 300 μmol/L are possible without dilution of the sample. By using the ordinary bottles used for oxygen samples to collect hydrogen sulphide samples, rapid work is possible with minimum risk of losses of sulphide and contact with oxygen.

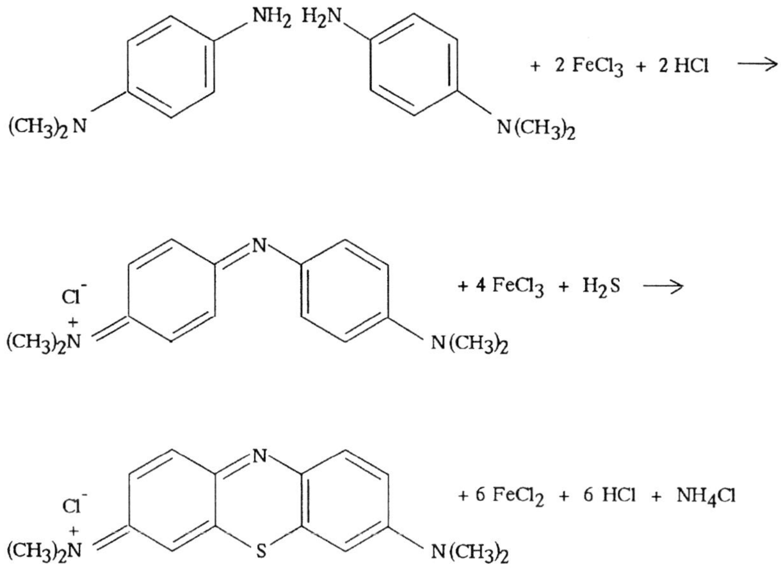

Fig. 5-1. Formation of methylene blue from dimethyl-*p*-phenylenediamine by hydrogen sulphide in an acidic medium using reaction with iron(*III*) chloride as oxidant.

5.3.1.1 Reagents

1. *N,N-dimethyl-p-phenylenediamine dihydrochloride solution*: 10 g of $(CH_3)_2N \cdot C_6H_4 \cdot NH_2 \cdot 2HCl$ (1,4) (analytical-reagent grade, a.g.) are dissolved in 500 mL of approximately 6 mol/L HCl. The acid is prepared by diluting concentrated HCl with an equal amount of distilled water. The reagent is stable for several months.

2. *Iron(III) solution:* 15 g of $FeCl_3$ (a.g.) are dissolved in 500 mL of approximately 6 mol/L HCl (prepared as above). The reagent is stable indefinitely.

3. *Zinc acetate solution:* 10 g of $ZnAc_2 \cdot 2H_2O$ (a.g.) are dissolved in 200 mL of distilled water.

4. *Oxygen-free water:* Oxygen-free distilled water is prepared by bubbling nitrogen (or argon) gas through a suitable volume (2–5 L) of distilled water for 30–60 minutes using a glass frit. The bottle is closed with a rubber stopper fitted with a siphon and a glass tube, both with stopcocks. This water is difficult to store properly and should be prepared and tested for oxygen just before use. The oxygen concentration should be < 0.2 mL/L.

5. *Sulphide stock solution:* A few crystals of sodium sulphide, $Na_2S \cdot 9H_2O$ (a.g.), are quickly washed with distilled water squirted from a plastic washing bottle. The crystals are dried immediately with filter paper and placed in a pre-weighed glass-stoppered weighing dish. About 1.50 g are weighed on an analytical balance and dissolved in oxygen-free distilled water to 1000 mL in a calibrated flask. The solution is not stable for long periods of time and should be used within a few days.

6. *Sulphide working solution:* 100 mL of the sulphide stock solution are pipetted into a 1000 mL calibrated flask containing oxygen-free distilled water; the flask is filled to the mark. This solution should be used as soon as possible. The solution contains approximately

$20 \,\mu g/mL \, S^{2-}$ (about $624 \,\mu mol/L \, S^{2-}$). The concentration is determined by titration as described in Section 5.3.1.8.

7. *Sodium thiosulphate solution (0.02 mol/L).*

8. *Potassium iodate solution (1.667 mmol/L).*

9. *Potassium iodide solution:* Dissolve 20 g KI (a.g.) in 100 mL of distilled water.

10. *Sulphuric acid solution (1 : 1 v/v).*

11. *Starch indicator*

Solutions 7, 8 and 10 are the same as some of the reagents described in Chapter 4 (Determination of oxygen).

5.3.1.2   Special apparatus

A spectrophotometer or a filter photometer at or close to 670 nm is used for the determination. Cuvettes with path lengths of 0.5, 1, 5 and 10 cm are used depending on the intensity of the colour development. Automatic piston pipettes adjusted to a volume of 0.5 mL are recommended for adding reagents. The absorbance readings should be recalculated according to the path length used for the calibration.

5.3.1.3   Sampling

Samples should be taken from the ordinary hydrocast samplers immediately after the oxygen sample and in the the same manner as these samples (see Chapter 4). The hydrocast samplers should preferably be all-plastic, since sulphide reacts with many metals. When no oxygen is present in the water, samples for hydrogen sulphide determination should be taken first (see also Chapter 2). We use 50 mL oxygen bottles. Hydrogen sulphide is probably present in the water when the manganese hydroxide precipitate in the oxygen sample is completely white instead of brownish.

5.3.1.4   Preservation of samples

Samples that cannot be analysed immediately on board ship may be preserved by adding zinc acetate (or chloride), which precipitates the sulphide as zinc sulphide. A 1 mL aliquot of zinc acetate solution is immediately added to a 50 mL sample with a piston pipette. Samples thus preserved may be stored for long periods of time if kept in a dark place. The other reagents are later added at the home main laboratory.

5.3.1.5   Procedure

The reagents, 0.5 mL of dimethyl-*p*-phenylenediamine and 0.5 mL iron(*III*) chloride solution, are added in exactly the same way as the reagents for determining oxygen, *i.e.,* simultaneously with piston pipettes or with dispensers. The tips of the pipettes/dispensers should touch the bottom of the oxygen bottles. No air bubbles must be trapped in the bottles. The colour develops within a few minutes. If other sizes of sample bottles are used, the amounts of reagents have to be adjusted accordingly. Note that knowledge of the exact volume of the sample bottles is not important, because concentrations are measured instead of total amounts.

5.3.1.6   Analysis

Absorbances are measured using a spectrophotometer at 670 nm or in a filter photometer with a filter close to 670 nm, using 0.5 or 1 cm cuvettes for high concentrations and 5 or 10 cm cuvettes for low concentrations. Measurements are performed no sooner than 1 h after addition of the reagents. Some workers have reported that the blue colour is not stable for more than 4 h, while others claim a stability of several days. We have measured solutions in distilled water; after 68 h an absorbance exceeding 0.8 in a 5 cm cuvette had increased by 2–3 %, whereas the lower concentrations remained unaltered (*Carlberg*, 1972). The samples are measured against distilled water; the blanks are found to be negligibly low, even with somewhat coloured diamine solutions. Hydrogen sulphide concentrations are determined from a standard curve as described in Section 5.3.1.9.

5.3.1.7   Dilution of samples

Samples containing high concentrations of hydrogen sulphide may have to be diluted before analysis, by precipitation with a zinc acetate solution (see under 'Reagent', Section 5.3.1.1) containing 2 g/L of gelatin. The solution is homogenized by vigorous shaking, and a suitable dilution is prepared to which the ordinary reagents are added (*Grasshoff and Chan*, 1971). However, the method by *Cline* (1969) (Section 5.3.2) should be given preference.

Note that by no means should the amounts of reagents added to the sample be doubled, if very high concentrations of hydrogen sulphide suggest that the reagent concentrations may be insufficient; this will change the pH of the sample leading to erroneous results even after dilution.

5.3.1.8   Standardization of the method

Standardization is carried out immediately after preparation of the working solution and the photometric standards. To each of six 200 mL Erlenmeyer flasks with glass stoppers are added 10 mL of potassium iodide solution and 10.00 mL of potassium iodate solution, exactly the same amount to each flask, and 1 mL of sulphuric acid; 50 mL of the sulphide working solution are pipetted into three of the flasks with a calibrated pipette, and 50 mL of oxygen-free water into the other three flasks. The flasks are set aside for about 10 min. Their contents are then titrated with thiosulphate solution using soluble starch as indicator:

$$8I^- + IO_3^- + 6H^+ \rightarrow 3I_3^- + 3H_2O$$
$$H_2S + I_3^- \rightarrow 3I^- + 2H^+ + S \tag{5-6}$$
$$I_3^- + 2S_2O_3^{2-} \rightarrow 3I^- + S_4O_6^{2-}$$

The titers of triplicates should agree to within 0.05 mL. The amount of H_2S present is calculated using the formula

$$\mu mol/L\ H_2S = 10^6 \cdot M \cdot (A - B) / (2V) \tag{5-7}$$

or

$$mL/L\ H_2S\quad = 22.41 \cdot 10^3 \cdot M \cdot (A - B) / (2V)$$

where

- A = mean of the titers of the three solutions without sulphide, in mL;
- B = mean of the titers of the three solutions with sulphide, in mL;
- V = mL of sulphide working solution added;
- M = concentration of the thiosulphate solution (mol/L).

5.3.1.9   Calibration of the method

A series of standards is prepared by adding decreasing volumes of working solution to 100 ml calibrated flasks by means of a pipette and dilution to volume with oxygen-free water *via* a siphon (see Table 5-1). Calculate the exact sulphide concentration using the formulae

$$\mu mol/L\ S^{2-}\quad =\quad C \cdot D/E \tag{5-8}$$

where

- C = mL of working solution;
- D = concentration of sulphide in the working solution, in $\mu mol/L$;
- E = volume of the flask + volume of added reagents, in mL.

Table 5-1.   Preparation of standard curve from working solution for low H_2S concentration range.

Volume diluted to 100 mL	H_2S concentration in $\mu mol/L$
50	306.0
40	245.0
30	183.8
20	122.5
10	61.3
8	49.0
4	24.5
2	12.3
1	6.1
0.5	3.1
0	0

Note that the concentrations given in Table 5-1 approximately correspond to a working solution containing $20\ \mu g/mL$ of sulphide. They have to be corrected to the value found by titration. With piston pipettes 1 mL each of both sulphide reagents are added simultaneously to the solutions immediately after preparation. The contents of the bottles are mixed well, and the standards are measured against distilled water at 670 nm in a 5 mm cuvette no sooner than 1 h after the reagent addition. (In modern spectrophotometers cuvettes with a path length of 5 mm can be used for all concentrations in the calibration curve.) From the results a standard curve is prepared which should pass through the origin, if the zero fraction of the absorbance is subtracted; see Fig. 5-2.

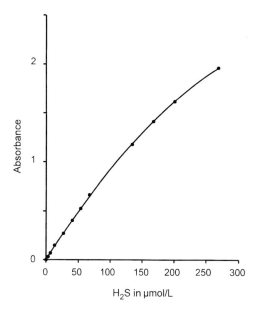

Fig. 5-2. Standard curve for hydrogen sulphide solutions according to Fonselius (at 670 nm with a 5 mm cuvette). The curve follows Beer's law up to 60 μmol/L, but can be used for concentrations up to around 300 μmol/L.

5.3.2   Method by Cline

A method similar to the method described in Section 5.3.1 has been developed and investigated by *Cline* (1969), also published by *Parsons et al.* (1984). For routine work at sea the method described in Section 5.3.1 appears to be simpler, especially when determining concentrations of hydrogen sulphide below 250 μmol/L. Cline's method should be used for determining high concentrations of hydrogen sulphide, of from 200 up to 1 000 μmol/L. The standards and samples must be diluted with distilled water after addition of the reagent; oxygen-free water is not needed. If the standards are diluted 1 : 25 and are measured in a 5 mm cuvette, a straight calibration curve obeying Beer's law is obtained. A small problem with this method in practical field work is the large volume of reagent used, 4 mL of reagent/50 mL of sample. It is not possible to use the bottles used for oxygen sampling. Special 50 mL bottles with long necks should be used instead, *e.g.,* calibrated flasks (see Section 5.3.2.3). A reduction in the reagent volume to 2 mL would require the use of fuming hydrochloric acid to keep the pH at the optimum value.

5.3.2.1   Reagents

Cline's reagent No. 4 for sulphide concentrations between 250 and 1 000 μmol/L is used.

Mixed diamine solution: Dissolve 20 g of *N,N*-dimethyl-*p*-phenylenediamine dihydrochloride and 30 g of iron(*III*) chloride, $FeCl_3$, in 500 mL of 6 mol/L HCl (cool, a. g.). The reagent is stable for several months.

5.3.2.2   Special apparatus

A spectrophotometer with 5 mm cuvettes is required.

5.3.2.3   Sampling

Samples are drawn as described in Section 5.3.1.3, but 50 mL calibrated flasks are used instead of oxygen bottles. The rubber or plastic tubing of the sampler should touch the bottom of the flask, while the flask is filled with sample water allowing at least one flask volume to overflow. The tube then is withdrawn carefully and the flask filled to the mark. Note that the calibrated flasks must have space for at least 4 mL of reagent above the mark.

5.3.2.4   Procedure

A 4 mL aliquot of the mixed reagent is added to the calibrated flasks with a piston pipette or a dispenser. The flasks are stoppered, turned upside down and shaken to mix the reagent with the sample. The samples are stored for at least 1 h before analysis.

5.3.2.5   Analysis

The samples are diluted 1 : 25 with distilled water no sooner than 1 h after addition of the reagent and are measured at 670 nm with the spectrophotometer using a 5 mm cuvette.

5.3.2.6   Standardization and calibration of the method

Standardization and calibration are performed as described in Sections 5.3.1.8 and 5.3.1.9, but the sulphide working solution is prepared by diluting 160 mL of sulphide stock solution to 1000 mL. A series of standards is prepared by adding the following volumes of working solution to 50 mL calibrated flasks with a pipette and diluting to volume with oxygen-free water using a siphon (see Table 5-2). Note that the calibrated flasks must have space for at least 4 mL of reagent above the mark. Calculate the exact sulphide concentration as described in Section 5.3.1.9.

Table 5-2.   Preparation of standard curve for H_2S concentrations according to Cline.

Volume diluted to 50 mL	H_2S concentration in $\mu mol/L$
50	926
40	741
30	556
20	370
10	185
0	0

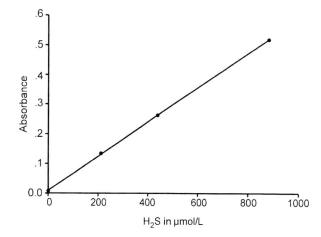

Fig. 5-3. Standard curve for hydrogen sulphide solutions according to Cline (at 670 nm with 5 mm cuvette). The curve follows Beer's law up to around 1000 μmol/L, but it is diffcult to measure low concentrations. Note that the hydrogen sulphide concentration is shown as the actual concentration of the sample before dilution (1:25).

Immediately after preparation of the dilutions, 4 mL of the mixed reagent are added with a piston pipette. The contents of the bottles are mixed well. The standards are diluted 1:25 with distilled water no sooner than 1 h after reagent addition. They are then measured against distilled water at 670 nm in a 5 mm cuvette (see Section 5.3.1.9). Note that the zero standard must also be diluted. From the results, a standard curve is prepared which should pass through the origin if the zero fraction of the absorbance is subtracted; see Fig. 5-3.

5.3.3 Titration methods

It is also possible to titrate the hydrogen sulphide content of samples according to the method described in Section 5.3.1.8. A simple method has been suggested by *Andersen and Føyn* (1969) which has the advantage that it only requires the reagents used for the determination of oxygen and therefore can be used when special reagents are not available. The hydrogen sulphide is reacted with the manganese(*II*) sulphate and the alkaline potassium iodide solutions (Winkler I and II). Manganese(*II*) sulphide forms and precipitates together with the hydroxide. The precipitate is allowed to settle as much as possible. The stopper is removed and 11 mL of the sulphide-free supernatant liquid is pipetted off. Then 10 mL of potassium iodate solution are added with a calibrated pipette and immediately afterwards 1 mL of sulphuric acid (50 % v/v). The bottle is re-stoppered and shaken to dissolve the precipitate. The solution is transferred quantitatively into a 200 mL Erlenmeyer flask and titrated with thiosulphate solution using soluble starch as the indicator. The reactions are:

$$Mn^{2+} + S^{2-} \rightarrow MnS \qquad (5\text{-}9)$$
$$8I^- + IO_3^- + 6H^+ \rightarrow 3\,I_3^- + 3H_2O$$
$$MnS + 2H^+ \rightarrow Mn^{2+} + H_2S$$
$$H_2S + I_3^- \rightarrow 3I^- + 2H^+ + S$$
$$I_3^- + 2S_2O_3^{2-} \rightarrow 3I_3^- + S_4O_6^{2-}$$

The concentration of hydrogen sulphide is calculated according to

$$\text{mL/L } H_2S = 22.41 \cdot 10^3 \cdot M \cdot (F-G) / (2 \cdot (U-R)) \tag{5-10}$$

where

$F =$ amount of thiosulphate consumed by titration of a mixture of 10 mL of potassium iodate, 1 mL of sulphuric acid and the volume of oxygen reagents used, in mL;

$G =$ amount of thiosulphate consumed for titration of the sample, in mL;

$U =$ volume of the bottle, in mL;

$R =$ volume of of the oxygen reagents added, in mL;

$M =$ concentration of the thiosulphate solution, in mol/L.

According to *Andersen and Føyn (1969)*, the method is not specific for H_2S since the results will also include the effects of other reductants. In fact, this may provide a better measure of the true oxygen deficiencies in the water studied and might therefore be of primary interest in conjunction with biological problems.

Note: For work in coastal areas, some oxygen titration methods include addition of sodium azide to the reagents to avoid effects of possible contaminants. In this case, the titration method for hydrogen sulphide using the oxygen reagents will not work, because the azide will react with hydrogen sulphide.

5.3.4 Methods using mercury compounds

Many other methods available for hydrogen sulphide determinations have not been applied in routine work. Potentiometric titration with $HgCl_2$ (*Boulègue*, 1981) or gravimetric methods, *e.g.*, precipitation of HgS *(Dyrssen et al.*, 1996) should be avoided because of detrimental effects on the environment. In Sweden such methods are not allowed in environmental routine work and will probably be illegal in other EU countries.

References to Chapter 5

Andersen, A.T., Føyn, L. (1969), in: *Chemical Oceanography*: Lange R. (Ed.). Oslo: Universitetsforlaget, 1969; pp. 129–130.

Boulègue, J. (1981), *J. Geochem. Explor.*, 15, 21.

Carlberg, S.R. (1972), *ICES Cooperat. Res. Rep. Ser. A*, 29, 30.

Cline, J. D. (1969), *Limnolog. Oceanogr.*, 14, 454.

Cline, J.D., Richards, F.A. (1972), *Limnol. Oceanogr.*, 17, 885.

Dyrssen, D.W., Hall, P.O.J., Haraldsson, C., Chierici, M. (1996), *Aquat. Geochem.*, 4, 1.

Fischer, E. (1883), *Chem. Ber.*, 26, 2234.

Fonselius, S. (1969), *Fishery Board of Sweden, Ser. Hydrography*, Rep. 23, 22.

Grasshoff, K, Chan, K.M. (1971), *Anal. Chim. Acta*, 53, 442.

Parsons, T.R., Maita, Y., Lalli, C.M. (1984), *A Manual of Chemical and Biological Methods for Seawater Analysis*. Oxford: Pergamon Press, 1984; pp.149–153.

Skopintsev, B.A., Karpov, A.V., Vershinina, O.A. (1959), *Transact. Mar. Hydrophys. Inst.*, Academy of Sciences of the USSR., Translated and produced by Scripta Technica Inc. for the American Geophysical Union; pp. 55–72.

6 Determination of thiosulphate and sulphur

D. Dyrssen, S. Fonselius and B. Yhlen

6.1 Introduction

When water containing sulphide is brought into contact with oxygen, the sulphide is rapidly oxidized by bacterial action. *Jannasch et al.* (1974) discussed the possiblility of a strong sulphide-oxidizing activity at the oxygen-hydrogen sulphide interface in the Black Sea where a substantial population of thiobacilli-type bacteria has been observed. The first oxidation product of sulphide is elemental sulphur (*Jørgensen et al.*, 1991) according to the reaction (*Goldhaber and Kaplan*, 1974).

$$2H_2S + O_2 \rightarrow 2S^0 + 2H_2O \tag{6-1}$$

Dyrssen et al. (1996) suggested that this reaction would dominate when oxygen was added to an excess of sulphide, *e.g.*, by interleaving of oxic water into anoxic water below the chemocline (redoxcline). Elemental sulphur has often been reported to occur in visible amounts in anaerobic-aerobic transition layers as dispersed or colloidal sulphur. It is also found at sediment surfaces where sulphide formed in the top sediment layers is oxidized whilst diffusing into the aerobic surface (*Troelsen and Jørgensen*, 1982). *Steudel* (1989) has shown that elemental S^0 in cultures of sulphur bacteria most likely consists of long-chain polythionates. The higher polythionates may be separated chromatographically (*Steudel*, 1987). Globules of S^0 form droplets which are deposited inside or outside the bacterial cells. These may be determined using an epifluorescence microscope according to *Hobbie et al.* (1977).

Further oxidation of elemental sulphur will lead to the formation of thiosulphate, polythionates, sulphite and finally sulphate:

$$2S^0 + O_2 + H_2O \rightarrow S_2O_3^{2-} + 2H^+ \tag{6-2}$$

It has also been suggested that the first product of biological sulphur oxidation is sulphite (*Goldhaber and Kaplan*, 1974), which reacts abiologically with the original sulphur to produce thiosulphate:

$$S^0 + O_2 + H_2O \rightarrow SO_3^{2-} + 2 H^+$$
$$SO_3^{2-} + S^0 \rightarrow S_2O_3^{2-} \tag{6-3}$$

It is still unknown how elemental sulphur is activated for further oxidation. The organisms largely responsible for the oxidation of sulphur belong to the genera *Chromatium* and *Chlorobium*. *Tuttle and Jannasch* (1972) suggested that heterotrophic oxidation might be more prevalent in marine environments. Later *Jannasch et al.* (1991) studied chemoauto-trophic sulphur-oxidizing bacteria from the Black Sea. The presence of green phototrophic sulphur bacteria in the Black Sea suggested that photosynthetic oxidation of sulphide took place (*Jørgensen et al.*, 1991). The main products were thiosulphate and sulphate, but evidence was also presented for the formation of elemental sulphur.

6.2 Principle of the determination of thiosulphate

Since thiosulphate is an intermediate oxidation product of sulphide in an alkaline or neutral medium, sulphide interferes with the determination of thiosulphate and, thus, any traces of sulphide should be removed by the precipitation of zinc sulphide with zinc acetate before the iodometric determination of thiosulphate. The solubility constant for ZnS has been given by *Dyrssen and Kremling* (1990). After precipitation of the zinc sulphide in a stoppered bottle full to the brim, a measured amount of the supernatant seawater sample is pipetted into a known amount of iodine. The latter oxidizes the thiosulphate to tetrathionate. The surplus iodine is then titrated by adding a thiosulphate solution of known concentration.

Ammonia will not be oxidized to nitrite before the reaction of iodine with the thiosulphate in the sample has taken place. Nitrite will also react with iodine; therefore, the surplus iodine must be determined rapidly in order to avoid formation of additional nitrite by oxidation of ammonia. The reactions are as follows:

$$
\begin{aligned}
&HS^- + Zn^{2+} \rightarrow ZnS(s) + H^+ \\
&IO_3^- + 8I^- + 6H^+ \rightarrow 3I_3^- + 3H_2O \\
&2S_2O_3^{2-} + I_3^- \rightarrow S_4O_6^{2-} + 3I^- \\
&NO_2^- + I_3^- + H_2O \rightarrow NO_3^- + 3I^- + 2H^+
\end{aligned}
\tag{6-4}
$$

6.2.1 Apparatus

The same instrumentation as used for the determination of dissolved oxygen is required for the determination of thiosulphate in seawater (see Chapter 4). The same type of sample bottles is used, but with a content of about 100 mL instead of 50 mL. The piston burette should be calibrated carefully and be readable to 0.01 mL.

6.2.2 Reagents

1. *Zinc acetate solution:* 18.4 g of zinc acetate (analytical-reagent grade, a.g.) are dissolved in 500 mL of distilled water. The solution is about 0.2 mol/L.
2. *Potassium iodate:* Exactly 356.7 mg of dried (180 °C) potassium iodate (a.g.) are dissolved and made up to 1000 mL. The solution is 1.667 mmol/L. (The reagent is the same as used for the determination of dissolved oxygen; see Chapter 4.)
3. *Potassium iodide:* 20 g of potassium iodide (a.g.) are dissolved in 500 mL of distilled water.

4. *Sodium thiosulphate:* A 0.02 mol/L thiosulphate solution is prepared by 1 : 10 dilution of the stock solution (49.5 g/L of $Na_2S_2O_3$). (This reagent is the same as used for the determination of dissolved oxygen; see Chapter 4.)

5. *Sulphuric acid:* An approximately 0.05 mol/L sulphuric acid solution is prepared by adding 3 mL of concentrated sulphuric acid (a.g.) to about 800 mL of distilled water. The solution is made up to 1000 mL with distilled water.

6. *Starch indicator solution:* 1 g of soluble starch is dissolved in 100 mL of distilled water, heated quickly to boiling and stored in a cool dark place (see also Chapter 4).

6.2.3 Sampling and storage

Samples for the determination of thiosulphate are taken in the same way as for oxygen. It is, however, only necessary to rinse the 100 mL bottle by allowing about 50 mL to overflow so that sample water is not wasted, as it might be needed for the determination of other constituents. The analysis must be performed within about 60 min after sampling, but the sulphide should be precipitated immediately (see Section 6.2.4.2).

6.2.4 Procedure

6.2.4.1 Standardization of the thiosulphate solution

About 50 mL of distilled water are filled into a 100 mL titration beaker; 5 mL of the sulphuric acid, 10.00 mL of the 1.667 mmol/L iodate standard and 1.0 mL of the potassium iodide solution are added. The liberated iodine is titrated to a light yellow. One ml of starch indicator solution is added, and the titration is carried to the colourless endpoint. At the approach of the endpoint the solution becomes 'cloudy' directly after the addition of an aliquot of the thiosulphate solution. The endpoint is reached when the cloudiness can no longer be observed. Diffuse illumination from below or moderate illumination facilitate detection of the endpoint.

6.2.4.2 Titration of the sample

Before stoppering, 1 mL of the zinc acetate solution is added to the sample and the bottle is shaken vigourously. The zinc sulphide is allowed to settle.

From the supernatant solution, 50 mL (*V* in Section 6.2.5) are pipetted carefully into the titration beaker. Care must be taken to avoid stirring up the precipitated zinc sulphide. Then 1 mL of the iodide solution and exactly 10 mL of the iodate standard solution are pipetted into the titration beaker followed by 5 mL of the diluted sulphuric acid. The sample is shaken gently and the titration beaker is covered with a watch glass or plastic film and set aside for approximately 5 min. The surplus of iodine not consumed by the thiosulphate initially present in the sample is then titrated with the stardardized thiosulphate solution. The endpoint is determined as described in Section 6.2.4.1.

6.2.5   Calculation of the thiosulphate content of the sample

A 10 mL volume of iodate standard solution corresponds to 5 mL of precisely 0.02 mol/L thiosulphate solution. If the volume of the thiosulphate solution consumed for its standardization is V mL, the factor f, with which this volume must be multiplied in order to obtain the corresponding volume of precisely 0.02 mol/L thiosulphate solution, is $f = 5/V$. The volume of thiosulphate solution used for the sample titration is corrected as follows:

$$V_{corr} \text{ (mL)} = V \cdot f \tag{6-5}$$

The amount of thiosulphate initially present in the sample then is

$$C = (5 - V_{corr})\, 0.02/50 \text{ mol/L} = (5 - V_{corr})\, 0.0004 \text{ mol/L}. \tag{6-6}$$

If the amount of thiosulphate is small, the calculation involves the subtraction of one large number from another. Therefore, the standardization and sample titration must be carried out with great care. See also Section 6.4 (Other methods).

6.2.6   Interferences

All substances dissolved in a seawater sample that react with iodine interfere with the determination of thiosulphate. The major interference, however, is caused by nitrite. If nitrite is present in the seawater sample it will be oxidized by iodine to nitrate, consuming two equivalents of iodine for the oxidation of one equivalent of nitrite (see Section 6.2). The amount of 'apparent' thiosulphate must, therefore, be corrected for the nitrite content of the sample by subtracting twice the amount of nitrite, the determination of which is described in Chapter 10 (Section 10.2).

6.3   Principle of the determination of sulphur

Two methods can be used for the determination of elemental sulphur.

Method A

Elemental sulphur can be determined simply by solvent extraction with chloroform according to *Boulègue and Popoff* (1979) and *Gagnon et al.* (1996). The chloroform phase is analysed by UV spectrophotometry at 270 nm. The operation must be carried out in an oxygen-free atmosphere. The oxidation of sulphide can also be avoided by precipitation with zinc acetate. In this case the ZnS (if not in colloidal form) may be filtered off and and extracted with chloroform since the elemental sulphur is coprecipitated with the zinc sulphide. If the seawater contains other substances that are extracted with chloroform and absorb UV light at 270 nm, one has to deal with these interferences separately (Section

6.3.6). Standard curves are prepared by dissolving elemental sulphur in chloroform. Other solvents such as hexane have also been used.

Method B

Troelsen and Jørgensen (1982) used the method of *Barlett and Skoog* (1954). In this method the interference of sulphide is also prevented by precipitation as zinc sulphide. The sediment sample or filtered water sample (see *Jørgensen et al.,* 1991) is extracted with CS_2. Phase separation is achieved by centrifugation. The CS_2 phase is evaporated and the S^0 is reacted with CN^- to form SCN (so-called cyanolysis). The thiocyanate is determined spectrophotometrically at 460 nm as an iron(*III*) complex. Standard curves are prepared by dissolving elemental sulphur in CS_2. Corrections may have to be made for blank absorption due to reagent colour ($FeCl_3$), pigments extracted and S^0 in the CS_2 solvent. *Troelsen and Jørgensen* (1982) found that elemental sulphur in the cells of sulphur bacteria were completely extracted with CS_2.

6.3.1 Apparatus

For both methods a UV spectrophotometer equipped with 1 cm cuvettes is required. A filter unit for 0.2 μm membrane filters (Nuclepore) is required for samples that contain S^0 in particulate form (*e.g.,* in bacteria or that co-precipitated with ZnS). Since phase separation is achieved by centrifugation a centrifuge is needed with 15 mL ground-glass stoppered bottles. Shake flasks should hold a seawater sample of 500 mL + the organic solvent. For 1 g sediment samples 10 mL ground-glass stoppered centrifuge tubes should be used.

6.3.2 Reagents

1. Chloroform (a. g.): Usually stabililized with ethanol. The ethanol does not have to be removed by shaking with water.
2. *Carbon disulphide:* Elemental sulphur may form slowly in the CS_2 solvent *(a.g.),* which should be discarded if the blank value becomes too high.
3. *Elemental sulphur:* Standard solutions are prepared by dissolving elemental sulphur (a.g.) in the organic solvent ($CHCl_3$ or CS_2). These standards must correspond to the levels of sulphur in the samples (500 mL of seawater or 1 g of sediment). The standard solutions are used to prepare standard curves.
4. *Zinc acetate solution*: 92 g of zinc acetate (a.g.) are dissolved in 500 mL of distilled water. This solution is about 1 mol/L. The solution is used to bind sulphide and to co-precipitate elemental sulphur. An excess of zinc is therefore required.
5. *Cyanolysis*: For the formation of thiocyanate, when applying the second method (B) a solution of 1 g/L of NaCN in a 19:1 mixture of acetone and distilled water is used.
6. *Iron(III) chloride:* For the formation of the iron thiocyanate complex a solution of 5 g/L of $FeCl_3 \cdot 6H_2O$ in a 19:1 mixture of acetone and distilled water is required.

6.3.3 Sampling and storage

When sampling seawater containing hydrogen sulphide extreme care must be exercised to avoid the oxidation of sulphide to elemental sulphur. Oxygen-free argon or nitrogen is also used to protect the sample when emptying the sampling bottle (*Dyrssen et al.,* 1996). In order to stabilize the sulphide, precipitation of ZnS with an excess of the zinc acetate solution is advisable. Elemental sulphur is co-precipitated and can be extracted with an organic solvent (here $CHCl_3$ or CS_2 are recommended).

The same protection is needed for a sediment sample, *e.g.,* coated with sulphur bacteria (*Troelsen and Jørgensen,* 1982).

6.3.4 Procedure

A sample volume of 500 mL is usually needed. If the elemental sulphur is co-precipitated with ZnS it can be removed by filtering through a 0.2 μm membrane filter. The filters, which can be stored, are extracted either with 5 mL of chloroform (method A) or 5 mL of carbon disulphide (method B). The S^0 in the chloroform is determined by UV spectrophotometry at 270 nm using the standard solutions (Section 6.3.2.3) as reference.

In the second method (B) the carbon disulphide is evaporated and the S^0 is redissolved in 5 mL of the cyanide reagent. After cyanolysis for 4 h at room temperature, 1.5 mL of the cyanolysate is mixed with 1.5 mL of the iron(*III*) chloride solution (Section 6.3.2.6). The absorption of this mixture is measured at 460 nm in a 1 cm cuvette. The standard solution in CS_2 (Section 6.3.2.3) is treated in the same way. The amount of S^0 added with the standard solution should be 5–50 μg.

For sediment samples *Troelsen and Jörgensen* (1982) used 1 g samples stabilized with 2 mL of 1 mol/L zinc acetate. The samples were shaken with 5 mL of CS_2 (a. g.) in stoppered centrifuge tubes. Phase separation was achieved by centrifugation. Zinc acetate in the top layer should be sucked off since Zn^{2+} interferes strongly with the subsequent cyanolysis.

6.3.5 Calculations

Method A

The concentration of S^0 in the sample extracted with chloroform is determined by using a standard calibration curve prepared by measuring the absorption (at 270 nm) at different concentrations of S^0 in chloroform (see Section 6.3.2.3). For 500 mL of seawater extracted with V mL of chloroform, the sample concentration of S^0 in μmol/L is given by $C \cdot V/500$, where C is the concentration of S^0 in μmol/L in V mL of chloroform. In order to obtain a value of V close to 10 mL one should compensate for the chloroform (about 3 mL) that dissolves in the seawater sample. It is advisable therefore to measure V.

Method B

If CS_2 and cyanolysis are used with subsequent reaction with iron(*III*), the S^0 concentration is determined from a standard additions curve (at 460 nm) by adding different amounts of sulphur using different volumes of the standard solution (Section 6.3.2.3). For procedure see Section 6.3.4.

6.3.6 Interferences

The seawater sample may contain substances that are extracted with chloroform and absorb light at 270 nm (method A). There are several ways to deal with this problem. One way is to use a less polar solvent, *e.g.,* petroleum ether or hexane. Another method is to evaporate the solvent and bind the sulphur by cyanolysis and then determine SCN^- by complexation with iron(*III*) (see Section 6.3.4, method B). Standard additions to the sample will not solve the problem, but running a full spectrum on both the standard solution and the sample extract might disclose what the interfering substance is.

In method B with CS_2, solvent corrections have to be made for blank absorption due to reagent colour ($FeCl_3$) and chlorophyll, and other pigments extracted from the sample. The last can be removed by passing the extract through a filter column packed with a few cm of Florisil (*Troelsen and Jørgensen,* 1982; *Chen et al.,* 1973).

6.4 Other methods

If the methods suggested here are not sensitive enough, one of the following methods may be tested.

Sulphite and thiosulphate have been determined by liquid chromatography after derivatization with 2,2'-dithiobis(5)nitropyridine (*Vairavanmurthy and Mopper,* 1990). At high levels of H_2S sulphide should be precipitated with $ZnCl_2$ (*Millero,* 1991).

Chen et al. (1973) have described a gas chromatographic method for the determination of sulphur, which is more sensitive than the two methods described earlier.

Sulphide, thiols, thiosulphate and sulphite can be titrated with mercury(*II*) chloride according to *Boulègue* (1981). The titration has been examined by *Dyrssen and Wedborg* (1986).

Luther et al. (1991) employed square wave voltammetry for the determination of thiosulphate, sulphite and tetrathionate. They used an EG & B Princeton Applied Research Model 383B-4 in conjunction with a Model 303 static dropping mercury electrode.

Note: Regarding the use of mercury compounds in environmental routine work, see Chapter 5 (Section 5.3.4).

References to Chapter 6

Barlett, J.K., Skoog, D.A. (1954), *Anal. Chem.*, 26, 1008.

Boulègue, J., Popoff, G. (1979), *J. Fr. Hydrol.*, 10, 83.

Boulègue, J. (1981), *J. Geochem. Explor.*, 15, 21.

Chen, K.Y., Moussavi, M., Sycip, A. (1973), *Environ. Sci. Technol.*, 7, 948.

Dyrssen, D., Wedborg, M. (1986), *Anal. Chim. Acta*, 180, 473.

Dyrssen, D., Kremling, K. (1990), *Mar. Chem.*, 30, 193.

Dyrssen, D., Hall, P.O.J., Haraldsson, C., Chierici, M. (1996), *Aquat. Geochem.*, 2, 111.

Gagnon, C., Mucci, A., Pelletier, E. (1996), *Mar. Chem.*, 52, 195.

Goldhaber, M.B., Kaplan, I.R. (1974), in: *The Sea:* Goldberg, E.D. (Ed.). New York: Wiley Interscience Publ., 1974, vol. 5; pp. 569–655.

Hobbie, J.E., Daley, R.J., Jasper, S. (1977), *Appl. Environ. Microbiol.*, 33, 1225.

Jannasch, H.W., Truper, H.G., Tuttle, J.H. (1974), in: *The Black Sea – Geology, Chemistry and Biology:* Degens, E.T., Ross, D.A. (Eds.). Tulsa: Am. Ass. Petrol. Geol., 1974; pp 419–425.

Jannasch, H.W., Wirsen, C.O., Molyneaux, S.J. (1991), *Deep-Sea Res.*, 38, 1105.

Jørgensen, B.B, Fossing, H., Wirsen, C.O., Jannasch, H. (1991), *Deep-Sea Res.*, 38, 1083.

Luther III, G.W., Church, T.M., Powell, D. (1991), *Deep-Sea Res.*, 38, 1121.

Millero, F. (1991), *Limnol. Oceanogr.*, 36, 1007.

Steudel, R. (1987), *Angew. Chem.*, 99, 143.

Steudel, R. (1989), in: *Autotrophic Bacteria:* Schlegel, H.G., Bowien B. (Eds.). Berlin: Springer Verlag, 1989; pp. 289–303.

Troelsen, H., Jørgensen, B.B. (1982), *Estuar., Coast. Shelf Sci.*, 15, 255.

Tuttle, J.H., Jannasch, H.W. (1972), *Limnol. Oceanogr.*, 17, 532.

Vairavanmurthy, A., Mopper, K. (1990), *Environ. Sci. Technol.*, 24, 333.

7 Determination of pH

M. Wedborg, D. R. Turner, L.G. Anderson and D. Dyrssen

7.1 Introduction

pH is often described as a 'master variable' in seawater and other aquatic systems since many properties, processes and reactions are pH-dependent. Examples include the surface charge of particulate matter (which in turn affects the adsorption of trace metals and other chemical species) and many metal complexation reactions. However, seawater pH is usually considered as part of the carbon dioxide system, which provides the major pH buffer in seawater. In recent years there has been somewhat of a renaissance in seawater pH measurement, primarily with the aim of using pH data to follow changes in the carbon dioxide system arising from physicochemical or biological processes.

7.2 List of symbols

A	=	Arbitrary fitting constant
A_1, A_2	=	Absorbance at absorption maxima of HI^- and I^{2-}, respectively
A_λ	=	Absorbance at wavelength λ
A_T	=	Total alkalinity
$a_{H^+}(F)$	=	Hydrogen ion activity on the 'free' concentration pH scale
$a_{H^+}(SWS)$	=	Hydrogen ion activity on the 'seawater scale' (including fluoride)
$a_{H^+}(T)$	=	Hydrogen ion activity on the 'total scale' (not including fluoride)
a_i, a_i^o	=	Activity of species i in actual and standard states, respectively
B	=	Arbitrary fitting constant
C	=	Arbitrary fitting constant
C_T	=	Total dissolved inorganic carbon
E	=	Electromotive force (emf), V or mV
E_1	=	Emf between glass electrode filling solution and internal reference electrode
E_2	=	Emf between glass electrode filling solution and glass membrane
E_3	=	Emf between glass membrane and test solution
$E_4 (E_J)$	=	Emf between test solution and reference electrode filling solution
E_5	=	Emf between external reference electrode and filling solution
E_B	=	Emf measured by pH cell in standard buffer

E_J	=	Liquid junction potential (E_4)
ΔE_J	=	Difference in E_J between buffer and sample
E_K	=	Arbitrary constant
E_X	=	Emf measured by pH cell in sample
ε_λ^I	=	Molar absorptivity of species I^{2-} at wavelength λ, $\lambda = 1, 2$ (see A_1, A_2)
ε_λ^{HI}	=	Molar absorptivity of species HI^- at wavelength λ, $\lambda = 1, 2$ (see A_1, A_2)
e_1	=	$\varepsilon_2^{HI}/\varepsilon_1^{HI}$
e_2	=	$\varepsilon_2^I/\varepsilon_1^{HI}$
e_3	=	$\varepsilon_1^I/\varepsilon_1^{HI}$
F	=	Faraday constant, 96 485 C/mol
f_H	=	'Apparent activity coefficient' of the hydrogen ion
γ_i	=	Activity coefficient of species i
I_T	=	Total indicator concentration
K_{HF}	=	Formation constant for HF
K_{HSO_4}	=	Formation constant for HSO_4^-
K_{HI}	=	Formation constant for HI^-
K_{H_2I}	=	Formation constant for indicator H_2I
l	=	Spectrophotometric cell path length
μ_i, μ_i^o	=	Chemical potential of species i in actual and standard states, J/mol
pH(F)	=	pH defined on the ionic medium 'free' scale
pH(NBS)	=	pH defined on the infinite dilution scale
pH(SWS)	=	pH defined on the 'seawater' scale (including fluoride)
pH(T)	=	pH defined on the 'total' scale (not including fluoride)
pH_B	=	pH measured in standard buffer
pH_P	=	pH at pressure P atm
pH_t	=	pH at temperature $t\,°C$
pH_{tris}	=	pH in tris buffer
pH_X	=	pH measured in the unknown solution
Q	=	Absorbance ratio A_2/A_1
R	=	Gas constant, 8.314 J/(K mol)
S	=	Salinity
T	=	Thermodynamic temperature, K
t	=	Temperature,°C
X	=	A_T/C_T

7.3   Definition of pH

Before a procedure for pH measurement in seawater is selected it is necessary to consider which activity definition should be applied and which pH scale should be used. The major reason for this is a practical one: the reliability and interpretation of the measurement depends on the selection of activity definition and pH scale.

The activity is defined from the difference between the chemical potential of the species considered in the sample solution and its chemical potential in a reference state, referred to as the standard state:

$$\mu_i - \mu_i^\circ = RT \ln \frac{a_i}{a_i^\circ} \qquad (7\text{-}1)$$

where μ_i and μ_i° are the chemical potentials in J/mol of species i in the actual and standard states, respectively, a_i is the activity of species i, a_i° is by definition unity, R is the gas constant in J/(K·mol) and T is the temperature in K. In aqueous solutions the standard state is normally based on a concentration scale, and Eq. (7-1) can then be written:

$$\mu_i - \mu_i^\circ = RT \ln a_i = RT \ln c_i \gamma_i \qquad \gamma_i \rightarrow 1 \text{ in pure water} \qquad (7\text{-}2)$$

where c_i is the concentration of i on an appropriate concentration scale and γ_i is the activity coefficient of i. The activity coefficient takes account of the interactions between the species i and other dissolved species. The activity coefficient is by definition unity in the standard state but varies with solution composition. This selection of pure water as the standard state in Eq. (7-2) is the traditional choice in solution chemistry. The pH is defined in turn from the activity of the hydrogen ion as

$$pH = -\log_{10} a_{H^+} \qquad (7\text{-}3)$$

It is not possible to measure pH as defined in Eq. (7-3) because solutions with zero ionic strength, corresponding to the standard state, cannot be prepared, and because single ion activity coefficients cannot be measured. Therefore, an operational definition based on potentiometric measurements and on an activity coefficient convention was introduced (*Bates and Guggenheim*, 1960; *Bates*, 1973) and recommended by NBS/NIST (National Bureau of Standards/National Institute of Standards and Technology) and IUPAC (International Union of Pure and Applied Chemistry). A series of standard buffer solutions were assigned pH values which were close to the best estimates of $-\log_{10} a_{H^+}$ and which were self-consistent. This scale became known as the NBS pH scale: pH values on this scale are close to, but not identical to, pH as defined by Eq. (7-3) with a pure water standard state. The careful work at NBS/NIST in preparing and assigning pH values to primary standard buffers based on the infinite dilution scale was a step forward but the problem of irreproducibility connected with practical measurement remained (see Section 7.5). For seawater, the use of the low ionic strength NBS buffers causes significant changes in the liquid junction potential between calibration and sample measurement, and unless this change is carefully characterized for each electrode system used, this introduces an error in the pH estimate which is larger than the desired accuracy of 0.01–0.001 pH units. This situation was greatly improved by the adoption of seawater as the standard state, coupled with the introduction of a new set of buffers based on synthetic seawater (*Hansson*, 1973). Since these buffers have a composition very close to that of the sample, the liquid junction potential changes between calibration and sample measurement are much reduced.

7.4 pH scales in seawater

Seawater pH scales are based on the adoption of seawater as the standard state, which means that Eq. (7-2) can be rewritten as:

$$\mu_i - \mu_i^o = RT \ln a_i = RT \ln c_i \gamma_i \qquad \gamma_i \to 1 \text{ in seawater} \qquad (7\text{-}4)$$

Concentration and activity are now identical by definition since the activity coefficient has been defined as unity in seawater. The hydrogen ion concentration in seawater is generally expressed in the mol/(kg-seawater) concentration scale. Three different seawater pH scales have been defined, based on differing ways of defining the hydrogen ion concentration.

The 'free' hydrogen ion concentration scale pH(F) uses the concentration of free hydrogen ion to define the hydrogen ion activity (*Bates and Culberson*, 1977)

$$a_{H^+}(F) = \left[H^+\right] \qquad (7\text{-}5)$$
$$pH(F) = -\log_{10} a_{H^+}(F) \qquad (7\text{-}6)$$

This is conceptually the simplest of the seawater pH scales, but suffers from the problem that the concentration of free hydrogen ion cannot be determined analytically, since when acid is added to seawater at low pH a proportion of the added acid is bound to sulphate and fluoride ions. Since fluoride is only a minor component of seawater (see Chapter 11), *Hansson* (1973) proposed fluoride-free synthetic seawater as the standard state, giving the 'total' hydrogen ion concentration scale pH(T)

$$a_{H^+}(T) = \left[H^+\right] + \left[HSO_4^-\right] = \left[H^+\right]\left\{1 + K_{HSO_4}\left[SO_4\right]_{tot}\right\} \qquad (7\text{-}7)$$

where

$$K_{HSO_4} = \frac{\left[HSO_4^-\right]}{\left[H^+\right]\left[SO_4^{2-}\right]}$$

$$pH(T) = -\log_{10} a_{H^+}(T) \qquad (7\text{-}8)$$

It was later suggested that fluoride be included in the buffer medium (*Dickson and Riley*, 1979; *Dickson and Millero*, 1987) which resulted in a slightly modified scale, known as the 'seawater' hydrogen ion concentration scale pH(SWS)

$$a_{H^+}(SWS) = \left[H^+\right] + \left[HSO_4^-\right] + \left[HF\right]$$
$$= \left[H^+\right]\left\{1 + K_{HSO_4}\left[SO_4\right]_{tot} + K_{HF}\left[F\right]_{tot}\right\} \qquad (7\text{-}9)$$

where

$$K_{HF} = \frac{\left[HF\right]}{\left[H^+\right]\left[F^-\right]}$$

$$pH(SWS) = -\log_{10} a_{H^+}(SWS) \qquad (7\text{-}10)$$

The naming of the total and seawater pH scales is far from consistent in the literature, so that it is advisable to check carefully which scale is actually used, *e.g.*, from the definition of a_{H^+} or from the calibration buffers used.

The NBS concentration scale has also been applied to seawater measurements. The measured values of pH(NBS) are, however, dependent on the change in liquid junction potential between the buffer(s) and sample, which needs to be taken into account if the measured pH values are to be used in thermodynamic calculations. This effect is often summarized as an 'apparent activity coefficient' f_H for the hydrogen ion:

$$f_H = \frac{10^{-pH(NBS)}}{\{[H^+] + [HSO_4^-]\}} \tag{7-11}$$

f_H must be determined in separate experiments for each electrode pair used.

Which one of the above-mentioned scales applies to a certain pH measurement depends on the composition of the buffer solution used for calibration of the electrodes and on the procedure for assigning pH to the buffer (potentiometric measurement) or by the pH scale used in the determination of the indicator protonation constant (photometric measurement). A long discussion has been devoted to the choice of pH scale to use for seawater (*Hansson*, 1973; *Bates*, 1982; *Covington and Whitfield*, 1988; *Dickson and Millero*, 1987; *Dyrssen and Wedborg*, 1987; *UNESCO*, 1987; *Dickson*, 1993a). The Carbon Dioxide Sub-Panel of the Joint Panel on Oceanographic Tables and Standards (JPOTS) has recommended buffers based on synthetic seawater be used instead of the NBS standard reference buffers (*UNESCO*, 1987) whereas the IUPAC Commission on Electroanalytical Chemistry recommends use of both types of buffers, supplemented with checks of the liquid junction potential error for the standard reference buffers, until standard saline buffers become commercially available (*Covington and Whitfield*, 1988). It has been suggested as an argument against the ionic medium scale that, for measurements in estuarine samples, it is necessary to use several buffers that reasonably match the salinity of the sample. In practice, however, this implies that a maximum of four buffers is needed for calibration to cover the whole salinity range in an estuarine environment. The practical disadvantage of this procedure is, in our opinion, less than the disadvantage arising from the need for frequent experimental characterization of the variable liquid junction potential error associated with each electrode pair when using low ionic strength standard buffers.

Our conclusion from this discussion is that synthetic seawater buffers containing sulphate but not fluoride, as originally suggested by *Hansson* (1973), should be used for pH measurement in seawater and estuarine water, *i.e.*, that the pH(T) scale should be used. With this choice the problems caused by the uncertainty in the stability constant for HF are avoided. Preparation of the buffers and assignment of pH is treated below.

7.5   Measurement of pH

7.5.1   Potentiometry

7.5.1.1   Theory

The potentiometric method is based on measurement of the cell emf in an electrochemical cell in which one of the electrodes is selective for hydrogen ions and the other electrode serves as a reference. An important consequence of this fact is that the change in potential on moving the electrodes from the buffer to the sample is the sum of all changes that occur in the contributions to the cell potential as shown in Fig. (7-1), where:

E_1 is the potential difference between the inner reference electrode (usually an Ag-AgCl electrode) and the inner reference solution (usually dilute hydrogen chloride solution);

E_2 is the potential difference between the inner reference solution and the inner surface of the glass membrane;

E_3 is the potential difference between the outer surface of the glass membrane and the sample/buffer solution;

E_4, often denoted E_J, is the liquid junction potential difference, which ideally depends only on composition of the outer reference and measurement solutions, but in reality also on the practical design of the liquid junction;

E_5 is the potential difference between the outer reference electrode and the outer reference solution.

Provided that the equipment works satisfactorily, and that temperature and pressure are maintained constant, the changes to be considered when the electrodes are moved from the buffer to the sample solution are the ones in E_3 (the desired pH difference signal) and E_J (caused by the unavoidable change in liquid junction potential). The pH estimate is based on the Nernst equation:

Standard buffer:
$$E_B = E_K - \frac{RT \ln 10}{F} \cdot pH_B \tag{7-12}$$

Sample:
$$E_X = E_K - \frac{RT \ln 10}{F} \cdot pH_X \tag{7-13}$$

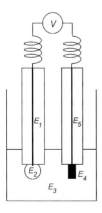

Fig. 7-1. Cell for potentiometric pH measurement.

where F is the Faraday constant, 96 485 C/mol and the Nernst slope $RT\ln10/F = 59.16$ mV per pH unit change at 25°C. Combination of Eqs. (7-12) and (7-13) gives:

$$pH_X = pH_B + \frac{(E_B - E_X)\cdot F}{RT\ln10} \qquad (7\text{-}14)$$

When E_B and E_X are measured and pH_B is known, the unknown pH_X is operationally defined according to Eq. (7-14). The weakness in this operational definition arises from the fact that E_K in Eqs. (7-12) and (7-13), which are assumed to be the same in both solutions above, in reality may differ between the buffer and sample measurements as a result of the difference ΔE_J in liquid junction potentials:

$$pH_X = pH_B + \frac{(E_B - E_X + \Delta E_J)\cdot F}{RT\ln10} \qquad (7\text{-}15)$$

The estimated pH_X will include ΔE_J which cannot be measured or calculated and which is not reproducible when the standard buffer is a low ionic strength solution. With the buffers suggested by *Hansson* (1973) this error is minimized.

7.5.1.2   Tris buffers

The buffer substance selected by *Hansson* (1973) is 2-amino-2-hydroxymethylpropane-1,3-diol, generally referred to as 'tris'. For assignment of pH to the synthetic seawater tris buffer solutions, Hansson used potentiometric titration with hydrochloric acid in which calibration of the electrodes was made in the region of excess of acid. This type of calibration is based on a mean value for E_K in Eq. (7-12) obtained in a region where the measured signal is very stable and changes in the liquid junction potential negligible. It follows that the pH values assigned to the Hansson buffers are very reliable.

Almgren et al. (1975) gave the following procedure for preparation of a Hansson buffer solution of salinity 35:

prepare solutions A and B according to Table (7-1).;

for a buffer volume of 1 L, dissolve 10.00 mmol of tris in 50.00 mL of solution A (Table 7-1) and bring the volume up to 1 L with solution B.

Table 7-1. Composition of solutions for preparation of Hansson tris buffers at 25 °C (*Almgren et al.*, 1975).

Dried salts/titrimetric standard solutions	Solution A mmol/L	Solution B mmol/L
NaCl	321.2	421.2
KCl	10.5	10.5
Na$_2$SO$_4$	28.9	28.9
MgCl$_2\cdot$6H$_2$O	54.4	54.4
CaCl$_2\cdot$6H$_2$O	10.6	10.6
HCl	100.00	–

The critical step is the addition of the right amounts of tris and hydrochloric acid, while the salt concentration is not very critical, and therefore magnesium and calcium chloride can be weighed. If greater accuracy is desirable, standardized stock solutions of $MgCl_2 \cdot 6H_2O$ and $CaCl_2 \cdot 6H_2O$ should be used instead (*Almgren et al.*, 1975). The water used in the preparation of solutions A and B should be free of carbon dioxide (boiled). Note that the concentrations in Table 7-1 are given in mmol/L at 25 °C which is practical for shipboard preparation of the buffers. These concentrations can be converted into the temperature and pressure independent mol/(kg-seawater) concentration scale by division with the density of seawater of salinity 35 at 25 °C (1.023 36 kg/L; see Table 4 in the Appendix of this book).

For preparation of buffers of salinities below 35 an appropriate volume of solution B is replaced with water (see Table 7-2).

Table 7-2. Volumes of solutions A and B to be used for the preparation of Hansson tris buffers of salinities < 35 ($t = 25°C$).

Salinity	Volume of A mL	Volume of B mL	Tris mmol	Water
5	50	92	10.00	up to 1 L
15	50	378	10.00	up to 1 L
25	50	664	10.00	up to 1 L
35	50	950	10.00	–

The pH of a tris buffer solution prepared as described above is a function of temperature and salinity and can be estimated from the following equation (*Almgren et al.*, 1975):

$$pH_{tris} = \frac{(4.5 \cdot S + 2559.7)}{T} - 0.01391 \cdot S - 0.5523 \tag{7-16}$$

which is the fitting equation for the experimental results of *Hansson* (1973) obtained for $S = 10–40$ and $T = 278.15–303.15$ K. Note that pH_{tris} in Eq. (7-16) is defined for hydrogen ion concentrations in mol/(kg-seawater).

7.5.1.3   Practical considerations

Sampling

Care must be taken not to bring a sample into contact with air, because exchange of CO_2 affects the pH. Precautions to be taken are the same as for oxygen and total carbonate sampling: the sample is drawn from the sampler through a silicon tube into a glass bottle while care is taken to avoid air bubbles; the sample bottle is rinsed and then slowly filled with the tube opening placed at the bottom; the sample is allowed to overflow by about one bottle volume; and finally the bottle is sealed so that no air is trapped. Cold water dissolves more air and when heated, air bubbles may be formed. Samples drawn from Niskin or other hydrographic bottles should be among the first drawn to minimize gas exchange. If storage is necessary, the bottles should be kept in the dark and cool but not for any length of time. Poisoning of samples for pH determination is not recommended as this will affect the pH. For samples that must be stored for longer times, it is recommended that they be poisoned and that instead A_T and C_T be determined (see Chapter 8).

Temperature control

It is essential to control the temperature because not only does the sample pH but also does the pH of the tris buffer and the Nernst slope of the electrodes depend on temperature. Ideally, pH measurements should be thermostated at the *in situ* temperature of the sample, which is not possible in practice. Very low temperatures, close to 0°C should be avoided because of limitations caused by the electronic equipment and the response time of the electrodes. A reasonable compromise is to maintain a temperature around 10 °C if some of the samples are cold, otherwise if all samples are warmer than 10 °C, a temperature somewhere within the range of the samples should be adopted.

Instrumentation

For measurement it is essential to use a voltmeter with a high input impedance because of the high inner resistance of the glass electrode (around $100\,M\Omega$). Instruments referred to as pH meters always fulfil this requirement, while general purpose digital voltmeters do not. If such an instrument is to be used, it must be combined with a voltage-follower-amplifier connected between the electrodes and the input. Electrodes to be used can either be separate glass and reference electrodes or a combination electrode. It is strongly recommended not to use electrodes with gel-filled reference half cells because of the badly characterized liquid junction. Although the ideal reference electrode would have an open free-flow liquid junction, a good practical compromise is to use a combination electrode with a sleeve-type junction.

Calibration and measurement

The calibration of the potentiometric cell should be made at the same temperature as the sample measurement. In principle, it is desirable to calibrate in direct connection with the sample measurement. When samples are continuously processed, a routine should be set up for calibration intervals (number of samples to be run between calibrations). The calibration interval depends on the stability of the potentiometric cell (the condition of the electrodes, the stability of the mains voltage supply, interference from other sources). For each calibration and sample measurement, the temperature is registered and E_B and E_X measured. The sample pH is estimated from the measured E_X and E_B, pH_B calculated with Eq. (7-16), which are all put into Eq. (7-14) for calculation of pH_X. If the measurements are made in an estuarine environment, the salinity of the sample must be taken from the CTD measurement for selection of the right buffer to use for the calibration.

Precision and accuracy

It should be remembered that the sample pH is operationally defined and as a consequence it is not possible to define how accurate the measured pH is. Clearly, the reliability of the pH estimate depends on the accuracy and precision of the preparation of the tris buffers. A precision of 0.01 pH unit is attainable with commercially available reference half cells without extra precautions, and down to 0.002 if special care is taken to design a free flow liquid junction (*Culberson*, 1981; *Butler et al.*, 1985; *Covington and Whitfield*, 1988).

7.5.2   Spectrophotometry

High precision spectrophotometric determination of pH using multi-wavelength combinations has been described and discussed by several workers (*e.g., Byrne and Breland,* 1989; *Clayton and Byrne,* 1993; *Dickson,* 1993b; *Zhang and Byrne,* 1996). Determination of the absorbance at several wavelengths eliminates the need to know the total concentration of indicator in the sample, thus making high precision determination possible.

7.5.2.1   Theory

Several indicators have been suggested for the determination of pH, but the most suitable are the sulphonphthalein indicators such as *m*-cresol purple, thymol blue and cresol red. They all exist in three forms, with the chemical equilibria following reactions (7-17) and (7-18)

$$I^{2-} + H^+ \leftrightarrow HI^- \qquad K_{HI^-} = \frac{\left[HI^-\right]}{\left[H^+\right]\left[I^{2-}\right]} \qquad (7\text{-}17)$$

and

$$HI^- + H^+ \leftrightarrow H_2I \qquad K_{H_2I} = \frac{\left[H_2 I\right]}{\left[H^+\right]\left[HI^-\right]} \qquad (7\text{-}18)$$

with the total indicator concentration being equal to

$$I_T = [I^{2-}] + [HI^-] + [H_2I] \qquad (7\text{-}19)$$

or

$$I_T = \left[I^{2-}\right] \cdot \left(1 + K_{HI^-} \cdot \left[H^+\right] + K_{HI^-} \cdot K_{H_2I} \cdot \left[H^+\right]^2\right) \qquad (7\text{-}20)$$

At seawater pH, H_2I can be neglected so that Eq. (7-17) dominates and thus

$$pH(T) = \log_{10} K_{HI^-}(T) + \log_{10}\left(\frac{[I^{2-}]}{[HI^-]}\right) \qquad (7\text{-}21)$$

where pH(T) is on the total seawater pH scale (see Eq. (7-8)) and $\log_{10} K_{HI^-}(T)$ is the first stability constant for the indicator (Eq. (7-17)) on the same pH scale.

When the absorbance at a wavelength λ (A_λ) is determined, Beer's law gives

$$\frac{A_\lambda}{l} = \varepsilon_\lambda^I \cdot [I^{2-}] + \varepsilon_\lambda^{HI} \cdot [HI^-] \qquad (7\text{-}22)$$

where l is the cell path length and the ε_λ values are the molar absorptivities at a given wavelength for the two indicator species.

When the absorbance is measured at two wavelengths (A_1 and A_2), the ratio of the two indicator species can be determined according to the following:

$$\frac{[I^{2-}]}{[HI^-]} = \left(\frac{A_2}{l \cdot [HI^-]} - \varepsilon_2^{HI}\right) \cdot \frac{1}{\varepsilon_2^I} \qquad (7\text{-}23)$$

$$[HI^-] = \frac{A_1}{l \cdot \left(\varepsilon_1^{I} \cdot \dfrac{[I^{2-}]}{[HI^-]} + \varepsilon_1^{HI} \right)}$$

(7-24)

Inserting Eq. (7-24) into (7-23) gives

$$\frac{[I^{2-}]}{[HI^-]} = \left\{ \frac{A_2}{A_1} \cdot \left(\varepsilon_1^{I} \cdot \frac{[I^{2-}]}{[III^-]} + \varepsilon_1^{HI} \right) - \varepsilon_2^{HI} \right\} \cdot \frac{1}{\varepsilon_2^{I}}$$

(7-25)

which can be rearranged to

$$\frac{[I^{2-}]}{[HI^-]} = \frac{\dfrac{A_2}{A_1} - \dfrac{\varepsilon_2^{HI}}{\varepsilon_1^{HI}}}{\dfrac{\varepsilon_2^{I}}{\varepsilon_1^{HI}} - \dfrac{A_2}{A_1} \cdot \dfrac{\varepsilon_1^{I}}{\varepsilon_1^{HI}}}$$

(7-26)

Substituting the following terms

$$Q = \frac{A_2}{A_1} \qquad e_1 = \frac{\varepsilon_2^{HI}}{\varepsilon_1^{HI}} \qquad e_2 = \frac{\varepsilon_2^{I}}{\varepsilon_1^{HI}} \qquad e_3 = \frac{\varepsilon_1^{I}}{\varepsilon_1^{HI}}$$

and inserting Eq. (7-26) into (7-21), gives

$$pH(T) = \log_{10} K_{HI^-}(T) + \log_{10} \left(\frac{Q - e_1}{e_2 - Q \cdot e_3} \right)$$

(7-27)

The highest precision is achieved when the two wavelengths are chosen at the maximum absorbances of the two species (Fig. 7-2).

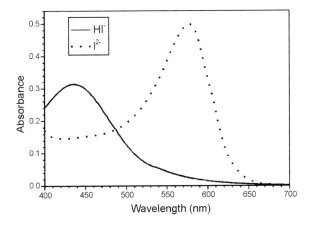

Fig. 7-2. Spectra of *m*-cresol purple for the two species HI^- and I^{2-}.

7.5.2.2   Indicator *p*K values for seawater

The accuracy of the spectrophotometric pH determination is set by the accuracies of the stability constant K_{HI} and of the different molar absorptivity ratios in Eq. (7-27). For cresol red, Eq. (7-27) was determined at 25°C to be

$$\text{pH} = 7.8164 + 0.004 \cdot (35- S) + \log_{10}\left(\frac{Q - 0.00286}{2.7985 - Q \cdot 0.09025}\right) \tag{7-28}$$

at salinities close to 34.5 where $Q=A_{573}/A_{433}$ (*Byrne and Breland*, 1989).

Clayton and Byrne (1993) determined pH as a function of Q for *m*-cresol purple at different temperatures and fitted this to an equation with the following result on the total hydrogen ion concentration scale (mol/(kg-seawater)), valid in the range $293 < T < 303$ K and $30 < S < 37$:

$$\text{pH} = \frac{1245.69}{T} + 3.8275 + 2.11 \cdot 10^{-3} \cdot (35 - S) + \log_{10}\left(\frac{Q - 0.0069_1}{2.222_0 - 0.133_1 Q}\right) \tag{7-29}$$

where $Q=A_{578}/A_{434}$.

pK_{HI} for *m*-cresol purple is on the low side for optimum use in surface waters at low latitudes. Hence, *Zhang and Byrne* (1996) fitted Eq. (7-27) for thymol blue in the range $278 < T < 308$ K and $30 < S < 40$, on the total hydrogen ion concentration scale (mol/(kg-seawater)), with Q being the ratio A_{596}/A_{435}.

$$\text{pH} = 4.706 \cdot \frac{S}{T} + 26.3300 - 7.17218 \cdot \log_{10} T + 0.017316 \cdot S + \log_{10}\left(\frac{Q - e_1}{e_2 - Q \cdot e_3}\right) \tag{7-30}$$

where

$$e_1 = -0.00132 + 1.600 \times 10^{-5} \cdot T$$

$$e_2 = 7.2326 - 0.0299717 \cdot T + 4.600 \times 10^{-5} \cdot T^2$$

$$e_3 = 0.0223 + 0.0003917 \cdot T$$

7.5.2.3   Measurement procedures

The absorbance ratio in Eq. (7-27) can be determined either by a manual or by an automatic method. The latter can be coupled to a continuous flow system and hence produce a quasi-continuous record. Regardless of which system that is used, a stock solution of the indicator has to be prepared. This is prepared by dissolving the sodium salt in deionized water to give a strength of 0.2 mol/L. Even if a small volume of the indicator stock solution is added to the seawater sample, it will perturb the pH of the sample. To minimize this perturbation it is recommended to adjust the pH of the stock solution as close as possible to that of the samples to be analysed. Even so, a minor correction might be necessary (see Section 7.5.2.3 – "Correction of pH due to the perturbation by the indicator solution"). As the indicator stock solution is not perfectly stable it should be stored in a dark bottle and its pH should be measured regularly. This is done by determining the absorbances at the appropriate wavelengths using a cell with a path length of 0.5 mm or less.

Instrumentation

The instrumentation needed includes a spectrophotometer, optical cells (at least two, one for determining pH in the stock solution and one for the sample) plus equipment for automation if this is desirable. The determination of the absorbance at several wavelengths can be achieved either by a scanning spectrophotometer or by a diode array spectrophotometer. The advantage with the latter instrument is that it measures the whole spectrum in a short time, making it suitable for automated measurements.

Manual system

In the manual system, a known volume of the indicator stock solution is added directly to the optical cell, after the cell has been filled with seawater sample. The volume of indicator stock solution added depends on its concentration and the cell volume. The best accuracy in the absorbance readings is achieved if they are in the range of 0.2–1 absorbance units. Typically, the optical cell volume during the manual procedure is about 30 mL with a path length of 100 mm. The cell should be temperature-controlled and the temperature has to be recorded at the time of the absorbance reading.

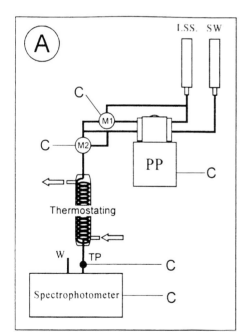
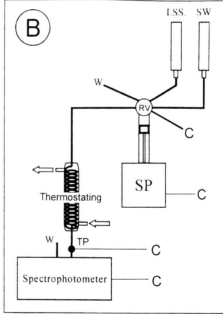

Fig. 7-3. Illustration of two automatic systems for the determination of pH by multi-wavelength measurements. In (A) the mixing ratio of the sample (SW) and indicator stock (ISS) solutions is determined by a peristaltic pump (PP), while in (B) it is determined by a syringe pump (SP). Other abbreviations are: M1 and M2 for magnetic three-way valves, RV for rotating valve, TP for temperature probe, C for computer control and W for waste.

Automatic method

In the automatic system, the sample and indicator can either be mixed in a flow system or in a syringe pump system, after which it flows through the optical cell (Fig. 7-3). The mixing system also determines the volume ratio of the two solutions, which is selected to give optimal absorbance readings. While absorbance readings are taken, the flow has to be stopped to minimize noise. As for the manual method, the temperature has to be controlled and recorded at the time of the absorbance reading. When using a flow-through optical cell, it is difficult to use a long path length and thus a higher concentration of indicator must be used in order to get the same absorbance as in the manual system. This means that the perturbation of the sample pH can be significant and has to be corrected for.

Correction of pH due to the perturbation by the indicator solution

This perturbation is an effect of the difference between the pH of the sample and that of the indicator stock solution. A correction can be made in two ways. One option is to add double volumes of indicator stock solution to seawater samples of varying pH. A plot is made of the absorbance ratio (Q) against indicator volume added, and the linear regression fit to this line can then be used to make the correction. Alternatively, the correction to be made can be read from an error diagram, computed by the theoretical titration of a sample by the indicator added, such as that shown in Fig. (7-4).

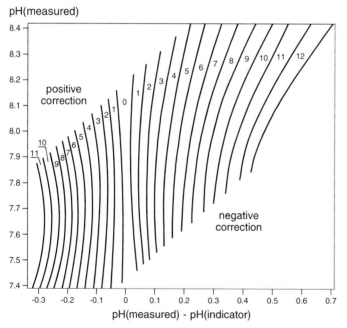

Fig. 7-4. Correction diagram for the perturbation of pH by the addition of indicator (*m*-cresol purple) at 15 °C and 35 salinity, using a 1 cm cell ($I_T = 0.02$ mmol/L). The corrections are in thousands of a pH unit, on the pH(T) scale (after *Chierici and Fransson*, personal communication).

7.5.3 Comparison of the various techniques

Glass electrode potentiometry was for many years the only practical method for pH measurement in seawater, and it is only within the last ten years that accurate spectrophotometric techniques have been developed and used. Both methods have been described in this chapter, and it is appropriate to consider the advantages and disadvantages of each.

Glass electrode potentiometry has the advantages of relatively simple instrumentation and a long history of development and usage. The disadvantages are that the glass and reference electrodes need careful handling and that the electrodes require regular calibration in seawater buffers, which are time-consuming to prepare. The accuracy of the measured pH depends strongly on the accuracy with which these buffers are prepared. Furthermore, for the glass and reference electrodes the operating range is practically restricted to $> 10\,°C$ due to slow electrode response at lower temperatures.

Indicator spectrophotometry requires somewhat more sophisticated instrumentation than potentiometry, but has the advantage that calibration depends largely on knowledge of the pK value(s) of the indicator selected. This 'calibration' can be carried out under well-controlled laboratory conditions rather than in the field, and in principle need only be done once. Furthermore, if more accurate pK values become available in the future, previously obtained absorption data can be recalculated to give improved pH values. A problem which requires careful attention is the change in pH caused by addition of the indicator.

Under optimum conditions a precision of 0.002 pH can be obtained with glass electrodes, compared with 0.0005 pH or better for spectrophotometry with a 100 mm path length. Systematic errors are currently estimated to be similar for both techniques, of the order of 0.004 pH. The systematic errors for spectrophotometric measurements are likely to decrease in the near future as further indicator pK determinations become available and as more accurate methods of correcting for the added indicator are developed. Spectrophotometry is therefore expected to become the method of choice for the most accurate and precise measurements of seawater pH.

7.5.4 Correction of pH to *in situ* conditions

The procedures described in this chapter are based on the collection of discrete samples of seawater followed by measurement of the pH in the sample. The temperature and pressure at which the measurement is made will in most cases differ from the temperature and pressure of the water from which the sample was taken, so that if the *in situ* pH is required, e.g., for thermodynamic calculations, then the measured pH must be corrected to the *in situ* conditions. This is best done by making use of a second measured CO_2 system parameter, normally alkalinity A_T or total inorganic carbon C_T (Chapter 8), and calculating both A_T and C_T on the mol/(kg-seawater) concentration scale. Since this scale is independent of both temperature and pressure, pH at the *in situ* temperature and pressure can then be calculated directly. For details of these calculations see Chapter 8 (Section 8.6). Alternatively, the empirical equations given below can be applied.

7.5.4.1   Empirical equations for correction to *in situ* temperature

Equations have been derived for correcting pH measured at 25 °C to an *in situ* temperature t °C both on the total pH scale (pH(T); *Millero*, 1979) and on the seawater pH scale (pH(SWS); *Millero*, 1995) as follows:

$$pH_t(T) = pH_{25}(T) + A \cdot (t - 25) + B \cdot (t - 25)^2 \qquad (7\text{-}31)$$

where

$$10^3 A = -9.296 + 32.505 \cdot \left[pH_{25}(T) - 8\right] + 63.806 \cdot \left[pH_{25}(T) - 8\right]^2$$
$$10^4 B = 3.916 + 23.000 \cdot \left[pH_{25}(T) - 8\right] + 41.637 \cdot \left[pH_{25}(T) - 8\right]^2 \qquad (7\text{-}32)$$

$$pH_t(SWS) = pH_{25}(SWS) + A + B \cdot t + C \cdot t^2$$

where

$$A = -2.6492 - 0.0011019 \cdot S + 4.9319 \times 10^{-6} \cdot S^2 + 5.1872 \cdot X - 2.1586 \cdot X^2$$
$$B = 0.10265 + 3.1618 \times 10^{-5} \cdot S - 0.20322 \cdot X + 0.084431 \cdot X^2$$
$$C = 4.4528 \times 10^{-5}$$
$$X = A_T / C_T$$

The accuracy of the correction is estimated as 0.003 pH in each case, based on the fitting of the equations to the pH values calculated using the appropriate thermodynamic constants

7.5.4.2   Empirical equations for correction to *in situ* pressure

The pressure dependence of the various carbonate equilibrium constants is less well defined than the temperature dependence, with the result that less attention has been devoted to empirical equations for the pressure dependence of pH. *Millero* (1979) quotes the following equation for the NBS pH scale as a function of pressure P:

$$pH_P(NBS) = pH_o(NBS) + A \cdot P \quad \text{(given in atm by Millero)} \qquad (7\text{-}33)$$

where

$$-10^3 A = 0.424 - 0.0048 \cdot (S - 35) - 0.00282 \cdot t - 0.0816 \cdot (pH_o(NBS) - 8)$$

No corresponding equation is available for the total and seawater pH scales, but given the uncertainties involved in these pressure corrections this equation can be used to give reasonable estimates of corrections to *in situ* pressures on these scales also.

References to Chapter 7

Almgren, T., Dyrssen, D., Strandberg, M. (1975), *Deep-Sea Res.,* 22, 635.

Bates, R.G. (1973), *Determination of pH. Theory and Practice.* New York: Wiley-Interscience.

Bates, R.G. (1982), *Pure. Appl. Chem.,* 54, 229.

Bates, R.G., Culberson, C.H. (1977), in: *The Fate of Fossil Fuel CO_2* in the Oceans: Anderson, N.R., Malahoff, A. (Eds.). New York: Plenum Press, 1977; pp. 45–61.

Bates, R.G., Guggenheim, E.A. (1960), *Pure. Appl. Chem.,* 1, 163.

Butler, R.A., Covington, A.K., Whitfield, M. (1985), *Oceanol. Acta,* 8, 433.

Byrne, R.H., Breland, J.A. (1989), *Deep-Sea Res.,* 36, 803.

Clayton, T.D., Byrne, R.H. (1993), *Deep-Sea Res.,* 40, 2115.

Covington, A.K., Whitfield, M. (1988), *Pure. Appl. Chem.,* 60, 865.

Culberson, C. (1981), in: *Marine Electrochemistry*: Whitfield, M., Jagner, D. (Eds.), Chichester: John Wiley, 1981; pp.187–261.

Dickson, A.G., Riley, J.P. (1979), *Mar. Chem.,* 7, 89.

Dickson, A.G., Millero, F.J. (1987), *Deep-Sea Res.,* 34, 1733.

Dickson A.G. (1993a), *Deep-Sea Res.,* 40, 107.

Dickson, A.G. (1993b), *Mar. Chem.,* 44, 131.

Dyrssen, D., Wedborg, M. (1987), *Anal. Chim. Acta,* 200, 261.

Hansson, I. (1973), *Deep-Sea Res.,* 20, 479.

Millero, F.J. (1979), *Geochim. Cosmochim. Acta,* 43, 1651.

Millero, F.J. (1995), *Geochim. Cosmochim. Acta,* 59, 661.

UNESCO (1987), *Thermodynamics of the carbon dioxide system in seawater.* Report by the Carbon Dioxide Sub-Panel of the Joint Panel on Oceanographic Tables and Standards, Techn. Pap. Mar. Sci., No. 51, Paris.

Zhang, H., Byrne, R.H. (1996), *Mar. Chem.,* 52, 17.

8   Determination of total alkalinity and total dissolved inorganic carbon

L.G. Anderson, D.R. Turner, M. Wedborg and D. Dyrssen

8.1   Introduction

The aim of this chapter is to describe both the high precision methods for the determination of total alkalinity A_T (potentiometric titration) and total dissolved inorganic carbon C_T (coulometry) that are the methods of choice today, and also a simpler method for total alkalinity (back titration) which does not require sophisticated computerised equipment. Thermodynamic calculations of carbonate speciation are also covered.

As suggested by *Dyrssen and Sillén* (1967) the concentration scale mol/(kg-seawater) is recommended. This scale is pressure and temperature independent.

The accuracy of the sophisticated methods is such that it is possible to follow shifts in alkalinity (A_T) and total dissolved inorganic carbon (C_T) caused by the following processes. Uptake of CO_2 or HCO_3^-, formation of carbohydrates:

$$CO_2 + H_2O \rightarrow CH_2O(org) + O_2; \qquad \Delta A_T = 0;\ \Delta C_T = -1 \qquad (8\text{-}1)$$

$$HCO_3^- + H_2O \rightarrow CH_2O(org) + OH^- + O_2; \quad \Delta A_T = 0;\ \Delta C_T = -1 \qquad (8\text{-}2)$$

Uptake of NO_3^- or NH_4^+, formation of organic nitrogen:

$$H^+ + NO_3^- + H_2O \rightarrow NH_3(org) + 2O_2; \qquad \Delta A_T = +1;\ \Delta C_T = 0 \qquad (8\text{-}3)$$

$$NH_4^+ \rightarrow NH_3(org) + H^+; \qquad \Delta A_T = -1;\ \Delta C_T = 0 \qquad (8\text{-}4)$$

Formation of calcium carbonate:

$$Ca^{2+} + 2HCO_3^- \rightarrow CaCO_3(s) + CO_2 + H_2O; \quad \Delta A_T = -2;\ \Delta C_T = -1 \qquad (8\text{-}5)$$

Decomposition of carbohydrates in oxic seawater:

$$CH_2O(org) + O_2 \rightarrow CO_2 + H_2O; \qquad \Delta A_T = 0;\ \Delta C_T = +1 \qquad (8\text{-}6)$$

Decomposition of organic nitrogen in oxic seawater:

$$NH_3(org) + 2O_2 \rightarrow H^+ + NO_3^- + H_2O; \qquad \Delta A_T = -1;\ \Delta C_T = 0 \qquad (8\text{-}7)$$

Decomposition of carbohydrates in anoxic seawater:

$$2CH_2O(org) + SO_4^{2-} \rightarrow HCO_3^- + CO_2 + HS^- + H_2O; \Delta A_T = +2;\ \Delta C_T = +2 \qquad (8\text{-}8)$$

Dissolution of calcium carbonate in deep water:

$$CaCO_3(s) + CO_2 + H_2O \rightarrow Ca^{2+} + 2HCO_3^-; \quad \Delta A_T = +2;\ \Delta C_T = +1 \qquad (8\text{-}9)$$

8.2   List of symbols

All concentrations and stability constants are on the mol/(kg-seawater) concentration scale unless otherwise stated.

A_T	=	Total alkalinity
A_C	=	Carbonate alkalinity
B	=	Borate contribution to H_T
B_T	=	Total borate
b	=	Blank coulometer reading, counts/min
c	=	Coulometer calibration factor
c_A, c_B	=	Acid and base concentrations in back titration method, mol/L
C	=	Carbonate contribution to H_T
c_C	=	Concentration of carbonate standard for coulometer calibration, mol/L
C_T	=	Total dissolved inorganic carbon
E	=	Electromotive force (emf), mV
E_K	=	Nernst equation intercept, mV
F	=	Faraday constant, 96 485 C/mol
F_1'	=	First Gran function
F_2'	=	Second Gran function
F_T	=	Total concentration of fluoride
Fl	=	Fluoride contribution to H_T
$f(CO_2)$	=	Fugacity of CO_2, atm or μatm
f_H	=	Apparent activity coefficient of H^+ (NBS pH scale)
$\{H^+\}$	=	Hydrogen ion concentration corresponding to pH scale definition, $\{H^+\}= 10^{-pH}$
H_b	=	Burette acid concentration in potentiometric titration, mol/L
H_T	=	Total concentration of hydrogen ion
K_0	=	CO_2 solubility constant, mol/((kg-seawater)·atm)
K_1	=	1st dissociation constant of carbonic acid
K_2	=	2nd dissociation constant of carbonic acid
K_{1P}	=	1st dissociation constant of phosphoric acid
K_{2P}	=	2nd dissociation constant of phosphoric acid
K_{3P}	=	3rd dissociation constant of phosphoric acid
K_B	=	Dissociation constant of boric acid
K_{HF}	=	Stability constant for HF
K_{HSO_4}	=	Stability constant for HSO_4^-
K_{Si}	=	Dissociation constant of silicic acid
K_W	=	Ionic product of water
N_s	=	Coulometer reading, counts
$n(CO_2)$	=	Number of moles CO_2 in coulometer calibration gas sample
P	=	Phosphate contribution to H_T
P_T	=	Total concentration of phosphate
pH(NBS)	=	pH defined on the infinite dilution scale
pH(SWS)	=	pH defined on the 'seawater' scale (including fluoride)
pH(T)	=	pH defined on the 'total' scale (not including fluoride)
R	=	Gas constant, 8.314 J/(K·mol)
ρ	=	Density of seawater sample, kg/L

Issue Receipt

University of Plymouth (3)

Date: Sunday, July 01, 2012

Time: 7:39 PM

Card number: 0021102290

Item ID: 9003898239

Title: Methods of seawater analysis / e

Due date: 02/07/2012 23:59

Total items: 1

Please keep your receipt until you

have checked your Voyager account

S	=	Salinity
Si	=	Silicate contribution to H_T
Si_T	=	Total concentration of silicate
Su	=	Sulphate contribution to H_T
Su_T	=	Total concentration of sulphate
T	=	Thermodynamic temperature, K
t	=	Temperature,°C
t_C	=	Time of coulometer sample determination, min
v_0	=	Sample volume, mL
v	=	Volume added during potentiometric titration, mL
v_A, v_B	=	Added volumes of acid and base solutions in back titration method, mL

8.3   Sampling and reference materials

8.3.1   Sampling

Care must be taken not to bring samples to be analysed for C_T into contact with air in order to avoid gain or loss of CO_2. Precautions to be taken are the same as for oxygen and pH sampling: the sample is drawn from the sampler through a silicone tube into a glass bottle while care is taken to avoid air bubbles; the sample bottle is rinsed and then slowly filled with the tube opening placed at the bottom; the sample is allowed to overflow by about one bottle volume and finally the bottle is sealed so that no air is trapped; cold water dissolves more air and when heated, air bubbles may be formed. Samples for C_T drawn from Niskin or other hydrographic bottles should be among the first drawn to minimize gas exchange; in contrast, A_T is not affected by CO_2 exchange so that A_T samples do not need to be among the first drawn. If the sample is likely to be exposed to varying temperatures before analysis, a head-space of 1 % of the bottle volume can be left without risk of significant error in C_T (*DOE*, 1994). Samples for A_T and C_T which will not be analysed within a short time after sampling should be poisoned by the addition of saturated mercury(II)-chloride (0.2 mL per litre of sample).

8.3.2   Standard reference materials

Reference materials with certified A_T and C_T values are now available from Scripps Institution of Oceanography. These materials are prepared from sterilized natural seawater, and are certified by titration against coulometrically standardized acid (A_T) and by manometric measurement of the CO_2 released on acidification (C_T). Regular analysis of these certified reference materials allows the analyst to check for bias and drift in the analytical technique. The use of certified reference materials in A_T and C_T analysis is discussed in detail in *DOE* (1994).

8.4   Total alkalinity

Two methods are described here: automated potentiometric titration and the simpler back titration method. Automated titration techniques using photometric detection have also been developed (*Anderson and Wedborg*, 1983), and have been compared with potentiometric titration (*Anderson and Wedborg*, 1985). The method by *Anderson and Wedborg* (1983) uses bromocresol purple as indicator and the titration is followed at 570 nm in the pH range 5.2–6.8. In this interval the titration of hydrogen carbonate to carbonic acid dominates and the evaluation is done by linear Gran functions for both equivalence points. The precision is slightly lower than for the potentiometric titration described in Section 8.4.2 and the photometric method is thus not described further here. New photometric techniques based on the high precision spectrophotometric measurement of pH are, however, expected to become available over the next few years (see Chapter 7).

8.4.1   Definition

The alkalinity is defined as the amount of hydrogen ions in moles required to neutralise the proton acceptors in 1 kg of seawater. *Dickson* (1981) suggested that bases formed from weak acids with p$K > 4.5$ (at 25° and zero ionic strength) be considered as proton acceptors, *i.e.*, defined as part of the alkalinity. This leads to the following definition of alkalinity in seawater:

$$A_T = \left[HCO_3^-\right] + 2\left[CO_3^{2-}\right] + \left[B(OH)_4^-\right] - \left[H^+\right] + \left[OH^-\right]$$
$$+ 2\left[PO_4^{3-}\right] + \left[HPO_4^{2-}\right] - \left[H_3PO_4^0\right] + \left[SiO(OH)_3^-\right] + ... \qquad (8\text{-}10)$$

where the small contributions (usually $< 1\,\mu mol/kg$) from hydroxide, phosphate, silicate and other bases can often be ignored.

8.4.2   Potentiometric titrations

In the potentiometric titration methods, the pH change during titration of a seawater sample with hydrochloric acid is followed with an electrochemical cell (described in Chapter 7). The titration serves to transfer the bases defined as part of the alkalinity to their acidic forms.

8.4.2.1   Instrumentation

Recommendations as to the electrochemical cell are given in Chapter 7. Because A_T is not affected by the exchange of carbon dioxide between sample and air, it is acceptable to use an open titration vessel if A_T only is to be estimated, while C_T or pH is measured separately. For the evaluation of C_T as well as A_T from the titration, it is necessary to use a closed titration vessel with as little head space as possible, because exchange of carbon dioxide between sample and air affects C_T as well as pH. Temperature should be controlled as in the case of pH measurements (see Chapter 7). The size of the titration vessel can be adjusted to the amount of sample available, typically between 10 and 150 mL.

Computer control

For good throughput of samples and optimum reproducibility it is recommended that the alkalinity titration be computer controlled. The computer is used for data collection from the voltmeter as well as for addition of acid from the burette. Software for PCs for this purpose is commercially available. A suitable resolution for the burette, which is generally of a motor driven piston type, is $1\,\mu L$.

Reagents

Hydrochloric acid (pro analysi): Concentration 0.01–0.1 mol/L, depending on the sample volume.

8.4.2.2   Analytical procedure

It is essential that the sample be stirred at a constant rate throughout a titration, because a change of stirring rate would lead to a small but possibly not negligible displacement of the measured emf. Before the titration is started, the sample should be thermostated to a temperature close to that *in situ*, if possible (see Chapter 7). Also, the electrochemical cell needs a few minutes to stabilise after the electrodes have been immersed in the sample. Each titration point should be taken as a mean of several readings, typically 5–10, and the two latest consecutive means compared until the difference is lower than a set limit. If the readings are noisy, this procedure may cause too large a variation in waiting times between points; in such cases, a fixed waiting time may be a better procedure. The titration should be stopped in the vicinity of pH 3.

8.4.2.3   Evaluation procedures

The mass balance equations to be used in the evaluation are given in the following. They are based on an exactly neutralized solution as a zero point, *i.e.*, the point in the titration where the amount of acid added equals the initial alkalinity, A_T, which is identical to $-H_T$ in equation (8-17).

$$C_T = [CO_3^{2-}] + [HCO_3^-] + [H_2CO_3^*] \tag{8-11}$$

The term $[H_2CO_3^*]$ includes $[CO_2(aq)]$ and $[H_2CO_3]$.

$$B_T = [B(OH)_3] + [B(OH)_4^-] \tag{8-12}$$
$$Si_T = [Si(OH)_4] + [SiO(OH)_3^-] \tag{8-13}$$
$$P_T = [PO_4^{3-}] + [HPO_4^{2-}] + [H_2PO_4^-] + [H_3PO_4] \tag{8-14}$$
$$Su_T = [SO_4^{2-}] + [HSO_4^-] \tag{8-15}$$

$$F_T = [F^-] + [HF] \tag{8-16}$$

$$H_T = [H^+] + [HSO_4^-] + [HF] - [HCO_3^-] - 2[CO_3^{2-}] - [B(OH)_4^-] - [SiO(OH)_3^-]$$
$$+ [H_3PO_4] - [HPO_4^{2-}] - 2[PO_4^{3-}] - [OH^-] \tag{8-17}$$

The initial analytical total concentration of hydrogen ion in the solution is thus the negative of the alkalinity, *i.e.*, $H_T = -A_T$.

The volume-corrected total concentrations must be calculated for each titration point and the concentrations of the complexes are expressed with the consecutive dissociation constants of the protonated forms. For each of the components, the contribution to the H_T equation (see Appendix 8A for derivation) is inserted according to:

$$\frac{v_0 \cdot H_T + v \cdot H_b}{(v_0 + v)} = [H^+] - \frac{K_w}{[H^+]} - \frac{v_0}{(v_0 + v)} \cdot (C + B + Si + P - Su - Fl) \tag{8-18}$$

where H_b is the concentration of acid in the burette, H_T is the total initial concentration of H^+ in the sample, v_0 mL is the sample volume, and v mL is the volume added from the burette at each titration point. The emf reading (E mV) has been used in Eq. (8-19) to calculate the hydrogen ion activity according to

$$[H^+] = 10^{\frac{(E-E_K) \cdot F}{RT \ln 10}} \tag{8-19}$$

The Nernst equation intercept, E_K, must be estimated in the evaluation according to the procedures below when the ionic medium 'total' scale is used, as recommended in Chapter 7. This replaces the less reliable calibration in a low ionic strength buffer.

Gran evaluation

In the Gran evaluation the mass balance equations given above are used together with the experimental variables volume of acid added (v mL) and measured emf (E mV). Linear functions, F_2' and F_1', which are valid for limited parts of the titration, are calculated from the titration variables and mass balance equations. Each of the two functions corresponds to a main reaction associated with an equivalence point which is mainly determined by the carbonate protonation:

$$F_2' \qquad HCO_3^- + H^+ \rightarrow H_2CO_3 \tag{8-20}$$

$$F_1' \qquad CO_3^{2-} + H^+ \rightarrow HCO_3^- \tag{8-21}$$

Function F_2' uses a solution of H_2CO_3 as the zero point for H_T and expresses the excess of acid added after v_2, the second equivalence volume. Analogously, F_1' uses a solution of HCO_3^- as the zero point for H_T and expresses the formation of H_2CO_3 after v_1, the first equivalence volume.

$$F_2' = (v_0 + v) \, ([H^+] + [HSO_4^-] + [HF] - [HCO_3^-] - 2[CO_3^{2-}] - [B(OH)_4^-]$$
$$- [SiO(OH)_3^-] + [H_3PO_4] - [HPO_4^{2-}] - 2[PO_4^{3-}] - [OH^-]) \tag{8-22}$$

$$F_1' = (v_0 + v) \, ([H^+] + [HSO_4^-] + [HF] + [H_2CO_3] - [CO_3^{2-}] - [B(OH)_4^-]$$
$$- [SiO(OH)_3^-] + 2[H_3PO_4] + [H_2PO_4^-] - [PO_4^{3-}] - [OH^-]) \tag{8-23}$$

The reactions with protolytes other than carbonate and the fact that the main reactions are not 100 % complete are taken into account in an iterative procedure which is based on Eq. (8-18) (*Hansson and Jagner*, 1973; *Bradshaw et al.*, 1981; *Anderson and Wedborg*, 1985; SOP3 in *DOE*, 1994). A flow chart of the iterative procedure is shown in Fig. 8-1.

For the purpose of estimation of A_T only, a rapid titration procedure based on a five point potentiometric titration with Gran evaluation has been suggested for large (40 mL) as well as small (1 mL) sample volumes (*Haraldsson et al.*, 1997).

Non-linear curve-fitting

For the non-linear curve-fitting evaluation one of the two experimental variables, v mL and E mV, must be selected as the dependent variable. In general, curve-fitting algorithms minimize the sum of the squares of the errors in the dependent variable only, whereas errors in the independent variable are disregarded. This indicates that in principle E mV would be the best choice of dependent variable, but this also leads to a non-explicit higher degree equation that must be solved at each step in the minimization procedure. From practical and theoretical tests of the procedures, *Johansson and Wedborg* (1982) concluded that it is acceptable to use v mL as the dependent variable. With this choice, Eq. (8-18) can be solved explicitly for v mL:

$$v = v_0 \cdot \frac{\left([H^+] - \dfrac{K_w}{[H^+]} - H_T - C - B - Si - P + Su + Fl \right)}{\left(H_b - [H^+] + \dfrac{K_w}{[H^+]} \right)} \tag{8-24}$$

The parameters to be estimated from the curve-fitting are C_T, A_T $(=-H_T)$ and the constant term E_K in the Nernst equation. An additional possibility is to include K_1 among the parameters. The remaining equilibrium constants must be taken from their salinity-temperature dependence equations. Minimization algorithms are now easily available in standard spreadsheet programmes such as the solver tool in Microsoft Excel.

Precision and accuracy

Procedures for quality assurance are discussed in *DOE* (1994), as well as suitable reference standards for checking the accuracy of the results (see Section 8.3.2). The precision, measured as repeatability, for A_T was reported to be slightly less than $\pm 1\ \mu$mol/L (*Johansson and Wedborg*, 1982; *Anderson and Wedborg*, 1985). *Haraldsson et al.* (1997) gave precisions of 0.05 % and 0.07 % for their large and small volume titrations, respectively.

It is essential to calibrate the volume of the titration vessel, or alternatively to use a calibrated pipette for the addition of the sample. Calibration should be made gravimetrically with air buoyancy correction (see SOP 21 in *DOE*, 1994).

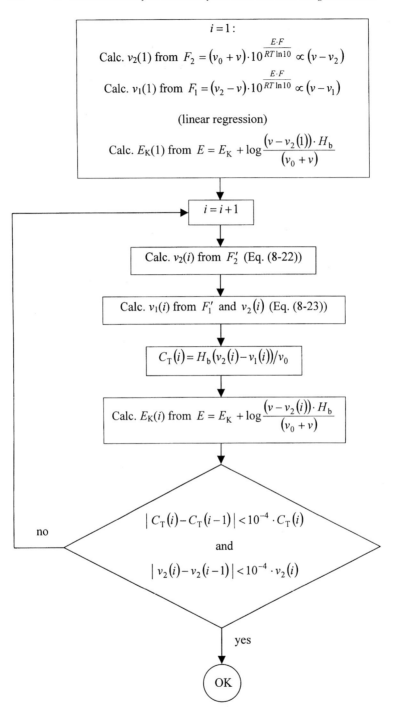

Fig. 8-1. Flow chart of the iterative Gran evaluation.

8.4.3 Back titration method

The back titration method can be used to measure total alkalinity with simpler equipment than that required for the potentiometric titrations described above, but at the cost of a poorer precision. The method described is that of *Gripenberg* (1936) as modified by *Koroleff* (1972). The sample is acidified with hydrochloric acid to a pH of about 3.5, and the carbon dioxide is driven off by boiling. The solution is then back-titrated to pH 6 with sodium hydroxide solution using bromothymol blue as indicator. The alkalinity is estimated simply as the difference between the amounts of hydrochloric acid and sodium hydroxide added.

8.4.3.1 Reagents

Hydrochloric acid (pro analysi), 0.013 mol/L.

Sodium hydroxide (pro analysi), 0.015 mol/L: Dissolve 20 g of sodium hydroxide in 20 mL of distilled water, keep in a closed flask several days and then centrifuge; transfer 2 mL of the clear supernatant to 2 L of CO_2-free distilled water. The sodium hydroxide solution should be standardized against hydrochloric acid by carrying out the analysis described in Section 8.4.3.2 for a distilled water sample.

Bromothymol blue indicator: Dissolve 100 mg of indicator in 1.6 mL of 0.1 mol/L sodium hydroxide solution and dilute to 100 mL.

Reference solution for endpoint determination: To 120 mL of seawater add 0.4 mL of bromothymol blue solution and adjust the pH (by measurement with a pH meter) to 6.0 (6.6 for Baltic Sea water) with boric acid.

8.4.3.2 Analytical procedure

The sample (100 mL) is pipetted into a 200 mL Erlenmeyer flask, and 15 mL of hydrochloric acid and 0.4 mL of bromothymol blue solution added. CO_2-free air, prepared by pumping air through a column of silica gel (to remove moisture) and Ascarite (to remove CO_2) is bubbled through the solution for 30 min before beginning the titration. The bubbling of CO_2-free air is maintained during the titration. The sodium hydroxide solution is titrated into the acidified sample until a pH of 6 is reached (6.6 for Baltic Sea water). The endpoint is assessed by comparison with a reference solution prepared as described.

8.4.3.3 Calculation of results

The alkalinity A_T mol/(kg-seawater) is calculated according to

$$A_T = \frac{(v_A c_A - v_B c_B)}{\rho v_o} \tag{8-25}$$

where v are the added volumes (mL) and c are the concentrations (μmol/L) of the hydrochloric acid (A) and sodium hydroxide (B), respectively, ρ is the density of the seawater sample (kg/L) and v_0 the sample volume (mL).

8.4.3.4   Precision and accuracy

Almgren et al. (1982) have reported a precision of 0.2 % for this method, but differences as high as 1 % between repeat determinations. As with the potentiometric titration methods, accuracy will depend to a large degree on the accuracy of the volume measurements and of the hydrochloric acid concentration.

8.5   Total dissolved inorganic carbon

8.5.1   Potentiometric titrations

Total dissolved inorganic carbon can also be evaluated from the potentiometric titration method described in Section 8.4.1, if it is performed in a closed cell. However, this technique does not give the highest precision possible and thus the coulometric method first described by *Johnson et al.* (1985) and outlined here is nowadays the method of choice. Nevertheless, C_T can be evaluated from the potentiometric titration curve, either by Gran function linearization (Section 8.4.2.3 'Gran evaluation') or by a curve fitting procedure (Section 8.4.2.3 'Non-linear curve-fitting').

8.5.2   Coulometric determination technique

A known volume of seawater is acidified and the carbon dioxide produced is stripped out of the sample by an inert gas. The gas is bubbled through a reagent containing ethanol-amine, which reacts with the carbon dioxide to produce hydroxyethylcarbamic acid (Eq. (8-26)). The latter is coulometrically titrated by the hydroxide ions generated at the cathode (Eq. (8-27)), and the pH in the reagent solution is monitored colorimetrically through the indicator thymolphthalein. At the anode silver is oxidised (Eq. (8-28)). The amount of electrons produced corresponds to the amount of carbon dioxide in the sample and can thus be converted into concentration by dividing by the sample volume.

$$HO(CH_2)_2NH_2 + CO_2 \ \rightarrow \ HO(CH_2)_2NHCOO^- + H^+ \tag{8-26}$$

$$H_2O + e^- \ \rightarrow \ \tfrac{1}{2}H_2(g) + OH^- \tag{8-27}$$

$$Ag(s) \ \rightarrow \ Ag^+ + e^- \tag{8-28}$$

A very detailed description of this technique and all its adherent fine points are given in *'Handbook of methods for the analysis of the various parameters of the carbon dioxide system in sea water,* version 2' (*DOE*, 1994). Here the main principles are discussed together with some of the critical issues to consider for achieving the best results.

8.5.2.1   Instrumentation

The principle of the instrumentation used for the coulometric determination of C_T is illustrated in Fig. (8-2). The system consists of different parts (numbered 1 to 6 in the figure) which preferably are controlled by a computer. All the plumbing should consist of non permeable tubing and the gas flow path is regulated by magnetic valves (normally three-way pinch valves) during the different steps of the analysis. The carrier gas has to be free of CO_2 and could either come from a gas cylinder (*e.g.*, 99.995 % pure nitrogen) or from a compressor fitted with a CO_2 remover (*e.g.*, Peak Scientific). It might also be necessary to add a CO_2 absorber (Ascarite or Malcosorb) between the gas source and the system to clean the carrier gas from minute residues of CO_2.

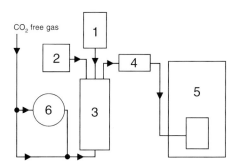

Fig. 8-2. Schematic illustration of the coulometric C_T determination setup. The different parts are: (1) seawater dispensing system, (2) acid adder, (3) extraction tower, (4) sample gas purifier, (5) coulometer and (6) gas loop calibration system.

The seawater dispensing system (1)

A known amount of seawater is dispensed from this part into the extraction tower. The volume should be in the range of 20–30 mL in order to pass enough carbon dioxide through the coulometer to achieve the required precision. In the laboratory it can be a syringe which is weighed before and after the sample is added. At sea it normally consists of a pipette which is temperature controlled by a water-jacket. The pipette has to be volume calibrated at appropriate temperatures before and after the cruise, and the temperature has to be controlled to within about 0.5 °C to achieve the accuracy and precision expected from this system.

Acid adder (2)

Acid is added to the extraction tower in order to acidify the sample and thus to degas all CO_2 more quickly. The acid addition can either be done by a simple burette or by pressurizing an acid bottle for a given time to push out the right amount of acid. To reach a favourable pH, 1–1.5 mL of 8.5 % phosphoric acid are added.

CO_2 extraction tower (3)

The extraction tower has to be constructed to degas all CO_2 efficiently from the acidified seawater sample, with a minimum loss of water vapour and unwanted acid gases to the coulometer. An example of its design is shown in Fig. (8-3). The ceramic frit should be selected to produce a flow of small bubbles, but without generating a significant pressure resistance. The top of the extraction tower (or the outlet from it) can also be fitted with a water-jacket, through which cold water is circulated. This will minimize the amount of water vapour that follows the gas stream to the coulometer. Once the analysis is finished, the sample has to be removed from the extraction tower. This can be done by pressurizing it from above at the same time as the waste outlet is opened. Another option is to push the sample down through the frit to a waste connection. This, however, requires a fairly coarse frit, which will produce large bubbles during the extraction step, and also any particles present in the sample will tend to collect in the frit, which might cause an increase in the coulometer signal, for example from metal carbonates that can slowly dissolve in the acidic environment during the extraction.

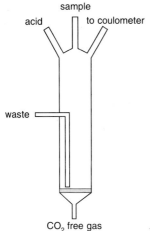

Fig. 8-3. The CO_2 extraction tower.

Sample gas purifier (4)

Even if a cold trap is used at the top of the extraction tower, a small amount of water vapour will pass through and if it reaches the coulometer it could affect the result. Hence, a cartridge of drying agent, *e.g.*, magnesium perchlorate, should be inserted in the gas stream. An alternative to the drying agent is a Teflon filter (typically 50 mm diameter) which will trap the water vapour without any risk of releasing particles as can happen with drying agents. Furthermore, the Teflon filters can be dried and re-used for a long time before the build up of salt crystals is so great that they have to be discharged.

In some oceanic environments, other reactive gases might be present in the sample and carried over into the gas stream. An activated silica gel trap can then be useful. If the sample contains hydrogen sulphide, a scrubber containing a silver sulphate solution has to be used.

Coulometer (5)

The coulometer normally used as detector of the degassed CO_2 is the UIC Inc. Model 5011. The coulometer is fitted with a titration cell, where the cathode solution contains a mixture of ethanolamine, tetraethylammonium bromide and thymolphthalein in a water-

DMSO (dimethyl sulphoxide) solution and the anode solution is a water-DMSO solution saturated with potassium iodide. The cathode is a platinum wire, while the anode is a silver rod. As the *pK* of the thymolphthalein indicator used to sense the pH in the titration cell is temperature-dependent, the titration cell should be kept at a constant temperature (within $\pm 0.2\,°C$) to achieve maximum precision. This can be done either by water-jacketing the titration cell, or by continuously blowing air of constant temperature around the cell.

Gas loop calibration system (6)

To calibrate the coulometric system, including the response factor of the coulometer, a gas loop calibration system can be used. In principle it consists of a chromatography valve with sample loops of different sizes that can be filled with a calibration gas of high purity ($> 99.99\,\%$ pure CO_2). Prior to injection of the calibration gas into the system, the temperature and pressure have to be measured accurately, as they have a significant impact on the amount of CO_2 in the gas loop.

Computer control

To achieve the best precision and accuracy in the determination a very reproducible procedure is necessary. This is best done if the system is computer controlled, including the handling of magnetic valves, temperature readings and reading of the coulometer. The UIC coulometer is equipped with both a parallel and serial (RS232) port and software to acquire the data in the serial mode is available from the manufacturer.

8.5.2.2 Analytical procedure

At the start of the analytical procedure the coulometer titration vessel is filled with fresh solutions. It has been shown that the background level will decrease if a few 'waste' seawater samples are run before the start of the real analysis. Also the background seems to decrease if the cell is left for several hours with nitrogen gas flowing through. Once the background level is stable at a low level (25 ± 10 counts/min) the coulometer should be calibrated. This can be done either by CO_2 gas, using the gas loop calibration system, or by analysing a carbonate solution or a reference material (Section 8.3.1).

Once the coulometer is calibrated the seawater samples can be analysed. An analysis is run by first adding the phosphoric acid and degassing it for half a minute to get rid of dissolved CO_2. After this, the seawater sample (with a known temperature) is added from the dispensing system at the same time as the coulometer reading is zeroed. The degassing of the CO_2 in the acidified sample is monitored by reading the coulometer output every 60 s. This is done until the readings of the last 2 min are constant and less than the maximum background. The average of the readings during the last 2 min are taken as the background and this value multiplied by the extraction time (minutes) is subtracted from the total coulometer reading. The stripping is interrupted and the extraction tower is emptied, and the system is then ready to analyse the next sample. After some 10–15 samples have been analysed a new calibration of the coulometer should be performed. At the end of each run with the same coulometer solutions a new analysis of the reference material is recommended.

During analytical work the gas flow has to be continuously checked. It should be constant, indicating no clogging problems, and around 150 mL/min to extract the CO_2 efficiently from the sample and at the same time to be slow enough for the cathode solution to absorb all CO_2 in the gas stream. The sample gas purifier also has to be checked continuously to ensure that its capacity has not been exceeded.

The maximum number of samples that can be determined without changing the coulometer solutions depends on the amount of CO_2 that has been titrated, which should not in total exceed 2 mmol. However, the solutions should also be changed after 12 h of use, or if the background increases significantly (above 25 counts/min), or if the calibration factor changes excessively (more than 0.1 %).

8.5.2.3   Calculation and expression of results

The total dissolved inorganic carbon concentration C_T μmol/(kg-seawater) is calculated as

$$C_T = \left(\frac{N_s - b \cdot t_c}{c} \right)\left(\frac{1}{v_0 \cdot \rho} \right) \tag{8-29}$$

where N_s is the coulometer reading (counts), b is the blank (counts/min), t_C is the time of extraction (min), c is the calibration factor, v_0 is the sample volume and ρ is the density of the seawater sample at the temperature measured in the sample dispensing system (Fig. 8-3, item 1). Calculation of the calibration factor c depends on the calibration technique used:

Standard reference material with certified concentration C_T(std) mol/(kg-seawater):

$$c = \left(\frac{N_s - b \cdot t_c}{C_T(\text{std})} \right)\left(\frac{1}{v_0 \cdot \rho} \right) \tag{8-30}$$

Carbonate solution of concentration c_C mol/L:

$$c = \left(\frac{N_s - b \cdot t_C}{c_C} \right)\left(\frac{1}{v_0} \right) \tag{8-31}$$

Gas loop calibration with $n(CO_2)$ moles of carbon dioxide:

$$c = \left(\frac{N_s - b \cdot t_C}{n(CO_2)} \right) \tag{8-32}$$

8.6 Thermodynamic calculations of the CO$_2$ system in seawater

8.6.1 Equations describing the CO$_2$ system in seawater

The equations given here include the contributions of only bicarbonate, carbonate, borate and hydroxide to the total alkalinity. As pointed out in Section 8.4.1, the contributions from other bases present in seawater are usually small. The system is now defined by three mass balance equations (8-33) to (8-35) and five stability constants (Eqs. (8-36) to (8-40)):

$$A_T = \left[HCO_3^-\right] + 2\left[CO_3^{2-}\right] + \left[B(OH)_4^-\right] - \left\{H^+\right\} + \left[OH^-\right] \tag{8-33}$$

$$\begin{aligned} C_T &= \left[H_2CO_3\right] + \left[CO_2(aq)\right] + \left[HCO_3^-\right] + \left[CO_3^{2-}\right] \\ &= \left[H_2CO_3^*\right] + \left[HCO_3^-\right] + \left[CO_3^{2-}\right] \end{aligned} \tag{8-34}$$

$$B_T = \left[B(OH)_3\right] + \left[B(OH)_4^-\right] \tag{8-35}$$

$$K_0 = \left[H_2CO_3^*\right] / f(CO_2) \tag{8-36}$$

$$K_1 = \left[HCO_3^-\right]\left\{H^+\right\} / \left[H_2CO_3^*\right] \tag{8-37}$$

$$K_2 = \left[CO_3^{2-}\right]\left\{H^+\right\} / \left[HCO_3^-\right] \tag{8-38}$$

$$K_B = \left[B(OH)_4^-\right]\left\{H^+\right\} / \left[B(OH)_3\right] \tag{8-39}$$

$$K_W = \left\{H^+\right\}\left[OH^-\right] \tag{8-40}$$

where $f(CO_2)$ is the fugacity of carbon dioxide, the term $[H_2CO_3^*]$ is a convenient short-hand for the sum of dissolved CO_2 and carbonic acid (Eq. (8-34)), and the meaning of $\{H^+\}$ is dependent on the pH scale in use, $\{H^+\} = 10^{-pH}$. For the total hydrogen ion concentration scale $\{H^+\} = [H^+] + [HSO_4^-]$ (see Section 7.4).

If all the stability constants, together with the total borate concentration B_T (*Uppström, 1974*), are known, then these eight equations have ten unknowns: A_T, C_T, $f(CO_2)$, $\{H^+\}$ $(= 10^{-pH})$ and the concentrations of dissolved CO_2/carbonic acid $[H_2CO_3^*]$, hydrogen carbonate, carbonate, boric acid, borate and hydroxide. It follows, therefore, that only two unknowns (*i.e.*, two of the measurable parameters A_T, C_T, $f(CO_2)$ and pH) need be determined in order to be able to calculate all the others. The values assigned to the stability constants are discussed in Section 8.6.2, and the total boron concentration can be calculated as $B_T = 0.00042 \cdot S/35$ mol/(kg-seawater) (*Whitfield and Turner, 1981*).

8.6.2 Selection of stability constants

Over the last 25 years, several groups have made reliable determinations of the stability constants defined in Eqs. (8-36) to (8-40). Unfortunately, these constants are not always directly comparable with one another since different groups have used different concentration scales and pH scales. The discussion will be restricted to the options recommended here for measurement and for reporting of results, *i.e.*, the mol/(kg-seawater) concentration scale and the total hydrogen ion concentration pH scale pH(T).

Table 8-1. Recommended stability constants on the mol/(kg-seawater) and pH(T) scales.

Constant	S range	t range /°C	σ^a	Reference
K_0	0–35	0–35	0.0013	*Weiss*, 1974
K_1	5–45	0–45	0.0021	*Roy et al.*, 1993; 1994
K_2	5–45	0–45	0.0030	*Roy et al.*, 1993; 1994
K_B	5–45	0–45	0.0018	*Dickson*, 1990
K_W	20–40	5–30	0.019^b	*Hansson*, 1973

[a] Standard deviation of individual values of $\log_{10} K$ from fitting Eqs. (8-41) to (8-45);
[b] Standard deviation about fitting equation given by *Millero* (1979), Eq. (8-45).

Table 8-1 lists the constants which are recommended for use on the mol/(kg-seawater) concentration scale and the total hydrogen ion concentration pH scale pH(T). The determinations by *Roy et al.* (1993; 1994) and *Dickson* (1990) listed in Table 8-1 supersede the careful determinations of *Hansson* (1973), which covered a more limited salinity range and that had a somewhat higher standard deviation. Careful stability constant measurements have also been made on the seawater pH(SWS) scale (K_1, K_2, *Goyet and Poisson*, 1989; K_W, *Dickson and Riley*, 1979) and on the pH(NBS) scale (K_1, K_2, *Mehrbach et al.*, 1973; K_W, *Culberson and Pytkowicz*, 1973). In addition, *Dickson and Millero* (1987) recalculated the K_1 and K_2 values of *Hansson* (1973) and *Mehrbach et al.* (1973) to the pH(SWS) scale and found no consistent difference between them.

However, conversion from the pH(NBS) scale depends on values of f_H for the electrode pair used, and conversion between the pH(T) and pH(SWS) scales depends on the values of the protonation constants K_{HSO_4} and K_{HF} (see Section 7.4). The parameter K_{HF} is not particularly well defined, and conversion between these scales is therefore not recommended for accurate calculations.

The references listed in Table 8-1 give the following equations for calculating these stability constants as their natural logarithms from values of salinity S and temperature T (K).

$$\ln K_0 = -60.2409 + 93.4517(100/T) + 23.3585 \cdot \ln(T/100)$$
$$+ S\left[0.023517 - 0.023656(T/100) + 0.0047036(T/100)^2\right]$$

(8-41)

$$\ln K_1 = 2.83655 - 2307.1266/T - 1.5529413 \cdot \ln T$$
$$+ \left(-0.20760841 - 4.0484/T\right)S^{0.5} + 0.08468345 \cdot S$$
$$- 0.00654208 \cdot S^{1.5} + \ln\left(1 - 0.001005\,S\right)$$

(8-42)

$$\ln K_2 = -9.226508 - 3351.6106/T - 0.2005743 \cdot \ln T$$
$$+ \left(- 0.106901773 - 23.9722/T\right) S^{0.5} + 0.1130822 \cdot S \tag{8-43}$$
$$- 0.00846934 \cdot S^{1.5} + \ln\left(1 - 0.001005\,S\right)$$

$$\ln K_B = 148.0248 + 137.1942 \cdot S^{0.5} + 1.62142 \cdot S$$
$$+ \left(- 8966.90 - 2890.53 \cdot S^{0.5} - 77.942 \cdot S + 1.728 \cdot S^{1.5} - 0.0996 \cdot S^2\right)/T \tag{8-44}$$
$$+ (-24.4344 - 25.085 \cdot S^{0.5} - 0.2474 \cdot S) \ln T + 0.053105 \cdot S^{0.5} \cdot T$$

$$\ln K_W = 148.9802 - 13847.26/T - 23.6521 \cdot \ln T$$
$$+ \left(- 97.9429 + 4149.915/T + 14.8269 \cdot \ln T\right) S^{0.5} - 0.023694 \cdot S \tag{8-45}$$

The final term in the expressions for $\ln K_1$ and $\ln K_2$ converts the constants from the mol/(kg-water) scale used by *Roy et al.* (1993) to mol/(kg-seawater).

8.6.3   Calculations with two measured CO_2 parameters

The system defined by Eqs. (8-33) to (8-40) can be solved analytically for all input combinations of two measured parameters with the exception of A_T, C_T, A_T and $f(CO_2)$, where an iterative approach is required. These solutions are summarized below: pH is represented by the corresponding hydrogen ion concentration $\{H^+\} = 10^{-pH}$. Full derivations of the solutions to the equations can be found in, *e.g., DOE* (1994).

8.6.3.1   Calculations with pH and A_T measured

$$A_C = A_T + \{H^+\} - K_W/\{H^+\} - B_T/\left(1 + \{H^+\}/K_B\right) \tag{8-46}$$

$$f(CO_2) = \frac{A_C\{H^+\}^2}{K_0 K_1\left(\{H^+\} + 2K_2\right)} \tag{8-47}$$

$$C_T = \frac{A_C\left(\{H^+\}^2/K_1 + \{H^+\} + K_2\right)}{\{H^+\} + 2K_2} \tag{8-48}$$

8.6.3.2   Calculations with pH and $f(CO_2)$ measured

$$C_T = K_0 f(CO_2)\left(1 + K_1/\{H^+\} + K_1 K_2/\{H^+\}^2\right) \tag{8-49}$$

$$A_T = \frac{K_0 K_1 f(CO_2)}{\{H^+\}}\left(1 + 2K_2/\{H^+\}\right) + B_T/\left(1 + \{H^+\}/K_B\right) + K_W/\{H^+\} - \{H^+\} \tag{8-50}$$

8.6.3.3   Calculations with pH and C_T measured

$$f(CO_2) = \frac{C_T\{H^+\}^2}{K_0\left(\{H^+\}^2 + K_1\{H^+\} + K_1 K_2\right)} \tag{8-51}$$

$$A_T = \frac{C_T K_1\left(\{H^+\} + 2K_2\right)}{\left(\{H^+\}^2 + K_1\{H^+\} + K_1 K_2\right)} + B_T/\left(1 + \{H^+\}/K_B\right) + K_W/\{H^+\} - \{H^+\} \tag{8-52}$$

8.6.3.4   Calculations with A_T and C_T measured

There is no analytical solution to this problem. The simplest method of solution is to solve Eq. (8-52) for $\{H^+\}$ by an iterative method (*e.g.*, Newton-Raphson; see Stephenson, 1969). Then $f(CO_2)$ can be calculated using equation (8-51).

8.6.3.5   Calculations with A_T and $f(CO_2)$ measured

This problem also lacks an analytical solution. The simplest method is again an iterative solution for $\{H^+\}$, this time solving Eq. (8-50), following which C_T can be calculated using equation (8-49).

8.6.3.6   Calculations with C_T and $f(CO_2)$ measured

This problem can be solved by rearranging Eq. (8-49) as a quadratic in $\{H^+\}$ with the solution:

$$\frac{1}{\{H^+\}} = \frac{-K_1 + \sqrt{K_1^2 - 4K_1 K_2\left(1 - C_T/[K_0 f(CO_2)]\right)}}{2K_1 K_2} \tag{8-53}$$

The alkalinity can then be calculated using Eq. (8-50).

8.6.4   Errors arising from the calculations

An important consideration when carrying out any of the calculations summarized in Section 8.6.3 is the anticipated uncertainty in the result. Since the equations themselves are non-linear, the uncertainties in the measured parameters and in the stability constants combine in a non-linear fashion. *Dickson and Riley* (1978) have compiled the partial derivatives which allow the errors associated with these calculations to be estimated. The values of σ in

Table 8-1 are used as estimates of the uncertainties in the constants, and the precisions of the individual measurement techniques are taken to be 0.002 pH (potentiometry) and 0.0005 pH (spectrophotometry) for pH, 2 μmol/kg for A_T, 1 μmol/kg for C_T and 0.5 μatm for $f(CO_2)$. The resulting precisions (random errors) in the calculated parameters are shown in Table 8-2.

Table 8-2. Estimated precision of calculated CO_2 parameters.

Measured parameter	Precision of calculated parameter
pH (pot.), A_T	2 μmol/kg C_T, 2.8 μatm $f(CO_2)$
pH (spect.), A_T	2 μmol/kg C_T, 2.1 μatm $f(CO_2)$
pH (pot.), $f(CO_2)$	16 μmol/kg C_T, 20 μmol/kg A_T
pH (spect.), $f(CO_2)$	12 μmol/kg C_T, 15 μmol/kg A_T
pH (pot.), C_T	2 μmol/kg A_T, 2.5 μatm $f(CO_2)$
pH (spect.), C_T	2 μmol/kg A_T, 2.1 μatm $f(CO_2)$
A_T, C_T	0.005 pH, 5.3 μatm $f(CO_2)$
A_T, $f(CO_2)$	0.002 pH, 2 μmol/kg C_T
C_T, $f(CO_2)$	0.002 pH, 2 μmol/kg A_T

The non-linearity in accumulation of errors can clearly be seen from Table 8-2. The measurement combinations A_T, C_T and pH, $f(CO_2)$ are obviously unsuitable for calculation of other CO_2 system parameters since the uncertainties in the measurements are magnified during the calculation. The results suggest, however, that with the right choice of measured parameters, A_T and C_T can be calculated with precision similar to that achieved in direct measurement (this also applies to pH measured by potentiometry). In contrast, $f(CO_2)$ cannot be calculated with the precision which can be achieved in direct measurement.

The estimates in Table 8-2 refer to precision, or random errors (see Chapter 1). The accuracy of the calculations, where systematic errors are also taken into account (see Chapter 1), can be expected to be somewhat poorer. Reasonable estimates for the accuracies of the current measurement techniques are 0.005 pH, 4 μmol/kg for A_T, 2 μmol/kg for C_T and 2 μatm for $f(CO_2)$. The resulting estimates of the accuracy of calculated CO_2 parameters are shown in Table 8-3.

Table 8-3. Estimated accuracy of calculated CO_2 parameters.

Measured parameter	Accuracy of calculated parameter
pH, A_T	4 μmol/kg C_T, 5.3 μatm $f(CO_2)$
pH, $f(CO_2)$	30 μmol/kg C_T, 40 μmol/kg A_T
pH, C_T	5 μmol/kg A_T, 5 μatm $f(CO_2)$
A_T, C_T	0.010 pH, 9 μatm $f(CO_2)$
A_T, $f(CO_2)$	0.003 pH, 4 μmol/kg C_T
C_T, $f(CO_2)$	0.003 pH, 2 μmol/kg A_T

The general pattern and conclusions remain the same: if only two CO_2 parameters are to be measured, then the best pairings are $f(CO_2)$ with C_T or A_T, followed by pH with C_T or A_T (but note that in this case the calculated values of $f(CO_2)$ will be subject to significant uncertainty: if reliable $f(CO_2)$ values are required this parameter should be measured directly).

The pairings of A_T with C_T and pH with $f(CO_2)$ should not be used for calculations of other CO_2 parameters. *Millero et al.* (1993) measured all four CO_2 parameters in the equatorial Pacific, and found differences between measured and calculated parameters in broad agreement with Table 8-3.

References to Chapter 8

Almgren, T., Dyrssen, D., Fonselius, S. (1983), in: *Methods of Seawater Analysis*, 2nd edn., Grasshoff, K., Ehrhardt, M., Kremling, K. (Eds.). Weinheim: Verlag Chemie, Weinheim; pp. 99–123.

Anderson, L.G., Wedborg, M. (1983), *Oceanol. Acta, 6*, 87.

Anderson L.G., Wedborg, M. (1985), *Oceanol. Acta, 8*, 479.

Culberson, C.H., Pytkowicz, R.M. (1973), *Mar. Chem., 1*, 309.

Bradshaw, A.L., Brewer, P.G., Shafer, D.K., Williams, R.T. (1981), *Earth Planet. Sci. Lett., 55*, 99

Dickson, A.G. (1981), *Deep-Sea Res., 28A*, 609.

Dickson, A.G. (1990), *Deep-Sea Res., 37*, 755.

Dickson, A.G., Millero, F.J. (1987), *Deep-Sea Res., 34*, 1733.

Dickson, A.G., Riley, J.P. (1978), *Mar. Chem., 6*, 77.

Dickson, A.G., Riley, J.P. (1979), *Mar. Chem., 7*, 89.

DOE (1994), *Handbook of Methods for the Analysis of the Various Parameters of the Carbon Dioxide System in Seawater*, Version 2: Dickson, A.G., Goyet, C. (Eds.), Report ORNL/CDIAC-74, Oak Ridge National Laboratory, Oak Ridge, TN, USA.

Dyrssen, D., Sillén, L.G., (1967), *Tellus, 21*, 113.

Goyet, C., Poisson, A. (1989), *Deep-Sea Res., 36*, 1635.

Gripenberg, S. (1936), *Comm. 108 V. Hydrolog. Conf. Baltic States*, Helsingfors.

Hansson, I. (1973), *Deep-Sea Res., 20*, 461.

Hansson, I., Jagner, D. (1973), *Anal. Chim. Acta, 65*, 363.

Haraldsson, C., Anderson, L.G., Hassellöv, M., Hulth, S., Olssson, K. (1997), *Deep-Sea Res., 44*, 2031.

Johansson, O., Wedborg, M. (1982), *Oceanol. Acta, 5*, 209.

Johnson, K.M., King, A.E., Sieburth, J.M. (1985), *Mar. Chem., 16*, 61.

Koroleff, F. (1972), in: *New Baltic Manual, Coop. Res. Rep. Ser. A.* (29): Carlberg, S.R. (Ed.).

Mehrbach, C., Culberson,C.H., Hawley, J.E., Pytkowicz, R.M. (1973), *Limnol. Oceaongr., 18*, 897.

Millero, F.J. (1979), *Geochim. Cosmochim. Acta, 43*, 1651.

Millero, F.J., Byrne, R.H., Wanninkhof, R., Feely, R., Clayton, T., Murphy, P., Lamb, M.F. (1993), *Mar. Chem. 44*, 269.

Roy, R.N., Roy, L.N., Vogel, K.M., Porter-Moore, C., Pearson, T., Good, C.E., Millero, F.J., Campbell, D.M. (1993), *Mar. Chem., 44*, 249.

Roy, R.N., Roy, L.N., Vogel, K.M., Porter-Moore, C., Pearson, T., Good, C.E., Millero, F.J., Campbell, D.M. (1994), *Mar. Chem., 44*, 337.

Stephenson, G. (1969), *Mathematical Methods for Science Students*. London: Longmans.

Uppström, L (1974), *Deep-Sea Res., 21*, 161.

Weiss, R.F. (1974) *Mar. Chem., 2*, 203.

Whitfield, M., Turner, D.R. (1981), in: *Marine Electrochemistry*: Jagner, D., Whitfield M. (Eds.). Chichester: John Wiley, 1981; pp. 3–66.

Appendix 8A

Here the details are given of the derivation of the contributions to H_T used in Eqs. (8-18) and (8-24).

Carbonate contribution C

$$\frac{v_0 \cdot C_T}{(v_0 + v)} = \left[CO_3^{2-}\right] \cdot \left(1 + K_{1C} \cdot \left[H^+\right] + K_{1C} \cdot K_{2C} \cdot \left[H^+\right]^2\right) \tag{8A-1}$$

The contribution C to H_T is therefore given by:

$$\left(\left[HCO_3^-\right] + 2\left[CO_3^{2-}\right]\right) = \left[CO_3^{2-}\right] \cdot \left(\frac{\left[H^+\right]}{K_2} + 2\right) =$$

$$\frac{v_0 \cdot C_T}{(v_0 + v)} \cdot \frac{\left(\dfrac{\left[H^+\right]}{K_2} + 2\right)}{\left(1 + \dfrac{\left[H^+\right]}{K_2} + \dfrac{\left[H^+\right]^2}{K_1 \cdot K_2}\right)} = \frac{v_0}{(v_0 + v)} \cdot C \tag{8A-2}$$

Borate and silicate contributions B and Si

The equations for these have the form given below where Bas_T and K_{Bas} denote the total concentrations, B_T and Si_T, and dissociation constants, K_B and K_{Si}, of boric and silicic acids, respectively)

$$\frac{v_0}{(v_0 + v)} \cdot Bas_T = \left[Bas^-\right] \cdot \left(1 + \frac{\left[H^+\right]}{K_{Bas}}\right) \tag{8A-3}$$

The contributions B and Si to H_T are therefore given by:

$$\left[Bas^-\right] = \frac{v \cdot Bas}{(v + v) \cdot \left(1 + \dfrac{\left[H^+\right]}{K_{Bas}}\right)} = \frac{v}{(v + v)} \cdot B(Si) \tag{8A-4}$$

Phosphate contribution P

$$\frac{v_0 \cdot P_T}{(v_0 + v)} = \left[PO_4^{3-}\right] \cdot \left(1 + \frac{\left[H^+\right]}{K_{3P}} + \frac{\left[H^+\right]^2}{K_{3P} \cdot K_{2P}} + \frac{\left[H^+\right]^3}{K_{3P} \cdot K_{2P} \cdot K_{1P}}\right) \tag{8A-5}$$

The contribution P to H_T is therefore given by:

$$\left(\left[HPO_4^{2-}\right]+2\left[PO_4^{3-}\right]-\left[H_3PO_4\right]\right)=\left[PO_4^{3-}\right]\cdot\left(\frac{\left[H^+\right]}{K_{3P}}+2-\frac{\left[H^+\right]^3}{K_{3P}\cdot K_{2P}\cdot K_{1P}}\right)=$$

$$=\frac{v_0\cdot P_T}{\left(v_0+v\right)}\cdot\frac{\left(\dfrac{\left[H^+\right]}{K_{3P}}+2-\dfrac{\left[H^+\right]^3}{K_{3P}\cdot K_{2P}\cdot K_{1P}}\right)}{\left(1+\dfrac{\left[H^+\right]}{K_{3P}}+\dfrac{\left[H^+\right]^2}{K_{3P}\cdot K_{2P}}+\dfrac{\left[H^+\right]^3}{K_{3P}\cdot K_{2P}\cdot K_{1P}}\right)}=$$

(8A-6)

$$=\frac{v_0}{\left(v_0+v\right)}\cdot P$$

(8A-6)

Sulphate and fluoride contributions *Su* and *Fl*

The equations for these have the form given below where HA_T and K_{HA} denote the total concentrations, Su_T and Fl_T, and the formation constants, K_{HSO_4} and K_{HF}, of hydrogen sulphate and hydrogen fluoride, respectively:

$$\frac{v_0\cdot HA_T}{\left(v_0+v\right)}=HA\cdot\left(1+\frac{1}{\left(\left[H^+\right]\cdot K_{HA}\right)}\right)$$

(8A-7)

The contributions Su and Fl to H_T are therefore given by:

$$HA=\frac{v_0\cdot HA_T}{\left(v_0+v\right)\cdot\left(1+\dfrac{1}{\left(\left[H^+\right]\cdot K_{HA}\right)}\right)}=\frac{v_0}{\left(v_0+v\right)}\cdot Su(Fl)$$

(8A-8)

For the evaluation, B_T, Su_T and F_T are calculated from their constant relationship to the salinity ($B_T = 0.00042 \cdot S/35$; $Su_T = 0.02824 \cdot S/35$; and $F_T = 0.000073 \cdot S/35$; all in mol/(kg-seawater)) and the stability constants for the protonation of the bases are calculated from their salinity dependence. Omission of Si_T from the equations has very little effect on the result for A_T, because the dominant portion of the silicate is present in the acidic form at the seawater pH. Omission of P_T has even less effect on the A_T estimate. If, however, C_T is also to be evaluated from the titration, omission of P_T in the evaluation equation introduces an overestimate of C_T, which is about equal to P_T (*Johansson and Wedborg, 1982*).

9 Determination of carbon dioxide partial pressure ($p(CO_2)$)

A. Körtzinger

9.1 Introduction

According to Dalton's law the total pressure of a mixture of gases behaving ideally is equal to the sum of the partial pressures p_i of all component gases i:

$$p = \sum_{i=1}^{k} p_i$$

The partial pressure of component i is defined as the product of its mole fraction x_i and the total pressure p of a gas mixture containing k components:

$$p_i = p \cdot x_i = p \cdot \frac{n_i}{\sum\limits_{j=1}^{k} n_j}$$

where n_i is the number of moles of component i.

The partial pressure of a volatile component assigned to a seawater sample more precisely denotes the partial pressure of this component in a gas phase that is in equilibrium with this sample. Hence, if the partial pressure is measured in a surface seawater sample and is also measured in the air actually overlying the seawater, it is possible to determine the momentary degree of saturation of this component in the water at the time of sampling. A partial pressure difference between the two phases (*i.e.*, the seawater being under- or supersaturated with respect to the air) is the thermodynamic driving force for any net gas exchange of this component. Together with a knowledge of the transfer coefficient, the partial pressure difference can be used to calculate the momentary net gas flux across the air-sea interface.

9.2 Principle of the measurement

The principle of the measurement of the partial pressure of CO_2 ($p(CO_2)$) in seawater is based on the determination of the CO_2 mixing ratio in a gas phase which is in equilibrium with a seawater sample at known temperature and pressure. The CO_2 mixing ratio can either be measured with a non-dispersive infrared analyser (NDIR) or with a gas chromato-graph (GC) using a flame ionisation detector after catalytic conversion of the CO_2 into methane. While the GC approach has a few advantages (*e.g.*, small sample volume) the more rugged infrared technique is more suitable for use at sea and allows measurements in a truly continuous fashion. Therefore only the NDIR approach will be described here. A fine example of a GC-based system can be found in *Weiss* (1981).

Depending on the sampling strategy (*discrete* or *continuous*) two different families of analytical systems have been developed. For the determination of the $p(CO_2)$ in air that is in equilibrium with a *discrete sample*, a known amount of seawater is isolated in a closed system containing a small known volume of air with a known initial CO_2 mixing ratio. For the determination of the $p(CO_2)$ in air that is in equilibrium with a *continuous flow* of sea-water, a fixed volume of air is equilibrated with seawater that flows continuously through an equilibrator.

Continuous $p(CO_2)$ systems are widely used in marine CO_2 research. They provide important information about the saturation state of seawater at the air-sea interface when operated on board research vessels with a continuous flow of seawater, usually obtained by means of a shipborne pumping system. This more frequently applied continuous or under-way mode is described here in detail. Three different designs of discrete $p(CO_2)$ systems can be found in *Wanninkhof and Thoning* (1993), *DOE* (1994) and *Neill et al.* (1997).

9.3   Apparatus for continuous mode of operation

9.3.1   The equilibrator

A great variety of effective designs of equilibrators for continuous $p(CO_2)$ systems have been described in the literature. Essentially three different families of equilibrator design can be distinguished:

1. the 'shower type' equilibrator (*e.g.*, *Keeling et al.*, 1965; *Inoue et al.*, 1987; *Robertson et al.*, 1993; *Goyet and Peltzer*, 1994);
2. the 'bubble type' equilibrator (*e.g.*, *Takahashi*, 1961);
3. the 'thin film type' equilibrator (*Poisson et al.*, 1993).

The common theme is to expose the well-mixed water phase to the likewise well mixed gas phase through a large surface area. This is accomplished by either showering the water into the gas phase (shower type) or by bubbling the air through the water phase (bubble type). The thin film type is of a somewhat different design in that the two phases move in laminar flows in opposite directions without generating turbulent mixing.

Furthermore, the published equilibrator designs not only vary in these basic function principles but also in many other aspects such as physical dimensions, flow rates and time constants. For example, a design described by *Copin-Montegut* (1985) combines aspects of the shower and bubble type. The equilibrator described here is a combination of the bubble and the thin film type equilibrator design (*Körtzinger et al.*, 1996). A schematic diagramme is shown in Fig. 9-1.

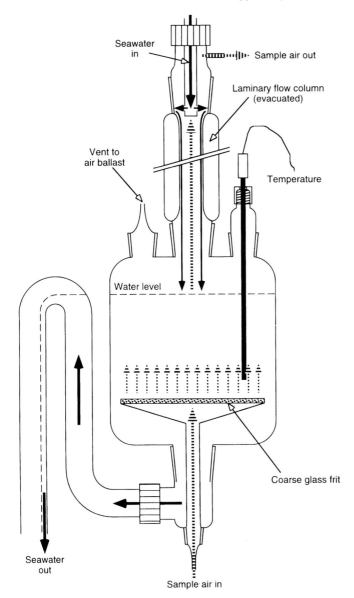

Fig. 9-1. Equilibrator for continuous measurement of $p(CO_2)$.

The continuous flow of seawater passes through a glass equilibrator, which is vented to the atmosphere (open system). Thus the water in the cell is always at ambient pressure. A fixed volume of air is recirculated continuously through the system so as to be in almost continuous equilibrium with the constantly renewed seawater phase. Two equilibration concepts are realized in two subsequent stages. One stage is a bubble type equilibrator: a chamber is filled with approximately 1 000 mL of constantly renewed water. The air enters from below through a coarse glass frit and bubbles through the water. The other stage is a laminar flow type equilibrator: a 45 cm glass column is centred on top of the water chamber. The seawater enters from the top and forms a laminar flow on the inner walls of the column, while the air coming from the water chamber below passes through the column before leaving the equilibrator at the column head. The countercurrent flow direction of seawater and air, as well as the large surface area, facilitate the establishment of equilibrium. An evacuated jacket minimises temperature changes in the water flow during passage through the column.

The flow rate of seawater is adjusted to 1.5–2.0 L/min. The air flow (total air volume approximately 400 mL) is maintained at a flow rate of about 1.0 L/min. The time constant of this equilibrator with respect to CO_2 is 75 ± 6 s, as determined from laboratory experiments under the above flow conditions (*Körtzinger et al.*, 1996). For measurements of seawater temperature, the equilibrator is equipped with a platinum resistance thermometer (Pt-100).

9.3.2   The analytical system

The schematic drawing in Fig. 9-2 depicts the general design of the underway $p(CO_2)$ system (modified after *Körtzinger et al.*, 1996). All numbers in the description below refer to numbers that appear in Fig. 9-2.

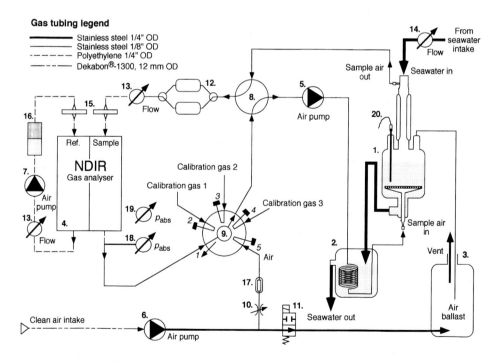

Fig. 9-2. Schematic diagramme of the continuous $p(CO_2)$ system. 1, Glass equilibrator; 2, seawater reservoir; 3, air ballast bottle; 4, non-dispersive infrared gas analyser; 5, 6 and 7, air pumps; 8, two-position valve; 9, multi-position valve; 10, needle valve; 11, solenoid valve; 12, drying unit; 13, gas flowmeter with flow regulator; 14, liquid flowmeter; 15, 1-μm PTFE membrane filter; 16, gas purification tube; 17, 2-μm particle filter; 18 and 19, pressure transducers; 20, platinum resistance thermometer.

After passage through the equilibrator (1) the air stream is pumped *via* a two-way valve (8), a gas drying unit (12), a flowmeter-regulator (13) and a 1-μm PTFE membrane filter (15) to the NDIR analyser (4), where the mole fraction of CO_2 ($x(CO_2)$) is measured relative to a dry and CO_2-free reference gas (absolute mode). The air stream is then recirculated *via* a multiport valve (9), valve 8 and an air pump (5) to a reservoir (2) which is flushed with outflowing water from the equilibrator. The gas tubing forms a coil in the reservoir to allow the air stream to adjust to the seawater temperature before reentering the equilibrator.

The present system features an LI-6262 CO_2/H_2O infrared gas analyser (LI-COR Inc., Lincoln, NE, USA), which is a dual-channel instrument that simultaneously measures the mole fractions of CO_2 and H_2O. With this particular NDIR instrument, the gas stream needs no drying prior to infrared gas detection, as the biasing effect of water vapour on the measurement of CO_2 can be eliminated based on the H_2O measurement. Appropriate internal algorithms correct for gas phase interactions of CO_2 with water vapour which cause a broadening of the absorption band of CO_2 (*McDermitt et al.*, 1993). However, this wet measurement approach frequently encounters operational problems from condensation of water in the gas tubing, if the seawater temperature is above ambient room temperature. The system described here therefore includes a gas drying unit (12), to make the design less restrictive with respect to the type of NDIR instrument and the acceptable range of surface seawater temperatures.

For the gas drying unit different approaches can be used. Peltier elements (dew point approximately –30 °C) can be used efficiently. A pair of two elements allows one element to be heated for thawing and drying while the other one is in use. A second option is chemical drying agents (see Section 9.4.2) which need regular replacement. Again a unit with paired drying tubes facilitates handling and avoids interruptions of the continuous measurements.

The reference gas circuit is a closed loop system which consists of a flowmeter-regulator (13), a miniature air pump (7), a gas purification tube (16) and a 1-μm PTFE membrane filter (15). This feature provides a constant supply of CO_2-free air as the reference gas and thus eliminates the need for compressed gas (*i.e.*, high-purity nitrogen) otherwise to be provided by the user.

For the measurement of atmospheric $p(CO_2)$, a diaphragm pump (6) continuously draws uncontaminated air from a clean air intake on the compass platform or foremast of the vessel through a Dekabon-type flexible tubing (Furon Dekoron Division, Aurora, OH, USA; 12 mm OD) to the $p(CO_2)$ system. While atmospheric $p(CO_2)$ is not being measured this air is used to flush the air ballast bottle (3) which provides a clean air buffer to the equilibrator vent. If any volume changes occur in the recirculated air, only clean outside air can invade the equilibrator through the vent line. For air measurements the normally-open solenoid valve (11) is closed and the air enters the analyser circuit *via* a needle valve (10) and a 2-μm particle filter (17).

Three different types of measurements (calibration, measurement of atmospheric air, measurement of seawater equilibrated air) are controlled with valves 8 and 9. Figure 9-3 shows a schematic drawing of the valve concept with the two different general states described below. Valve 8 separates the equilibrator circuit from the measuring circuit. Thus, during calibration (ports 2–4 of valve 9) or measurement of atmospheric air (port 5 of valve 9) the air pump (5) keeps the short-circuited equilibrator circuit in progress (Fig. 9-3, right), while the separated analyser circuit is flushed with ambient air or calibration gases. Valve 9 selects the gas to be measured.

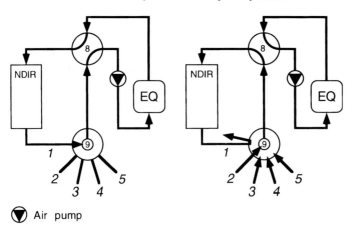

▼ Air pump

Fig. 9-3. Schematic diagramme of the valve concept.

Two high-accuracy pressure transducers are used to monitor the pressure in the NDIR cell (18) as well as the barometric pressure (19). The whole system is computer-interfaced and operated with purpose-made software. All data generated by the system are averaged and logged at user-chooseable intervals. A typical routine comprises a data interrogation interval of 6 s and an averaging interval of 1–2 min. The fully accessible duty cycle is carried out by remote control of the valves. The duration and interval of measurement states (calibration, measurement of atmospheric or seawater equilibrated air) can be chosen from a set-up menu. A typical routine consists of calibration, 55 min measurement of $p(CO_2)$ in seawater and 5 min measurement of $p(CO_2)$ in air alternating, and re-calibration after 4–6 cycles (*i.e.*, approximately 4–6 h). A delay interval (typically 1–2 min) can be defined to avoid logging of data during a short time interval after valve switching.

9.4   Reagents

9.4.1   Calibration gases

Griffith et al. (1982) have demonstrated the effect of the gas matrix on the response of NDIR instruments. This introduces some error if different gas matrices are involved in the calibration (*e.g.*, CO_2 in pure nitrogen) and the measurement (CO_2 in air). Therefore, a set of three calibration gases with known CO_2 concentrations in air is used for the calibration of the NDIR instrument. All mixtures are based on the 'natural' air concept and contain nitrogen, oxygen and argon in natural proportions (*i.e.*, 780 ‰ N_2, 210 ‰ O_2, 9.3 ‰ Ar). The CO_2 concentrations of the calibration gases should be chosen to span the range of $x(CO_2)$ values that can be expected in the field. Typically these are in the range of 200–500 μmol/mol, but may be as low as 100 μmol/mol and as high as 800 μmol/mol in extreme situations.

9.4.2   Gas purification reagents

1. *CO$_2$ removal agent*: A CO$_2$ scrubber is used in the purification tube (16) to remove any traces of CO$_2$ from the reference gas. Soda lime and Ascarite (sodium hydroxide coated silica; Thomas Scientific Inc., Swedesborough, NJ, USA) are good choices.
2. *Drying agent*: A desiccant is used in the gas purification tube (16) to remove any traces of water. It can also be used in the drying unit (12) if chemical drying is preferred. Magnesium perchlorate and Aquasorb (phosphorus pentoxide on solid support; Mallinckrodt Chemical Inc., Paris, KY, USA) can be used for this purpose.

9.5   Calculation of results

The calculation of final $p(CO_2)$ values from the raw voltage readings v' (mV) of the NDIR analyser involves a number of steps which are described below. The response of NDIR instruments is non-linear. Furthermore, the detector signal depends on the number of CO$_2$ molecules in the optical path which in turn is mainly a function of pressure and temperature for a given CO$_2$ mixing ratio. The calculation procedure therefore requires temperature and pressure corrections as well as a calibration function. The following procedure is only valid for operation of the NDIR analyser in absolute mode, where the sample is measured against a CO$_2$-free reference gas.

In the first step, raw voltage readings v' (mV) are corrected from the measured NDIR cell pressure p (kPa) during CO$_2$ measurement *versus* the standard barometric pressure p° (101.325 kPa). It has been found empirically (*Welles and Eckles*, 1991) that pressure affects the voltage signal in a linear fashion:

$$v = v' \cdot \frac{p^\circ}{p}$$

The CO$_2$ mole fraction $x(CO_2)^*$ ($\mu mol/mol$ = ppmv) is then calculated from the pressure-corrected voltage readings v (mV) on the basis of the calibration polynomial:

$$x(CO_2)^* = a_1 \cdot v + a_2 \cdot v^2 + a_3 \cdot v^3$$

where a_1, a_2 and a_3 are the calibration coefficients calculated from the calibration runs using a least-squares procedure.

In the next step the CO$_2$ mole fraction $x(CO_2)^*$ is linearly corrected for any deviation of the cell temperature T (K) during the CO$_2$ measurement from the calibration temperature T° (K). *Welles and Eckles* (1991) have shown that the mole fraction is scaled linearly with the inverse of the absolute temperature:

$$x(CO_2) = x(CO_2)^* \cdot \frac{T}{T^\circ}$$

with $x(CO_2)$ representing the final value of the mole fraction of CO$_2$ in dry sample air. As the air equilibrated with the seawater can be assumed to be at 100 % humidity a correction has to be applied to account for the increase in the CO_2 mole fraction due to the removal of water vapour prior to the infrared measurement. In marine sciences the $p(CO_2)$ is traditionally expressed as microatmospheres (1 atm = 101.325 kPa). For better comparability with published data we also use this unit here. Thus the $p(CO_2)$ (μatm) at 100 % humidity is given by

$$p(CO_2) = x(CO_2) \cdot (p - p(H_2O))$$

where p (atm) is the ambient (= equilibrator) pressure and $p(H_2O)$ (atm) is the saturation vapour pressure of water, which can be calculated from the following equation after *Weiss and Price* (1980):

$$\ln p(H_2O) = 24.4543 - 67.4509 \cdot \frac{100}{T} - 4.8489 \ln \frac{T}{100} - 0.000544\,S$$

where T (K) is the seawater temperature in the equilibrator and S is the seawater salinity.

For very accurate measurements the non-ideal behaviour of CO_2 has to be taken into account, *i.e.*, fugacity has to be used instead of partial pressure. This is the case if the results are to be used to calculate other parameters of the CO_2 system in seawater. The fugacity can be calculated from a knowledge of the virial expression of the equation of state for CO_2. For the binary mixture CO_2-air the fugacity of CO_2 ($f(CO_2)$) is given by

$$f(CO_2) = x(CO_2) \cdot (p - p(H_2O)) \cdot \exp\left(p\frac{(B + 2\delta)}{RT}\right)$$

The first virial coefficient of CO_2, B (cm^3/mol), can be calculated using a power series given by *Weiss* (1974) for the temperature range $265 < T < 320$ K:

$$B = -1636.75 + 12.0408\,T - 3.27957 \cdot 10^{-2}\,T^2 + 3.16528 \cdot 10^{-5}\,T^3$$

The parameter δ (cm^3/mol) is the cross virial coefficient B_{12} for interaction between gases 1 and 2 minus the mean of B_{11} and B_{22} for the two pure gases (CO_2 and air). The temperature dependence of δ in the binary mixture of CO_2 and air can be calculated after *Weiss* (1974) for the temperature range $273 < T < 313$ K:

$$\delta = 57.7 - 0.118T$$

The fugacity coefficient, *i.e.*, the ratio between fugacity and partial pressure of CO_2, is of the order 0.996–0.997 under typical conditions ($p = 1$ atm, $T = 270$–300 K, $p(CO_2)$ = 350 μatm).

As the $p(CO_2)$ in seawater strongly varies with temperature, a correction is necessary to compensate for any difference between the equilibration temperature and the *in situ* seawater temperature. Different equations have been proposed for the temperature dependence of CO_2 partial pressure/fugacity in seawater (*e.g.*, *Gordon and Jones*, 1973; *Weiss et al.*, 1982; *Copin-Montegut*, 1988, 1989; *Goyet et al.*, 1993; *Takahashi et al.*, 1993). As temperature deviations are generally $<1\,°C$ in underway work the correction is small and the choice from the above suite of equations is not critical. The following equation given by

Weiss et al. (1982) is valid for ranges of 0–36 °C in temperature, 30–38 in salinity and 80–2000 μatm in $f(CO_2)$:

$$\frac{\partial \ln f(CO_2)}{\partial t} = 0.03107 - 2.785 \cdot 10^{-4}\ t - 1.839 \cdot 10^{-3}\ \ln f(CO_2)$$

where t (°C) is the seawater temperature measured in the equilibrator.

For larger temperature deviations (>1 °C) more robust equations which include salinity should be used (*e.g., Copin-Montegut*, 1988, 1989; *Goyet et al.*, 1993). This is especially true for discrete $p(CO_2)$ measurements, where much larger temperature corrections of up to 20 °C and more have to be applied.

Measurements of the $p(CO_2)$ difference $\Delta p(CO_2)$) between surface seawater and air can be used to calculate momentary net fluxes of CO_2 across the air-sea interface after the following equation:

$$F = k\ K_0\left(\Delta p(CO_2)\right)$$

where F is the net flux, k is the transfer velocity and K_0 is the solubility coefficient of CO_2 in seawater. The transfer velocity k mainly depends on wind speed and seawater temperature while K_0 is a function of temperature and salinity (*Weiss*, 1974). Details of these flux calculations and the inherent problems are beyond the scope of this chapter. An introduction can be found in *Liss and Merlivat* (1986). *Watson et al.* (1995) gave a review of the *status quo*.

9.6 Accuracy

NDIR instruments can usually be calibrated to yield CO_2 measurements with an accuracy of 1 μmol/mol or better. Barometric pressure should be measured with an accuracy of 50 Pa. This translates to an error in $f(CO_2)$ of about 0.05 % (or 0.15 μatm at 300 μatm). Considerably more important are accurate measurements of the seawater temperatures (*in situ* and in the equilibrator). As the $f(CO_2)$ strongly varies with temperature an error in temperature of 0.1 °C is equivalent to an absolute error in $f(CO_2)$ of 0.4 % (or 1.2 μatm at 300 μatm). Therefore temperature measurements have to be accurate to at least 0.05 °C. The resulting accuracy of the final $f(CO_2)$ values is typically of the order of 2 μatm (or 0.5 % under typical surface ocean conditions).

References to Chapter 9

Copin-Montegut, C. (1985), *Mar. Chem.*, 17, 13.

Copin-Montegut, C. (1988), *Mar. Chem.*, 25, 29.

Copin-Montegut, C. (1989), *Mar. Chem.*, 27, 143.

DOE (1994), *Handbook of Methods for the Analysis of the Various Parameters of the Carbon Dioxide System in Seawater*: Dickson, A.M., Goyet, C. (Eds.), Report ORNL/CDIAC-74, Oak Ridge National Laboratory, Oak Ridge, TN, USA.

Gordon, L.I., Jones, L.B. (1973), *Mar. Chem.*, 1, 317.

Goyet, C., Peltzer, E.T. (1994), *Mar. Chem.*, 45, 257.

Goyet, C., Millero, F.J., Poisson, A., Shafer, D.K. (1993), *Mar. Chem.*, 44, 205.

Griffith, D.W.T., Keeling, C.D., Bacastow, R.B., Guenther, P.R., Moss, D.J. (1982), *Tellus*, 34, 385.

Inoue, H., Sugimura, Y., Fushimi, K. (1987), *Tellus*, 39B, 228.

Keeling, C.D., Rakestraw, N.W., Waterman., L.S. (1965), *J. Geophys. Res.*, 70, 6087.

Körtzinger, A., Thomas, H., Schneider, B., Gronau, N., Mintrop, L., Duinker, J.C. (1996), *Mar. Chem.*, 52, 133.

Liss, P.S., Merlivat, L. (1986), in: *The Role of Air-Sea Exchange in Geochemical Cycling*: Buat-Ménard, P. (Ed.). Dordrecht: Reidel, 1986, pp. 113–127.

McDermitt, D.K., Welles, J.M., Eckles, R.D. (1993), Publication 116, LI-COR, Lincoln, NE, USA.

Neill, C., Johnson, K.M., Lewis, E., Wallace, D.W.R. (1997), *Limnol. Oceanogr.*, in press.

Poisson, A., Metzl, N., Brunet, C., Schauer, B., Bres, B., Ruiz-Pino, D., Louanchi, F. (1993), *J. Geophys. Res.*, 98, 22 759.

Robertson, J.E., Watson, A.J., Langdon, C., Ling, R.D., Wood, J.W. (1993), *Deep-Sea Res.*, 40, 409.

Takahashi, T. (1961), *J. Geophys. Res.*, 66, 477.

Takahashi, T., Olafsson, J., Goddard, J.G., Chipman, D.W., Sutherland, S.C. (1993), *Glob. Biogeochem. Cycl.*, 7, 843.

Wanninkhof, R., Thoning, K. (1993), *Mar. Chem.*, 44, 189.

Watson, A.J., Nightingale, P.D., Cooper, D.J. (1995), *Phil. Trans. R. Soc. Lond.*, B, 348, 125.

Weiss, R.F. (1974), *Mar. Chem.*, 2, 203.

Weiss, R.F. (1981), *J. Chromatogr. Sci.*, 19, 611.

Weiss, R.F., Price, B.A. (1980), *Mar. Chem.*, 8, 347.

Weiss, R.F., Jahnke, R.A., Keeling, C.D. (1982), *Nature* (London), 300, 511.

Welles, J., Eckles, R. (1991), *LI-COR 6262 CO_2/H_2O Analyzer Operating and Service Manual*, Publication 9003–59, LI-COR, Lincoln, NE, USA.

10 Determination of nutrients

H.P. Hansen and F. Koroleff

10.1 Introduction

10.1.1 Oceanic distributions of nutrients

A nutrient element is one which is involved in the production of organic matter by photosynthesis in the upper ocean. Traditionally, in chemical oceanography the term has been applied almost exclusively to phosphorus, inorganic nitrogen compounds and silicon, but strictly a number of the major constituents of seawater, together with a number of essential trace metals, are also nutrient elements. With regard to the latter group, for example, strong evidence has been obtained from recent studies that in certain areas of the world ocean iron plays an important role in controlling the growth of marine plants (*e.g., Martin et al.,* 1994; see also Chapter 12). This chapter will be restricted to the determination of 'traditional' nutrient elements (P, N, Si). The section outlined here is confined to a brief consideration of the abundance and chemistry of phosphorus, inorganic nitrogen and silicon in the sea.

These constituents tend to be efficiently stripped from the surface waters through incorporation into the cells, tissues and extracellular structures of living organisms. Only a relatively small portion of the biogenic particles moving downwards in the water column is deposited in sediments. The average atom is released and returned to the surface layer many times before it is finally removed from the water column. The upward transport of dissolved matter by turbulent mixing is much less effective than the gravity driven downward flux of particles; *i.e.*, the 'biological pump' causes pronounced concentration increases with depth.

Some typical vertical profiles of phosphate, nitrate and silicate in three major ocean basins are shown in Fig. 10-1.

It has been generally accepted that a major part of the deep ocean waters has a common origin in surface waters of the North Atlantic (*Broecker and Peng*, 1982). Surface cooling increases the density and initiates thermal convection. The initial surface water is forced downward where it propagates as a deep current towards the Pacific Ocean. The particle flux from the surface gradually changes the chemical composition of the deep water resulting in an increase in nutrients with age (Fig.10-1). The rather slow decomposition of silica skeleton material as compared with soft tissue, which contains the major part of nitrogen and phosphorus, leads to a more gradual enrichment of silicon with depth, while phosporus and nitrogen have been already released in the upper layers.

In areas of lower depths (< 1000 m) increased amounts of phosphorus and nitrogen containing particles reach the sea floor. In addition, topographic conditions (fjords, trenches or land-locked basins) may restrict the horizontal transport (exchange) of deep waters and

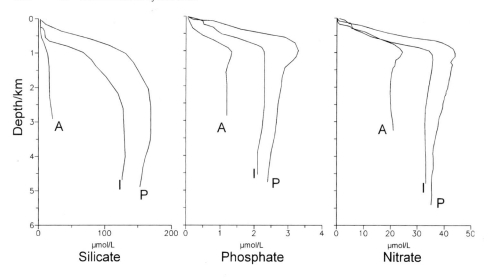

Fig. 10-1. Typical vertical distributions of silicate, phosphate, and nitrate in the North Atlantic (A), Indian Ocean (I) and the Central Pacific Ocean (P) based on GEOSECS data 1977/78.

haline stratification may block the vertical exchange causing stagnant deep water and anoxic conditions (*e.g.*, the Black Sea or Baltic Sea).

Consequently, the concentrations of phosphate in the deep water may increase to very high levels towards the bottom (*e.g.*, $> 8 \mu$mol/L, Baltic Sea; *Hansen*, 1990). Nitrate concentrations here are $< 0.1 \mu$mol/L, because most of the nitrate is converted into ammonia under anoxic conditions while a considerable part is removed by denitrification at low oxygen concentrations (see also Chapter 5).

10.1.2   Chemistry of nutrients in the marine environment

Phosporus exists in the sea in the form of ions of (ortho) phosphoric acid, H_3PO_4. About 10 % of the inorganic phosphate in seawater is present as PO_4^{3-} and practically all the remaining phosphate exists as HPO_4^{2-}. The condensed phosphoric acids, *i.e.*, diphosphoric acid, $H_4P_2O_7$, and all polyphosphoric acids with P-O-P linkages, have so far not been detected in open seawaters but are known to occur in estuarine and coastal waters as a result of pollution with detergents.

A variety of organic phosphorus compounds is present in the upper layers of the sea. These compounds are decomposition and excretion products of organisms and, therefore, phospholipids, phosphonucleotides and their derivatives may be found in seawater.

Phosphorus may be deposited in sediment as iron(*III*), calcium or magnesium phosphate (Stumm and Morgan, 1981). The reduction of iron(*III*) to iron(*II*) at the turn to anoxia in water or sediment may release phosphate from the sediment and cause elevated phosphate concentrations in bottom waters (*Hansen*, 1990).

Nitrate is the final oxidation product of nitrogen compounds in seawater. Nitrogen exists in the sea as elementary dissolved nitrogen and at different oxidation levels from -3 to $+5$.

The redox potential of seawater (pE) determines the form that has dominant stability. Elementary nitrogen prevails from a pE of -4 to $+12$. All other oxidation levels are products of biological processes. Below a pE of -4, NH_4^+ dominates while beyond a pE of $+12$ and a pH of about 7, NO_3^- is the most stable species. The fact that N_2 is not entirely converted into NO_3^- indicates a lack of efficient biological mediation of the reverse reaction also, for the mediating catalyst must operate equally well in both directions. This also suggests that the reduction of NO_3^- to N_2 must follow an indirect path as, *e.g.*, reduction of NO_3^- to NO_2^- and consequent con-proportioning of NO_2^- and NH_4^+ to N_2 (*Stumm and Morgan*, 1981). In the world oceans dissolved elementary nitrogen is the most abundant form of nitrogen (about $22 \cdot 10^{15}$ times all other forms) followed by nitrate, ammonia (inorganic and organic compounds), nitrite and gaseous dinitrogen oxide.

A considerable part of the nitrogen loads of the oceans comes *via* the atmosphere as wet or dry deposition of ammonia and nitrogen oxides. In the vicinity of highly industrialized and agricultural areas as, *e.g.*, in Central Europe, the atmospheric load can be as much as 45 % of the river discharge with nearly equal amounts of ammonia and oxidized forms of nitrogen (*HELCOM*, 1996, 1997).

Nitrite occurs in seawater as an intermediate product in microbial redox processes of nitrate (denitrification) at low oxygen levels. In addition, nitrite may be excreted by phytoplankton especially during periods of luxury feeding, *i.e.*, when a surplus of nitrate and phosphate stimulates a heavy bloom of plankton (*Martin*, 1968; *Grasshoff*, 1967). The natural level of nitrite in seawater is usually very low ($<0.1\,\mu$mol/L), but in transition zones, between oxic and anoxic layers, thin layers of nitrite concentrations of up to $2\,\mu$mol/L may occur (*Hansen*, unpublished). Nitrite is formed at low oxygen levels. In the presence of higher oxygen concentrations nitrite should be oxidized to nitrate according to the thermodynamic energy levels. However, nitrite concentrations of $>3\,\mu$mol/L associated with 80 % oxygen saturation at 4.2 °C have been observed in a Kattegat (Baltic Sea) water layer which displayed no noticeable decrease in nitrite for 7 days (*Hansen and Grasshoff*, unpublished).

Ammonium (NH_4^+) concentrations in the sea show considerable variations and can change rapidly. The amounts rarely exceed $5\,\mu$mol/L in oxygenated, unpolluted waters, but in anoxic deep stagnant water, such as in the Black Sea, the concentration of ammonium can be as high as $100\,\mu$mol/L. Since the acid-base pair, NH_4^+–NH_3, has a pK_a value of about 9.3, the ammonium ion (NH_4^+) is the dominant species in natural waters with a pH of 8.2 or less (*Stumm and Morgan*, 1981). It must be borne in mind that it is NH_3 which is toxic to fish and other marine organisms and not the NH_4^+ ion.

Ammonium is often the most abundant form of inorganic nitrogen in the surface layers after phytoplankton blooms have removed the greater part of the nitrate and phosphate. In the assimilation processes of phytoplankton, ammonium is preferentially used for synthesizing proteins. When nitrate is incorporated it must first be reduced to ammonia before it can be transferred into amino acid compounds. However, there is no indication that growth rates are particularly increased by either form.

Soluble and particulate organic nitrogen compounds resulting from decaying organisms, together with those excreted by plants and animals, are rapidly broken down to NH_3 by various species of proteolytic bacteria. Ammonia is excreted directly by animals together with urea and peptides.

Silicon is, after oxygen, the most common of all elements in the upper lithosphere. During weathering of silicate material, silicon is brought into solution, probably in the form of alkali salts of orthosilicic acid, $Si(OH)_4$.

Silicon enters the sea by glacial weathering of rocks in Antarctica and through rivers. Silicon concentrations in rivers very much depend on the geological formation of the respective area. The highest concentrations are found in volcanic areas (350–550 μmol/L). The low nutrient concentrations of oceanic surface waters apply also to silicate. Deep water concentrations range from about 25 μmol/L (Atlantic) to 170 μmol/L (Pacific).

Seawater is undersaturated with silicate with respect to the solubilities of marine silicon minerals.

The concentration of suspended particulate silicone in seawater also varies from trace concentrations to several mg/L. About half of this matter is of inorganic origin, mainly from clays.

Silicon is a major constituent of diatoms, which form a large proportion of marine phytoplankton. Some other algae, fungi and the siliceous sponges also have structural parts consisting of silica. The diatoms and radiolaria can excrete silica in the form of opal (amorphous silica, $SiO_2 \cdot nH_2O$).

10.2   Analytical methods

Principally, there are three groups of analytical methods for the determination of nutrients:
- manual methods, where each sample is treated individually and manually for each variable;
- automated methods, which are generally automated versions of the manual methods which may be combined to provide simultaneous multi variable analyses;
- sensors, which provide a physical signal representing the analyte concentration on contact with the seawater sample, preferably without prior chemical treatment.

Obviously, sensors would be ideal nutrient 'analysers'. There are several promising approaches for chemical sensors, *e.g.*, solid-state Chem-Fets (*Wohltjen*, 1984), ion-selective electrodes, bio-sensors (*Schultz*, 1991), coated fibre optic sensors (*Saari*,1987; *Barnard and Walt*, 1991) and several others. At the time of writing none of these sensors was accurate enough for direct detection of nutrients at natural seawater concentrations. General drawbacks are limited sensitivity and stability. As some sensors are available to measure nutrients in waste waters or other high concentration samples, and, considering the rapid pace of technical progress, during the following decade suitable sensors for marine applications might become available.

At the time of the first edition of this book the automated chemical analysis of nutrients, introduced into oceanography by *Weichart* (1963) and *Grasshoff* (1964), was still a relatively new technique but now it is state-of-the-art in most oceanographic laboratories.

Commercial nutrient analysers as well as standardized components for laboratory-designed analyser setups are available from various manufacturers and dealers of laboratory equipment.

The majority of instruments designed for use in land and ship laboratories are based on continuous flow-analysis (on-line wet chemical treatment and spectrophotometric detection of the sample) and variations hereof and can determine several variables simultaneously at rates of up to more than 100 samples per hour. However, they have to be operated and maintained by qualified personnel. International monitoring programmes and the establishment of monitoring networks (monitoring of pollution and environmental changes) have stimulated the development of analysers capable of unattended autonomous operation (remote, *in situ*). Such instruments (*e.g.*, the automated pumped analyser, APP, Meerestechnik – Elektronik, Trappenkamp, Germany, and the Nutrient Monitor, W.S. Ocean Systems Ltd., UK) generally are slightly less sensitive and accurate and cannot replace the man-operated laboratory analysers. Several *in situ* flow-systems have been described in the literature (*e.g.*, *Johnson et al.*, 1986, 1989; *Jannasch et al.*, 1994). These instruments are still prototypes and up to now used by the constructors in research projects only.

The standard setups for nutrient analysis are either manual or automated air-segmented-flow-analysis versions of a wet chemical treatment of water samples to convert the desired nutrient into a coloured compound with high molar absorptivity in a suitable range of the light spectrum. Flow-injection analysis is based on identical chemical procedures but requires different reagent recipes. A more detailed introduction into automated flow-analysis is presented in Section 10.3 at the end of this chapter.

As the outlines of methods are nearly identical for 'manual' analysis and automated air-segmented-flow-analysis (with regard to chemistry, sample pretreatment, preparation of standards and reagents), we have combined the analytical procedures for each nutrient.

All nutrient analyses in this book are performed by spectrophotometry, a well established analytical method which does not require repetition here. However, it is important to remember some of the particulars of the method and some implications of Beer-Lambert's law, which is the basis of spectrophotometric determinations.

The monochromatic light, either emitted by a light emitting diode (LED) or a lamp and optical filter, passes through the cuvette containing the sample. The logarithm of the ratio of light intensities before and after passage through the sample solution is called absorbance (A) and is a linear function of the concentration of the absorbing component according to Beer-Lambert's law

$$\log \frac{I_0}{I} = \varepsilon \cdot c \cdot d$$

where I_0 is the initial light intensity, I is the intensity after passage through the sample solution, ε and c are the molar absorptivity and the concentration of the absorbing compound, respectively, and d is the optical path length (cuvette length).

Like many physical laws this is an idealised formulation of a natural process. In fact, however, the relationship between absorbance and concentration is linear only to about 1–1.5 absorbance units (depending on the chemical component). An absorbance of 1.5 means that only 3 % of the incident light passes through the cuvette.

In special applications non-linear calibration curves may be used to expand the measuring range, however the relative precision of the spectrophotometer decreases considerably at higher absorbance ranges. Consequently an analytical setup for the determination of a wide concentration range of the desired component does not allow the maximum analytical precision.

Spectrophotometer cuvettes of nominally equal sizes are generally not identical, *i.e.*, a dye solution may result in slightly deviating absorbances in different cells. If the cells of one set do not match exactly, individual cell absorbances (A_{cell}) have to be determined and considered in the calculation of the results (see Section 10.2.4).

10.2.1   Pretreatment of samples

The procedures of sampling and storage have been thoroughly described in Chapters 1 and 2. Therefore, only some particular problems with respect to the determination of nutrients will be considered here.

It is obvious that any treatment of the sample, *i.e.*, transfer from one container to another, chemical treatment, dilution, filtration, preservation or storage bears a contamination risk or may otherwise modify the sample. Thus the optimal treatment would be transfer from the water sampler into the final analytical sample or reaction container and instantaneous analysis (within about 1 h).

If the water from the sampler can be used without prior filtration, the desired sample volumes for manual analysis (generally about 50 mL) may be poured into the final reaction bottles using the simple tool shown in Fig. 10-2. A hole of about 10 mm diameter is drilled into a 100 mL graduated cylinder of polypropylene just above the desired volume mark. The lower bend of the hole is adjusted (with a file) so that the cylinder in the vertical position retains the desired volume of seawater. The accuracy of 40 mL sample volumes thus collected was found to be better than ±0.2 mL causing analytical errors of less than ±0.5 % (less than the method precision).

Fixed-volume sampling is more accurate when the water is subsampled into calibrated reaction bottles as, *e.g.*, for oxygen, and excess sample water is then displaced by reagent addition. This procedure considerably reduces the contamination risk and generally improves the determination of ammonia.

Subsamples for automated analysis should be taken directly from the water samplers into subsample bottles of the desired volumes, which can be accommodated by the autosampling device of the nutrient analyser.

Nalgene 60 mL polyethylene bottles, which also allow preservation of the samples by deep freezing, fit into several commercial autosamplers (see also Section 10.3.2).

Before the analysis the sample is homogenized (by shaking). The bottle caps are replaced by a seal of polyethylene or aluminium foil, which the sampler needle inlet easily penetrates. The sample is thus protected until required for analysis, and the volume is sufficient for repeated analyses.

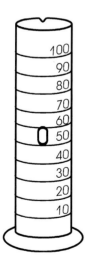

Fig. 10-2. A simple tool for fixed volume sampling.

10.2.2   The matrix

The sample matrix may influence the analytical signal (absorbance). If the matrices of reference standard and sample differ, the calibration may be inaccurate. The reason can be constituents in the sample, which increase or quench the analytical signal or which interfere with the chemical reaction forming the signal-producing compound.

Turbidity has to be removed by filtration or, in case of minor turbidity signals, compensated by using differential spectrophotometers with the chemically untreated water sample as reference. Correction of the results for a measured turbidity absorbance is another option.

The use of differential spectrophotometers also removes interferences by absorbing materials as, *e.g.*, humic substances.

Seasalt also may cause errors in the spectrophotometric signal. These 'salt effects' are either a suppression of the analyte absorbance (*e.g.*, in the determination of silicate and phosphate) by the ions of seawater or an effect of the buffer capacity of seawater (*e.g.*, shifts in the reaction pH interfere with the determination of ammonia).

Salt effects may be ignored if standards and blanks are prepared with nutrient-free or low-nutrient seawater (LNSW) or artificial seawater (ASW). Salinities should be equal to those of the samples. This is the best solution for most analyses of waters where salinity variations are small and a LNSW can be found or prepared. In estuaries or semi-enclosed brackish seas (*e.g.*, the Baltic Sea) salinity variations generally are too large for the us of a single standard salinity LNSW or ASW. In the following we describe and discuss the preparation and quality of the different 'water standards' in some more detail and also outline a procedure for how to compensate for salt effects.

Pure water

Doubly distilled water (DDW) or deionized water (DIW) is usually pure enough for nutrient analysis. Distillation is a time consuming procedure with considerable risk of contamination by gaseous components (N-oxides, ammonia) and silicic acid from the quartz apparatus. Deionized water is easier to prepare, but the exchange resin(s) must be freshly regenerated. On board research vessels the fresh water supply is notoriously contaminated and a rather poor basis for the preparation of pure water. It is better to carry a sufficient amount of water from the pure water supply of the onshore laboratory as feed water for the still or the ion exchange apparatus. Pure water for the analyses should be prepared on board ship no more than 2 days before use either by re-distillation or by ion exchange. A small deionizer cartridge if used to re-purify already demineralized water lasts for months and hundreds of litres, and guarantees the best water quality. Storing pure water is always a risk because glass containers may release silicic acid, and gaseous components pass through the walls of most plastic containers.

Artificial seawater (ASW)

As a matrix for standards and blanks in nutrient analyses three properties of an artificial seawater are important, as they may affect the analytical result:
– ionic strength;
– salinity (density, absorbance, refractive index, *etc.*);
– buffer capacity.

A sodium chloride solution of a g/kg concentration equal to the sample salinity usually satisfies the first and second requirements. Problems encountered with potential impurities in the chemicals used for the preparation of artificial seawater according to the protocol in the Appendix of this book are likely to introduce errors rather than improve the analysis. In flow-analysis of nitrate, nitrite, phosphate and silicate we could not detect analytical differences between using ASW or sodium chloride solutions (*Hansen*, unpublished).

Chemical procedures which are affected by the buffer capacity of the seawater, *e.g.*, determination of ammonia, either require an adjustment of the matrix for standards or the analytical results must be corrected for a salinity error.

If ASW is available with a certified quality, it may be used as standard/blank matrix. For most applications, however, we recommend either the use of LNSW or a salinity adjusted sodium chloride solution and correction of the salinity error according to the salinity effect determined during calibration (see Section 10.2.4).

Low-nutrient seawater (LNSW)

The best matrix for the preparation of standards is a natural seawater without, or low in, nutrients, such as oceanic surface waters in late spring or summer. Silicate concentrations should be lower than 5 μmol/L and nitrate preferably not higher than 1 μmol/L. The water should be filtered through a 10 μm pore size filter to remove particles and stored in the dark for a couple of weeks (see Chapter 2). Traces of ammonia may be removed by sparging the LNSW for several hours with air freed of ammonia by passing it through dilute sulphuric acid.

Correction of salt effects

To compensate for matrix effects either LNSW (preferable), ASW or a sodium chloride solution is recommended with a salinity equal to the mean of the (expected) sample salinities (*Hansen*, 1994). For all components showing salt effects for the expected salinity range, correction functions have to be established. In most cases simple linear corrections are sufficient.

The procedure outlined below should be followed to determine the required correction terms for compensating for the salt effects.

Prepare a series of five dilutions of a basic matrix (LNSW, ASW or NaCl) from lowest to highest expected sample salinities. Spike aliquots of the 'zero' matrix with individual or mixed nutrient standard solutions to give five concentrations of each series from 0 to a high standard concentration (just above the expected maximum sample concentration).

Analyse the samples and calibrate against the high-nutrient standard solution with medium salinity. Plot the *deviations* of the resulting concentrations from the nominal (spike) concentrations *versus* the spike concentrations (Fig. 10-3a). The plot should display a series of five straight lines (one regression line per salinity). If the basic matrix contains traces of nutrients, these concentrations or the respective dilutions must be added to the determined and the nominal concentrations. The medium (standard) salinity line is parallel to the abscissa in concentration deviation zero and the higher and lower salinity lines are symmetrically below and above. The plot indicates the linearity of the determination and shows whether the correction for a salt effect can improve the analytical accuracy.

The salinity correction factor F_{salt} is determined as the slope of the regression line of the relative concentration deviations between the high standard in salinity S and medium sali-

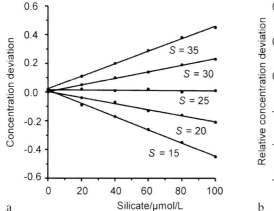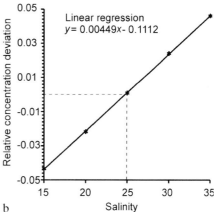

Fig. 10-3. Determination of the salt-effect (a) and correction factor (b) for silicate determination 0–100 μmol/L, salinities 15–35 and zero water salinity of 25. The salinity correction factor (F_{salt}) is 0.0045 (b).

nity *versus* the corresponding salinity. The correction factor may be calculated or determined graphically (Fig.10-3b).

The salinity error of a nutrient determination (N_{unc}) is corrected by a linear correction equation including terms for the differences in the sample salinity S and the standard salinity S_{std} which is used as the blank, wash and standard matrix.

$$N_{cor} = N_{unc} \cdot [1 + F_{salt} \cdot (S - S_{std})]$$

In cases where samples are diluted prior to the analysis (*e.g.*, analysis of pore waters) the resulting salinities have to be estimated or measured to about ± 1.

10.2.3 References and standard materials

The preparation of nutrient standards should not cause any severe problems for a chemical laboratory with standard equipment, *i.e.*, analytical balance, calibrated glassware, pure water and access to analytical grade chemicals. The recipes are given with the reagent recipes for the determination of the respective components (Sections 10.2.5–10.2.19). All primary standards (stock solutions) are prepared with pure water at concentrations of 10 mmol/L. The pure water may be filtered through < 0.2 μm membrane filters to remove bacteria. Polycarbonate is the best material for bottles to store primary and diluted standards. It is gas and vapour tight, does not release any interfering material and is nearly unbreakable. Glass bottles (preferably Duran, Pyrex or equivalent) may also be used except for silicate which must be stored in polycarbonate, polyethylene or polypropylene bottles. Rechecks of 10 mmol/L standards which had been stored in sealed standard glass ampoules for more than 10 years revealed no changes in composition (*Wenck*, unpublished). Even the increase in silicate was almost negligible. It is extremely important to cross-check any new batch of standards, both internally and externally. For this purpose, carefully seal and store aliquots of each batch of standards and compare new standards against several 'old' standards. External checks may be performed against standards from other laboratories, in combination with intercalibration/intercomparison exercises, or against commercial standards (*e.g.*, Sagami Research Centre, Tokyo). The permanent use of commercial standards, which are relatively expensive, is not necessary.

If stored dark and cool (in a refrigerator) in polycarbonate bottles, mixed primary standards of nitrate, nitrite, ammonia, phosphate and silicate in pure water or other matrices are stable for months. Mixed working standards, *i.e.*, dilutions in the range of natural concentrations, can be used but should be prepared freshly at least every day.

10.2.4   Calibration, blank determination and calculation procedures

As previously discussed, all nutrients are quantified by spectrophotometry. Calibration in the context of spectrophotometric analysis means comparison of the sample absorption after chemical reaction with the absorption of a standard (*i.e.*, an artificial sample) of known concentration, which has been treated in exactly the same manner. The reliability of calibrations decreases with increasing concentration differences between sample and standard. Consequently, the best calibration and thus the best analytical results are obtained with exactly matched sample and calibration ranges. Since the methods for calibration, blank determination and calculation of results follow the same procedures for all nutrients discussed here, the different steps to be taken simultaneously are outlined in the following.

Starting the analysis with a new analytical setup requires a full range (initial) calibration including zero (blank) determination and linearity test. The operational calibration performed later on at regular intervals is a reduced two- or three-point calibration.

The general calibration and calculation procedure consists of six steps.

1. Calibration always starts with an estimation of the expected sample concentrations, salinities and possible interfering components. Calculate the required cuvette length based on the expected maximum concentration and the molar absorptivity to give absorbances according to Beer-Lambert's law (generally less than 0.8–1). Preferably only one cuvette size should be used for measuring a sample sequence. Change of cuvette sizes requires re-calibration starting at step 3.
2. Prepare an adequate supply of zero water (ZW), *i.e.*, LNSW, ASW or a sodium chloride solution of medium sample salinity (see Section 10.2.2). If the expected variation of salinity is $< \pm 2$, salinity corrections are generally not required.
3. Prepare a sequence of standards (minimum 5) from zero to a concentration slightly beyond the expected maximum sample concentration by pipetting 1–5 aliquots of a standard stock solution in pure water into calibrated flasks and fill up with ZW. Treat at least three sample aliquots per standard (including ZW as zero standard) as described for the respective analytical method and measure the absorbances (absolute or relative units). In manual analysis use ZW in the reference cell. A plot of the nominal standard (spike) concentrations *versus* absorbances (or the equivalent analogue signal, *e.g.*, voltage, recorder pen deflection) should look like Fig. 10-4.

The plot indicates a linear concentration/absorbance relationship, *i.e.*, calibration has to consider a calibration factor (slope of the regression line, $F = 5.1286$) and a non-zero blank absorbance of the ZW ($A_{zero} = 0.0369$ equivalent to a concentration of $0.0318 \, \mu mol/L$). The zero offset is caused by small nutrient concentrations in the ZW, absorbances of the reagents and a cell blank (A_{cell}) resulting from small optical differences between sample and reference cuvettes (see Section 10.2). Because the absorbance caused by nutrient traces in the ZW is a property of the ZW and the standards, but not the seawater samples, only the blank absorbance ($A_{blank} = A_{reag} + A_{cell}$) has to be corrected for in the determinations.

4. In the manual procedure, the blank absorbance (A_{blank}) is determined by measuring a sample volume of pure water plus reagents (as for sample treatment) against a pure water reference without reagents. In analyses which use acid reagents, the pure water in the reference cell is acidified to the sample pH after reagent addition. This compensates for a possible absorbance caused by turbidity, which is usually reduced by acid addition. In flow-analysis, the absorbance difference is determined by a pure water sample with and without reagent addition (pure water instead of reagents).

Reagent absorbances should generally be negligible, *i.e.*, non-zero absorbances of ZW are mostly due to small amounts of nutrients in the ZW, in particular if LNSW is used for the preparation of blanks and standards.

5. The calibration and the blank determination yield the terms required to calculate sample concentrations (C) from the measured sample absorbances (A_m) with the equation

$$C = F \cdot (A_m - A_{cell})$$

where F is the factor (slope) of the calibration curve.

Example:
The equation of the calibration curve (Fig. 10-4) is

$$C = 5.1286 \cdot A - 0.0318 \ (F = 5.1286)$$

Assume a (rather high) blank absorbance of 0.005, which, multiplied by F is equivalent to a concentration of 0.0256 μmol/L and is included in the constant term 0.0318. Subtraction (0.0318 − 0.0256) gives a ZW concentration of the corresponding nutrient of 0.0062 μmol/L.

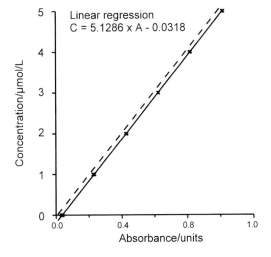

Linear regression
C = 5.1286 x A - 0.0318

Fig. 10-4. Calibration curve of nominal standard concentrations (C) *versus* absorbances (A). Marks and solid line refer to the measurements (triplicates) and the linear regression. The dashed line is the C/A relationship corrected for nutrient traces in the zero water.

The calculation formula for sample concentrations from absorbances in the example is

$$C = 5.1286 \cdot (A_m - 0.005)$$

or, using the blank concentration instead of blank absorbance

$$C = 5.1286 \cdot A_m - 0.0062$$

This formula is only valid for one specific analytical setup, *i.e.*, spectrophotometer, cuvette and cuvette length, and reagents used for the calibration procedure. If the concentration range of the samples requires the use of different cuvette sizes, individual calibrations must be performed (steps 3–5) for each concentration range and cuvette size.

Note that the precision of figures in the calculations is only reasonable up to a maximum of four decimal places.

6. Without changes to the analytical setup and reagents only occasional re-checks of the factor are required. In flow methods every 1–2 h (20–40 samples) an on-line two-point calibration is necessary. Use ZW as the low standard and the highest standard of the dilution sequence as the high standard. The low standard concentration is the nutrient content of the ZW as determined from the difference of the 'zero' absorbance (offset) minus the reagent absorbance divided by the regression factor. In our example (Fig. 10-4) the ZW concentration is 0.006 μmol/L. The high standard is the nominal concentration plus the ZW concentration (5 + 0.006 = 5.006 μmol/L).

The on-line calibration procedure of flow-analyses (baseline and sensitivity drift compensation) is described in Section 10.3.6.

10.2.5   Determination of dissolved inorganic phosphate

10.2.5.1   Principle of the method

All methods for the determination of inorganic phosphate in seawater are based on the reaction of the ions with an acidified molybdate reagent to yield a phosphomolybdate heteropoly acid, which is then reduced to a highly coloured blue compound. In early work, tin(II) chloride was used as the reductant in flow-analysis (*Hager et al.*, 1968). However, this reductant has several disadvantages, including the appreciable temperature dependence of the reduction rate and the pronounced salt error.

Hydrazine (*Bernhard and Wilhelms*, 1967) or ascorbic acid are the common reductants in the flow-analysis of phosphate. The hydrazine method seems to reduce dye coating of the flow cell window but is equivalent to the ascorbic acid method described below. In continuous flow analysers (CFA) the same manifold can be used for both methods.

Ascorbic acid was first used in the analysis of seawater by *Greenfield and Kalber* (1954), but present methods essentially follow the procedure by *Murphy and Riley* (1962), in which trivalent antimony ions are components of a single solution reagent. This reagent reacts rapidly with phosphate ions to yield a bluish complex containing antimony in a 1 : 1 atomic ratio to phosphorus. The method described here is a modification of the *Murphy and Riley* procedure as given by *Koroleff* in the first edition of this book. Two solutions are used instead of a single reagent: the first contains sulphuric acid, ammonium molybdate and anti-

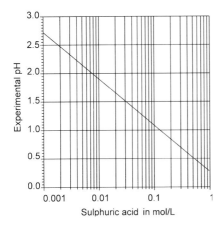

Fig. 10-5. Experimental pH values of sulphuric acid concentrations at 25 °C.

mony ions, and the second solution contains the ascorbic acid. Furthermore, the final concentrations of acid and molybdate are decreased from 0.2 to 0.1 mmol/L and from 0.78 to 0.39 mmol/L, respectively. A lower final acidity is also used by *Strickland and Parsons* (1968). Two separate reagent solutions are more stable and the compensation for turbidity is easier.

To obtain rapid colour development and to depress the interference of silicate, it is of fundamental importance that the final reaction pH is ≤ 1, and that the molar ratio of sulphuric acid to molybdate is 230–330. *Gripenberg* (1929) drew attention to these facts and a reinvestigation was carried out by *Koroleff* (1968).

As the pH of the reaction is mainly determined by the sulphuric acid concentration, the relationship between acid concentrations from 0.001 to 1 mol/L and pH at 25 °C is depicted in Fig. 10-5.

10.2.5.2   Range and precision

The molar absorptivity of the partly reduced heteropoly acid is 22 700 at 880 nm, which means that the net absorbance for a 1.0 μmol/L phosphate sample is 0.227 measured in a 10 cm cuvette. With a good spectrophotometer an absorbance of 0.002 can be read, and the smallest concentration of phosphate-phosphorus that can be determined directly thus is 0.01 μmol/L. The absorbances are a linear function of the concentrations up to 28 μmol/L, equal to a net absorbance of approximately 0.630 in a 1 cm cell.

A spectrophotometer wavelength of 660–700 nm can be used for the determination of high phosphate concentrations, or if required by the spectrophotometer design (*e.g.*, an LED photometer). This results in a loss of sensitivity of about 30 % (see Section 10.3.4.4, Fig. 10-24).

In the report of the ICES/SCOR nutrient calibration, *Koroleff and Palmork* (1977) give a relative precision for phosphate of ± 15 % at the low level (0.2 μmol/L), at the medium level (0.9 μmol/L) ±5% and at the high level (2.8 μmol/L) ±2 %. At the Baltic Intercalibration Workshop (Kiel, 1977) nearly identical results were obtained with natural samples having salinities of from 5 to 34. The Fifth ICES Intercomparison Exercise (*Aminot and Kirkwood*,

1995) confirmed these results. Standard deviations of 0.03–0.09 μmol/L were found for phosphate concentrations up to 3 μmol/L. This is considerably higher than the reproducibility in the laboratory of 0.02–0.03 μmol/L (*Riley et al.*, 1972; *Strickland and Parsons*, 1968).

10.2.5.3 Interferences

The greatest advantage of using ascorbic acid as the reductant is that the blue phosphomolybdic complex formed is stable for hours, and that the colour intensity is not influenced by variations in salinity. However, the influence of some other ions occurring in natural waters must be considered:

Silicate. As stated by *Murphy* and Riley (1962) there is no effect of silicon up to 350 μmol/L. *Koroleff* (personal communication) has found this to be valid only if the absorbance is measured after about 5 min. During prolonged standing, a blue silicomolybdic complex is gradually formed. The increase in colour intensity with time is fairly linear during the first hour; thereafter the increase is small. The effect of silicate depends also on the acidity of the reaction; the increase in colour is somewhat smaller in 0.2 than in 0.1 mol/L sulphuric acid. If the colour is measured after 10 min, there is practically no interference caused by silicate at natural seawater concentrations (up to 200 μmol/L). If measurements are performed after about half an hour, 200 μmol/L silicate will increase the net absorbance by 0.003 in a 10 cm cell, which still is almost negligible.

Arsenate. Arsenate ions produce a similar colour to phosphate because they also form heteropoly acids. However, since their concentration in natural seawater is only 0.01–0.03 μmol/L, they will not interfere seriously in the determination of phosphate unless high precision measurements of low phosphate concentrations are performed.

The formation of the blue arsenomolybdic complex has been studied by *Koroleff* (unpublished), who found that in the absence of phosphate after a reaction time of 30 min, less than 24 % (both in pure water and seawater) of the arsenate originally present had reacted. However, the reaction speed was catalysed by phosphate ions, and even at 0.5 μmol/L phosphate all the arsenate may react. The effect of arsenic can be eliminated by reduction to arsenite as given in Section 10.2.6.

Hydrogen sulphide. The deep water in stagnant basins is often anoxic and contains dissolved hydrogen sulphide in amounts of up to 600 μmol/L (the Black Sea). In some low-salinity estuaries the concentration of sulphide may exceed 4 mmol/L. When the acid molybdate reagent is added to such waters, colloidal sulphur is formed resulting in a greenish colour. *Koroleff* (unpublished) has investigated this effect and found that up to *ca.* 60 μmol/L sulphide does not interfere with phosphate determinations. Studies by *de Jonge and Villerius* (1980) have confirmed this for phosphorus values lower than 12 μmol/L. Since high sulphide concentrations are most often associated with high phosphate content, the effect of sulphide can be eliminated by simple dilution of the sample with distilled water. If the phosphate concentration is so low that dilution is not advisable, the sulphide ions should be oxidized with bromine water (reagent 2) added to an acidified sample (0.2 mL of 4.5 mol/L acid to 100 mL of sample). Excess of bromine is removed by sparging with air. The hydrogen sulphide may also be driven off by acidifying the sample and passing a stream of nitrogen through the sample for about 15 min.

Other interfering compounds. Other compounds interfering with the determination of phosphate are not found in seawater. However, in certain chemical waste waters a copper content exceeding $160\,\mu$mol/L has been found to decrease the colour intensity; also chromium(*VI*) in amounts greater than $40\,\mu$mol/L and chromium(*III*) in a concentration exceeding $600\,\mu$mol/L have been found to interfere. By adding the phosphate reagents in the reverse order, phosphate may be determined in the presence of up to about $770\,\mu$mol/L of Cr^{6+}. The effect of iron ions causes a small increase in colour intensity of about 1 % per $180\,\mu$mol/L.

Nitrate concentrations in waste water exceeding 2 mmol/L can also affect the determination of phosphate. The interference from fluoride is more serious, and this ion in concentrations > 10 mmol/L totally inhibits the colour development. However, up to 1.6 mmol/L F^- has no effect.

In pulp and paper mill effluents, according to *Buchanan and Easty* (1973), lignin sulphonates at concentrations greater than 200 mg/L may decrease the colour.

10.2.5.4 Reagents

1. *Sulphuric acid, 4.5 mol/L*: Carefully add 250 mL of concentrated sulphuric acid ($d = 1.84$ g/mL) to 750 mL of pure water. Allow to cool and dilute to 1 L. Store in a polyethylene bottle.
2. *Bromine water*: Add 0.9 mL of bromine, Br_2 ($d = 3.11$ g/mL) to 100 mL of water. Store the saturated solution in the dark in an amber glass bottle.
3. *Phosphate standard solution*: Potassium dihydrogen phosphate, KH_2PO_4 (relative molecular mass 136.09), is dried in an oven at $110\,°C$, then placed in a desiccator. Exactly 136.09 mg are dissolved in pure water to which has been added 0.2 mL of sulphuric acid (reagent 1) and made up to 100 mL. Stored cold in a glass bottle the solution is stable for months. This standard stock solution contains 10 mmol/L phosphate .
4. *Ascorbic acid solution (manual method)*: Dissolve 10 g of ascorbic acid, $C_6H_8O_6$ in 50 mL of pure water, then add 50 mL of sulphuric acid (reagent 1). Stored dark in a brown bottle at $< 8\,°C$, the reagent is stable for several weeks as long as it remains colourless.
5. *Ascorbic acid solution (flow-analysis)*: Dissolve 6 g of ascorbic acid, $C_6H_8O_6$, in pure water. Add 20 mL of glycerol, 50 mL of methanol and 2 mL of a surfactant (Levor-IV or Aerosol-22 diluted 1:10 with pure water). Make up to 2 L with pure water. Stored dark in a brown bottle at $< 8\,°C$, the reagent is stable for about one month as long as it remains colourless.
6. *Mixed reagent (manual method)*: Dissolve 12.5 g of ammonium heptamolybdate tetrahydrate, $(NH_4)_6Mo_7O_{24} \cdot 4H_2O$, in 125 mL of pure water. Also dissolve 0.5 g of potassium antimony tartrate, $K(SbO)C_4H_4O_6$, (with or without $\frac{1}{2}H_2O$), in 20 mL of pure water. Add the molybdate solution to 350 mL of sulphuric acid (reagent 1) while continuously stirring. Add the tartrate solution and mix well. Store in a laboratory glass bottle. This mixed reagent is stable for several months.
7. *Mixed reagent (flow-analysis)*: Dissolve 10 g of ammonium heptamolybdate tetrahydrate, $(NH_4)_6Mo_7O_{24} \cdot 4H_2O$, in 600 mL of pure water. Carefully add 120 mL concentrated of sulphuric acid ($d = 1.84$ g/mL) and mix. Add 20 mL of a solution of 2.3 g potassium antimony tartrate, $K(SbO)C_4H_4O_6 \cdot nH_2O$ in 100 mL of pure water. Mix well and make up to 2 L. This mixed reagent is stable for several months.

10.2.5.5   Analytical procedures

Preparation

Clean all glass- and plastic-ware with a brush and a detergent solution followed by a thorough rinse with water. The detergent must be stringently tested for phosphate. Glassware used for the colour development should be reserved exclusively for phosphate determination.

The blue phosphomolybdic complex, being a colloid, has a tendency to stick as a thin film on the walls of a cuvette. This film can be removed by rinsing with dilute sodium or potassium hydroxide solutions followed by a thorough rinse with water.

No salt correction is required.

Manual determination

Add 1 mL of ascorbic acid (reagent 4) and 1 mL of mixed reagent (reagent 6) to 50 mL of sample. Mix thoroughly after each reagent addition.

The absorbance of the blue phosphorus complex reaches a maximum in a few minutes and stays almost constant for many hours. However, the absorbances should be read within 10–30 min, using a wavelength of 880 nm. This determination, on an unfiltered sample, gives the amount of dissolved inorganic phosphate ions in true solution and probably also includes a small fraction of those ions that are adsorbed onto particles and subsequently dissolved by the acid in the mixed reagent. The latter fraction is not included when a filtered sample is analysed.

Flow-analysis (for operation see Section 10.3)

The manifold in Fig. 10-6 requires a sample time of about 2.5 min (4 mL sample) to provide a steady-state peak. Heating of the 24 turns reaction coil to 37 °C accelerates the reaction independent of the environmental temperature. This allows reduction of the reaction coil to 19 turns or the sample time to about 2 min (3.2 mL sample volume).

Using microtube systems, all pump tubes may be reduced to half their flow rates. The air injection has to be adjusted to provide sufficient liquid segment volumes for the spectrophotometric operation.

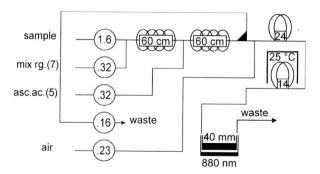

Fig. 10-6. Schematic CFA manifold for dissolved inorganic phosphate.

10.2.6   Determination of dissolved inorganic phosphate in the presence of arsenic

Arsenic occurs in seawater at very low concentrations, not exceeding $0.05 \, \mu$mol/L in unpolluted waters. Since the arsenate ion is isomorphic with the phosphate ion, it forms similar yellow and blue molybdate complexes and will accordingly be included in the phosphate determination.

10.2.6.1   Principle of the method

For the determination of phosphate in the presence of arsenic, the arsenate ion is reduced with thiosulphate to arsenite which does not form complexes with molybdate. Arsenate up to $3 \, \mu$mol/L is quantitatively reduced within 15 min. The blue phosphomolybdic complex is developed using the reagents given in Section 10.2.5.4. Calibration must be performed as described because thiosulphate reduces the colour of the blue phosphorus complex by *ca.* 10 %. Range and precision are the same as for the main procedure.

10.2.6.2   Reagents

Sodium thiosulphate solution: Dissolve 2.4 g of sodium thiosulphate pentahydrate, $Na_2S_2O_3 \cdot 5H_2O$, in 100 mL of pure water. Add and dissolve *ca.* 50 mg of Na_2CO_3 as a preservative. Stored cold in an amber glass bottle the reagent is stable for several weeks.

10.2.6.3   Analytical procedures

Calibration and measurement are performed as in Section 10.2.5.5. Add 1 mL of ascorbic acid (reagent 4, Section 10.2.5.4) and 0.5 mL of the thiosulphate solution. Mix and wait for 15 ± 1 min before adding 1 mL of the mixed reagent (reagent 6, Section 10.2.5.4). Measure the absorbance after 15–20 min. In pure water samples colloidal sulphur may appear after about 25 min, seawater samples remain clear for several hours.

10.2.7   Determination of dissolved inorganic phosphate by an extraction procedure (high-sensitivity method)

The detection limit of the main procedure (Section 10.2.5) for the determination of dissolved phosphate in seawater is about 0.01–$0.02 \, \mu$mol/L, a sensitivity that is satisfactory for most marine investigations. In special biological studies, however, the determination of phosphate below this level is sometimes necessary. Increased sensitivity can be obtained by solvent extraction of the phosphomolybdic acid, followed by reduction in the organic phase or by extraction of the already reduced blue complex. *Stephens* (1963) combined the single solution method by *Murphy and Riley* (1962) with solvent extraction. This method has been adopted by *Strickland and Parsons* (1968).

10.2.7.1   Principle of the method

This is a modification of the procedure originally described by *Stephens* (1963), using the same reagents as in the normal procedure (Section 10.2.5). Isobutanol is replaced by n-hexanol, since the former is somewhat more soluble in water. Also the amount of solvent is increased to an amount sufficient to fill an ordinary 10 cm cuvette.

10.2.7.2   Range and precision

Using 350 mL of sample, the smallest amount of phosphate phosphorus that may be detected with certainty is about 0.004 μmol/L, corresponding to a net absorbance of about 0.005 in a 10 cm cell. Beer-Lambert's law is valid between 0 and 0.3 μmol/L.

10.2.7.3   Interferences

Silicate. As the blue phosphomolybdic complex is extracted after 10 min, silicate does not interfere.

Arsenate. In samples with a phosphate content of $<0.5 \mu$mol/L, usually but traces of arsenate are present, and of that amount only about 20 % reacts within about 10 min.

Hydrogen sulphide. Samples to be analysed according to this procedure are very seldom anoxic. If sulphide ions are present, a pre-treatment with bromine (see Section 10.2.5.4) is recommended.

10.2.7.4   Reagents

In addition to the reagents listed in Section 10.2.5.4, n-hexanol and absolute ethanol are required.

10.2.7.5   Analytical procedure

Measure 350 mL of sample or standard with a graduated cylinder into a clean 500 mL separatory funnel. Add 7 mL of the mixed reagent (reagent 6, Section 10.2.5.4) and 7 mL of the ascorbic acid (reagent 4, Section 10.2.5.4). Mix between additions and allow to stand for 10 min. Add from a graduated cylinder 40 mL of n-hexanol and shake vigorously for at least 1 min. Allow to stand for a few minutes and drain off the lower aqueous phase. Rotate the funnel once more and remove the small volume of water still in the funnel. Run 35 mL of the organic solvent into a dry 50 mL graduated cylinder. Add 1 mL of absolute ethanol to the cylinder to dissolve micro-droplets of water. Measure the absorbance at 690 nm against a reference of n-hexanol.

10.2.8. Determination of nitrite

10.2.8.1 Principle of the method

The standard method for the determination of nitrite in seawater is based on the reaction of nitrite with an aromatic amine leading to the formation of a diazonium compound which couples with a second aromatic amine. The product is an azo dye which is quantified by spectrophotometry. Early methods were modifications of the Griess-Ilosvay procedure where the nitrite diazotized with sulphamic acid. The diazonium salt then coupled with 1-naphthylamine. The method described below is generally accepted for seawater analysis and is based on the method proposed by *Shinn* (1941) and adapted for seawater by *Bendschneider and Robinson* (1952). This method is very sensitive and is unaffected by the presence of other constituents normally occurring in seawater. Sulphanilamide hydrochloride is used as the amino compound, which after diazotization is coupled with *N*-(1-naphthyl)-ethylenediamine dihydrochloride.

The reaction leading to the formation of the azo dye can be formulated as follows:

Formation of the diazonium ion.

Coupling with *N*-(1-naphthyl)-ethylenediamine.

The amount of the azo dye formed is proportional to the initial concentration of nitrite over a wide range of concentrations (0–10 μmol/L).

10.2.8.2 Range and precision

The molar absorptivity of the azo dye at 540 nm is very high (about 46 000) making the determination of nitrite one of the most sensitive spectrophotometric methods.

Beer-Lambert's law is obeyed up to $10\,\mu mol/L$ with this method, but sample concentrations exceeding $3\,\mu mol/L$ should be diluted with the respective water matrix because of the high absorptivity (see Section 10.2.2).

The precision of the method is $\pm 0.02\,\mu mol/L$. With the appropriate analytical setup and sample pretreatment the sensitivity, however, allows the detection of much lower concentrations. We have tested the flow-analysis version of this method and found that a well calibrated system can analyse spiked natural (filtered) seawater samples from 0 to 100 nmol/L with a resolution of 1 nmol/L and a reproducibility of ± 3 nmol/L (*Hansen and Johannsen, unpublished*).

10.2.8.3 Interferences

At 540 nm light scattering in a turbid sample may cause considerable absorption and thus a considerable systematic error especially for low concentrations. Filtration through cellulose acetate filters of $2–5\,\mu m$ generally removes the turbidity. However, the required pore size of the filter may have to be less than $2\,\mu m$ (depending on the type of turbidity). In flow-analysis the filter can be connected to the entrance of the sample tube.

Intercalibration tests have demonstrated that impurity of the nitrite standard or partial decomposition of the standard solutions are the most likely sources for deviations between different laboratories (*Koroleff and Palmork*, 1977).

Sulphide ions have been reported to interfere with the determination of nitrite. Thus, when hydrogen sulphide is suspected to be present in a sample, the gas should be expelled with nitrogen after the addition of the acid sulphanilamide reagent.

10.2.8.4 Reagents

1. *Sulphanilamide (manual method)*: 10 g of crystalline sulphanilamide are dissolved in 100 mL of concentrated hydrochloric acid in about 600 mL of pure water. (Moderate heating accelerates the dissolution.) After cooling, the solution is made up to 1 L with pure water. Store in the dark at $<8\,°C$. The reagent is stable for at least one month.
2. *Sulphanilamide (flow-analysis)*: 5 g of sulphanilamide are dissolved in 72 mL of hydrochloric acid (25 %) in 300 ml of pure water and made up to 1 L.
   Store in the dark at $<8\,°C$. The reagent is stable for at least one month.
3. *N-(1-naphthyl)-ethylenediamine dihydrochloride* (NED, *manual method*): 0.5 g of the amine dihydrochloride is dissolved in 500 mL of pure water. The solution should be stored in a brown bottle at $<8\,°C$. The reagent is stable for more than a month and can be used until a brown discolouration occurs.
4. N-*(1-naphthyl)-ethylenediamine dihydrochloride (*NED, *flow-analysis*)*: 0.5 g of the amine dihydrochloride is dissolved in 1 L of pure water together with 1 mL of a surfactant (Brij 35, 10 %). Use of another surfactant might cause problems and is not recommended. (**Warning**: Do not use Levor-IV here, because it reacts with the amine.) Store in the dark at $<8°C$. The reagent is stable for at least one month.
5. *Nitrite standard solution*: Anhydrous sodium nitrite ($NaNO_2$) is dried at $100\,°C$ for 1 h and 0.690 g is dissolved in 1 L of pure water. The solution contains 10 mmol/L of nitrite and should be stored cool and dark (a refrigerator is not required but is preferable).

Commercially available sodium nitrite, even analytical grade, may contain $<100\%$ $NaNO_2$. Aged solid reagents should, therefore, not be used for the preparation of the nitrite standard.

10.2.8.5 Analytical procedures

Manual determination

High bacterial activity may rapidly produce nitrite in the sample. The determination of nitrite should, therefore, be carried out without delay, *i.e.*, the reagents should be added to the sample within 30 min of subsampling.

A 50 mL volume of sample, with or without prior filtration, is transferred into a reaction bottle (60–100 mL volumes) and 1 mL of sulphanilamide (reagent 1) is added. After mixing and about 1 min reaction time, 1 mL of aromatic amine (reagent 3) is added. The flask is shaken and the azo dye is allowed to develop for at least 15 min. The absorbance is measured in 1–5 cm cuvettes at 540 nm against a reference of the respective ZW within 1 h after the addition of the reagents. The dye should not be exposed to bright daylight. Calculate the concentrations as described in Section 10.2.4.

For precision measurements it is necessary to correct the absorbance of the sample for turbidity (even if samples have been filtered). This is done by adding the sulphanilamide reagent to 50 mL of the sample. It is necessary to do this because, for example, some of the turbidity may be caused by calcareous particles that are dissolved by the acid in the reagent and, thus, would not contribute any further to the turbidity. The turbidity absorbance (A_T) is measured using the same cuvette size as for the samples against ZW.

Calculate the concentration from the absorbances of sample, turbidity and reagents and the established calibration factor (see Section 10.2.4) by

$$C = F \cdot (A_s - A_T - A_{reag})$$

No salt effect has to be considered.

Flow-analysis

The manifold in Fig. 10-7 requires a sample time of only 1 min (1.6 mL sample) to produce a steady state plateau. The 25 °C coil could be reduced to about seven turns. However, economic reasons suggest the use of only one type of jacketed coil (the standard is 14 turns).

Microtubing systems allow 50 % flow rates for the pump tubes. Mixing is considerably reduced and thus sample times of less than 1 min may be achieved (about 800 μl sample volume).

The manifold can be simplified if one mixed reagent is used (0.32 mL/min) instead of two. The mixed reagent is prepared from 1:1 volumes of sulphanilamide (SA, reagent 2) and *N*-(1-naphthyl)-ethylendiamine (NED, reagent 4). The mixed reagent, if stored in the dark and at <8 °C is stable for about 5 d.

The filter at the sample inlet is optional and only required in cases of considerable sample turbidities. A pore size of 1–5 μm and a filter area of about 5 cm^2 is required. Most filter types and materials are acceptable, *e.g.*, Sartorius Minisart SM 17593 or 17594 cellulose acetate syringe filters (see also Chapter 2).

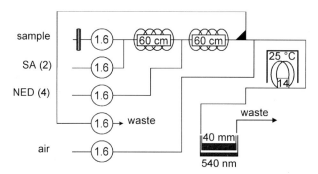

Fig. 10-7. Schematic CFA manifold for nitrite.

The precision of the automated method equals that of the manual method, *i.e.*, $\pm 0.02\,\mu$mol/L.

10.2.9   Determination of nitrate

10.2.9.1   Principle of the method

Five different methods have been used for the determination of nitrate in natural waters. Most of these are either not sensitive enough to allow determination of the low concentrations usually found in surface waters, or they suffer from serious interference by other constituents present in seawater, or a time-consuming slow reaction step is involved in the method.

The determination by means of nitration of an organic compound such as 2,6-xylenol (*Hartley and Asai*, 1963) is subject to interference from large amounts of chloride in seawater and is not sensitive enough.

A polarographic method based on the catalytic reduction of nitrate in the presence of uranyl ions (*Barnes*, 1959) is not suitable for ship-board analysis, even if a rapidly dropping mercury electrode is applied (*Grasshoff*, unpublished).

Some organic compounds form coloured products if they are oxidized by nitrate ions in acid solution. Examples are brucine (Jenkins *and Medsker*, 1964), diphenylbenzidine (*Atkins*, 1954), reduced strychnidine (*Dal Pont*, 1962), diphenylamine (*Isaeva*, 1958) and resorcinol (*Costa*, 1951). However, no reliable trouble-free method has been developed based on this technique. Another spectrophotometric method based on the formation of nitrosyl chloride in seawater containing 50 % concentrated sulphuric acid has been developed by *Armstrong* (1963), but this method cannot be recommended because of the difficulties connected with handling concentrated sulphuric acid solutions at sea, and because it is relatively insensitive.

Specific ion electrodes have been used successfully in the determination of nitrate in fresh water (*Langmuir and Jacobson*, 1970) but their application to the determination of nitrate in seawater is limited by interferences from most common anions, especially from chloride. For a detailed discussion of the nitrate electrode see *Whitfield and Jagner* (1981).

The most sensitive and generally applied methods for the determination of nitrate in seawater are based on the reduction of nitrate to nitrite, which is then determined *via* the formation of an azo dye. The reduction can be performed as a homogeneous or a heterogeneous reaction. A method based on the homogeneous reduction of nitrate by hydrazine in the presence of copper ions as catalyst has been proposed by *Mullin and Riley* (1955b) but the reaction is time-consuming, not quantitative and highly dependent on external conditions (see also *Dal Pont et al.*, 1963).

The reduction of nitrate in a heterogeneous system using either metallic zinc or cadmium in granules, dust or filings has been used by several workers (*Bray*, 1945; *Føyn*, 1951; *Price and Priddy*, 1959; *Chow and Johnson*, 1962; *Skougstad and Fishman*, 1963; *Morris and Riley*, 1963; *Grasshoff*, 1964; *Wood et al.* 1967).

Precision and accuracy of all reduction methods also are highly dependent on the experimental conditions, and many of the heterogeneous reduction methods cannot be recommended for routine work at sea.

The method described here has been developed from an automated procedure (*Grasshoff*, 1970) and has proved to be both reliable and useful for work at sea.

Nitrate prevails as an ion in seawater and is neither bound nor complexed (see Section 10.1.2). For the determination, it is reduced in a reductor (see Fig. 10-8) filled with copper-coated cadmium granules. The conditions of the reduction are adjusted so that nitrate is almost quantitatively converted into nitrite and not reduced further:

$$NO_3^- + Me(s) + 2H^+ \rightarrow NO_2^- + Me^{++} + H_2O \qquad (10\text{-}1)$$

The yield of the reduction of nitrate to nitrite depends upon the metal used in the reductor, on the pH of the solution and on the activity of the metal surface. Reaction solutions that are too alkaline or inactive metal surfaces result in only partial reduction; solutions that are too acidic, the use of highly electronegative metals or too active surfaces result in the nitrate being reduced further than the nitrite step. In both cases the analysis will result in nitrate values that are too low. The standard electromotive forces E_0 of the two reactions

$$NO_3^- + 3H^+ + 2e^- \rightarrow HNO_2 + H_2O \; (E_0 = 0.94 \text{ V}) \qquad (10\text{-}2)$$

and

$$NO_3^- + 4H^+ + 3e^- \rightarrow NO + 2H_2O \; (E_0 = 0.97 \text{ V})$$

are similar. In a neutral or alkaline solution the standard EMF is

$$NO_3^- + H_2O + 2e^- \rightarrow NO_2^- + 2OH^- \; (E_0 = 0.015 \text{ V}) \qquad (10\text{-}3)$$

Copper-coated cadmium filings or fine granules are very suitable for the heterogeneous reduction, but in a neutral or weakly alkaline medium the cadmium ions formed during the reduction of the nitrate react with hydroxyl ions and form a precipitate. Furthermore, the reduction potential needed for the reduction of nitrate to nitrite depends on the hydrogen ion activity according to the Nernst equation:

$$E = E_0 \frac{RT}{nF} \cdot \ln \frac{a_{NO_2^-} \cdot a^2_{OH^-}}{a_{NO_3^-}} \qquad (10\text{-}4)$$

This implies that the pH is changed if the solution is not buffered, especially in the vicinity of the metal surfaces. Because the buffer capacity of seawater is not sufficient ammonium chloride is added both as a complexant and as a buffer:

$$2NH_4^+ \leftrightarrow 2NH_3 + 2H^+ \tag{10-5}$$

$$Cd^{++} + 2NH_3 \rightarrow [Cd(NH_3)_2]^{++} \tag{10-6}$$

As can be seen from Eqs. (10-3) and (10-5), the two hydroxyl ions formed are balanced and the ammonia is bound in the diamine complex (10-6).

The time during which the nitrate (nitrite) is in contact with the metal must, however, still be controlled within certain limits.

Under controlled conditions, the nitrite originally present in the seawater sample passes the reductor without further reduction. This amount of nitrite must be considered and subtracted from the total amount of nitrate measured. After the reduction procedure, the nitrate determination is continued exactly as for the determination of nitrite in seawater.

10.2.9.2   Range and precision

With the same reductor the maximum deviation between duplicate samples (including reduction and determination of the nitrite) is $\pm 0.1 \, \mu mol/L$ in the range 0–5 $\mu mol/L$, ± 0.2 in the range 5–10 $\mu mol/L$, and ± 0.5 in the higher concentration range. If different reductors are used, the agreement strongly depends on the reductor efficiencies. The agreement should be less than double the values given here for the use of the same reductor.

For highest accuracy, the nitrate analysis should be performed without delay after sampling. If no systematic sampling errors are involved, the c.v. (coefficient of variation; see Chapter 1) of the nitrate determinations can be assumed to be $\pm 3 \%$ in the range 0–10 $\mu mol/L$.

10.2.9.3   Interferences

The determination of nitrate in seawater is not subject to interferences. It has been claimed that nitrate might occur together with small amounts of hydrogen sulphide. For thermodynamic reasons this can only be true for waters from the transition layers between oxic and anoxic environments where intense vertical turbulent mixing processes occur. In this case the hydrogen sulphide is precipitated on top of the reductor as copper or cadmium sulphide and does not interfere in the nitrate analysis. Nitrate values observed together with hydrogen sulphide should be interpreted with care.

10.2.9.4 Reagents

1. *Ammonium chloride buffer (manual method)*: 10 g of ammonium chloride are dissolved in 1 L of pure water. The pH is adjusted to 8.5 with about 1.5 mL of 25 % ammonia solution.
2. *Ammonium chloride buffer (flow-analysis)*: 75 g of ammonium chloride are dissolved in 5 L of pure water. The solution is adjusted to pH 8.5 with about 12 mL of 25 % ammonia solution. The buffer is stable.
3. *Sulphanilamide solution (manual method)*: 10 g of sulphanilamide are dissolved in a mixture of 100 mL of concentrated hydrochloric acid and about 500 mL of pure water and made up to 1 L with pure water. The solution is stable for several months.
4. *Sulphanilamide (flow-analysis)*: 5 g of sulphanilamide and 50 mL of concentrated hydrochloric acid are transferred into 500 mL of pure water, dissolved and made up to 1 L with pure water. The reagent is stable for at least one month.
5. N-*(1-naphthyl)-ethylenediamine dihydrochloride (NED, manual method)*: 0.5 g of the amine dihydrochloride is dissolved in 500 mL of pure water. The solution should be stored in a brown glass bottle and is usually stable for about one month.
6. N-*(1-naphthyl)-ethylendiamine dihydrochloride (NED, flow-analysis)*: 0.5 g of the amine dihydrochloride is dissolved in 1 L of pure water together with 1 mL of a surfactant (Brij 35, 10 %). Use of another surfactant might cause problems and is not recommended. Stored in a brown bottle the reagent is stable for about one month.
7. *Reductor filling*: Commercially available granulated cadmium (*e.g.*, Merck) is sieved and the fraction between 40 and 60 mesh is retained and used.
8. *Copper sulphate*: 1 g of copper sulphate pentahydrate ($CuSO_4 \cdot 5H_2O$) is dissolved in about 100 mL of pure water.
9. *Nitrate standard solution*: 1.011 g of dry potassium nitrate (KNO_3) is dissolved and made up to 1 L with pure water. The stock solution contains 10 mmol/L nitrate and is stable.

The sulphanilamide reagent and the NED are the same as the reagents used to determine nitrite. In multi-channel flow-systems common reservoirs may be used for the corresponding reagents. As in the nitrite determination the sulphanilamide and NED reagents may be combined (1 : 1) to a mixed reagent and applied as such. The mixed reagent should be stored in the dark and at < 8 °C for a maximum of 5 d.

The absorbance is linear up to 20 μmol/L.

10.2.9.5 Preparation of the reductor

The reductor column is a glass tube of 10–25 cm length and 3–5 mm inner diameter. It can be either straight, U-shaped or formed otherwise (Fig. 10-8) depending on the mechanical mounting and flow requirements.

The required reactor path length (contact path of cadmium surface and sample) depends on the sample volumes and allowed reduction times. For example, 20 cm of a 4 mm i.d. (internal diameter) reductor allow flow rates of about 6–8 mL/min, *i.e.*, 50 mL of sample are reduced in 6–8 min. The same reductor may be used for manual and flow-analysis. In flow-analysis with flow rates of about 2 mL/min reduction is close to 100 % with a reductor 10 cm long (4 mm i.d.). Excessive length increases the flow resistance and the risk of blocking and

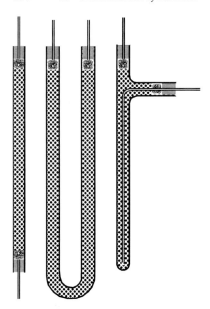

Fig. 10-8. Reductor types for manual and flow analyses.

'over reduction'. The reductor reaction consumes some cadmium and it has to be repacked as soon as loose packing, gaps or channels are visible. A short reductor, more frequently renewed, is preferable to a long one which might last longer.

Vertical or sloped mounting of the reactor, preferably with the inlet at the lower end, avoids bubbles being trapped during operation and provides optimal flow.

The cadmium granules are freed from adhering cadmium dust and oxides by washing with about 2 mol/L HCl and subsequently with pure water.

A small plug of glass wool is inserted in the lower end of the reductor, the tube is closed and filled with water. The cadmium granules are then slowly poured through a funnel into the reductor. Dense packing is supported by vibrating or gently tapping the reductor with a pencil. A glass wool plug is inserted in the upper end of the tube.

By means of a syringe 30 mL of the copper sulphate solution are slowly passed through the reductor. The reductor is washed (using a syringe or peristaltic pump) until the outflow is clear and free from particles. The reductor is activated by passing about 250 mL of buffer solution through it containing about 100 μmol/L of nitrate. After rinsing thoroughly with buffer solution, the reductor is ready for use.

The reductor should never become dry, small air bubbles, however, passing through the reductor during operation do no harm.

The procedure to determine the efficiency of the reductor is described in Section 10.2.9.6.

The efficiency should be close to 100 %. If the efficiency drops to less then 90 %, the reductor has to be reactivated or renewed.

The activation should be repeated after periods of no operation or determining low nitrate concentrations. When not in use, the reductor is stored filled with the ammonium chloride buffer solution. In the same way prefabricated reductors with or without copper coating may be kept as spare parts.

10.2.9.6 Analytical procedures

Manual determination

As the nitrate concentration in seawater may vary from 0 to 40 μmol/L, the calibration factor should be determined for a low and a high range (undiluted and diluted samples).

A 50 mL volume of the unfiltered sample is pipetted into a 100 mL calibrated flask and filled up with buffer solution. If nitrate concentrations in excess of 15 μmol/L are expected, the samples should be diluted 1 : 2 or 1 : 5 by correspondingly increased buffer volumes; *i.e.*, only 25 mL or 10 mL sample (instead of 50 mL) should be filled up to 100 mL with buffer solution.

A reductor setup for manual analysis using a peristaltic pump is shown in Fig. 10-9.

After thorough mixing the buffered sample is pumped at about 5–8 mL/min through the reductor into a 50 mL graduated cylinder. The first 30 mL are used to flush the reductor and the cylinder and are then discarded. Another 25 mL are then passed through the reductor and transferred into the reaction bottle. Instead of the graduated cylinder, a reaction bottle with a 25 mL mark can be used.

A 1 mL aliquot of the sulphanilamide (reagent 3) is added to the sample and mixed. After a reaction time of about 1 min, 1 mL of the amine (NED, reagent 5) is added, mixed and a reaction time of 15 min allowed. The dye should not be exposed to bright daylight.

The absorbance is measured within 1 h at a wavelength of 540 nm against the respective ZW.

The 5.0 μmol/L standard and the blank are measured in a 1 or a 5 cm cell, whereas the 40.0 μmol/L standard and blank are measured in the 1 cm cuvette only. The procedure is the same as for the determination of nitrite (see Section 10.2.8.5).

The nitrate concentrations are calculated according to Section 10.2.4.

Assuming a 100 % reduction efficiency, the calibration factors F should be close to 8 and 40 for 5 and 1 cm cuvettes, respectively. As the 40 μmol/L standard is diluted 1 plus 3, F will be about 80 for a 1 cm cell. If the factors are considerably lower, the efficiency of the reductor should be checked.

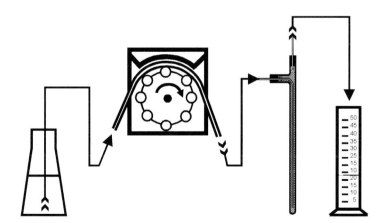

Fig. 10-9. Reductor setup for manual nitrate determinations.

The apparent nitrate values must be further corrected for the nitrite measured in the same sample. Again assuming a 100 % efficiency of the reductor, the nitrite concentration is subtracted from the nitrate value. If the reductor efficiency is significantly less than 100 %, the nitrite correction may be performed as follows.

(*i*) Subtract half of the nitrite absorbance (compensating for the the 1+1 dilution of nitrate samples with buffer) of the samples from the nitrate absorbance. Preferably use identical cells for nitrate and nitrite; if not, calculate the absorbances for identical cells. Convert the resulting corrected absorbance into nitrate concentrations using the nitrate calibration factor.

(*ii*) Subtract the nitrite concentration of the sample considering the reductor efficiency (E %) and the buffer dilution of the nitrate samples (D_B) according to

$$Nitrate_{cor} = Nitrate_{unc} - Nitrite \cdot D_B / E$$

D_B is equal to the sample volume which is made up to 100 mL with buffer, *i.e.*, 50 for an 'undiluted' sample, 25 for a 1:2 and 10 for a 1:5 sample dilution, *etc.*).

Efficiency test of the reductor

Using an identical analytical setup for the determinations of nitrite and nitrate (identical cuvette lengths and reagents), the calibration factor (see Section 10.2.4) for nitrate should be 0.48 times that of nitrite. A 50 % reduction of the absorbance/nitrate ratio is caused by the 1+1 sample/buffer dilution in the nitrate determination. The addition of 2 mL of reagents (2·1 mL) to a 25 mL nitrate sample as opposed to 2 mL of reagents to a 50 mL nitrite sample causes another 2 % loss in the absorbance/nitrate ratio.

Prepare standard samples in ZW (pure water or ASW, free from nitrite and nitrate) containing (*i*) 2 and 20 μmol/L nitrate and (*ii*) 2 and 20 μmol/L nitrite in triplicate. Analyse the samples in the setup for nitrate as described above. The efficiency of the reductor (E %) for the two concentration levels is given by the ratio of the corresponding mean absorbances of the nitrate and the nitrite samples.

$$E = 100 \cdot \frac{[A_{nitrate(1)} + A_{nitrate(2)} + A_{nitrate(3)}]}{[A_{nitrite(1)} + A_{nitrite(2)} + A_{nitrite(3)}]}$$

An effective reductor should reduce nearly 100 % of the nitrate of the sample. If an efficiency drop < 90 % is observed, the reductor has to be reactivated or renewed.

Flow-analysis

The manifold consists of two parts (see Fig. 10-10). In the first part the sample stream is mixed with the buffer solution and passed through the reductor. The second part is the same manifold as for the determination of nitrite. Thus, both nitrate and nitrite may be determined using the same manifold. By means of a three-way valve the reductor may be bypassed. By pumping pure water instead of sample (system wash) the reductor is filled with slightly diluted clean buffer. Switching the valve to 'bypass' puts the reductor into 'sleep mode'.

Most seawaters, however, contain more nitrate than nitrite (about ten times as much). The determination of nitrite and nitrate with one analytical manifold, therefore, might require different spectrophotometer adjustments.

Whenever beginning a series of analyses, the reductor should be reactivated by passing a high concentration nitrate standard (>20 μmol/L) through it for several minutes. If the standard peak of the on-line calibration is reduced by more than 10 %, the reductor should be reactivated or replaced.

The efficiency of the reductor is determined in the same way as for the manual procedure.

The water used as ZW and for the preparation of the nitrite and nitrate standards should be free from nitrate or nitrite. The efficiency determination may be performed using pure water for ZW and standards.

Analyse triplicates of the two nitrite and nitrate standards (2 and 20 μmol/L, as outlined before) without any modifications to the nitrate manifold. The efficiency is calculated from the mean absorbances (or equivalent spectrophotometer readings) as in the manual method above.

There is no salt effect and no interference from the normal constituents of seawaters. Hydrogen sulphide, however, will form some cadmium or copper sulphide precipitations on the cadmium reductor. Unless high amounts of sulphide are passed through the column repeatedly, the reductor will stay active.

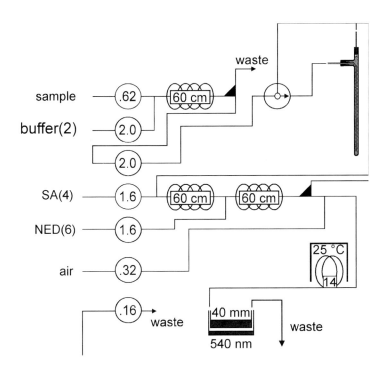

Fig. 10-10. Schematic CFA manifold for nitrate.

10.2.10   Determination of ammonia

In all methods for the determination of ammonia the sum of NH_4^+ and NH_3 is measured. Most of the earlier procedures also included varying amounts of labile organic nitrogen compounds such as trimethylamine and the amino acids in the determination. These methods have been reviewed by *Riley* (1975).

The blue colour of indophenol formed by phenol and hypochlorite in the presence of NH_3 was first reported by *Berthelot* (1859). About 30 applications of this reaction have been adopted for the determination of ammonia in various media. To be reasonably sensitive, the reaction requires elevated temperature or a catalyst. A number of transition metals ions have been used as catalyst, including: Mn^{2+}, Ag^+, Fe^{2+}, Cu^+, $[Fe(CN_6)]^{4-}$, $[Fe(CN)_5NO]^{2-}$. The last was suggested by *Lubochinsky and Zalta* (1954) and seems to yield the highest sensitivity. In this ion NO has a positive charge and Fe is divalent. *Mann, Jr.* (1963) has applied the Lubochinsky-Zalta technique for NH_3 determination after a micro Kjeldahl digestion but experienced difficulties with the Hg ion used as a catalyst in the digestion.

Sagi (1966) introduced the indophenol blue method with a nitroprusside catalyst for the direct measurement of ammonia in seawater.

Koroleff (1969, 1970) examined this procedure more closely and suggested the method as described below.

10.2.10.1   Principle of the method

Ammonia reacts in moderately alkaline solution with hypochlorite to give monochloramine which, in the presence of phenol, catalytic amounts of nitroprusside ions and excess of hypochlorite, gives indophenol blue. The reaction mechanism is complicated and cannot be fully explained. Quinone chloramine is possibly formed in one of the intermediate stages. The formation of monochloramine requires a pH of between 8 and 11.5. At higher pH ammonia is incompletely oxidized to nitrite, as in the method by *Richards and Kletsch* (1964). The indophenol formed has a pK value of about 8.9 and is fully oxidized at pH 10.8. The ratio phenol/hypochlorite must be fairly constant at about 25 (mass ratio) of phenol/ available chlorine, where 'available chlorine' is the total amount of chlorine at all oxidation levels. Hypochlorite solutions (*e.g.*, commercial bleaching agents) tend to give rather unstable dilute solutions with rapidly decreasing contents of available chlorine (about 3.5 %). *Grasshoff and Johannsen* (1972) introduced dichloro-s-triazine-2,4,6-(1H,3H,5H)-trione sodium salt (Trione, DTT) as an alternative hypochlorite donor. It has been used by *Dal Pont et al.* (1974) and *Liddicoat et al.* (1975) among others. This reagent has the advantage of being a stable solid, and the formation of hypochlorite when hydrolysed is rapid. The acid form contains 32 % of positive monovalent chlorine which is equivalent to 64 % available chlorine in a bleaching solotuion (28 and 58 % if the DTT dihydrate is used).

In seawater at a pH > 9.6, Mg and Ca ions may precipitate as hydroxides and carbonates, but these ions can be held in solution by complexation with citrate as suggested by *Solorzano* (1969). The efficiency of the citrate buffer has been improved considerably by the addition of a small amount of the disodium salt of EDTA (ethylenediaminetetraacetic acid) (*Ryle et al.*, 1981).

With the reagent concentrations suggested by *Koroleff* (1969), the formation of indophenol blue takes several hours at room temperature. The reaction can be accelerated (*i*) by increasing the concentration of the reagents as has been done in the majority of recent procedures, (*ii*) by increasing the reaction pH to more than 12, (*e.g., Scheiner*, 1975), (*iii*) by increasing the reaction temperature as in most automatic determinations and (*iv*) by irradiation of samples with long-wave ultraviolet light (*Liddicoat et al.*, 1975).

All these alternatives have been studied by *Koroleff* (unpublished), who found that a three-fold increase of the main reagent concentrations is advantageous. Also, it was observed that a reaction pH > 11.0 must be avoided, otherwise erratic blank values with greenish shades are obtained, and finally, that a reaction temperature of 37–40 °C is better than irradiation for complete colour formation within 30 min. The reaction time is 2–6 h at 20 °C depending on the salinity of the sample.

In the flow-analysis method described in Section 10.3, maximum sensitivity is obtained by heating to about 40 °C. Higher temperatures do not noticeably increase the sensitivity but enhance the tendency of turbidity formation by Ca and Mg hydroxides.

10.2.10.2 Range and precision

The molar absorptivity of the indophenol blue is about 20 000. Thus, the absorbance for 1 μmol/L is 0.200 measured in a 10 cm cell. Taking an absorbance of 0.010 as the detection limit, 0.05 μmol/L can be observed. The chemical process allows the determination of up to 150 μmol/L, but such samples must be diluted as Beer-Lambert's law is obeyed only to a concentration of about 40 μmol/L (absorbance 0.8 in a 1 cm cell). Dilution of the coloured solution after reaction is possible.

Analysis of filtered Baltic Sea samples of about 0.5 μmol/L spiked to nominally 1, 2 and 3 μmol/L ammonia and analysed in three independent analytical runs resulted in a mean standard deviation of ±0.092 μmol/L or ±2.7 % (*Hansen and Johannsen*, unpublished). Similar results have been reported by *Riley et al.* (1972) and *Solorzano* (1969). The recent ICES intercomparison exercise (*Aminot and Kirkwood*, 1995) showed an overall relative standard deviation of more than 20 %, indicating that, despite good precision of ammonia measurements within one laboratory, the inter-laboratory precision is comparatively poor. This is probably due to the ease of contamination in preparations of zero water and standards as well as in the handling of samples for ammonia determinations.

10.2.10.3 Interferences

The possibility of interference from amino acids and urea has been examined by several workers (*e.g., Zadorojny et al.* ,1973), using similar concentrations of phenol and hypochlorite and amino acid concentrations of about 14 μmol/L. Of 19 amino acids, *L*-phenylalanine caused the greatest interference and decreased the ammonia recovery by 13 %. For 14 other acids the interferences fluctuated around ±2 % of the 'true' value. The overall content of amino acids in seawater generally is about 0.5 μmol/L, therefore interferences by these substances may be neglected. No interference has been observed for urea.

Zadorojny et al. (1973) tested 25 possibly interfering inorganic constituents of seawater and observed no effects up to levels of 5 mg/L. At the 10 mg/L level a 7–14 % positive error was estimated for cyanide, thiocyanate and sulphide, and a 6 % negative effect by mercury (*II*) ions.

Koroleff (1969, 1970) found that mercury(*II*) ions at a concentration of 10–200 μmol/L (polluted waters in the vicinity of pulp and paper industry) decrease the indophenol blue by about 20 % and that samples containing more than 60 μmol/L of sulphide should be diluted.

The indophenol blue produced by the same amount of ammonia is less in seawater than in pure water. This salt effect is caused by (*i*) magnesium ions, as also stated by *Grasshoff and Johannsen* (1974) and (*ii*) by buffering capacity of seawater. Increasing amounts of Mg^{2+} ions and increasing buffer capacity decrease the final reaction pH. Consequently, the salt effect can be assumed to be caused by the pH. The chloride ion has no influence.

If the reagents are adjusted to establish a pH of 11.0 in a pure water sample, the resulting pH in natural seawater samples of salinities S is approximately

$$pH = 11.0 - 0.500 \cdot S + 0.00045 \cdot S^2$$

Samples from brackish waters, with a wide salinity range, have to be corrected with respect to the salt error. Standards are prepared using a zero water (ZW) of medium sample salinity (S_0).

The experimental linear salinity correction

$$NH_{3(cor)} = [1 + 0.0073 \cdot (S_s - S_0)] \cdot NH_{3(unc)}$$

where S_s is the salinity of the actual sample, corrects the salinity error to within ± 1 %. Considering the methodical standard deviation of ± 3–5 % (see Section 10.2.10.2), this is sufficient for most applications.

For manual analyses, standards may be prepared in pure water and the term $(S_s - S_0)$ in the above equation is then displaced by S_s.

10.2.10.4   Reagents

All pure water, either distilled or deionized, should be passed through an ion exchange column immediately before use. Cation-exchange resin (*e.g.*, Permutit RSB 100) is preferable, but 'mixed bed' cation-anion resins are also sufficient. As only traces of ions have to be removed, columns containing about 500 mL of resin are capable of deionizing more than 100 L of pure water. In flow-systems, a small cation-exchange column (about 10 cm long with 1 cm i.d.) can be inserted in the pure water lines, and the segmentation air is aspirated through a small wash bottle containing dilute sulphuric acid.

1. *Sodium hydroxide, 1.0 mol/L*: Dissolve 40 g of sodium hydroxide (NaOH) in pure water and dilute to 1 L. Store in a well-stoppered polyethylene bottle.
2. *Sodium hydroxide, working solution*: Add 2 mL of phenol reagent (reagent 3) and 1 mL of citrate solution (reagent 8) to 50 mL of pure water. Titrate with the NaOH (reagent 1) to a pH of 11.0 using a pH meter. Dilute the 1 mol/L NaOH solution so that the pH is 11.0 when 2 mL are added. The solution thus obtained, contains about 0.8 mol/L NaOH and is used for preparing the hypochlorite reagent. Store in a tightly closed polyethylene bottle.
3. *Phenol reagent (manual method)*: Dissolve 80 g of colourless phenol (C_6H_5OH) in 300 mL ethanol and add 600 mL of pure water. Dissolve 600 mg of disodium nitroprus-

side dihydrate [$Na_2Fe(CN)_5NO \cdot 2H_2O$] in 100 mL of water. Add this solution to the phenol solution and store in a tightly closed amber glass bottle at $<8\,°C$. The reagent is stable for several months.

4. *Phenol reagent (flow-analysis)*: 35 g of phenol and 0.4 g of disodium nitroprusside dihydrate [$Na_2Fe(CN)_5NO \cdot 2H_2O$] with 100 mL of ethanol are diluted and made up to 1 L with pure water. This stock solution is stable for about 4 weeks if kept at $<8\,°C$. The working solution is prepared by diluting 350 mL of the stock to 1 L with pure water and is stable for about one week if kept at $<8\,°C$.

5. *Hypochlorite solutions*: Hypochlorite solutions, usually containing 1 mol/L (3.5 %) of available chlorine in *ca.* 0.1 mol/L NaOH, can be obtained from some manufacturers. Their available chlorine content may be determined as follows. Dissolve approximately 0.5 g of potassium or sodium iodide in 50 mL of about 0.5 mol/L sulphuric acid. Add 1.0 mL of hypochlorite solution and titrate the iodine liberated with 0.1 mol/L thiosulphate solution in the usual way. Always check the stock solution before use. Thus 1 mL of the thiosulphate solution is equivalent to 0.1 mmol or 1.77 mg of positive monovalent chlorine (equivalent to 3.54 mg of available chlorine).

6. *Hypochlorite reagent (manual method)*: Dissolve 0.5 g of dichloro-s-triazine-2,4,6-(1H, 3H, 5H)-trione sodium salt (Trione, DTT) in 100 mL of the working NaOH (reagent 2) or dilute the equivalent hypochlorite solution containing 300 mg or 8.47 mmol of available chlorine to 100 ml with the working NaOH (reagent 2). Store at $<8\,°C$ in an amber glass bottle. The reagent is stable for at least 3 weeks.

7. *Hypochlorite (flow-analysis)*: The stock solution contains 20 g of NaOH and 2 g of Trione or the equivalent hypochlorite solution (containing 1.2 g or 34 mmol of available chlorine) in 1 L of pure water. The stock solution is stable for about 6 weeks if kept at $<8\,°C$. Working solutions are 350 mL of the respective stock solutions made up to 1 L with pure water. Working solutions are stable for about one week if kept at $<8\,°C$.

8. *Citrate buffer (manual method)*: Dissolve 240 g of trisodium citrate dihydrate ($C_6H_5Na_3O_7 \cdot 2H_2O$) and 20 g of EDTA (disodium salt) in about 600 mL of pure water. Make the solution alkaline with 10 mL of NaOH solution (reagent 1). Add boiling chips and remove any ammonia by boiling until the volume is less than 0.5 L. Cool and dilute to 500 mL with pure water. Store in a well-stoppered polyethylene bottle. The solution is stable.

9. *Citrate buffer (flow-analysis)*: 120 g of trisodium citrate, 1.5 g of NaOH and 10 g of EDTA (disodium salt) are dissolved in 1 L of pure water. Boil the solution down to about half the original volume. Allow to cool and make up to 1 L with pure water.

10. *Standard stock solution*: Dry ammonium chloride (NH_4Cl) at $100\,°C$. Dissolve 53.5 mg (or 66.07 g of NH_4SO_4) in pure water and dilute to 100 mL. Add a drop of chloroform as preservative. Kept in a glass bottle at $<8\,°C$ the solution is stable for several months. The standard contains 10 mmol/L ammonium.

10.2.10.5 Analytical procedures

Manual method

Ammonia must be determined in a well-ventilated room where no ammoniacal solutions are stored. Use only pure water which has been repurified as described in Section 10.2.10.4.

As open-system filtration exposes the samples to the environment and bears the risk of contamination, samples which display a visible turbidity should be centrifuged prior to manual analysis.

The main manual procedure is performed in stoppered 50 mL flasks or centrifuge tubes (reserved solely for this determination). Flasks and tubes should be cleaned with acid, rinsed well with pure water and kept closed between analyses. Heating the bottles does not always remove traces of ammonia from the walls.

Standard solutions for the determination of the calibration factor should be prepared on the day of analysis. The pure water used as the blank should be taken from the same batch used for the preparation of the working standards. The calibration factor for the determination in a 10 cm cell is about 4.8. Samples resulting in absorbances of more than 0.8 in a 1 cm cuvette have to be diluted, as Beer-Lambert's law is not obeyed beyond this value.

To 50 mL of sample add 2 mL of phenol reagent (reagent 3), 1 mL of citrate solution (reagent 8) and 2 mL of hypochlorite reagent (reagent 6). Mix well by swirling between additions. Close the bottles and put them for 30 min in a water bath maintained at $37 \pm 1\,°C$. The water level in the bath should be about the same as the sample level in the bottles. (If a thermostatically controlled water bath is not available, a well insulated box of suitable height, filled with water at 40 °C, may be used.) Next, put the bottles on the bench and allow them to cool for a further 30 min. Measure the absorbances in the appropriate cell at 630 nm against a reference of acidified pure water.

For work exclusively on oceanic waters, the calibration factor is determined using standard stock solutions diluted with natural seawater low in ammonia. Proceed as described above. In this case, the value of F should be close to 6.0 with a 10 cm cell.

Flow-analysis

In flow-analysis determinations a 5 μm filter may be inserted at the sample inlet.

The schematic layout of the manifold is illustrated in Fig. 10-11. The air should be scrubbed prior to introduction into the system by passing it through a gas-washing flask containing dilute sulphuric acid in order to remove any ammonia. The phenol reagent should be filtered through a glass-fibre filter before use.

To accelerate the indophenol blue formation, the final reaction requires a temperature of from 37 to 41 °C.

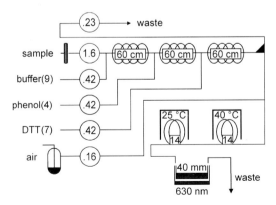

Fig. 10-11. Schematic CFA manifold for ammonia.

It is important to introduce the citrate buffer into the system flow before the first sea-water sample. Contact of seawater and reagents without buffer will immediately block the system by heavy precipitation of magnesium hydroxide. At intervals, precipitated hydroxides must be removed with some dilute acid.

The total system time in plateau-mode is about 5 min, minimum sample time is 2 min and wash time 0.2 min.

10.2.11 Determination of dissolved inorganic silicate

10.2.11.1 Principle of the method

The determination of dissolved silicon compounds in natural waters is based on the formation of a yellow silicomolybdic acid when an acid sample is treated with a molybdate solution. According to *Chow and Robinson* (1953), all soluble silicates in seawater can be determined in this manner, even those with sodium oxide/silicon dioxide ratios of less than one. Colloidal silicic acid in seawater quickly changes from its colloidal state and is detectable spectrophotometrically. Only long chain polymers containing three or more silicic acid units do not react at any appreciable speed.

Depending on the pH, the yellow silicomolybdic complex exists in two isomeric forms which differ only in their degree of hydration. The α-isomer is formed at pH 3.5–4.5 and is very stable once formed; the β-silicomolybdic acid is formed rapidly in the pH range 0.8–2.5, but is much less stable. According to *Grasshoff* (1964), the molar absorptivity of the β-acid at 390 nm is 1.7 times that of the α-form.

In early work silicate was exclusively determined by visual or photometric estimation of the β-complex. The literature and methodology have been reviewed by *Mullin and Riley* (1955a) and *Isaeva* (1958). In 1968 a careful study of the optimum conditions for this technique was performed by a research group at the Sagami Chemical Centre in Japan. Using their modified method, the colour of the complex was stable for 5–20 min at room temperature in both distilled water and in seawater. The salt effect for ocean water was reduced to 3 %.

Grasshoff (1964) developed a method based on the formation of the α-silicomolybdic acid in seawater and claimed the colour to be stable for days after the first 2 h and no salt error. Since the molar absorptivity is only 1200, 5 or 10 cm cuvettes are usually needed, which in turn requires a reference sample.

The absorptivity for the β-isomer method is 3300 at 380 nm (calculated from the Sagami Centre method), and, therefore, long absorption cells and reference samples would also be required. The stability of the colour is also a limiting factor when serial analyses are to be performed.

Since both of the yellow silicomolybdic acids have only low intensity colours, several methods have been developed in which the complexes are reduced to intensely coloured blue complexes. These heteropoly acids are well-defined soluble compounds and not colloidal products as are the blue phospho- and arsenomolybdic complexes. The most common reducing agents are metol (*p*-methylaminophenol sulphate) and sulphite (*Strickland and Parsons*, 1968), and ascorbic acid (*Koroleff*, 1971). The manual method and flow-analysis described below use ascorbic acid as the reductant.

The optimum acidity for the formation of the yellow β-silicomolybdic acid is 0.035–0.065 mol/L, and the molar ratio of acid and ammonium heptamolybdate is 17–23. It may be recalled that in the phosphate determination this ratio is 230–330 with an acidity of 0.1–0.2 mol/L.

After a mean reaction time of 10 min, oxalic acid is added for two reasons: (*i*) to avoid reduction of the excess molybdate and (*ii*) to eliminate the influence of phosphate present in the sample. As the silico complex has only a limited stability in the presence of oxalic acid, the reductant (ascorbic acid) is added immediately after the oxalic acid. The amount of ascorbic acid required is one quarter of that in the phosphate determination. No further acid need be added, as is the case in the metol-sulphite reduction. The blue silicomolybdic complex is formed within 30 min and is stable for several hours.

10.2.11.2   Range and precision

The molar absorptivity of the partly reduced heteropoly acid is about 22 000 (810 nm) in distilled water and about 19 000 in ocean water. Beer-Lambert's law is valid for up to 200 μmol/L. However, such samples must be diluted (either original samples or after colour development), as 100 μmol/L gives a net absorbance of *ca.* 1.9 in ocean water in a 1 cm cell. With a 5 cm cell, 0.010 absorbance units are equal to about 0.1 μmol/L.

High concentrations of silicate may be measured using the absorbance of the blue complex at 660 nm, which is about 40 % of the 810 nm absorbance. The absorbance of 100 μmol/L at 660 nm is *ca.* 0.85 in ocean water in a 1 cm cell, thus no dilution is necessary.

Much less attention has been paid to the determination of silicate in recent intercomparisons as compared with phosphate and the nitrogen components. The 5th ICES Intercomparison Exercise (*Kirkwood et al.*, 1991) revealed large standard deviations. Most of results from more than 30 participating laboratories agreed to within about ± 10 % for low- and high-level concentrations. However, a large number of results were a long way off ± 10 %.

Early nutrient calibration exercises (*Koroleff and Palmork*, 1977; *Grasshoff*, 1977), where most laboratories used the reduced silicomolybdic acid method for the determination of silicate with either metol-sulphite or ascorbic acid as the reductant determined precisions of ± 4 % at 4.5 μmol/L, ± 2.5 % at 45 μmol/L and ± 6 % at the 150 μmol/L level. These figures probably represent the attainable level of precision in joint experiments including an intercalibration exercise.

10.2.11.3   Interferences

Hydrogen sulphide affects the isomerisation and reduction process of the heteropoly acid, and a greenish colour may be observed initially. This has no effect on the final colour intensity as long as the formation of colloidal sulphur is suppressed. About 150 μmol/L can be tolerated. When higher amounts are present, the sample should be diluted or, preferably, the sulphide oxidized with bromine (see Section 10.2.5).

In the presence of oxalic acid, and with the reductant added separately, phosphate ions do not interfere. If, however, the oxalic acid and the reductant are added together (as in the metol-sulphite method), a slight increase in colour intensity can be observed; 1 μmol/L of phosphate corresponding to about 0.05 μmol/L of silicate.

As in the phosphate determination, fluoride of > 2.5 mmol/L decreases the colour intensity of the blue silico complex. Complexing the ion with boric acid reduces the interference

(*Case*, 1944). It should be noted that ocean water contains about 0.07 mmol/L of fluoride (see Chapter 11).

Transition metals, such as Fe, Cu, Co and Ni, in high concentrations interfere owing to the absorbance of their ions. In this case, a reference sample with 0.5 mL of sulphuric acid (reagent 1) and 1 mL of ascorbic acid per 25 mL sample water should be used.

In brackish waters, spanning a wide range of salinities, precise silicate measurements must be corrected for a salt error. For the method described here a linear correction

$$Si_{cor} = [1 + 0.0045 \cdot (S_s - S_0)] \cdot Si_{unc}$$

where S_s and S_0 are the salinities of the sample and the zero water (ZW). ZW of medium sample salinity is used for standards and blanks. The salinity error after correction is $< 0.2\%$.

10.2.11.4   Reagents

As a precaution, the silicate content of the pure water should be checked according to the described procedure at frequent intervals. Silicate-free water is best prepared by passing water through a three-column ion exchanger (anion, cation, mixed bed). If the deionizer is well regenerated and the tap water is not extremely high in silicate (as some harbor fresh-water supplies are) the deionized water is sufficiently low in silicate. Double distillation does not noticeably improve the quality. However, it should be noted that silicate is the first ion to 'break through' an anion- or mixed-bed exchange column. Deionizers are generally supplied with a conductivity meter. Conductivities of less than $0.5\,\mu S/cm$ are acceptable for pure water in nutrient analysis. The specific conductivity of silicate ions is low. Therefore, $0.5\,\mu S/cm$, if caused entirely by silicate ions, indicate low water quality. Frequent tests for silicate contents are highly recommended.

Use only plastic containers (polycarbonate, polyethylene or polypropylene).

1. *Sulphuric acid, 4.5 mol/L*: Add 250 mL of concentrated acid ($d = 1.84\,g/mL$), slowly while stirring, to 750 mL of water in a 1 L plastic beaker. Allow to cool and make up to 1 L with pure water in a graduated cylinder.
2. *Acid molybdate reagent (manual method)*: Dissolve 38 g of ammonium heptamolybdate tetrahydrate, $(NH_4)_6Mo_7O_{24} \cdot 4H_2O$, in 300 mL of pure water. Add this solution to 300 mL of sulphuric acid (reagent 1). Do not add acid to molybdate! If stored protected from direct sunlight, the reagent is stable for several months.
3. *Molybdate reagent (flow-analysis)*: 7 g of sodium molybdate dihydrate, $Na_2MoO_4 \cdot 2H_2O$, and 21 mL of sulphuric acid (reagent 1) are dissolved and made up to 1 L with pure water.
4. *Oxalic acid solution (manual method)*: Dissolve 10 g of oxalic acid dihydrate, $(COOH)_2 \cdot 2H_2O$, in 100 ml of pure water. Store this saturated solution in a plastic bottle at room temperature. It is stable indefinitely.
5. *Oxalic acid (flow-analysis)*: Dissolve 6 g of oxalic acid dihydrate, $(COOH)_2 \cdot 2H_2O$, and make up to 1 L with pure water.
6. *Ascorbic acid solution (manual method)*: Dissolve 2.8 g of ascorbic acid, $C_6H_3O_6$, in 100 mL of pure water. Store in an amber glass bottle at $< 8\,°C$. The reagent is effective as long as it remains colourless.

7. *Ascorbic acid solution (flow-analysis)*: Dissolve 16 g of ascorbic acid, $C_6H_3O_6$, and 1 mL of surfactant (Levor-IV) in 1 L of pure water. Store in an amber glass bottle at $< 8°C$. The reagent is effective as long as it remains colourless.

8. *Bromine water*: Prepare a saturated solution by adding 0.9 mL of bromine ($d = 3.11$ g/mL) to 100 mL of pure water. Store the solution in a thick-walled polyethylene bottle in the dark and at $< 8°C$.

9a. *Standard stock solution (I, preferable)*: Pure quartz powder, SiO_2, is heated at 1000 °C for 1 h to constant mass and then cooled in a desiccator. Weigh 300.5 mg, the theoretical amount for 5 mmol Si, or 301.7 mg of a 99.6 % purity quartz into a platinum crucible and add 1.5 g of anhydrous sodium carbonate, Na_2CO_3, mix with a metal spatula and fuse until the melt is quiescent. Keep the melt at 1000 °C for a few minutes until it is clear. Allow to cool and dissolve in several portions of hot pure water. Transfer the portions quantitatively into a 500 mL calibrated flask and dilute to the mark. Transfer the solution immediately into a polycarbonate bottle (or high pressure polyethylene or polypropylene). This solution contains 10 mmol/L Si. The standard is stable for at least one year.

9b. *Standard stock solution (II)*: Dry good quality disodium hexafluorosilicate, Na_2SiF_6, at 105 °C in a Pt or Ni crucible. Calculate from the assay of Si given by the manufacturer the amount equal to 5 mmol Si (theoretically $5 \cdot 188.06$ mg). Dissolve the salt in about 100 mL of pure water in a plastic beaker. Warm carefully if necessary.

Transfer the solution immediately into a polycarbonate bottle (or high pressure polyethylene or polypropylene). This solution contains 10 mmol/L. The standard is stable for at least one year.

Calibration solutions are prepared by dilution with pure water or seawater. These working standards should be used on the day of preparation.

10.2.11.5 Analytical procedures

Manual method

If a sample is visibly turbid, it must be centrifuged or filtered through a well-rinsed 0.4 μm Nuclepore membrane filter. Samples containing more than 150 μmol/L of sulphide are treated as follows.

To 50 ml of sample are added 0.1 mL of sulphuric acid (reagent 1) and bromine water (reagent 8) drop by drop until the solution remains pale yellow for about 5 min. The excess bromine is removed by a fast stream of air from a capillary positioned above the surface of the solution. For precision measurements, correct for the dilution caused by the added sulphuric acid and bromine water (about 0.4 %).

To 50 mL of sample in a plastic reaction flask add 2 mL of the molybdate reagent (reagent 2), swirl and allow to stand for 10–20 min for freshwater and 5–10 min for seawater samples. Next, add 2 mL of oxalic acid (reagent 4) followed immediately by 1 mL of ascorbic acid (reagent 6). Mix gently between additions.

With amounts of up to about 30 μmol/L, the colour reaches its maximum after *ca.* 20 min and is then stable for several hours. If the colour is dark blue, indicating a silicate content of more than about 80 μmol/L, the reduced sample is diluted with 50 mL of acidified ZW

(1 mL of sulphuric acid, reagent 1, per 100 mL) or surface water from the same station. The silicate in the dilution water does not react because oxalic acid is present in the sample.

Measure the absorbances in a cell of suitable length at 810 nm (or 660 nm for high concentrations) after a reduction time of 30–60 min against a cell of equal length filled with ZW.

To calculate the concentration of the diluted sample, multiply the absorbance (sample absorbance – blank absorbance) by the dilution factor of 1.91 (105/55) and the calibration factor.

Flow-analysis

The flow-analysis of silicate is rather sensitive to temperature changes.

The effect tends to increase with reduced system lengths (capillary system) and is probably caused by isomerization of the β-isomer of silicomolybdic acid. Temperatures of ships laboratories often are notoriously variable. If the ambient temperature varies by more than about ± 3 K within a calibration interval, the manifold should be thermostatically controlled. This may be done by either using a thermostated test-tube as a coil core for the mixing and reaction coils (shaded rectangle in Fig. 10-12) or deploying the coils in a water bath.

The wide variation of silicate concentrations in natural seawater (0–$300\,\mu$mol/L) requires an adaptation of the system to the actual concentration range, if high-precision measurements are performed. Selection of the spectrophotometer wavelength (810 nm for maximum absorbance and 660 for about 40 %) and cuvette sizes from 1 to 4 cm allow a sensitivity adjustment of from 1 to 10. Samples may be diluted on-line without modifications of manifold or reagents by using two sample inlets instead of one and adding a three-way-valve which enables the selection of either sample or ZW in one of the inlets. The total sample (respective sample + ZW) flow must be 0.6–0.9 mL/min to keep the acid and reagent concentrations within the proper ranges. The errors, introduced by dilution are acceptable up to a sample to ZW ratio of about 1 : 5. Higher dilution ratios are less accurate and stable due to differences of the tube geometries. Silicate concentrations beyond 150 μmol/L stringently require dilution or the application of a non-linear calibration, as Beer-Lambert's law is only obeyed up to 150 μmol/L.

There is no interference from compounds commonly present in seawater other than phosphate amounts above 5 μmol/l.

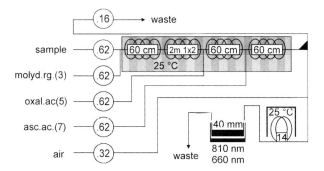

Fig. 10-12. Schematic CFA manifold for dissolved inorganic silicate.

The salt error is $< 3\%$ for salinities from 25 to 35 if sodium chloride solutions or LNSW of medium sample salinities are used as ZW. The linear salinity correction (see Section 10.2.11.3) may be applied if required.

10.2.12   Determination of nitrogen, phosphorus and silicon in particulate and dissolved organic matter

The nutrient elements N, P and Si are metabolised between their dissolved inorganic forms, bound in dissolved organic matter or incorporated in organisms. The last are not detected by the standard nutrient determination procedures. However, if not deposited permanently by sedimentation they are decomposed by biochemical processes to their inorganic dissolved forms. They thus represent a nutrient pool which is only temporarily unavailable.

As the determination of nutrients is an elementary part of any seawater analysis, the required equipment should be available in every marine chemical laboratory. Consequently, only a decomposition procedure and subsequent determinations of the inorganic components is required. For analysis, the organic material in the samples must be decomposed before measurements. The result is the total amount (inorganic and organic) of the respective nutrient. The total organic amount is calculated by subtracting the inorganic from the total. Dissolved and particulate organic forms may be discriminated by parallel analysis of filtered and unfiltered samples (see Chapter 2) .

Koroleff (unpublished) has reviewed a number of methods to oxidize organic material in seawater using peroxodisulphate or hydrogen peroxide as oxidants supported by either heat or UV irradiation. He introduced a method combining persulphate oxidation and autoclave heating. The method was slightly modified by *D'Elia et al.* (1977). A buffer was added to the system keeping the pH at 9.7 at the beginning and 4–5 at the end (*Koroleff*, 1977). The oxidizing agent thus breaks down organic components, releases phosphorus as phosphate and oxidizes nitrogen components to nitrate. The peroxodisulphate oxidation is supported by heating in an autoclave to about 115 °C.

With minor modifications the method can be applied for the determination of total nitrogen, phosphorus and silicate individually, or nitrogen and phosphorus simultaneously.

A persulphate oxidizing agent for the decomposition of organic material containing N and P is commercially available (Oxisolv, Merck).

Though manual determinations of nutrients are still in use, the majority of samples are analysed by automated flow-analyses. The oxidation procedures for the determination of total nitrogen, phosphorus and silicon as presented in the previous editions of this book, have thus been adapted to subsequent flow-analysis of inorganic nutrients without modifications to the manifolds or reagents.

A calibration and calculation procedure is suggested in Section 10.2.12.2, which is applicable to all total N, P and Si determinations both in manual and flow-analyses. A calibrated analytical setup is required for the respective inorganic component (nitrate, phosphate, silicate).

The desired sample volumes for the determination of total N, P or Si are poured into the oxidation bottles immediately after sampling. If the bottles are closed tightly, the samples may be stored without further precautions. Total Si samples should only be stored in plastic bottles.

10.2.12.1 Equipment

Polyethylene, polypropylene or Teflon 60 mL bottles with screw caps are used.

For the determination of phosphorus and nitrogen glass bottles (Pyrex or Duran) may be used. If the glass bottles are provided with a metal screw cap, a liner must be inserted in the caps which is cut from a 2–3 mm thick colourless silicon rubber sheet. The polypropylene and polyethylene bottles must be cooled and decompressed very carefully, because they are easily deformed at elevated temperatures.

Before use, containers must be tested to ensure that they seat well.

Materials other than the above mentioned must be tested for leaching of N, P or Si.

The flasks are filled with pure water and the reagents and autoclaved for 30 min at about 115 °C. Check whether the water volume in the flasks has changed (weighing) after autoclaving and cooling and analyse for nitrate, phosphate and silicate. After several repetitions of the procedure, the flasks are ready for use.

Between determinations the bottles are stored filled with *ca.* 0.1 mol/L HCl.

If deionized water is used, repeat the above test with distilled water, because deionized water may contain organic contaminations (in particular N and Si components).

Instead of an autoclave, a standard kitchen pressure cooker may be used. Most cookers are set to about 115 °C boiling temperature.

10.2.12.2 Calibration and calculation of total and organic nutrients

The total concentration of a nutrient element is determined after a decomposition step (oxidation, hydrolysis, carbonate fusion, *etc.*). The difference between concentrations before and after decomposition is the concentration of the nutrient element bound organically or in other non reactive forms.

According to the general calibration procedures for nutrient determinations (see Section 10.2.4), blanks and standards have to undergo the same treatment as the samples.

The determination of total and organic nutrients in general is less precise than the determinations of the respective inorganic dissolved component. This is due to inhomogeneities of samples containing inorganic and organic particles, and also a consequence of the additional sample handling and treatment.

In many applications the calibration procedure may be simplified without (or with tolerable) loss of the analytical precision.

Provided the digestion step does not change the total concentration of the respective nutrient element by contamination, evaporative losses or any other process, the original sample content can be calculated from the determined concentration by simple calculation of all dilution or concentration steps involved. This assumption has to be verified at the very beginning of an analytical procedure, *i.e.*, by running the specified standards through the entire procedure and calculation of the concentrations (1) with respect to the original standards and blanks and (2) with respect to the calibration standards of the dissolved inorganic nutrient determination and the sample dilutions throughout the entire procedure. The sample concentrations calculated by (1) and (2) should match within the precision determined for a series of replicate samples.

For routine analyses the correct performance of the method is checked by regular determinations of blank and standard samples. The autoclave, for example, is commonly filled with some 50 samples, which should be accompanied by triplicate blanks and standards.

The dilution factors (F_D), or the respective calculation procedures, in case the factors are dependent on variable sample sizes or dilutions, are given with the respective determination procedures. Correct the concentration determined for the respective constituents after conversion into the inorganic component (NUT_{inorg}) by

$$NUT_{total} = F_D \cdot NUT_{inorg}$$

to give the concentration of the total nutrient in the sample.

10.2.13   Determination of total and organic phosphorus by acid persulphate oxidation

The concentration of organic phosphorus in seawater commonly ranges from 0 to about 2 μmol/L. After the oxidation, concentrations are determined with the same analytical setup as for dissolved phosphate but without dilution.

The range of the procedure, the precision and interferences are the same as for inorganic phosphate (see Section 10.2.5).

Hydrogen sulphide is oxidized to sulphate. However, pre-oxidation with bromine water is recommended if the sulphide concentrations exceed 20 μmol/L.

10.2.13.1   Reagents

In addition to the reagents for the phosphate determination (see Section 10.2.5.4) either one of the following alternative oxidation reagents may be used.

Potassium persulphate solution: Dilute 5 mL of sulphuric acid (4.5 mol/L; reagent 1 of the phosphate determination in Section 10.2.5.4) with water to 100 mL. Add and dissolve 5 g of potassium peroxodisulphate, $K_2S_2O_8$. Store this solution at room temperature in a polyethylene bottle protected from direct sunlight. The reagent is stable for about one week.

Potassium peroxodisulphate, $K_2S_2O_8$, solid.

Standard solutions are prepared from the phosphate standard (see Section 10.2.5.4; reagent 3).

10.2.13.2   Analytical procedure

To 50 mL of the sample is added either 4 mL of peroxodisulphate solution or 0.2 mL of sulphuric acid (reagent 1, Section 10.2.5.4) and 220 $\pm$ 20 mg of solid potassium persulphate. Dissolve the reagent by swirling. Close the bottles and place them in the autoclave, which has been filled with about 200 mL of water. Autoclave the samples for 30 min. Cool to below 50 °C and decompress the autoclave carefully, because the inner pressure of the hot sample bottles may pop the caps (if polyethylene or polypropylene bottles are used, do not decompress the autoclave before the content has cooled to room temperature!). After opening, check that the stoppers are tight and the volumes of the samples are unchanged.

Free chlorine formed is reduced by first adding ascorbic acid in the subsequent phosphate analysis. The flow-analysis manifold for phosphate may be modified accordingly, *i.e.*, ascorbic acid solution is introduced as the first reagent, or 0.5 ml of ascorbic acid (Section 10.2.5.4; reagent 4) is added to the sample and mixed before the flow-analysis.

Transfer the required volumes into the reaction flasks for phosphate determination or into sample containers for the flow-analyser.

Dilution factors (*see* 10.2.12.2):

a) Sample + 4 mL oxid. reagent $F_D = (\text{mL sample} + 4.0) / \text{mL sample}$
b) Sample + 0.2 mL sulph. ac. + solid $F_D = (\text{mL sample} + 0.2) / \text{mL sample}$
   persulphate

If ascorbic acid is added prior to the phosphate determination, add the respective volume to the sample volume in the numerator.

In near-shore waters a reference may be necessary to compensate for turbidity. The reference is made from a second portion of the sample, by omitting the reagents after the oxidation step.

Proceed as described in Section 10.2.5.5, determination of dissolved inorganic phosphate (note the change of reagent sequence or additional ascorbic acid addition!), and in Section 10.2.12.2, calibration and calculation of total and organic nutrients.

10.2.14 Determination of total and organic phosphorus by alkaline persulphate oxidation

It was observed that organically bound phosphorus is completely decomposed to phosphate when oxidized with persulphate in an alkaline medium. Furthermore, more than 60 % of condensed phosphates are hydrolyzed. As concentrations of polyphosphates are negligible in most natural waters, a simultaneous oxidation procedure for organic phosphorus and nitrogen compounds has been developed by *Koroleff* (1977). *Valderrama* (1981) compared the procedure with former methods using separate determinations. In the simultaneous oxidation, the pH of the reaction starts at *ca.* 9.7 and ends at 4–5. These conditions are obtained by a boric acid-NaOH system. In seawater samples no precipitation is formed when the oxidation mixture is added. At elevated temperatures a precipitation is formed, which, however, dissolves as oxidation proceeds.

Range, precision and interferences are identical to those outlined in Section 10.2.13.

10.2.14.1 Reagents

One of the two oxidizing reagents below is required in addition to the reagents listed for phosphate (see Section 10.2.5.4).

Oxidizing reagent: Dissolve 50 g of potassium peroxodisulphate and 30 g of boric acid in 1 L of 0.375 mol/L (15.00 g/L) NaOH. Store this solution at room temperature in a tightly closed polyethylene bottle wrapped with aluminium foil. The mixture is stable for several

weeks. *Note*: If total nitrogen is also to be determined, NH_3-free water and persulphate low in nitrogen must be used (see Section 10.2.16).
   Oxidizing agent Oxisolv, Merck.

10.2.14.2   Analytical procedure

To 50 mL of standard or sample are added either 5 mL of the oxidizing solution or 1 dispensing spoon of Oxisolv, Merck. Close the flask and autoclave for at least 30 min. When the autoclave is opened, check that the stopper sits tight and swirl the bottle to dissolve any small precipitation on the bottom. Then allow the flask to cool to about room temperature. Withdraw from the flask 5.0 mL with a bulb pipette for separate determination of the total nitrogen as nitrate. Transfer the remaining 50 mL (45 mL, respectively, if the Oxisolv reagent is used) to a phosphate reaction flask.

Add ascorbic acid as first reagent in the subsequent phosphate analysis. The flow-analysis manifold for phosphate may be modified accordingly, *i.e.*, ascorbic acid solution is introduced as the first reagent, or 0.5 ml of ascorbic acid (see Section 10.2.5.4; reagent 4) is added to the sample and mixed before the flow-analysis.

Proceed as described in Sections 10.2.5.5 and 10.2.12.2 (note the change of reagent sequence or additional ascorbic acid addition!).

10.2.14.3   Dilution factors (see Section 10.2.12.2)

a) Sample + 5 mL oxid. reagent $F_D = (\text{mL sample} + 5.0) / \text{mL sample}$
b) Sample + Oxisolv $F_D = 1\ [F_D = (\text{mL sample} + 0) / \text{mL sample}]$

If ascorbic acid is added prior to the phosphate determination, add the respective volume to the sample volume in the numerator.

10.2.15   Determination of polyphosphates

This determination may be of interest in near-shore and estuarine water possibly contaminated with traces of polyphosphate-based detergents. In the method, any polyphosphate is converted into phosphate by acid hydrolysis.

10.2.15.1   Analytical procedure

To 100 mL of the sample is added 0.5 mL of sulphuric acid (reagent 1 of the phosphate determination, Section 10.2.5.4). Two 50 mL portions are taken from this solution and poured into two oxidation flasks. After heating and transferring to phosphate reaction flasks, add to both flasks 1 mL of ascorbic acid solution (see Section 10.2.5.4; reagent 4) and to only one flask 1 mL of the mixed reagent (see Section 10.2.5.4; reagent 6).

Continue as for phosphate, but measure against a reference (or subtract the absorbance) of the sample with ascorbic acid only.

This determination on an unfiltered sample yields the amount of total hydrolysable phosphorus (polyphosphate) plus inorganic phosphate in true solution and also the fraction of these ions adsorbed onto particles. The latter fraction is not included when analyses are performed using filtered samples.

The dilution factor (F_D; see Section 10.2.12.2) is 1.035.

10.2.16 Determination of total and organic nitrogen after persulphate oxidation

The determination of total and organic nitrogen has been one of the most difficult tasks in marine chemistry. The oldest procedure is the Kjeldahl digestion, which is still in use on a microscale. The Kjeldahl method and a method by *Krogh and Keys* (1934) decompose and reduce the nitrogen components to ammonia which is then determined.

Most modern methods oxidize the nitrogen material with an oxidant aided by heating or UV light at a wavelength below 250 nm (*Armstrong et al.*, 1966).

New methods use catalytic pyrolysis in an oxygen stream and detection of the nitrogen oxides by appropriate detectors or trapping in oxidizing aqueous solutions and subsequent determination as nitrate. The method and associated analytical problems have been discussed by, *e.g.*, *Williams et al.* (1993).

The oxidation of organic nitrogen compounds with potassium peroxodisulphate was introduced by *Koroleff* (1969, 1973) and has subsequently become a recommended standard method.

Organic phosphorus and arsenic compounds are easily broken down to inorganic compounds by boiling with peroxodisulphate in acid solution. Under the same conditions nitrogen compounds in freshwater are converted into a mixture of ammonia and nitrate, the respective amounts formed depending upon the composition of the organic compounds present. Unfortunately, when the oxidation is performed on a seawater sample, losses occur, possibly due to the formation of elementary nitrogen or as a result of the volatilization of compounds together with the free chlorine generated.

Koroleff (1969) studied the oxidation in an alkaline medium and obtained quantitative recoveries as nitrate from a variety of organic nitrogen compounds including waste waters with high organic nitrogen contents; the results compared favorably with Kjeldahl digestions. In some cases, nitrogen in heterocyclic five-membered rings was not released; the amounts of such compounds in natural waters, however, are so small that this may be neglected.

The oxidation of ammonia to nitrate is spontaneous and no losses are observed when the oxidation is performed in a sealed bottle; 80 % of the ammonia is retained after boiling in an open vessel. The nitrate content of the sample after oxidation is finally determined after conversion into nitrite and, since the method for the determination of nitrite is very sensitive, the volume of the initial sample required is only 5–20 mL.

Solorzuno and Sharp (1980) found that the recovery of urea from seawater is only 75 % when a sample volume larger than 10 mL is analysed. They suggested an increase in the amount of NaOH to a level well over that needed to precipitate all Mg^{2+} ions in an oceanic sample.

Koroleff (1977) has simplified the original procedure by introducing a buffer system that results in an initial pH of 9.7, and then decreases to 4–5 at the end. In seawater samples no precipitate is formed when the oxidation mixture is added. At elevated temperature a precipitate appears, which, however, dissolves as oxidation proceeds.

Two alternatives are given: (1) for total nitrogen only and (2) for a larger volume of sample allowing simultaneous oxidation with phosphorus (see Section 10.2.17).

10.2.16.1   Range and precision of the method

The complete oxidation of organic material with persulphate needs a ten-fold excess of oxygen. Using 10 mL of the oxidizing solution (100 mg of $K_2S_2O_8$) and 10 mL of sample, the total carbon content of the sample should not exceed 20 mg/L or a Chemical Oxygen Demand (COD) of 50 mg/L. With a sample volume of 10 mL, an amount of nitrogen up to about 75 μmol/L can be determined using a calibration factor (*i.e.*, linear relationship between concentration and absorbance). The detection limit for a 10 mL sample and a 1 cm cuvette is about 0.2 μmol/L.

Several intercalibration exercises have been carried out by the Nordic Council for Applied Sciences (Nordforsk). The results have been summarized by *Dahl* (1974), who showed that at a level of about 50 μmol/L the c.v. was ±8.3 %. Included in the investigations were samples of waste water, freshwater and various seawaters.

Ions or compounds normally present in clean or slightly polluted natural waters do not interfere with the method. Hydrogen sulphide present in anoxic water causes no interference up to 150 μmol/L. In cases of higher concentrations, *e.g.*, more than 500 μmol/L as found in the Black Sea, the amount of persulphate should perhaps be increased, but this has not yet been investigated.

10.2.16.2   Reagents

In addition to the reagents for the nitrate determination (see Section 10.2.9.4) the following reagents are required.

Sodium hydroxide, 0.375 mol/L: Dissolve 15.0 g of NaOH with < 0.001 % nitrogen in pure water and dilute to 1 L. Store in a tightly stoppered polyethylene bottle.

Sodium hydroxide, 0.075 mol/L: Dissolve 3.0 g of NaOH low in nitrogen in pure water and dilute to 1 L. Store in a well-stoppered polyethylene bottle.

Potassium peroxodisulphate, $K_2S_2O_8$: The persulphate used in this procedure should have a low nitrogen content. Suitable reagents are from Merck (No. 1.050 92 with a maximum of 0.001 % N) and BDH 'Analar' (No. 10218 with a maximum of 0.0005 % NH_3). Both reagents have been tested and found to contain *ca.* 0.00045 % N. One or two recrystallizations will improve the quality of analytical grade reagents to the above level or better. Dissolve 16 g of the persulphate in 100 mL of pure water at 70–80 °C. Cool the clear solution to almost 0 °C and filter. Dry the recrystallized salt in a desiccator over anhydrous $CaCl_2$. The recovery is about 80 %.

Oxidizing solution: Dissolve 10 g of the purified potassium peroxodisulphate ($K_2S_2O_8$), and 6 g of boric acid, H_3BO_3, in 1 L of the 0.075 mol/L NaOH reagent. Store the oxidant in a tightly closed polyethylene bottle in the dark and at < 8 °C. The solution is stable for several weeks.

Standard stock organic N solution: Disodium-EDTA, $(HOOCCH_2)_2N(CH_2)_2N(CH_2-COONa)_2 \cdot 2H_2O$, is a well-defined and stable reagent. Dissolve 186.2 mg of the reagent in 100 mL of water. The solution is stable for several months if stored in the dark and at $< 8\,°C$ in a glass bottle. The standard contains 10 mmol/L.

10.2.16.3   Analytical procedure

The pH value of samples to be analysed must be between 5 and 9. Estimate the amount of nitrogen in the sample and withdraw a test portion of 5–20 mL, observing that the amount of nitrogen per sample volume should not exceed 0.75 μmol (75 μmol/L,) if a calibration factor is to be used. Sample volumes are measured with bulb pipettes, as is the 10 mL of the oxidizing solution (see reagents). The commercial oxidation reagent Oxisolv (Merck) may be used instead. One portioning spoon of the solid reagent is sufficient for the oxidation of a 50 mL sample. Close the oxidation flask immediately after the addition of the slightly alkaline persulphate solution. Autoclave for 30 min. Cool the autoclave to less than $50\,°C$ and decompress carefully. Take out the bottle and swirl to dissolve any small amount of precipitate on the bottom. Release any over-pressure and cool to room temperature. Next, transfer the sample into a flask marked at 50 mL. Rinse twice with pure water and add the rinses to the sample, which is finally made up to 50 ml with pure water.

Determine the nitrate concentrations manually or by flow-analysis (see Section 10.2.9.6). The dilution factor (see Section 10.2.12.2) is:

$$F_D = (\text{mL sample}) / 50$$

The analysis of unfiltered samples yields the total amount of nitrogen. By subtracting the concentration of nitrate + nitrite and ammonia determined separately, the amount of organically bound nitrogen is obtained.

The determination of total N in a filtered sample excludes the particulate organic nitrogen. With extreme care, particulate nitrogen can be determined as the difference between total N in filtered and unfiltered samples.

10.2.17   Simultaneous oxidation of nitrogen and phosphorus compounds with persulphate

This procedure was introduced by *Koroleff* (1977) and has been found to be very useful in routine work. It is described in 10.2.14 on phosphorus and differs from the 'single' total nitrogen determination only in the use of a more concentrated oxidizing reagent.

10.2.17.1   Range and precision of the method

In this procedure a total carbon content of 10 mg/L is allowed for. Because the sample is diluted ten-fold in the final stage, the total nitrogen content may be as high as 150 μmol/L

(2.1 mg/L N), with the calibration factor still valid. Usually a dilution this high permits the use of a 5 cm cell for seawater samples.

10.2.17.2   Oxidizing reagent

Dissolve 5 g of purified $K_2S_2O_8$ and 3 g of H_3BO_3 in 100 mL of the 0.375 mol/L NaOH (see Section 10.2.14). Store at room temperature in a tightly stoppered polyethylene bottle wrapped with aluminium foil. The reagent is stable for at least a week.

10.2.17.3   Analytical procedure

A 5 mL volume of the oxidizing reagent (or 1 portioning spoon of the commercial oxidation reagent Oxisolv, Merck) is added to 50 mL of sample and the mixture is autoclaved for 30 min. From the total volume of 55 mL (or 50 mL if Oxisolv is used) of oxidized sample, 5.0 mL are withdrawn for the determination of total nitrogen as nitrate (nitrite), so that 50 mL (45 mL, respectively) remain for the determination of total phosphorus as phosphate. The 5 mL portion is diluted to 50 mL, and analysed for nitrate manually or by flow-analysis (see Section 10.2.9.6).

The dilution factors (see Section 10.2.12.2) are:

a) Sample + 5 mL oxid. reagent
   Total–P: $F_D = 1.10$ total–N: $F_D = 11$
b) Sample + Oxisolv
   Total-P: $F_D = 1$ total–N: $F_D = 10$

10.2.18   Determination of total silicon

In seawater silicon occurs predominantly as reactive silicate, the dissolved ions of orthosilicic acid, but also as inorganic and organic fractions of suspended material, particularly originating in plankton diatoms (see Section 10.1). The concentration of suspended silicon is very small, usually not exceeding a few μmol/L. Treating with an alkaline persulphate solution will break down these silicon complexes and convert them into reactive silicate. Only when large amounts of clay material are present a carbonate fusion is needed.

10.2.18.1   Principle of the method

The methodology of alkaline, borate buffered persulphate oxidation (*Koroleff*, 1977), is also applicable to the decomposition of suspended silicon. The procedure given here is similar to that for total phosphorus (see Section 10.2.14), *i.e.*, the sample is autoclaved with peroxodisulphate at pH 9.7 in the beginning and at about 5 at the end of the oxidation. Silicate is then determined according to Section 10.2.11.

Range, capacity, precision and interferences are the same as outlined for inorganic silicate (see Section 10.2.11).

10.2.18.2 Reagents

In addition to the reagents for the determination of silicate (see Section 10.2.11.4) the following reagents are needed.

Sodium hydroxide, 0.375 mol/L: Dissolve 15.0 g of NaOH, preferably with a SiO_2 content of less than 0.001 %, in pure water and dilute to 1 L. Store in a polyethylene bottle.

Oxidizing reagent: Dissolve 5 g of potassium peroxodisulphate, $K_2S_2O_8$, and 3 g of H_3BO_3 in 100 mL of the hydroxide solution (reagent 1). Store this solution at room temperature in a tightly-closed polyethylene bottle wrapped with aluminium foil. The reagent is stable for several weeks.

Compensating salt solution, CSS: 20 mL of the oxidizing reagent is autoclaved as described. After cooling, add 30 mL of water. Transfer into a plastic bottle for storage.

10.2.18.3 Analytical procedure

Pour 25 mL portions of the samples into oxidation flasks. Add 2 mL of oxidizing solution with a plastic pipette. Close the bottles and autoclave for *ca.* 30 min. Allow to cool to below 50 °C, decompress carefully, take out the bottles and swirl them to dissolve any small amount of precipitate on the bottom. Allow to cool to room temperature. Pour a sample into a graduated cylinder, rinse with 2 mL of water, adjust the volume to 30 mL and transfer into a reaction flask. Proceed with manual or flow-analysis as described for the determination of silicate (see Section 10.2.11).

The dilution factor (see Section 10.2.12.2) is: $F_D = 1.2$.

The oxidizing solution causes a salt effect which must be compensated for. Aliquots of 25 mL of the blanks and standards are oxidized and treated exactly the same as the samples, or 5 mL of the CSS solution is added to 25 mL blanks and standards, omitting the oxidation step. The salt effect due to salinity differences between samples and standards is corrected as described for the determination of silicate (see Section 10.2.11).

The absorbance of the blank (after oxidation or CSS addition) includes the absorbance caused by traces of Si in the ZW and the reagent blank.

Anoxic samples with sulphide in excess of 150 μmol/L should be aerated with air or nitrogen for about 30 min prior to the determination.

The determination on an unfiltered sample yields the sum of dissolved silicates, organically bound silicon and the bulk of the inorganic particulate silica.

When particulate silicon is to be estimated, samples should not be stored for more than a few hours because organic material may be partly dissolved and converted into reactive silicate.

The difference between total silicon and dissolved inorganic silicate (see Section 10.2.11) is usually equal to the particulate silicon in a sample. As different salt factors are used in the two procedures, the accuracy of the estimation may be affected. A better procedure is recommended by the following 'difference' method.

Total silicon is determined as described above, but the amount of reactive silicate is obtained from another 25 mL sample after addition of 5 mL of CSS solution. Now the same calibration, salt factors and reagent blanks can be used for both determinations.

10.2.19   Determination of total silicon by carbonate fusion

Only evaporation and fusion with alkali carbonate release larger amounts of silica from suspended inorganic material and make it available for spectrophotometric analysis.

10.2.19.1   Reagents

Sodium carbonate solution: Dissolve 10 g of anhydrous sodium carbonate, Na_2CO_3, in pure water and dilute to 100 mL. Store in a polyethylene bottle.

Dinitrophenol indicator: Prepare a 0.1 % solution of 2,4- or 2,5-dinitrophenol in 70 % ethanol. The indicator is colourless at pH < 2 and yellow at pH > 4.

Sulphuric acid, 0.45 mol/L: Dilute 10 mL of the 4.5 mol/L sulphuric acid (reagent 1 of the silicate determination, see Section 10.2.11) to 100 mL with pure water. Store in a polyethylene bottle.

10.2.19.2   Analytical procedure

Pour 25 mL of sample (using a bulb pipette) into a platinum crucible and evaporate to about 5 mL at a low temperature (about 100 °C) on a sand bath or in an oven. With a plastic pipette add 5 mL of carbonate solution (see reagents). Evaporate further to dryness and fuse until the melt is quiescent and clear. Allow to cool and leach the melt with two 5 mL portions of hot water. Transfer quantitatively into a 100 mL plastic beaker. Add 0.2 mL of the indicator (see reagents) followed by dilute sulphuric acid (see reagents) until colourless (about 11 mL). Then add a further 4 mL. Transfer into a 50 mL plastic graduated cylinder, with a stopper, and dilute with water to 40 mL. Allow to stand for at least 1 h and then use 25 mL for analysis. Proceed as described for the determination of silicate (see Section 10.2.11). Multiply the resulting silicate concentrations with a dilution factor $F_D = 1.6$ to correct for the dilution during the carbonate fusion procedure.

10.3   Automated nutrient analysis

Automated methods not only increase the amount of chemical data obtainable during a cruise, they also avoid some of the main pit falls of manual analytical procedures. Most of the standard manual analytical methods include multiple handling of samples in open vessels while adding reagents or during titration, *etc.* The resulting risk of sample contamination, though tolerable in a clean laboratory, becomes incalculable under the somewhat provisional conditions in a ship's laboratory even if the floor is not in constant motion.

However, it seems necessary to state that automation does not mean progress *per se.* While the sources of errors in the application of approved manual analytical methods are well known and most of the occurring faults are obvious to the operator, malfunctions of an automated system might pass completely unnoticed by the operator as long as the data do not show significant anomalies. Thus, calibration and control procedures are the dominating problems in automated analysis and will be discussed in detail.

The large amount of data produced by an automated analytical system has to be processed by an on-line computer. Owing to the miniaturization of analytical and electronic components multi-channel nutrient analysers may be designed as rather compact and handy units, even fit to be used on small vessels. However, they are very complex and by no means 'switch on-measure-switch off' machines. Well trained personnel, skilled in analytical chemistry, mechanics and data processing, and a great deal of experience is required to reliably operate nutrient analysers. Automated analysis is most profitable in every day routine measurements of large sample numbers. A few occasional nutrient samples, however, are better analysed by manual methods.

Since the introduction of automated chemical nutrient analysis a number of methods for the determination of different compounds in seawater and modifications of the methods have been developed. With few exceptions the continuous flow method of the Technicon AutoAnalyzer is the fundamental principle of these methods.

In addition to the determination of the 'traditional' nutrients, flow methods have been reported for a number of other components using either flow spectrophotometers or other types of detectors. Flow methods can be easily combined in multi-variable systems. Despite the occasionally low accuracy, they can provide useful additional information about the analysed samples. As an example, consider conventional water sampling in an estuary using a small vessel and sample bottles on a hydrographic wire. To confirm that a sample has really been taken at the depth indicated by the cable length and sampler position on the cable, it is necessary to measure the salinity. Instead of using sample water for separate salinity determinations with a salinometer (increasing the required sampler size), a small conductivity cell in the flow analyser would provide this information.

10.3.1 Principle of automated analysis

All modern automated analytical systems are wet chemical analysers based on the Continuous Flow Analysis (CFA, Technicon) introduced by *Weichart* (1963) and *Grasshoff* (1964). For seawater analysis a continuous stream of sample water is taken either from sample bottles or from a direct seawater intake. All operations such as the addition of reagents, heating, dialysing or phase transfer are performed in a closed tubing system between the inlet and flow-through cell of the detector system (usually a spectrophotometric cell).

Most chemical reactions employed are based on those used in the manual methods. Modifications aim at accelerating reactions and reducing the sample volume in order to save time and chemicals. The analyte is converted into a light absorbing compound whose absorbance is measured with a flow-through spectrophotometer.

The absorbance can be measured before the analyte has been converted quantitatively into the detectable compound if the conditions for the reaction (time and temperature) are kept constant and the system is 'zeroed' by washing with a blank or pure water before the next sample is introduced. In this case, the reacted portion of the analyte may be related to its concentration (peak-detecting method). The time required to analyse one sample including washing time is about 1–2 min. Allowing for quantitative reaction (steady-state method) takes about twice as long. However, no intermediate washing is required, because the spectrophotometer cell is washed out completely by the next sample. The use of a capillary tubing system can reduce the time per sample to equal that of the peak-detecting method.

Steady-state and peak-detecting methods may both be used in single- and multichannel systems. Advantages of the peak-detecting mode are increased analytical speed and considerably reduced sample volume. However, the advantages are offset by increased sensitivity to environmental changes (temperature, flow changes, *etc.*) and more complicated data evaluation.

This is also valid for a modification of the CFA where fixed sample volumes are injected at regular intervals into a flow of premixed reagents (Flow Injection Analysis, FIA). The sample spreads in the reagent stream, and the reaction products form a zone of absorbance which is recorded by the flow photometer as a peak. Flow Injection Analysis requires only about 30 s for one sample (15 s sample plus 15 s zero water) which is 6–10 times faster than CFA. However, the advantage of high analytical speed is compensated for by high reagent consumption and increased sensitivity of the system. The dye-forming reaction in FIA-systems is not quantitative (generally stopped at about 20 %) and therefore depends significantly on the system temperature. The resulting peaks show a pointed top instead of a plateau and are more difficult to quantify and sensitive to noise (electrical or other). If the expected number of samples per time does not demand FIA, we recommend the use of CFA. We also recommend beginning nutrient analyses with manual analysis or CFA before attempting the more complicated FIA. For more details on flow-analysis methods see, *e.g.*, *Stockwell* (1996). A good compilation of FIA methods and techniques is given by *Růžička and Hansen* (1988).

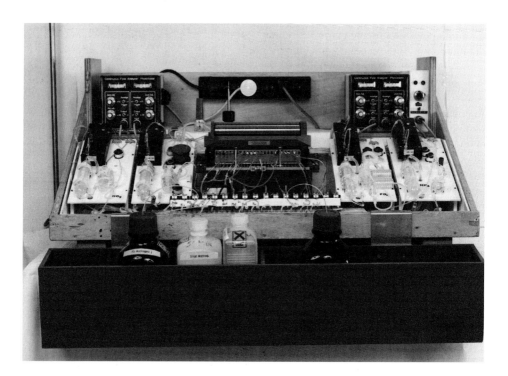

Fig. 10-13. Mobile 4-channel Continuous Flow Analyser for shipboard use (Institut für Meereskunde, Kiel).

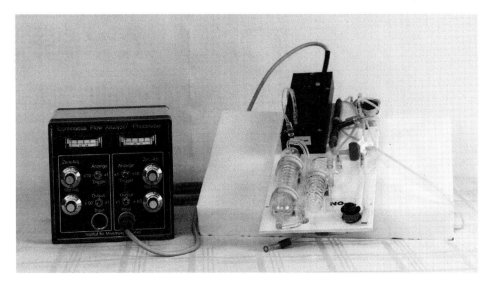

Fig. 10-14. CFA manifold for nitrate with LED-photometer and twin photometer amplifier (Institut für Meereskunde, Kiel).

The automated methods presented here are CFA methods which may be operated in the steady-state or peak-detection mode.

Analyser systems are available commercially but may also be constructed from single components according to individual requirements. A typical system for the automated analysis of seawater constituents consists of a sampler, proportioning pump, the analytical manifold (a delay and reaction system), a flow-through spectrophotometer and a data acquisition system. Fig. 10-13 shows a 4-channel CFA system built and used in the Institute of Marine Research in Kiel (Germany). One example manifold (nitrate) including a flow-through-spectrophotometer is displayed in Fig.10-14.

10.3.2   The sampler

Sample water is pumped into the analytical system either from a direct seawater inlet (continuous recording) or from individual sample bottles by means of a sampler. Most samplers consist of a stepwise rotating plate carrying cups, tubes or bottles with individual samples (Fig. 10-15).

Random position samplers are also available, which can randomly select, under programme control, sample containers in racks or on trays. The sizes of the sample plate and containers depend on the amount of sample required for the determination of each individual compound. Using the peak-recording principle 2–3 mL of sample usually is needed per analysis of each compound while the steady-state method uses twice the amount. A pick-up device controlled by a mechanical or electrical timer directs a thin stainless steel tube into the sample vessel and the sample is pumped into the analytical system for a pre-determined time interval. The pick-up is then lifted and moved into a rinsing solution reservoir filled with the appropriate 'zero water'. Using the peak-detecting method, the rinsing time equals

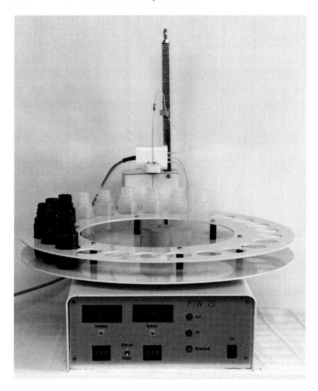

Fig. 10-15. Sampler for 24 sample bottles (50 to 100 mL), MA-RON GmbH, Germany.

the sampling time while it may be reduced considerably or dropped completely in the steady-state method. For the determination of micro-nutrients or other trace constituents of seawater, a sample plate is preferable that allows the direct insertion of the vessels used for subsampling, because any additional sample transfer increases the risk of contamination. Sealing or covering the sample bottles with aluminium or polyethylene foil, which may be penetrated by the pick-up tube, is recommended to avoid sample contamination by dust from the laboratory air.

The choice of the sample containers is a compromise between minimizing the sample volumes and required volumes for the analyses. Small sample cups of 2–5 mL volumes allow small samplers with large numbers of samples, however, at least one additional sample transfer is required between water sampler and analyser sample container and the surface/volume ratio is rather high for small cups. Considering that the contamination risk is proportional to time and the area of sample contact with the environment, sample containers which contain enough sample for repeated analysis (if required) taken directly from the water sampler are preferable to small cups.

An example of a commercial sampler with rotating sample plates is shown in Fig. 10-15.

A sampler for CFA should position the sample inlet for pre-set time intervals (0.1–5 min) in the desired sample (sequentially or randomly) and in an intermediate wash container. Some samplers are equipped to select additional standard and/or blank reservoirs by mechanical positioning through solenoid valves. The common procedure, however, is insertion of the desired standards or blanks in the same way as samples.

The sampler must communicate with the data acquisition either as master (sending a start signal on sample change) or slave (change sample on receipt of a command signal).

10.3.3 The proportioning pump

The heart of the system for automated analysis of seawater constituents is the proportioning pump. Though reagents are added to the sample in excess, changes in proportioning ratios may alter the concentration of the analyte by varying the dilution. Various types of peristaltic proportioning pumps are commercially available. A roller, or roller chain, acts on a single or multiple pumping tubes.

The rollers move at constant speed and the flow rate is determined by the inner diameter of the pumping tubes. The tubes are usually made of a special PVC (Tygon), but for solvents that might destroy the Tygon tubes silicone rubber or neoprene are used. The ratio of the tube wall thickness and the inner diameter of the tube is such that the thicknesses of the squeezed tubes are identical for all tube sizes ranging from inner diameters of 0.005–0.11 in corresponding to flow rates of 0.015–3.9 mL/min at standard pump speed. For details of these tubes see Tab. 10-1. These analytical peristaltic pumps are standardised, *i.e.,* they fit the same pumping tubes. The pump speed is either fixed or variable. Figure 10-16 illustrates the principle of a peristaltic pump.

A commercial pump supporting 24 tubes (Ismatec, Switzerland) is shown in Fig. 10-17.

This peristaltic pump uses a roller wheel with a planet drive, *i.e.,* each roller is driven actively to reduce the frictional wear on the tubes. The pump allows the pressure of the roller to be adjusted individually for each tube. The lifetime of a tube is 100–200 h of continuous operation depending on the corrosive action of the reagent on the tubing material. As the frictional wear acts mainly on the roller side of the tubes, turning them over after about half the lifetime may reduce the risk of the tube breaking.

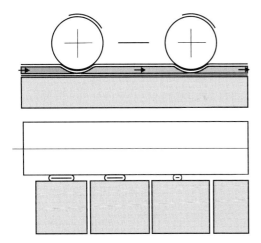

Fig. 10-16. Principle of the peristaltic multi-channel pump.

Table 10-1. Tygon pumping tubes (according to Technicon).

Colour code	Inner diameter (in)	Flow rate mL/min
Macro pumping tubes		
black	0.030	0.32
orange	0.035	0.42
colourless	0.040	0.60
red	0.045	0.80
grey	0.050	1.00
yellow	0.056	1.20
yellow–blue	0.060	1.40
blue	0.065	1.60
green	0.073	2.00
purple	0.081	2.50
purple–black	0.090	2.90
purple–orange	0.100	3.40
purple–colourless	0.110	3.90
Micro pumping tubes		
orange–black	0.005	0.015
orange–red	0.0075	0.030
orange–blue	0.010	0.048
orange–green	0.015	0.096
orange–yellow	0.020	0.159
orange–white	0.025	0.235

Remarks: The pumping volumes depend on the speed of the peristaltic pump. The values listed refer to a 'standard speed' which is the fixed speed (or one of those selected) of most peristaltic pumps which these tubes fit geometrically.

Fig. 10-17. 24-channel peristaltic pump by Ismatec (Switzerland).

10.3.4 The analytical manifold

As in the manual determination methods, one or more reagents are added sequently to the sample and mixed with it. After formation of the coloured component, the solution passes through a spectrophotometer cuvette.

Instead of individual addition of defined reagent volumes to a defined volume of sample, fixed volume flows of sample and reagents are combined in the CFA by means of multi-channel flow-calibrated pumps and a system of tubes, coils, fittings, *etc.*, in the manifold. Figure 10-18 shows a schematic diagram of a CFA system. Without additional measures, however, the mixing zone between two consecutive samples in the flow system would spread due to retention of part of the first sample on the tube walls and mixing into the following sample.

The 'trick' of the CFA is the insertion of air bubbles into the flow which separate the liquid phase into small segments. These liquid segments act as small subsample containers and considerably reduce the mixing. The mixing zone between conscutive samples is restricted to 5–10 liquid segments.

The ratio of air to liquid volumes depends on the type of manifold tubing and the flow-through spectrophotometer. The air has to be removed before entering the spectropho-tometer cuvette. Some flow-through spectrophotometers do not require removal of the air bubbles, because they are able to detect the segmenting air bubbles in the cuvette and only read the absorbance during phases of bubble-free filling. The reading is then stored until the next bubble has passed through the cuvette.

Recommended air/liquid ratios are about 1:4 for standard tubing systems with air being removed before entering into the spectrophotometer and about 1:10 for capillary systems and systems without bubble separators. The liquid segments must be long enough to fill the spectrophotometer cuvette bubble-free for a minimum spectrophotometer reading interval.

The air used for segmentation should be filtered and washed through dilute sulphuric acid to remove particles and ammonia. In special cases inert gas or air free from carbon dioxide is required. Air and reagents are injected by means of specially designed glass fittings or injector blocks whose inner diameters differ depending on whether standard or capillary systems are used. The importance of constant flow rates of the individual tubes to preserve the sample/reagent ratio is obvious. The number of required injection blocks is given by the number of reagents for the specific reaction. Each injection block (except the air injection) is followed by a mixing coil to homogenize the sample and reagent. After all reagents have been added to the sample, a sufficient length of delay coil has to be added to allow for the completion of the chemical reaction (or to reach a defined state of reaction in peak-detecting systems). In multi-channel systems additional delay coils may be needed to

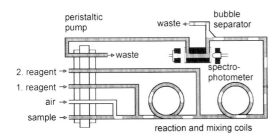

Fig. 10-18. Schematic diagramme of a continuous flow system.

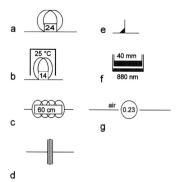

Fig. 10-19. Components used in the schematic manifold drawings. (a) Glass coils of x turns (2 mm i.d., coil diameter 17 mm, 180 μL per turn); (b) same as (a) but heated to t °C (25 °C stands for room temperature); (c) coil of x cm polyethylene tube, 0.5 mm i.d. and 'wild' mounting if not specified otherwise; (d) filter unit as specified in text; (e) debubbler; (f) flow cuvette, 2 mm i.d, x mm path length and y nm spectrophotometer wavelength; and (g) pump tube on peristaltic pump with flow medium at x mL/min flow rate at standard pump speed.

adjust the total system times of the individual channels to a phased simultaneous output. This considerably simplifies the design of the data acquisition system. However, the length of the system should be as short as possible, because mixing effects from one segment into the following increase with a long tubing system causing tailing effects and reducing the separation of individual samples. Usually the required reaction times in autoanalyser systems are shorter than those of the corresponding manual methods, the capillary system reaction time being even shorter. This may be due to wall-effects of the reaction tubing, which seem to accelerate the chemical processes involved.

The schematic manifolds presented in Sections 10.2.5–10.2.11 use symbols for the manifold components. The symbols are explained in Fig. 10-19.

10.3.4.1 Standard manifold components

All manifold components are commercially available. Injectors, either h- or Y-shaped tubes or straight glass tubes with a small platinum injection tube, are the most common components. Fitting two tube components together is simple, but requires the observance of a few rules. The entire manifold must guarantee a smooth flow, and dead volumes, gaps, cracks, *etc., i.e.,* spaces in the tubing which are remote from the general flow and could trap parts of the air bubbles, must be avoided. Identical glass tubes are fitted tightly end to end, the ends are moistened with cyclopentanone and fixed with a jacket of PVC (Tygon) tube. Cyclopentanone or cyclohexanone are good solvents for PVC tubes and allow their gluing to nearly almost any surface and 'welding'.

All parts of the analytical path should have identical inner diameters. If, nevertheless, components of different sizes have to be connected, the transition must be smoothed by insertion of coned glass tubes. Figure 10-20a illustrates the bubble splitting at an improper fitting of tubes of different diameters and their proper fitting by means of a small glass cone (Fig. 10-20b). The proper connection of a group of two injectors combining sample flow, segmenting air and a reagent is shown in Fig. 10-20c.

Coils for mixing and flow delay are the second dominating components in a CFA manifold.

While in the standard system the mixing processes are driven by differences in the specific gravities of sample and reagent solution and on frictional turbulence at the coil walls, only the latter is effective in capillary systems.

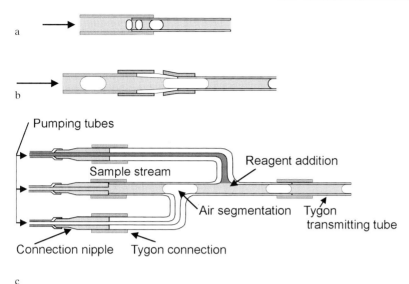

a

b

Pumping tubes

Reagent addition

Sample stream

Air segmentation   Tygon
transmitting tube

Connection nipple   Tygon connection

c

Fig. 10-20. (a) Bubble splitting at improper connection; (b) proper fitting with glass cone; and (c) CFA triple fitting with air segmentation and reagent addition.

Using capillary mixing coils (mostly polyethylene), the segmentation air is inserted after the last mixing coil, *i.e.*, mixing is performed in a non-segmented liquid stream. The loops of standard mixing coils should be oriented vertically as indicated in Fig. 10-21, capillary coil loops may be oriented in any desired position. In some applications plaiting of the capillary coil is recommended to avoid effects of the coil orientation on the mixing efficiency. The lengths of the mixing coils are determined by the times required for complete mixing. In standard systems small density differences between sample and reagent require long mixing coils and *vice versa*.

Delay coils are generally glass coils. Their number and length is determined by the time required to achieve the desired reaction after addition of all reagents.

The calculation of the delay coils required for the determination of phosphate may serve as an example.

The required reaction time for the manual method is about 10 min, which suggests about 5 min for the steady state and 3 min for the peak-detection flow method. The slowest step of the reaction is the formation of the phosphomolybdate complex which starts after the addition of the first reagent (mixed reagent). Consequently mixing and system flow after the first reagent addition (about 1.5 min) already contribute to the reaction time leaving a necessary delay of about 3.5 min for steady state and 1.5 min for peak detection mode.

Standard glass coils (see Fig. 10-10) have a volume of about 170 μL per loop, which results in a delay of 1 min per 14 turns of delay coil (standard coil length) for a total flow rate (sample + reagent + air) of 2.45 mL/min, *i.e.*, 49 coil loops (21 respectively) are required as the delay coil.

The last coil in front of the spectrophotometer should be 14 loops, jacketed to allow cooling or thermal equilibration.

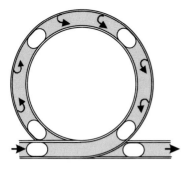

Fig. 10-21. Glass mixing coil (vertical mounting). Glass tube of about 2 mm i.d. and 3.5 mm o.d., coil diameter at tube centre 17 mm.

10.3.4.2   Heating and cooling

Chemical reactions are usually accelerated by heating. According to van't Hoffs rule, a temperature increase of 10 K increases the reaction speed by a factor of 2. The reaction coil may be heated by immersion in a thermostated bath or by inserting a regulated heating core into the centre of the coil thermally insulated on the outside. In capillary systems the mixing and reaction coil (preferably Teflon) may be wrapped around a temperature controlled electrically heater.

When part of the manifold is heated, it is necessary to cool the flow to room temperature before entering the cuvette. Even without any heating devices in the system, it is recommended to equilibrate the liquid flow thermally before it enters the cuvette. Thermal inhomogeneities are also inhomogeneities of optical density (refractive index and absorptivity). These contribute to the noise of the spectrophotometer signal which distorts the analysis especially at low concentrations. Replacing the (last) delay coil by a jacketed coil and circulating water from a room temperature reservoir of about 2 L through the equilibration coil jacket considerably reduces the 'optical' noise of the spectrophotometer.

In the peak-detecting mode, the heating procedure is more delicate than in the steady-state mode, because small changes in temperature may drastically modify the state of reaction achieved.

Seawater samples are often collected at temperatures below the laboratory temperature. Consequently, small gas bubbles tend to separate out while warming up. These bubbles enter the analytical manifold and disturb the spectrophotometric detection. If the air segments are removed in front of the cuvette, the random micro-bubbles will also be removed. When a spectrophotometer is used which allows the air segments to pass through the cuvette, a bubble separator has to be inserted at the very beginning of the manifold just after the sample inlet. The bubble separator removes random and micro-bubbles before the defined air segmentation.

A bubble separator fitted to a cuvette inlet is shown in Fig. 10-22.

The liquid/air phase enters the device through the horizontal tube, a defined volume flow is pumped through the vertical outlet and the cuvette. The upper outlet removes the air and a small amount of liquid phase while slightly less from the original liquid flow enters the cuvette. The fitting is wider at the top to provide a reservoir or trap for larger air bubbles. It depends on the overall manifold whether the air outlet or the liquid phase are actively pumped. Pumping the liquid phase through the cuvette and allowing the remaining liquid/air phase to flow into waste usually produces a less noisy spectrophotometer signal.

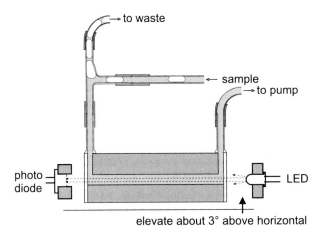

Fig. 10-22. Debubbler fitting and flow cell mounting in spectrophotometer.

10.3.4.3   Special devices

Attempts have been made to include a variety of sample treatments as on-line devices into flow systems. Separation of a gaseous component and re-introduction into a liquid flow system has been described by *Grasshoff and Hansen* (1979) for the determination of total carbon dioxide and by *Ehrhardt* (1969) for the determination of total dissolved organic carbon.

A dialyser introduced by Skalar in the flow-analysis of sulphate, permits an exchange of sulphate from the sample into a reagent flow through a membrane. Contact with sample components other than sulphate are thus avoided and sulphate enters the analytical flow at a high dilution.

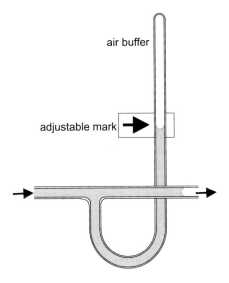

Fig. 10-23. System pressure monitoring device.

As previously described, the determination of nitrate requires reduction to convert nitrate into nitrite. Reductors which can be used in manual as well as flow analyses are described in Section 10.2.9.5. A three-way valve allows bypassing of the reduction.

Blocking of a tube (in particular capillary tubes) by particles is one of the major malfunctions of flow systems. As a consequence the system pressure increases causing variations in pump and flow rates and finally a blow-off of tube connections.

A valuable device to monitor the system flow, *i.e.*, detect system blockages or tube breaking, has been introduced by H. Johannsen (Institute for Marine Research, Kiel). The device (Fig. 10-23) monitors the pressure of the flow system at the beginning of the tubing system.

The mark is set to the initial level. The air-water interface oscillates around the mark due to pump pulses. Detection of increasing system pressure may help to solve the problem before the entire system breaks down.

10.3.4.4 The flow-through spectrophotometer

Owing to the rapid development of optical and electronic components (light transducers, integrated circuits, *etc.*) a large number of flow-through spectrophotometers are available commercially for automated marine chemical application. Miniaturization of the components has lead to the use of small spectrophotometer modules, one per analytical channel, rather than bulky multi-channel spectrophotometers. Other than a conventional laboratory spectrophotometer commonly used for manual determinations of nutrients, a flow-through spectrophotometer for CFA can be much simpler and, therefore, cheaper and smaller. For each analyte only one wavelength is required in the visible to the infrared spectral range. Most spectrophotometers use a standard tungsten or halogen lamp and interference filters of about 15 nm band width. During the last 10 years light emitting diodes (LEDs) have been introduced as light sources in spectrophotometers, which emit light of a specific constant wavelength. As the intensity of the emitted light also is extremely stable LEDs are ideal light sources for miniature spectrophotometers. Another advantage is that they convert nearly all the input energy into the output of one spectral wavelength, while tungsten lamps output more than 99 % of 'waste' light and a lot of heat.

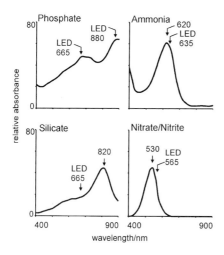

Fig. 10-24. Relative spectral absorbances of the analytical dyes, conventional (filter) wavelength selection and available LED. Except for nitrate/nitrite and the 820 nm peak of silicate, LEDs within ± 15 nm of the desired wavelengths are available.

LEDs are available for several discrete wavelengths closely matching the optimum wavelengths for the analytical methods (Fig.10-24). Minor losses in sensitivity due to mismatch of LED and optimal wavelength are generally acceptable. As the choice of available LED wavelengths is permanently increasing this problem will most probably be solved in the near future. The advantages of LEDs are obvious: small size, low energy consumption, mechanically robustness, extremely cheap as compared with expensive interference filters.

The standard spectrophotometer setup requires separation of the air bubbles segmenting the sample stream before entering the cuvette to produce a constant spectrophotometer signal. Figure 10-22 shows the debubbler device and the cuvette. Only part of the liquid volume of the sample is pumped through the cuvette, while the air bubbles are guided into the waste outlet. The cuvette is mounted with a slope in the direction of flow to avoid trapping of small bubbles produced by degassing of the sample. Flow cells with optical path lengths of 1–10 cm and inner diameters of 1–3 mm are used, depending on the molar absorptivity of the detectable component solution and the sensitivity of the optical system.

An air-water interface in the cuvette (bubble) blocks the light beam resulting in an infinite absorbance reading. As absorbances are not valid beyond about 1.5 (according to Beer-Lambert's law), this effect can be used to discriminate between valid and non-valid optical status of the spectrophotometer. Some spectrophotometers are equipped with an electronic amplifier setup which selects the absorbance reading during the valid state (cuvette completely filled with liquid sample). The reading is stored until the next bubble has passed the cuvette and the signal is valid again. This method has been introduced by Technicon. The air bubbles clean the cuvette as they remove particles which may have accumulated in it. Manifolds for this method use a debubbler only at the sample entrance to the system to trap micro-bubbles formed by degassing but not in front of the spectrophotometer. As debubblers represent considerable dead volumes omitting them improves the sample/segment separation (reduced peak tailing). Finally, the spectrophotometer signal is less noisy, because during reading the cuvette is filled with the homogeneous liquid of a single subsample (liquid segment between two air bubbles). Debubbling in front of the cuvette always mixes two or more consecutive subsamples of differing optical properties in the cuvette.

10.3.5   Data acquisition

The most simple data acquisition device, which has been used by generations of automated nutrient analysts, is an analogue multi-pen recorder connected with the photometer voltage output(s). The recorded peaks can be evaluated graphically. Adjusting the delay coils of the manifolds (in multi-channel operation) to identical system times facilitates the sample/peak correlation. This method may be old-fashioned but still is an acceptable setup for testing and developing analytical manifolds.

An analogue record of the spectrophotometer signals, or a corresponding hard copy generated by whatever acquisition system is used, provides a most valuable documentation of the performance of the analytical system and is the best insurance policy against data loss due to computer failures (a very common problem in ship operation of analysers).

Commercial analysers generally include the hard- and software for data acquisition. However, any standard personal computer (PC, preferably colour display and printer) combined with an analogue-to-digital converting interface (ADC) are sufficient to provide the analyst with an acceptable multi-purpose data acquisition system. The system described here is used by the author for data acquisition of the Kiel-Analysers.

The AD-interface connecting the analyser and the data unit is an 8-channel serial AD-interface with 12-bit resolution. The ADC is based on a Linear Technology chip (LTC 1290) and requires only a few additional components (a print board, complete with all components can be purchased for about 50 US$ from Conrad Electronic, D-92240 Hirschau). This interface is connected to a maximum of 8 spectrophotometer analogue outputs (or other instruments, *e.g.*, an attached conductivity sensor) and to a serial port of the PC. One of the analogue inputs reads the sampler status. A serial ADC allows longer distance between 'wet' analysis and 'dry' data treatment (about 15 m). If parallel ADC interface cards mounted in the computer are used, the wet-analyser and computer have to be close together (less than 5 m). No additional hardware is required.

The critical task for any nutrient analyser software is the determination of the peak height representing the component concentration. The peak shape depends on the sample volume and the mixing in the manifold.

The sample volume (at a given flow rate and system length) determines whether the reaction is complete and its products present in the spectrophotometer for some length of time (steady-state mode) or just a short moment of maximum non-quantitative reaction (Fig. 10-25).

Reduced flow-system lengths and, consequently, reduced flow-mixing generally cause steeper slopes of the peak fronts in the peak detection mode.

In the steady-state mode the peak height is determined by averaging about 10 s of spectrophotometer signal well before the peak ends (synchronised by the sampler signal). It is important to exclude the very peak end, *i.e.*, the last few seconds, because small density differences between subsequent samples or sample and wash solution often produce a small additional signal at the peak end and a negative one at the peak start (Fig. 10-26). In the peak-detection mode the real peak top has to be determined.

A particular problem is the carry-over of samples. A low-concentration sample may result in a slightly elevated signal if preceded by a high concentration sample, *i.e.*, the history of a sample in the analytical series may influence the result. A mathematical procedure to eliminate carry-over errors has been given by *Zang* (1997).

The peak height is calculated by subtracting a zero reading from the determined peaks. In the steady-state mode the zero value is determined by additional blank samples after an

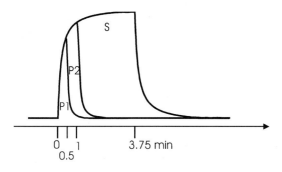

Fig. 10-25. Peak forms for steady-state mode (S) and two different sample volumes in peak mode (P1 and P2). Sample volumes at sample flow of 1.6 mL/min are 6 mL (S), 1.6 mL (P1) and 0.8 mL (P2).

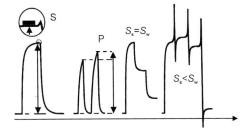

Fig. 10-26. Determination of the peak height in steady-state (S) and peak detection (P) mode and steady-state peaks for sample and wash salinities $S_s = S_w$ and $S_s < S_w$.

appropriate number of samples (Section 10.2.4), while in the peak-detecting mode an equally long 'washing' with zero water follows each sample.

The design of the software very much depends on the attempted 'operating comfort'. Basically the software has to determine the peaks in the spectrophotometer signals, assign the peaks to the correct samples and produce an output of the results. Good software should consider that the system flow, and consequently phasing between samples as well as between samples and sampler synchronization, may vary during the analysis. Adjustments should be possible without interrupting the analysis.

As mentioned before, we recommend a hardcopy record of the spectrophotometer signals in addition to a mere printing of the analytical results of the samples, to monitor the system performance and as a safety backup, independent of electronic data storage.

10.3.6 System calibration

A basic calibration procedure is required to adjust the spectrophotometer amplifiers to convert the two measured light intensities, I (sample) and I_0 (optical or electronic reference), into the correct solution of the term $\log (I_0 / I)$, *i.e.*, the absorbance as defined by Beer-Lambert's law (see Section 10.2). Mechanical or electronic simulation of an intensity $I = I_0$ must result in a zero photometer output. The linearity of the absorbance output should be checked by measuring a standard dilution sequence (see Section 10.2.4).

The system is calibrated by determining the correlation between concentration and spectrophotometer output for a zero sample and a sample of known concentration (standard).

The concentration of the standard should be just above the expected maximum sample concentration. If intermediate sample concentrations are expected, calibration between a low and high standard may be preferable. The calibration has to be repeated at regular intervals of from one to several hours depending on the overall performance of the analytical system.

Aging of the peristaltic pump tubes may modify the sample dilution. Some of the analytical channels tend to form precipitates on the cuvette walls, thus changing the spectrophotometer response, especially after a period of analysing high compound concentrations. Temperature effects may alter the equilibrium state of the formation of the spectrophotometrically detected compound, and, finally, the stability of some of the reagents is limited.

Zero water (wash and blanks) and standards should use a seawater matrix identical (or at least similar) to the samples (Section 10.2.2). Analysers employing steady-state mode are calibrated with a low standard (or zero) and a high standard, which are inserted in the same way as samples. A third, medium standard may be added if a linearity test is desired. After about 30–40 samples (about 2 h) a sequence of zero, zero, standard, zero samples (or low standard instead of zero) is inserted. A signal decrease from the first to the second zero at the beginning of the calibration sequence usually indicates dye precipitation in the cuvette which is now washed off.

In the peak-detecting mode the insertion of zero samples is not necessary because peaks are always followed by a zero (wash) reference. However, to detect possible contamination of the wash solution, a zero check is recommended at intervals (*i.e.*, every 3–4 calibrations). Standards should be inserted more frequently as compared with the steady-state mode, *i.e.*, about every 10th or 20th sample.

As the final spectrophotometer reading (represented either by recorder pen deflection or by a digital value read into the computer from the ADC) is calibrated against standard concentrations, no calibration procedures are required to produce exact absolute absorbance outputs and absolute analogue to digital conversion. Any arbitrary signal output is acceptable as long as it is linearly related to the absorbance.

The final sample concentration C_S is calculated as

$$C_S = C_L + \frac{(C_H - C_L) \cdot (A_S - A_L)}{A_H - A_L}$$

where C_H and C_L are the concentrations of the high and low (zero) standards, A_S, A_H and A_L are the absorbances (or the corresponding arbitrary readings) of sample, (high) standard and zero (low standard), respectively. Standard and zero readings are taken from the standard/zero sequences inserted in the analysis at the beginning and the end of a sample sequence.

This calibration assumes linear relationships between concentrations and spectrophotometer readings (validity of Beer-Lambert's law) and negligible shifts of baseline (zero line) and sensitivity between two subsequent calibrations.

If the baseline shift or the change of sensitivity detected by two consecutive zero/standard samples cannot be neglected, several compensation methods may be applied. Figure 10-27 shows a record of ten CFA samples in steady-state mode preceded and followed by zero/standard sequences. Both, baseline and sensitivity, differ considerably between the first and the final zero/standard sequence.

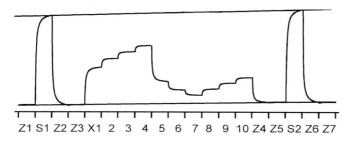

Z1 S1 Z2 Z3 X1 2 3 4 5 6 7 8 9 10 Z4 Z5 S2 Z6 Z7

Fig. 10-27. Compensation of baseline and sensitivity drift (lower and upper line) in steady-state flow analysis Z1 to Z7 and S1 to S2 are the zero and standard plateaus, respectively; X1 to X10 are samples.

The analyst has to decide whether:

(I)   the first zero/standard peaks are correct, the second is not valid, or
(II)   the second zero/standard sequence is correct, the first is not valid, or
(III) both zero/standards are equaly acceptable in which case the average is used for calculations, or
(IV)   assume linear drifts of baseline and standard levels and calculate accordingly.

The first three options simply use the first, the last or the mean of first and last zero/standard readings in the formula given to calculate the concentrations. Note that the choices (I)–(III) are poor compromises because they ignore that the analysis is performing badly.

The last option (IV) is the most common and optimal method.

As the sample-to-sample time interval is constant, the sample number can be used as a time basis for the assumed linear drifts.

The start and end of baseline and sensitivity drift are assumed at the beginning of the first and the end of the last sample. The zero line (lower) and the sensitivity line (upper) in Fig. 10-27 at the time of the respective peak-end represent the terms A_L and A_H in the concentration calculation formula.

Using the sample number (N = actual sample, N_T = total number of samples) as the time scale, A_L and A_H can be calculated from the initial zero and standard levels (Z_I and S_I) and the end levels (Z_E and S_E):

$$A_L = Z_I + (Z_E - Z_I) \cdot N / N_T \text{ and } A_H = S_I + (S_E - S_I) \cdot N / N_T$$

The final formula to calculate concentrations from absorbances or equivalent signals including baseline and sensitivity drift correction, therefore, is:

$$C_S = C_L + \frac{(C_H - C_L) \cdot [A_S - Z_I + (Z_E - Z_I) \cdot N / N_T]}{[S_I + (S_E - S_I) \cdot N / N_T] - [Z_I + (Z_E - Z_I) \cdot N / N_T]}$$

Z_E and Z_I are either the zero values adjacent to the samples or the means of the zero readings of the two zero standard sequences.

In some special analyses it may be necessary to extend the absorbance measurements beyond the linear range (*i.e.*, beyond the validity of Beer-Lamberts's law). In these cases a non-linear calibration curve has to be generated by measuring concentrations *versus* spectrophotometer readings. The sample concentrations are then calculated according to the non-linear calibration function or corresponding correction terms applied to the linearly calculated sample concentrations.

The use of non-zero water (*e.g.*, LNSW) as ZW and for standard preparation requires determination of the nutrient content of the respective matrix and a determination of reagent blanks. Reagents in nutrient analysis do not absorb at the analytical spectrophotometer wavelength unless contaminated or old. Thus, a reagent blank is mostly caused by traces of the corresponding nutrient contained in chemicals or in the pure water.

The ZW nutrient content is determined as described with the calibration procedure in Section 10.2.4.

In flow-analysis the following three-step procedure establishes the linearity and provides the reagent blank and the zero water nutrient concentration.

1. Prepare a sequence of five standards by adding the appropriate stock standard amounts to the ZW used for analysis. Plot the nominal concentrations (C) *versus* the spectrophotometer output signals (A, absorbance or arbitrary units). Calculate the concentration/absorbance regression

$$C = F \cdot A + O$$

for the calibration factor (F) and offset concentration (O).

2. Without any electrical adjustments switch the sample inlet to pure water (or insert pure water samples) and read the spectrophotometer (A_{wr}, pure water plus reagents). Then connect the reagent inlets to pure water and read the spectrophotometer (A_{ww}, pure water only).

3. The reagent absorbance is $A_{wr} - A_{ww}$ and the corresponding concentration

$$C_{reag} = F \cdot (A_{wr} - A_{ww})$$

The offset concentration after subtraction of the reagent blank (C_{reag}) is the nutrient concentration C_{zero} of the ZW. In the formula for the determination of sample concentrations use the nominal standard concentrations plus C_{zero} for C_H and C_L.

References to Chapter 10

Aminot, A., Kirkwood, K. (1995), *ICES Coop. Res.Rep.*, 213, 79.

Armstrong, F.A.J., Williams, P.M., Strickland, J.D.H. (1966), *Nature*, 211, 481.

Armstrong, F.A. (1963), *Anal. Chem.*, 35, 1292.

Atkins, W.R.G. (1954), *J. Cons. inst. Explor. Mer.*, 22, 271.

Barnard, S.S., Walt, D.R. (1991), *Nature*, 353, 338.

Barnes, H. (1959), *Apparatus and Methods of Oceanography*, Part I. London: Allen and Unwin, 1959.

Bendschneider, K., Robinson, R. (1952), *J. Mar. Res.*, 2, 1.

Bernhard, H., Wilhelms, A. (1967), *Technicon Symp.*, 1967, Vol. I., 386.

Berthelot, M.E.P. (1859), *Repert. de Chim. Appliq.*, 284.

Bray, R.H. (1945), *Soil. Sci.*, 60, 219.

Broecker, W.S., Peng, T.H. (1982), Tracers in the Sea, New York, Columbia University, 1982.

Buchanan, M.A., Easty, D.B. (1973), *Tappi*, 56(5), 127.

Case, O.P. (1944), *Ind. Eng. Chem., Anal. Ed.*, 16, 309.

Chow, D.T-W., Robinson, R.J. (1953), *Anal. Chem.*, 25, 646.

Chow, T.J., Johnson, M.S. (1962), *Anal. Chim. Acta*, 27, 441.

Costa, R.L. (1951), *Bol. Inst. Esp. Oceanogr.*, 43, 13.

Dahl, I. (1974), *Vatten*, 30, 180.

Dal Pont, G. (1962), *CSIRO Australia Oceanogr. Cruise Rept.*, 4. Oceanographic Obs. in the Indian Ocean in 1960.

Dal Pont, G., Newell, B., Staniford, J. (1963), *Austr. J. Mar. Freshwater Res.*, 14.

Dal Pont, G., Hoggan, M., Newell, B. (1974), *CSIRO, Div. Fish. Oceanogr. Rep.* 55.

De Jonge, V.N., Villerius, L.A. (1980), *Mar. Chem.*, 9, 191.

D'Elia, C.E, Stendler, P.A., Corwin, N. (1977), *Limnol. Oceanogr.*, 22(4), 760.

Ehrhardt, M. (1969), *Deep-Sea Res.*, 16, 393.

Føyn, E. (1951) *Rep. Norweg. Fish. Invest.*, 9, No. 14.

Grasshoff, K. (1964), *Deep-Sea Res.*, 11, 597.

Grasshoff, K. (1967), *Meteor Forsch. Ergeb.* Reihe A.

Grasshoff, K. (1970), *Technicon Paper* (691–57).

Grasshoff, K., Johannsen, H. (1972), *J. Cons. Int. Explor. Mer.*, 34, 516.

Grasshoff, K, Johannsen, H. (1974), *J. Cons. Int. Explor. Mer.*, 36, 90.

Grasshoff, K. (Compiler), (1977), *Rep. Baltic Intercalibr. Workshop*, Kiel.

Grasshoff, K., Hansen, H.P. (1979), *Vom Wasser*, 53, 73.

Greenfield, L.J., Kalber, F.A. (1954), *Bull. Mar. Sci. Gulf Carib.*, 4, 323.

Gripenberg, S. (1929), *Det 18. Skand. Naturforskarmode*, Copenhagen.

Hager, S.W., Gordon, L.I., Park, P.K. (1968), *Final Report to the Bureau of Commercial Fisheries.* Contract 14-17-0001-1759. Oregon State University, Dept. of Oceanography, Ref. No. 68-33, 31.

Hansen, H.P. (1990), Phosphate anomaly at the oxic-anoxic interface revealed by high resolution profiles. *Proc. of the 17th Conference of Baltic Oceanographers*, Norrköping, 1990.

Hansen, H.P. (1994), *HELCOM Baltic Sea Environmental Proceedings*, 58, 48.

Helcom (1996), *Baltic Sea Environmental Proceedings*, 64b.

Helcom (1997), *Baltic Sea Environmental Proceedings*, 69.

Hartley, A.M., Asai, R.I. (1963), *Anal. Chem.*, 35, 1207.

Isaeva, A.B. (1958), *Trud. Inst. Okeanol. Adad. Nauk S.S.S.R.*, 26, 234.

Jannasch, H.W., Johnson, K.S., Sakamoto, C.M. (1994), *Anal.Chem.*, 66, 3352.

Jenkins, D., Medsker, L.L. (1964), *Anal. Chem.*, 36, 611.

Johnson, K.S., Beehler, C.L., Sakamoto-Arnold, C.M. (1986), *Anal. Chim. Acta*, 179, 245.

Johnson, K.S., Sakamoto-Arnold, C.M., Beehler, C.L. (1989), *Deep-Sea Res.*, 36(9), 1407.

Kirkwood, D., Aminot, A., Perttilä, M. (1991), *ICES Coop. Res. Report*, 174.

Koroleff, F. (1968), *ICES, CM 1968/C: 33.*

Koroleff, F. (1969), *ICES, C.M. 1969/C: 9.*

Koroleff, F. (1970), *ICES, Interlab. Rep.*, 3, 19.

Koroleff, F. (1971), *ICES, C.M. 1971/C: 43.*

Koroleff, F. (1972), *ICES, C.M. 1972/C: 21.*

Koroleff; F. (1973), in: *Interkalibrering av metoder for bestämning av nitrat och totalnitrogen.* Nordforsk, Miljövardssekretatiated, 1973; 3, 31.

Koroleff, F (1977), in: Report of the Baltic Intercal. Workshop, Kiel; Grasshoff, K. (Compiler).

Koroleff, F., Palmork, K.H. (1977), *ICES Coop. Res. Rep.*, 67.

Krogh, A., Keys, A. (1934), *Biol. Bull. Woods Hole*, 67, 132.

Langmuir, D., Jacobson, R.L. (1970), *Environ. Sci. Technol.*, 4, 834.

Liddicoat, N.J., Tibbits, S., Butler, E.J. (1975), *Limnol. Oceanogr.*, 20(I), 131.

Lubochinsky, B., Zalta, J. (1954), *Bull. Soc. Chim. Biol.*, 36, 1363.

Mann Jr., LT. (1963), *Anal. Chem.*, 35 (13), 2179.

Martin, J.H. (1968), *Limnol. Oceanogr.*, 13, 63.

Martin, J.H., Coale, K H., Johnson, K.S., Fitzwater, S.E., Gordon, R.M., Tanner, S.J., Hunter, C.N., Elrod, V.A., Nowicki, J.L., Coley, T.L., Barber, R.T., Lindley, S.,Watson, A.J., Van Scoy, K., Law, C.S., Liddicoat, M.I., Ling, R., Stanton, T., Stockel, J., Connins, C., Anderson, A., Bidigare, R., Ondrusek, M., Latasa, M., Millero, F.J., Lee, K., Yao, W., Zhang, J.Z., Friedrich, G., Sakamoto, C., Chavez, F., Buck, K.,Kolber, Z., Greene, R., Falkowski, P., Chisholm, S.W., Hoge, F., Swift, R., Yungel, J.,Turner, S., Nightingale, P., Hatton, A., Liss, P., Tindale, N.W. (1994), *Nature*, 371, 123.

Morris, A.W, Riley, J.P. (1963), *Anal. Chim. Acta*, 29, 272.

Mullin, J.B., Riley, J.P. (1955), *Anal. Chim. Acta*, 12, 162.

Mullin, J.B., Riley, J.P. (1955b), *Anal. Chim. Acta*, 12, 464.

Murphy, J., Riley, J. P. (1962), *Anal. Chim. Acta*, 27, 31.

Price, J.B., Priddy, R.R. (1959), *Bull. Mar. Sci.*, 9, 310.

Richards, F.A., Kletsch, R.A. (1964), in: *Recent Researches in the Fields of Hydrosphere, Atmosphere and Nuclear Geochemistry*: Miyake, Y., Koyama, T. (Eds.). Tokyo: Maruzen Co. Ltd.; 1964, pp.65.

Riley, J.P. (1975), in: *Chemical Oceanography:* Riley, J.P., Skirrow, G. (Eds.). London: Academic Press, 2nd. edn., 1975; Vol. 3, pp. 193–514.

Riley, J.P., Grasshoff, K., Voipio, A. (1972), in: *A Guide to Marine Pollution:* Goldberg, E.D. (Ed.). New York: Gordon and Breach Science Publishers Inc., 1972; pp. 81–110.

Růžička, J., Hansen, E. (1988), *Flow Injection Analysis. Chemical Analysis.* New York: John Wiley & Sons, 2nd. edn.

Ryle, V.D., Muller, H.R., Gentien, P. (1981), *Automated Analysis of Nutrients in Tropical Sea Waters,* Australian Inst. of Mar. Sci., Ref.: AIMS.OS.81.2.

Saari, L.A.(1987), *Trends Anal.Chem.*, 6, 85.

Sagi, T. (1966), *The Oceanogr. Magazine*, 18, 43.

Scheiner, D. (1975), *Water Res., 10*, 31.

Schultz, J.S. (1991), *Sci. Am.*, 64.

Shinn, M.B. (1941), *Ind. Eng. Chem., Anal. Ed.,* 13, 33.

Skougstad, M.W., Fishman, M. (1963), *Anal. Chem.,* 35, 190.

Solorzano, L. (1969), *Limnol. Oceanogr.,* 14, 799.

Solorzano, L., Sharp, J.H. (1980), *Limnol. Oceanogr.,* 25 (4), 751.

Stephens, K. (1963), *Limnol. Oceanogr.*, 8, 361.

Stockwell, P. (1996), *Automatic Chemical Analysis.* London: Taylor & Francis, 1996.

Strickland, J.D.H., Parsons, T.R. (1968), *A practical Handbook of Seawater Analysis*, Fish. Res. Bd. Can. Bull. 167.

Stumm, W., Morgan, J.J. (1981), *Aquatic Chemistry.* New York: John Wiley & Sons, 2nd. edn., 1981; Chap. 7, 418.

Valderrama, J.C. (1981), *Mar. Chem.*, 10, 109.

Weichart, G. (1963), *Dtsch. Hydrogr. Z.,* 16 (6), 272

Whitfield, M., Jagner, D. (Eds.) (1981), *Marine Electrochemistry.* New York, John Wiley & Sons, 1981.

Williams, P.M., Bauer, J.E., Robertson, K.J., Wolgast, D.M., Occelli, M.L. (1993), *Mar.Chem.,* 41, 271.

Wohltjen, H. (1984), *Anal. Chem.,* 56, 87A.

Wood, E.D., Armstrong, F.A.J., Richards, F.A. (1967), *J. Mar. Biol. Ass. U.K.,* 47, 23.

Zadorojny, C., Saxton, S., Finger, R. (1973), *Journal WPCF,* 45, 905.

Zang, J.-Z. (1997), *J. Autom. Chem.,* 19 (6), 193.

11 Determination of the major constituents

K. Kremling

11.1 Introduction

Only 11 elements can be considered major components of seasalt: the cations sodium, potassium, magnesium, calcium and strontium, and the anions chloride, sulphate, bromide, hydrogen carbonate (carbonate), borate (borid acid) and fluoride. These major dissolved constituents (concentrations > 1 mg/kg in ocean waters) make up > 99 % of the soluble ionic species of seawater. The elemental ratios are relatively constant throughout the world ocean, and their concentrations change due to the addition or substraction of water only (concept of 'conservatism'). Therefore, it is possible to characterize the composition by determining only one constituent that is easy to measure and is conservative in its behaviour. An example is chlorinity (Cl, as defined in Section 11.2.4).

Table 11-1 lists the concentrations and their ratios to chlorinity considered to be representative of average seawater. With the exception of calcium, all oceanic chlorinity ratios studied show little or no variation with depth. The increase of the Ca^{2+}/Cl values in deep waters of most oceans (on average 0.3–0.5 %, with maximum deviation in deep North Pacific waters by as much as 1.3 %) can be explained by (a) calcium extraction from surface waters by biological activity, (b) decomposition of organic material in deeper layers and (c) the increased solubility of calcium carbonate at the lower temperature and higher pressure.

Although the concentration ratios of the major constituents of seawater are relatively constant, a number of factors may cause considerable regional variations. Major processes and areas to be considered are, for example, (a) estuaries (with high river inputs of varying composition, *e.g.*, as in the Baltic Sea); (b) anoxic basins due to bacteria using SO_4^{2-} as a source of O_2 leading to lower SO_4^{2-}/Cl values (*e.g.*, the Black Sea); (c) polar regions where freezing can cause anomalies in the relative composition of the major components; (d) hot brines with their highly saline waters are very different from average seawater (*e.g.*, the Red Sea); or (e) interstitial waters which differ appreciably from seawater in some major constituents. An individual consideration of the particular regions or processes, however, is beyond the scope of this book (for a brief review see *e.g., Millero and Sohn*, 1992).

This chapter presents the analytical procedures for calcium, magnesium, potassium, chlorinity, sulphate, bromide, boron and fluoride, selected as the most interesting major constituents. (The methods for determination of the CO_2 species are thoroughly outlined in Chapters 8 and 9.) It describes the methods, firstly from a theoretical and, secondly from a practical point of view. To make meaningful studies of the relative variations of these ions in the sea it is important to apply analytical procedures capable of high precision. The methods reported here have been tested by the author in large series and found to be among the most reliable. Most of the procedures have been subject to several modifications since their first applications. A short outline and critical discussion of an alternative method is given for some components. (For chemical methods of interstitial water analysis refer to *Gieskes et al.*, 1991). For aspects of sampling and storage see Chapter 2.

Table 11-1. Concentration and relative composition of the major components of average seawater at $S = 35.000$, pH = 8.1 and $t = 25\ °C$ (compiled from *Culkin and Cox*, 1966; *Morris and Riley*, 1966; *Riley and Tongudai*, 1967; *Carpenter and Manella*, 1973; and *Millero*, 1982).

Species	Concentration		Ratio	
	g/kg	mol/kg[a]	g_i /Cl	mol_i /Cl
Na^+	10.7816	0.468 97	0.5565	0.024 21
Mg^{2+}	1.2837	0.052 82	0.066 26	0.027 26
Ca^{2+}	0.4121	0.010 28	0.021 27	0.000 5307
K^+	0.3991	0.010 21	0.020 60	0.000 5269
Sr^{2+}	0.0079	0.000 09	0.000 41	0.000 0047
Cl^-	19.353	0.545 88	0.998 91	0.028 1759
SO_4^{2-}	2.7124	0.028 24	0.1400	0.014 574
HCO_3^-	0.1135	0.001 86	0.005 86	0.000 0960
Br^-	0.0672	0.000 84	0.003 47	0.000 0434
$B(OH)_3$	0.0203	0.000 33	0.001 05	0.000 0170
CO_3^{2-}	0.0116	0.000 19	0.000 60	0.000 0100
$B(OH)_4^-$	0.0066	0.000 08	0.000 34	0.000 0043
F^-	0.0013	0.000 07	0.000 067	0.000 0035

[a] To convert to molar units (mol/L) multiply by the density of samples (see Table 4 in the Appendix).

11.2   Analytical methods

11.2.1   Calcium

Complexometric titration with ethylenedioxy-bis-(ethylenenitrilo)tetraacetic acid (EGTA) has been used for the direct determination of calcium in seawater for several years. A very accurate and simple procedure by visual titration using glyoxal-bis(2-hydroxyanil) (GBHA) as metal indicator has been developed by *Tsunogai et al.* (1968) and will be described with slight modifications in this text.

A spectrophotometric method using the same complexone but with zincon and zinc-EGTA complex as an indirect indicator (*Culkin and Cox*, 1966) is outlined in Section 11.2.1.7.

11.2.1.1   Principle of the method

Amino polycarboxylic acids with the characteristic grouping $-N(CH_2COOH)_2$ form strong chelates with many polyvalent cations (*Schwarzenbach et al.*, 1945). Elements such as calcium, strontium or magnesium are therefore determined by titration with the solution of a chelating agent (complexone).

The special complexone EGTA, developed by *Schwarzenbach et al.* (1957) shows a difference between the stability constants of the calcium and magnesium complex of more than five orders of magnitude (log $K_{Ca} = 11.0$, log $K_{Mg} = 5.2$) allowing almost selective titration of calcium in the presence of large amounts of magnesium. In the presence of magnesium concentrations of > 27 mmol/L, however, slightly lower values for calcium are observed (by about 0.2 %). Strontium is partially titrated and must also be corrected for.

The endpoint of the above procedure is detected by the metal indicator GBHA, which forms a red calcium chelate at pH 11.7, very different in colour from free GBHA and less stable than the metal-complexone complex. The red complex is extracted into an organic phase, but the titration is performed without separation of the two layers. When the titration proceeds, calcium ions are complexed as much stronger EGTA complexes, and at the stoichiometric endpoint the red Ca-GBHA complex is replaced by the colourless free indicator.

11.2.1.2 Reagents and equipment

1. *EGTA stock solution*: $C_{14}H_{24}N_2O_{10}$, 0.1 mol/L of the sodium salt. Dissolve 38.0 g of ethylenedioxy-bis-(ethylenenitrilo)tetraacetic acid in 300 mL of 1 mol/L sodium hydroxide (NaOH) solution and dilute to 1000 mL.
2. *EGTA working solution*: 0.01 mol/L (or 0.005 mol/L for samples with a salinity of < 20). Take aliquots of the stock solution and dilute to 0.01 mol/L. The empirical titre is determined by standardization with the calcium standard solution.
3. *Borate buffer:* Dissolve 10 g of sodium tetraborate (analytical-reagent grade, a.g.), $Na_2B_4O_7 \cdot 10\ H_2O$ and 30 g of sodium hydroxide (a.g.) in distilled water and dilute to 500 mL.
4. *GBAH indicator solution:* Glyoxal-bis(2-hydroxyanil), $C_{14}H_{12}N_2O_2$, 0.05 % in *n*-propanol.
5. n-*Butyl alcohol.* 1-Butanol (a.g.).
6. *Standard calcium solution,* 0.0103 mol/L Ca^{2+}: Dissolve exactly 1.0309 g of pure calcium carbonate ($CaCO_3$, spectrographically standardized) in a few mL of dilute hydrochloric acid (HCl) and dilute to 1000 mL after addition of 13.670 g of magnesium nitrate, $Mg(NO_3)_2 \cdot 6H_2O$ (*e.g.*, 'Suprapur', Merck), 0.0240 g of strontium chloride (a.g.), $SrCl_2 \cdot 6H_2O$ and 27.400 g of sodium chloride (*e.g.*, NaCl, 'Suprapur', Merck). The concentrations correspond to a salinity of about 35.0.
7. *Accurate burette:* For example Metrohm piston burette, calibrated, with subdivisions of 0.005 mL.
8. 100 mL *conical flasks.*
9. *Magnetic stirrer.*

11.2.1.3 Standardization of the EGTA solution

When strontium and magnesium are present in the same proportions and amounts as in ocean waters, the titre of calcium is increased by about 0.54 %. This means the EGTA titre must be corrected for these interferences by multiplying with a factor of 0.9946, or the EGTA must be standardized with a standard calcium solution containing strontium and magnesium in the same ratios to calcium as in the seawater samples to be analysed. The latter method of empirical standardization is preferable in this case.

Transfer 10.00 mL of the 0.0103 mol/L standard calcium solution into a 100 mL conical flask and titrate with EGTA as described in Section 11.2.1.4.

$$\frac{M_{Ca} \cdot a}{b} = M_{EGTA}$$

where

M_{Ca} =   molarity of the calcium standard solution
a =   mL of calcium standard used
b =   mL of EGTA solution required
M_{EGTA} =   molarity of the EGTA solution

11.2.1.4   Analysis of the sample

Transfer about 10 mL of the sample into a 100 mL beaker covered with a watch glass and weigh accurately. Add 95 % of the standardized EGTA solution necessary for the end-point (roughly calculated from the salinity/chlorinity of the sample; see Table 11-1). Stir the solution with the magnetic stirrer, add 4 mL of 0.05 % GBHA and 4 mL of the buffer solution. Stir for another 3 min to develop the calcium-GBHA complex, then add 7 mL of *n*-butyl alcohol to extract the red complex into the organic layer. Immerse the capillary outlet of the burette directly into the solution and commence the titration immediately. Stir vigorously and observe the organic layer occasionally, after having stopped the stirrer. At the endpoint, which can be detected to within 0.005 mL of a 0.01 mol/L EGTA solution, the butanol layer changes from red to colourless (best observed by placing a sheet of white paper behind the beaker). Since the calcium-complex is relatively unstable, finish the titration within 15 min.

11.2.1.5   Calculation of results

The concentration of calcium is calculated from

$$Ca\,(g/kg) = \frac{a \cdot M \cdot 40.08}{w}$$

where

a =   mL of EGTA solution consumed
M =   molarity of EGTA solution
w =   mass of seawater sample (g); with correction for buoyancy

11.2.1.6   Precision

The precision of the method was estimated by *Tsunogai* et al. (1968) to be better than 0.1 %. This figure has been confirmed by this author from recent measurements of Baltic

Sea water samples (with a mean calcium concentration of about 3 mmol/L). The coefficient of variation has been found to be 0.06 % (*Kremling and Wilhelm, 1997*).

11.2.1.7 Spectrophotometric EGTA titration with zincon and Zn-EGTA as indirect indicator

Culkin and Cox (1966) applied spectrophotometric titration with 0.01 mol/L EGTA using zincon (1-carboxy-2'-hydroxy-5'-sulphoformacylbenzene) and zinc-EGTA as an indirect indicator to seawater analysis. The titration is performed with 10 g of seawater at pH 9.5 and at a wavelength of 500 nm. Interference of magnesium and strontium is compensated for by empirical standardization of EGTA. The method is as accurate as the procedure described above but requires additional equipment (spectrophotometer).

11.2.2 Magnesium

The method described here is based on the difference between measurements of total alkaline earths by complexometric titration with EDTA (ethylenediamine-N,N,N',N'-tetra-acetic acid) and selective measurement of calcium described in Section 11.2.1. The simultaneous EDTA titration of calcium, strontium and magnesium involves Eriochrome Black T (EBT) as indicator and was originally applied to seawater analysis by *Voipio* (1959) and *Pate and Robinson* (1961). To eliminate subjective errors in the determination of the endpoint, *Culkin and Cox* (1966) used photometric endpoint detection. A slight modification of this procedure, including the standardization of EDTA by magnesium is reported here.

A method for the isolation of magnesium from the bulk of sea salts by ion exchange and subsequent EDTA titration (*Greenhalgh et al., 1966*) is outlined in Section 11.2.2.7.

11.2.2.1 Principle of the method

For general references of complexometric titrations with aminopolycarboxylic acid see Section 11.2.1. The chelates formed by calcium, strontium and magnesium with EDTA are very stable, with logarithmic stability constants of 10.70, 8.63 and 8.69, respectively (at 20 °C and an ionic strength of 0.1). The metallochromic indicator EBT is used for endpoint detection. This compound, an o,o'-dihydroxy-azonaphthalene derivative, is blue in the pH range 7-11 and changes to a brilliant red after addition of polyvalent cations such as alkaline earths. Although the colour of the magnesium-EBT complex (1:1 ratio), which is more stable than the calcium-EBT complex (the log K of Mg and Ca at pH 10 are 5.4 and 3.8, respectively), changes rather abruptly at the endpoint, considerably greater precision can be obtained by photometric titration at 640 nm. It is important, as was demonstrated by *Carpenter and Manella* (1973), that the EDTA solution used to titrate the total alkaline earths is standardized against a solution of a magnesium salt. If a calcium solution is applied instead, then the photometric endpoint as used by *Culkin and Cox* (1966) is computed to be about 1 % beyond the equivalence point of the titration. Very probably this error is also a major reason for discrepancies in the Mg/Cl ratios of ocean water data sets (*Millero and Sohn, 1992*).

11.2.2.2   Reagents and equipment

1. *EDTA solution:* $C_{10}H_{14}N_2Na_2O_8 \cdot 2H_2O$, 0.025 mol/L for samples with $S < 20$. Dissolve 18.61 g of ethylenediaminetetraacetate (disodium salt, dihydrated, analytical-reagent grade, a.g.) dried at 50 °C for 1 h) in distilled water and dilute to 2000 mL. Determine the exact titre by standardization with the magnesium standard solution.
2. *Buffer solution:* Dissolve 36 g of ammonium chloride (a.g.), NH_4Cl, in 300 mL of concentrated ammonia solution, NH_4OH ($d = 0.91$), and dilute to 500 mL.
3. *EBT indicator solution:* Dissolve 0.2 g EBT (3-hydroxy-4-[(1-hydroxy-2-naphthyl)azo]-7-nitro-1-naphthalene-sulphonic acid, sodium salt) in 100 mL methanol. Filter the solution and store between 0 and 5 °C.
4. *Standard magnesium solution,* 0.025 mol/L: Dissolve exactly 0.6076 g of spectrographically standardized magnesium crystals (Mg) in a few mL of dilute hydrochloric acid (HCl) and dilute to 1000 mL with distilled water.
5. *Spectrophotometric titrator:* For detection of the titration endpoint two instrumental modifications are possible.
   a) A commercial spectrophotometer may be converted into a titrator by replacing the cell holder with a titration cell made of glass or acryl glass with a (suitable) volume of about 500 mL. Also, a normal beaker may be employed as the titration cell. A specially constructed cover is necessary with two holes to accommodate the tip of the burette and the shaft of a stirrer. Care should be taken to prevent light entering the cell compartment through the holes in the light-proof lid.
   b) A modern version recommended here for automatic titration is a spectrophotometer with a light guide titration cell which allows direct measurement of the endpoint (*e.g.,* by Model 662 from Metrohm AG, Herisau, Switzerland).
6. *Accurate burette:* For example, a Metrohm piston burette, calibrated in subdivisions of 0.005 mL, needed only if the titration curve is performed by conventional procedure (*e.g.,* as described under 5a of this section).

11.2.2.3   Standardization of EDTA solution

To standardize the EDTA solution, transfer 20.00 mL of the standard magnesium solution into the titration cell, add 5 mL of buffer solution, 1 mL of EBT indicator, dilute to 350 mL and titrate with EDTA as described in Section 11.2.2.4.

$$\frac{M_{Mg} \cdot a}{b} = M_{EDTA}$$

where

M_{Mg}	=	molarity of the magnesium standard solution
a	=	mL of magnesium standard solution used
b	=	mL of EDTA solution
M_{EDTA}	=	molarity of the EDTA solution

11.2.2.4 Analysis of the sample

Weigh accurately 10 mL of the seawater sample in an Erlenmeyer flask covered with a watch glass and quantitatively transfer into the titration vessel. Add 5 mL of buffer solution, 1 mL of EBT indicator and dilute with distilled water to 350 mL. When applying the conventional spectrophotometer method proceed as follows. Position the titration vessel in the cell compartment and cover it with the light-proof lid accommodating the stirring rod and the burette tip. Draw out the capillary end of the burette to a sufficient length so that the tip is always below the surface of the solution in the titration vessel.

Stir vigorously and adjust the spectrophotometer slit so that the instrument reads zero absorbance at 640 nm. Add EDTA solution continuously until the first faint colour change occurs. As the endpoint is approached, add aliquots of 0.005 mL, wait for equilibrium, and record the absorbance.

Plot the absorbance *versus* added mL of EDTA (see Fig. 11-1) and determine the volume of EDTA needed to reach the endpoint from the resulting titration curve. The intersection of the straight lines indicates the endpoint.

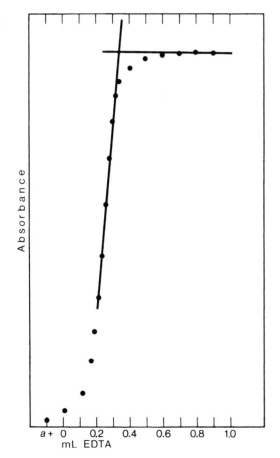

Fig. 11-1. A spectrophotometric titration curve of Mg + Ca + Sr with EDTA at pH 10 using Eriochrom Black T as indicator (at 640 nm).

In the case where an automatic titration system is applied (see Section 11.2.2.2), the volume of EDTA needed to reach the endpoint is reported in a print-out together with registration of the titration curve. It is important with this system to pre-select the increments of volume with which the titration should proceed (*e.g.*, by 0.1 mL steps in this case).

11.2.2.5 Calculation of results

Calculate the total amount of calcium, strontium and magnesium (in mol/kg) using the relationship

$$\text{mol/kg (Ca + Sr + Mg)} = \frac{a}{w} \cdot 2.5 \cdot 10^{-2} \tag{11-1}$$

where

a = mL of EDTA solution consumed
w = mass of seawater sample (g); with correction for buoyancy

The value $2.5 \cdot 10^{-2}$ is derived from the relationship:

$$1 \text{ ml } 0.025 \text{ mol/L EDTA solution} \equiv 2.5 \cdot 10^{-5} \text{ mol (Ca + Sr + Mg)} \tag{11-2}$$

Subtract the calcium concentration (in mol/kg), as determined by EGTA titration (Section 11.2.1), from the result of Eq. (11-1), with

$$1 \text{ mol Ca} = 40.08 \text{ g Ca} \tag{11-3}$$

Finally, subtract the strontium concentration as calculated from the relationship (see Table 11-1)

$$\text{Sr (mol/kg)} = 0.47 \cdot 10^{-5} \text{ Cl} \tag{11-4}$$

11.2.2.6 Precision

The relative standard deviation for measurements of the total alkaline earths in ocean waters is reported to be less than 0.1 % (*Pate and Robinson*, 1961; *Culkin and Cox*, 1966). In the case where an automatic titration system is applied, the author found a comparable coefficient of variation (c.v.) of 0.15 %. However, for water of lower salinity with an average magnesium content of 0.205 g/kg, a c.v. of only 0.65 % was measured by the author. Although the determination of the magnesium concentration is based on the difference between titrations, errors caused by the measurements of calcium and strontium can be neglected, when the analytical procedure described in Section 11.2.1 is used.

11.2.2.7 Ion-exchange separation of magnesium

To separate magnesium from seawater for subsequent individual spectrophotometric titration *Greenhalgh et al.* (1966) developed a cation-exchange scheme using Amberlite CG 120 resin. The required sample mass is equivalent to about 30 mL of seawater with a salinity of *ca.* 35; a suitable flow rate is about 20 mL/h. After sodium and potassium, magnesium is then eluted with 450 mL of 0.35 mol/L ammonium chloride solution and titrated using Eriochrome Black T as indicator and EDTA as titrant. However, standardization of the complexone (performed with magnesium metal) should be slightly modified from the original procedure to ensure that the concentration of ammonium chloride in the standard magnesium solution is the same as in the solution eluted from the ion-exchange column (*Carpenter and Manella*, 1973).

The ion-exchange method is highly reproducible with a c.v. of 0.03 % but is rather time-consuming.

11.2.3 Potassium

The gravimetric determination of potassium as its barely soluble tetraphenylborate is recommended and described here. This method has also been applied to seawater for an ocean-wide survey of the element (see Table 11-1). The procedure is preferred because of its high accuracy and precision and the minimum amount of equipment required to perform the analysis. Basically, the method outlined here is that developed by *Wittig* (1950) and *Raff and Brotz* (1951). A brief description of a potentiometric method is presented in Section 11.2.3.6.

11.2.3.1 Principle of the method

Sodium tetraphenylborate ('Kalignost') reacts with potassium according to

$$[B(C_6H_5)_4]^- + K^+ \rightarrow K[B(C_6H_5)_4] \downarrow$$

The solubility of the precipitate is lowest between pH 4 and 6. Its solubility product is around $2.3 \cdot 10^{-8}$ (at 20 °C). If a minor excess of the precipitating agent (*i.e.,* sodium tetraphenylborate) is available in solution, the solubility of the precipitate is negligible. Co-precipitation of calcium and magnesium ions may lead to serious errors when they are first precipitated. This interference is minimized, however, as carbonates which, in contrast to potassium tetraphenylborate, are soluble in acetic acid. Potassium tetraphenylborate is crystalline and starts to decompose when heated above 100 °C. Other cations which form stable precipitates with tetraphenylborate under the conditions applied are rubidium, cesium, ammonium, mercury, thallium(*I*) and silver (see Section 11.2.3.6). In natural seawaters, however, the concentrations of these constituents are so low as to be negligible.

11.2.3.2 Reagents and equipment

1. *Sodium tetraphenylborate* or 'Kalignost', $Na[B(C_6H_5)_4]$: Dissolve 34.2 g in around 900 mL of distilled water and add 5 g of aluminum oxide (Al_2O_3, analytical-reagent grade, a.g.) to

reduce eventual turbidity. Shake for 5 min and then dilute to 1000 mL with distilled water. Afterwards filter through a dense filter. The clear solution is stable for about 2 weeks. If the solution is stored in a glass container, then turbidity may occur due to a release of potassium from the glass walls. In this case the solution must be filtered before use.

2. *Dilute 'Kalignost' solution:* Dissolve 50 mg of 'Kalignost', $Na[B(C_6H_5)_4]$, in 500 mL of distilled water, add 0.5–1 g of Al_2O_3, shake and filter as described for reagent 1.
3. *Sodium carbonate*, Na_2CO_3 (a.g.), 5 % (w/v).
4. *Glacial acetic acid*, CH_3COOH (a.g.).
5. *Acetone*, CH_3COCH_3 (a.g.).
6. 250 mL beakers, with watch glasses.
7. *Glass frit filter*: G4, with porosity 10–16 μm.
8. *Drying-oven.*
9. *Desiccator.*

11.2.3.3   Analysis of the sample

Transfer about 100 mL of the seawater sample (for ocean water use 50 mL) into a 250 mL beaker covered with a watch glass and weigh to ± 0.02 g. Add 10 mL of the Na_2CO_3 solution (5 %) while stirring, then add 10 mL of the 'Kalignost' solution (reagent 1). Carefully add about 1 mL of glacial acetic acid to a pH of 4–6 (check by pH paper). Take care to lose no solution by 'CO_2-splashes'. Wait for 10 min, with stirring from time to time.

Collect the precipitate on a glass frit filter which has been heated to 120 °C (for 2 h) in the drying-oven and weighed before use. Wash the precipitate with several portions of a few mL of the dilute 'Kalignost' solution (a total of about 15–20 mL), then rinse the precipitate with 3–6 mL of distilled water. Finally, dry the precipitate for 2–3 h in a drying-oven at maximum temperature of 100 °C. After cooling in a desiccator, reweigh the glass frit filter.

Note: The frit is best cleaned by rinsing with distilled water and about 5 mL of acetone to dissolve the residual precipitate.

11.2.3.4   Calculation of results

The concentration of potassium is calculated from

$$K \, (g/kg) = \frac{a \cdot F \cdot 1000}{w}$$

where

a =   mass of precipitate (g); with correction for buoyancy
w =   mass of seawater sample (g); with correction for buoyancy
F =   stoichiometric conversion factor $K/K[B(C_6H_5)_4]$ with 0.1091

11.2.3.5   Precision

The method allows very precise determinations of even low potassium concentrations, which is mainly due to the high relative molecular mass of the precipitate relative to that of potassium (39.098). This author found recently that the c.v. of more than 100 analyses of Baltic Sea water on average was 0.26 %.

11.2.3.6 Potentiometric method

After precipitation of the halogen ions with $AgNO_3$, potassium is precipitated with sodium tetraphenylborate as described. The excess of tetraphenylborate is then determined by $AgNO_3$ titration, using a silver electrode for endpoint detection. The method has been applied to seawater samples as small as 1 mL as well as for interstitial waters, with a precision of better than 1 %. The procedure saves the tedious filtration and drying process but offers a somewhat lower precision than the gravimetric method. A detailed description is presented by *Marquis and Lebel* (1981).

11.2.4 Chlorinity (high precision method)

The chlorinity (Cl, in ‰) was originally defined as the mass of chlorine in 1 kg of seawater which is equivalent to the total amount of halides measured by titration with $AgNO_3$. Since changes occured in the relative atomic masses of Ag and Cl, the chlorinity was redefined by *Jacobsen and Knudsen* (1940) as the mass in grams of pure silver necessary to precipitate the halogens in 328.523 grams of seawater sample; *i.e.*, $Cl(‰) = 0.328\,523\,Ag\,(‰)$, where $Ag(‰)$ is grams of silver per kilogram of seawater. The 1969 relative atomic masses yield a value of 0.328 670, thus $0.328\,670/0.328\,523 = 1.000\,45$ times the $Cl(‰)$. For example, in seawater with a chlorinity of 19.374 (corresponding to a salinity of 35), the concentration of chloride is 19.353 g/kg (Table 11-1). The other halides converted into chloride add another 0.030 g/kg of chlorine equivalents. The difference of 0.009 g of chloride per kilogram of seawater beween the chlorinity and the total halide expressed as chloride reflects the corrections of relative atomic masses, which have occurred since standard seawaters were first introduced by *Forch et al.* (1902).

11.2.4.1 Principle of the method

As discussed, the chlorinity is determined by titration with silver nitrate solution precipitating the halides dissolved in seawater as insoluble silver halides with silver ions:

$$Ag^+ + Cl^-, Br^-, I^- \rightarrow AgCl(s), AgBr(s), AgI(s)$$

The titration endpoint is detected potentiometrically using an Ag/AgCl electrode, preferably by an automatic method. The procedure described here is based on an execptionally precise method outlined by *Hermann* (1951)

In this method, the chlorinity is first determined approximately by a modified Mohr-Knudsen titration to an accuracy of about ± 0.02‰ *(Grasshoff and Wenck*, 1972). Next, for the precise determination, a suitable amount of concentrated silver nitrate solution (standardized by comparison with batches of Standard Seawater) is weighed in a glass beaker. Then slightly more than the equivalent amount of sample seawater is weighed into the same beaker. The sample is titrated to the stochiometric endpoint with dilute silver nitrate solution.

11.2.4.2   Reagents and equipment

1. *Concentrated silver nitrate solution:* Dissolve 48.4 g of $AgNO_3$ (analytical-reagent grade, a.g.) in 1000 mL of distilled water. Store the solution in a light-shielded bottle (preferably amber glass or painted black).
2. *Dilute silver nitrate solution:* Dilute the concentrated $AgNO_3$ solution (see reagent 1) by 1 : 40 with distilled water. Store this solution in a light-shielded bottle also.
3. *Potassium chromate solution:* Dissolve 8 g of K_2CrO_4 (a.g.) in 100 mL of distilled water. This solution is stable indefinitely if kept in a ground-stoppered glass bottle.
   Remarks: Solutions (as under 1–3) must be made up in distilled water completely free of traces of halides.
4. *Standard Seawater, SSW* (as from Standard Seawater Service, Wormley, UK.; see also Chapter 3).
5. *Dilute SSW:* Weigh about 10 mL of SSW (as under 4) within to ±0.1 mg and dilute to 500 mL with distilled water. Then 1 mL of this solution corresponds to

$$g\,Cl = \frac{a \cdot b}{10^3 \cdot 500} \tag{11-5}$$

where

> a = chlorinity of the SSW (*e.g.,* 19.374 ‰)
> b = mass in g of the SSW volume (10 mL) used for the 'Dilute SSW'

6. *Automated titration system:* For example, Metrohm SM-Titrino 702, with subdivision of 0.005 mL; connected to a combined Ag/AgCl electrode. It is also possible to plot the titration curve by measuring the potential (E) with a digital voltmeter using two silver wires as electrodes; one placed in the burette (*i.e.,* in dilute $AgNO_3$ solution) maintaining a constant potential and the other wire is immersed in the seawater sample (as the AgCl electrode). This electrode must be cleaned from time to time.
7. *Accurate burette:* For example, Metrohm Dosimat, 20 mL, used for the Mohr-Knudsen titration. The connections and the piston burette should be either amber glass or painted black to avoid any photochemical reaction of the silver ions (as for apparatus under 6.).
8. 100 mL *conical flasks:* Of amber glass as titration beakers.
9. *Magnetic stirrer.*

11.2.4.3   Standardization of the silver nitrate solutions

The dilute and concentrated $AgNO_3$ solutions are calibrated with SSW (as prepared under Section 11.2.4.2). The estimation of the chlorinity is based on the salinity using the relationship $S = 1.806\,55\,Cl$; *i.e.,* a SSW batch with a salinity of 35.000 (or a K_{15} value of 1.000 00; see also Chapter 3) has a chlorinity of 19.374 ‰.

1. *Calibration of dilute AgNO₃ solution:* Transfer exactly 10 mL of the dilute SSW (see Section 11.2.4.2) into a titration beaker and titrate with the dilute $AgNO_3$ solution to the end-

point. An automated titration system will registrate the titration curve together with the amount of AgNO$_3$ consumed (in mL) to reach the equivalence point. In case where the more 'primitive' but much cheaper voltmeter system is applied (see Section 11.2.4.2), it is preferable to plot the first derivative of the function manually, $E = f [Ag^+]$. The dilute AgNO$_3$ should then be added in increments of 0.1 mL. According to Eq. (11-5), 1 mL of 'dilute AgNO$_3$ solution' corresponds to

$$g\ Cl = \frac{a \cdot b \cdot 10}{10^3 \cdot 500 \cdot c} \tag{11-6}$$

where

    c =   consumption of 'dilute AgNO$_3$ solution' for 10 mL of 'dilute SSW'

2. *Calibration of the concentrated AgNO$_3$ solution:* Transfer 10 mL of SSW into a titration beaker, weigh and dilute with about 25 mL of halide-free distilled water. Then add 6 drops of the potassium chromate solution. Adjust the speed of the stirrer bar to ensure thorough mixing. Titrate with concentrated AgNO$_3$ solution (*e.g.*, by means of a piston burette) until the first indication of a constant colour change of the precipitate (due to precipitation of Ag$_2$CrO$_4$). Rinse any droplets from the walls of the beaker. Then slow down the piston speed and continue the titration. The endpoint is reached when the colour has changed from yellow to a shade of rather dirty orange.

Weigh another 10 mL of SSW in a titration beaker and add approximately 0.2 mL less than the amount of concentrated AgNO$_3$ solution (with a piston burette) as determined above (no chromate indicator is added at this time). Then weigh again and titrate with the 'dilute AgNO$_3$ solution' (potentiometrically) until the endpoint as described for the 'calibra-ation of dilute AgNO$_3$ solution'. Take care that no air bubbles are trapped in the burette while titrating.

According to Eq. (11-5), the weight of SSW (b' in grams) used for this calibration corre-sponds to

$$g\ Cl = \frac{a \cdot b'}{10^3} = A \tag{11-7}$$

From this figure must be subtracted the amount of gCl which was consumed by the 'dilute AgNO$_3$ solution' to reach the endpoint (in v mL). This corresponds to (see Eq. (11-6))

$$g\ Cl = \frac{a \cdot b \cdot 10 \cdot v_{AgNO_3}}{10^3 \cdot 500 \cdot c} = B \tag{11-8}$$

If the weight of the added 'concentrated AgNO$_3$ solution' (in grams) is C and considering results of Eqs. (11-7) and (11-8), then 1 g of this solution corresponds to

$$g\ Cl = \frac{A - B}{C} \tag{11-9}$$

11.2.4.4 Titration of the sample

Transfer 10 mL of the seawater sample (or more, if the chlorinity of the sample is expected to be much lower than 'Standard Seawater') into a titration beaker, weigh and proceed in exactly the same way and with the same precautions as described under the standardization procedure (*i.e.*, start with Mohr-Knudsen titration *etc.*).

11.2.4.5 Calculation of the chlorinity

Example calculation:

Mass of the sample		10.0342 g
Mass of the conc. $AgNO_3$		9.0872 g
Consumption of dilute $AgNO_3$		4.25 mL
1 g conc. $AgNO_3$	$\equiv$	$0.920\,05 \cdot 10^{-2}$ g Cl
1 mL dil. $AgNO_3$	$\equiv$	$2.012 \cdot 10^{-4}$ g Cl
Then		
$9.0872 \cdot 0.920\,05 \cdot 10^{-2}$	$=$	$8.3607 \cdot 10^{-2}$ g Cl
and $4.25 \cdot 2.012 \cdot 10^{-4}$	$=$	$8.55 \cdot 10^{-4}$ g Cl
	$=$	$8.3607 \cdot 10^{-2} + 8.55 \cdot 10^{-4}$ g Cl
	$=$	$8.4462 \cdot 10^{-2}$ g Cl

The chlorinity of the sample is then

$$\frac{8.442 \cdot 10^{-2} \cdot 10^3}{10.0342} \quad = \quad \underline{8.417\ \text{\textperthousand}}$$

11.2.4.6 Precision

The method described here covers practically the entire range of chlorinities occurring in natural seawater. With correct calibrations and some experience, the standard deviation is ± 0.001 Cl ‰ or better.

11.2.5 Sulphate

In the classic gravimetric method the sulphate is precipitated and weighed as barium sulphate. Although it has been applied to seawater for more than 150 years, the method is still preferred because of its high precision and the minimum amount of equipment required to perform the determination. The procedure by *Bather and Riley* (1954) is described here in detail.

A potentiometric back-titration method of excess barium ions following precipitation as $BaSO_4$ (*Mucci*, 1991) is outlined in Section 11.2.5.6.

11.2.5.1 Principle of the method

When barium sulphate ($BaSO_4$) is precipitated directly out of seawater, the serious errors encountered are due to the co-precipitation of sodium, potassium, calcium and strontium ions. The interference by alkali metals is minimized in the presence of picric acid. Lowering the pH by a suitable addition of HCl will reduce the co-precipitation of calcium and strontium sulphate. However, if the solution is too acidic, low results are obtained because solubility of barium sulphate increases with decreasing pH. The concentration of the acid should therefore be optimized for quantitative recovery (*Bather and Riley*, 1954). For good crystallization and filtration, precipitation should be performed from a hot solution.

11.2.5.2 Reagents and equipment

1. *Barium chloride dihydrate*, $BaCl_2 \cdot 2H_2O$, (analytical-reagent grade, a.g.), 10 % (w/v).
2. *Hydrochloric acid*, HCl, (a.g.), concentrated.
3. *Picric acid:* 2,4,6-Trinitrophenol, (a.g.), saturated aqueous solution.
4. *600 mL beakers.*
5. *Sintered silica filter, low porosity.*
6. *Muffle furnace.*
7. *Desiccator.*

11.2.5.3 Analysis of the sample

Transfer about 50 mL of a seawater sample into a 600 mL beaker covered with a watch-glass and weigh to ± 0.02 g. Dilute with 235 mL of distilled water and add 10 mL of saturated picric acid solution and 5 mL of concentrated hydrochloric acid. Heat the solution to about 90 °C, taking care to prevent boiling and add 10 mL of a hot 10 % solution of barium chloride while stirring vigorously. Allow the precipitate to settle whilst the solution is kept hot.

Collect the precipitate on a sintered silica filter of low porosity which has been ignited and weighed before use. Transfer the barium sulphate (after filtration of the supernatant solution) to the crucible with a jet of hot water and wash with portions of a few mL of warm water until the filtrate gives no opalescence with silver nitrate solution.

Dry the crucible for 30 min at 110 °C and then heat for 2 h at 800 °C in a muffle furnace. After cooling for at least 1 h in a desiccator, weigh the filter again.

11.2.5.4 Calculation of results

The concentration of sulphate is calculated from

$$SO_4^{2-} \text{ (g/kg)} = \frac{a \cdot F \cdot 1000}{w}$$

where
  a = mass of precipitate in g; with correction for buoyancy
  w = mass of seawater sample in g; with correction for buoyancy
  F = stoichiometric conversion factor $SO_4/BaSO_4$ (0.4116)

11.2.5.5   Precision

With seawater from the Irish Sea the precision of the method has been established to be 0.14 % (*Bather and Riley*, 1954). A similar c.v. (0.16 %) has been found by *Kremling* (1969) for a low salinity Baltic Sea water sample with a sulphate concentration of 1.18 g/kg.

11.2.5.6   Potentiometric back-titration method

Sulphate concentrations may also be determined accurately by potentiometric back-titration of excess Ba^{2+} with a mercury electrode following the precipitation of $BaSO_4$ (*Mucci*, 1991). The seawater sample is freed from most seasalt cations with an ion-exchange-column. Then the eluate is reacted with an excess of barium, and after filtration of precipitated $BaSO_4$, the solution is titrated potentiometrically with an EGTA solution (see also Section 11.2) to the endpoint. The method applies over a wide range of salinities and sulphate concentrations in 1 mL or less of seawater and marine pore water samples, however, it is somewhat less precise (c.v. of about 0.6 %) than the 'simple' gravimetric procedure described.

11.2.6   Bromide

The iodometric method, principally developed by *Kolthoff and Yutzy* (1937) and first applied to seawater analysis by *Thompson and Korpi* (1942) is the most accurate procedure for the estimation of bromide in seawater.

11.2.6.1   Principle of the method

In a slightly acidic solution, bromide is oxidized to bromate with hypochlorite which is followed by its reaction with iodide and acid according to the equation:

$$BrO_3^- + 9I^- + 6H^+ \rightarrow 3I_3^- + Br^- + 3H_2$$
$$I_3^- + 2S_2O_3^{2-} \rightarrow 3I^- + S_4O_6^{2-}$$

(11-10)

The reaction is relatively slow but can be catalysed with molybdate ions. The high sensitivity of the method is based on the fact that 1 mole of bromate produces 3 moles iodine (I_2) which are then reduced by 6 moles of thiosulphate.

11.2.6.2   Reagents and equipment

1. *Phosphate buffer*, $NaH_2PO_4 \cdot H_2O$ (analytical-reagent grade, a.g.): Dissolve 25 g of sodium dihydrogen phosphate in about 250 mL of distilled water and dilute to 500 mL.
2. *Sodium chloride solution*, NaCl, (a.g.), 10 % (w/v).
3. *Sodium hypochlorite solution*, NaOCl, 1 mol/L in 0.1 mol/L NaOH: Prepare a 1 mol/L solution of sodium hypochlorite (low in bromine) in a 0.1 mol/L sodium hyroxide solution (a.g.).

4. *Sodium formate solution*, HCOONa, 50 % (w/v): Dissolve 30 g of sodium hydroxide (a.g.), NaOH, in about 50 mL of distilled water. Cool the solution in an ice bath and add 32 mL of 90 % formic acid (a.g.), HCOOH, while stirring. Dilute to 100 mL.
5. *Ammonium molybdate solution*, $(NH_4)_6 Mo_7O_{24} \cdot 4H_2O$, (a.g.), 3 % (w/v).
6. *Sulphuric acid*, H_2SO_4 (a.g.), 3 mol/L.
7. *Starch solution* (see Chapter 4).
8. *Potassium iodide*, KI (a.g.), crystals.
9. *Sodium thiosulphate solution*, $Na_2S_2O_3$, 0.002 mol/L: Prepare daily an approximately 0.002 mol/L solution of a 0.1 mol/L standard stock solution (*e.g.*, 'Titrisol', Merck). A 1000 mL stock solution is stabilized by addition of 1 mL of amyl alcohol. The thiosulphate solutions should be kept in the dark. The exact titre of the 0.002 mol/L solution is determined as described.
10. *Accurate burette*: (For example a piston burette), calibrated with subdivisions of 0.01 mL.
11. *Erlenmeyer flask*: Stoppered, 250 mL.

11.2.6.3 Standardization of the sodium thiosulphate solution

Standardize the 0.002 mol/L thiosulphate solution with a potassium bromate solution prepared by dissolving 2.7834 g of $KBrO_3$ (recrystallized 3–4 times from hot water and dried at 180 °C) in distilled water and diluted to 1000 mL, giving a $1.667 \cdot 10^{-2}$ mol/L solution. Prepare the working standard ($3.334 \cdot 10^{-4}$ mol/L) by accurate dilution with distilled water 1 : 50. This solution has an oxidation concentration of 0.0020 mol/L of electrons. Transfer 40.00 mL aliquots (with an accurate burette, *e.g.*, piston burette) of the working standard into an Erlenmeyer flask, add about 0.25 g of potassium iodide, 4 drops of 3 % ammonium molybdate, 10 mL of 3 mol/L sulphuric acid and titrate in the same manner as described for the procedure in Section 11.2.6.4.

$$M_{Na_2S_2O_3} = \frac{3.334 \cdot 10^{-4} \cdot 6 \cdot a}{b}$$

where

a =   mL potassium bromate standard solution used
b =   mL thiosulphate solution required
$M_{Na_2S_2O_3}$ =   molarity of the thiosulphate solution.

11.2.6.4 Analysis of the sample

Approximately 10 g seawater are weighed into a 250 mL stoppered Erlenmeyer flask. Add 10 mL of sodium chloride, 10 mL of phosphate buffer and 2 mL of hypochlorite. Dilute to about 40 mL and heat the solution on a hot plate or with a flame. Boil for a maximum of 5 s, remove from the hot plate and carefully add 5 mL of sodium formate solution whilst stirring. Rinse 'CO_2-splashes' from the flask with distilled water and cool to room temperature.
Then add 0.25 g of potassium iodide, 4 drops of molybdate solution, 10 mL of 3 mol/L sulphuric acid and freshly prepared starch solution. Start the titration after 30 s with standardized 0.002 mol/L sodium thiosulphate solution until the solution is colourless.

The endpoint detection is possible within to ±0.01 mL. Carry out the titration without interruption, thus preventing photooxidation of iodide and evaporation of the liberated iodine.

The reagent blank can be fairly large but is, from the author's experience, highly reproducible. The value is established from several runs, substituting the seawater sample with distilled water.

11.2.6.5 Calculation of results

Concentrations of bromide in seawater are estimated from:

$$Br^-(g/kg) = \frac{a \cdot M_{Na_2S_2O_3}}{w} \cdot \frac{79.904}{6}$$

where

a	=	volume in mL of thiosulphate solution consumed in the titration, less the reagent blank
$M_{Na_2S_2O_3}$	=	molarity of the thiosulphate solution
w	=	mass of seawater sample in g; with correction for buoyancy

11.2.6.6 Precision

The precision of the procedure has been investigated by *Morris and Riley* (1966) who carried out 10 replicate analyses on an average bromide concentration of 0.06058 g/kg. The computed coefficient of variation was 0.15 %.

Reproducibility in the low salinity range has been found to be much worse. Replicate measurements (10 analyses) on a Baltic Sea sample (0.014 23 g/kg Br⁻) resulted in a c.v. of 0.57 % (*Kremling*, 1969).

11.2.7 Boron

For determination of the total boron content, the spectrophotometric curcumin method is described, which was first applied to seawater by *Greenhalgh and Riley* (1962) but was modified and simplified by *Uppström* (1968). An outline of the mannitol-boric acid method is presented in Section 11.2.7.7.

11.2.7.1 Principle of the method

With curcumin (1,7-bis(4-hydroxy-3-methoxyphenyl)-1,6-heptadiene-3,5-dione) boron forms two different complexes; both procedures are known in the literature as 'the curcumin method'. Rubrocurcumin is formed from boric acid, curcumin and oxalic acid in a 1 : 1 : 1 complex; rosocyanin forms in the presence of strong sulphuric acid by the reaction between boric acid and curcumin, probably in a ratio of 1 : 2. An extensive study of the chemistry has been given by *Dyrssen et al.* (1972). Rosocyanin is the preferred complex for boron determinations because of its high molar absorptivity (at 545 nm) and excellent stability against hydrolysis. The elimination of water is very important for the dye formation; in the method of *Uppström* (1968) it is effected by propionic acid anhydride catalysed with oxalyl chloride. The colour of the rosocyanin complex is completely stable for about 2 h, after which time it starts to fade slightly. After 5 d the absorbance decreases by only about 10 %. Salinity does not interfere.

11.2.7.2 Reagents and equipment

1. *Curcumin-acetic acid reagent:* Dissolve 0.125 g of curcumin, $C_{21}H_{20}O_6$, in 100 mL of glacial acetic acid, CH_3COOH. The efficiency of the reagent remains unchanged for at least 2 weeks.
2. *Sulphuric-acetic acid reagent:* Mix equal volumes of conc. sulphuric acid, H_2SO_4 ($d = 1.84$) and glacial acetic acid.
3. *Propionic acid anhydride,* $(CH_3CH_2CO)_2O$, > 97 %.
4. *Oxalyl chloride,* $(COCl)_2$.
5. *Buffer solution:* Mix 135 mL of glacial acetic acid with 90 mL of ethanol (96 %) and 180 g of ammonium acetate (CH_3COONH_4) in a 1 L flask and dilute to volume with distilled water.
6. *Standard stock solution of boric acid:* Dissolve 0.5715 g of boric acid (H_3BO_3) in distilled water and dilute to 1000 mL, to give 0.1 mg/mL B.
7. *Spectrophotometer* with 1 cm *cells.*
8. *Micropipette,* 0.500 mL.
9. 50 mL *polyethylene beaker.*

11.2.7.3 Analysis of the sample

To 0.500 mL of the seawater sample (with a boron content of 0–5 mg/L) in the polyethylene beaker add 1.0 mL of glacial acetic acid and 3.0 mL of propionic acid anhydride. Mix by swirling and then add dropwise 0.25 mL of oxalyl chloride and wait at least 15 min for it to cool to room temperature.

Add 3.0 mL of the sulphuric-acetic acid mixture and 3.0 mL of curcumin reagent to the solution, mix thoroughly and wait for at least 30 min.

Finally, add 20 mL of the buffer solution. After cooling to room temperature, measure the absorbance of the orange solution at 545 nm in 1 cm cells against a reagent blank, obtained in the same manner with distilled water.

The rosocyanin complex is insensitive to light. Because of hydrogen chloride formation the reaction should be performed in a fume hood.

11.2.7.4 Calibration

To compensate for slight fading of the complex colour, determine a calibration curve with each set of seawater samples.

Prepare boron standards of 1, 2, 3, 4 and 5 mg/L B from the boric acid stock solution and follow the procedure described in Section 11.2.7.3. Plot absorbances against corresponding boron concentrations.

11.2.7.5   Calculation of results

The boron concentration in the sample is calculated from

$$B\ (g/kg) = \frac{C}{1000 \cdot d}$$

where

C = concentration of boron from calibration curve in mg/L B
d = density of the seawater sample

11.2.7.6   Precision

Uppström (1968) determined the c.v. for standard boric acid solutions of 0.5, 3.0 and 8.0 mg/L B as 0.9, 1.2 and 1.6 %, respectively. Replicate measurements of Baltic Sea water samples with about 2 mg/L B yielded a reproducibilty of ± 0.87 % (*Kremling*, 1972).

11.2.7.7   Mannitol-boric acid method

The method is based on the tendency of organic polyhydroxy-compounds with two or more hydroxyl groups in the *cis*-position to form complexes with boric acid (*Gripenberg*, 1966). When about 5 g of mannitol or another hexose is added to 100 mL of neutralized (pH 7) seawater, the acidity increases to a pH near 3–4. The amount of alkali used to bring the sample back to neutrality is then equivalent to the boric acid content.

With boron-free glassware and potentiometric titration, a c.v. of about 1 % may be obtained. A detailed procedure, but more time-consuming than the curcumin method, has been presented by *Noakes and Hood* (1961).

11.2.8   Fluoride

The spectrophotometric procedure with alizarin complexone is the most widely used method for the determination of fluoride in seawater. Small modifications have been made to the method first proposed by *Greenhalgh and Riley* (1961).

11.2.8.1   Principle of the method

In the presence of fluoride, the wine-red chelate fromed at pH 4.5 between alizarin complexone (1,2-dihydroxyanthraquinonyl-3-methylamine-*N'N*-diacetic acid) and lanthanum is converted into a stable blue ternary complex. The complex whose structure has been investigated by *Langmyhr et al.* (1971) apparently contains fluoride and lanthanum alizarin complexone in a 1 : 1 molar ratio. Highest absorbances, measured at 622 nm, are obtained at a pH of about 4.50. There is a linear relationship between fluoride content and absorbance up to *ca.* 25 μg of F. The calibration curve must be corrected for the salt error due to the presence of magnesium.

11.2.8.2   Reagents and equipment

1. *Combined lanthanum alizarin complexone reagent:*
   a) Dissolve 0.0479 g of alizarin complexone ($C_{19}H_{15}NO_8 \cdot 2H_2O$) in 0.1 mL of concentrated ammonia solution (NH_4OH) and 1.0 mL of 20 % (w/v) ammonium acetate (CH_3COONH_4) together with a few mL of distilled water.
   b) Filter the solution into a 200 mL calibrated flask containing 8.2 g of anhydrous sodium acetate CH_3COONa (analytical-reagent grade, a.g.), 6.0 mL of glacial acetic acid (CH_3COOH) and enough distilled water to dissolve the solids. Wash the filter with a small volume of distilled water. Slowly add 100 mL of acetone (CH_3COCH_3) while swirling.
   c) Dissolve 0.105 42 g of lanthanum nitrate ($La(NO_3)_3$, spectrographic grade) in 2.5 mL of 2 mol/L hydrochloric acid (HCl), warming gently to aid dissolution. Then mix with the aqueous acetone solution. Dilute to 200 mL with distilled water. Mix well and readjust the volume after about 30 min. The reagent is stable for at least 1 week.
2. *Sodium fluoride solution,* NaF: Prepare a stock solution of 1 mg/mL F by dissolving 2.210 g of sodium fluoride (a.g., dried at 105 °C for 2 h) in 50 mL of distilled water plus 1 mL of 0.1 mol/L sodium hydroxide solution and dilute to 1000 mL.
   A working standard containing 10 μg/mL F is freshly prepared by dilution of the stock solution.
3. *Artificial seawater:* Prepare an artificial seawater solution, but without sodium fluoride (see Tables 1 and 2 in the Appendix).
4. *Acetic acid,* CH_3COOH (a.g.), 6 mol/L.
5. *Spectrophotometer* with 1 cm *cells.*
6. 25 mL *flasks,* calibrated.
7. *Accurate burette:* (For example, piston burette), calibrated, with subdivisions of 0.01 mL.

11.2.8.3   Analysis of the sample

Weigh approximately 15 mL of seawater into a 25-mL volumetric flask. Add 8.0 mL of the combined alizarin reagent, dilute to volume, and mix well. When diluted to volume, the solution should have a pH of 4.50±0.02.

Samples with chlorinities considerably less than 20 ‰ require pH adjustment with (0.4– 0.02x) mL of 6 mol/L acetic acid, where x is the chlorinity of the water which should be known to the nearest 5 units (*i.e.,* to 0, 5, 10 or 15 ‰).

Prepare a reagent blank with pH 4.50±0.02 by adding the same reagent volume to 15 mL of distilled water containing 0.4 mL of 6 mol/L acetic acid.

Measure the absorbance after 30 min at 622 nm in a 1 cm cell against the reagent blank.

11.2.8.4   Calibration

To compensate for the salt error, calibrate the method as described in Section 11.2.8.3 using the artificial seawater. To 15 mL aliquots add 0.5, 1.0, 1.5 and 2.0 mL of the fluoride working standard and 8.0 mL of the alizarin reagent. For seawater samples with different chlorinities, prepare individual calibration curves with chlorinities of 0, 5, 10, 15 and 20 ‰.

11.2.8.5   Calculation of results

The fluoride concentration is calculated as:

$$F\,(g/kg) = \frac{a}{w \cdot 1000}$$

where

a =   amount of F in μg from the calibration curve
w =   mass of seawater sample in g; with correction for buoyancy

11.2.8.6   Precision

Greenhalgh and Riley (1961) determined the c.v. with a fluoride content of 1.37 mg/kg by eleven replicate measurements as 0.28 %. The precision is much less in the low chlorinity range. Recently *Kremling and Wilhelm* (unpublished) determined from replicate measurements of Baltic Sea water samples a mean c.v. of 1.1 %.

References to Chapter 11

Bather, J.M., Riley, J.P. (1954), *J. du Conseil,* 20, 145.

Carpenter, J.H., Manella, M.E. (1973), *J. Geophys. Res.,* 78, 3621.

Culkin, F., Cox, R.A. (1966), *Deep-Sea Res.,* 13, 789.

Dyrssen, D.W., Novikow, Y.P., Uppström, L.R. (1972), *Anal. Chim. Acta,* 60, 139.

Forch, C., Knudsen, M., Sorensen, S.P.L. (1902), *Kgl. Dan. Vidensk. Selsk. Roekke naturvidensk., og. methem. Afd XII, I., Skrifter,* 6, 151pp.

Gieskes, J.M., Gamo, T., Brumsack, H. (1991), *Joides Resolution Ocean Drilling Program,* Texas A & M University, Technical Note 15.

Grasshoff, K., Wenck, A. (1972), *J. Cons. Perm. Int. Explor. Mer,* 34, 522.

Greenhalgh, R., Riley, J.P. (1961), *Anal. Chim. Acta,* 25, 179.

Greenhalgh, R., Riley, J.P. (1962), *Analyst,* 87, 970.

Greenhalgh, R., Riley, J.P., Tongudai, M. (1966), *Anal. Chim. Acta,* 36, 439.

Gripenberg, S. (1966), *Societas Scientiarum Fennica Comment. Phys.-Math.,* 32, 1.

Hermann, F. (1951), *J. du Conseil,* 17, 223.

Jacobsen, J.P., Knudsen, M. (1940), *Int. Union Geodesy Geophys.-Assoc. Phys. Oceanogr., Publ. Sci.,* 7, 38pp.

Kolthoff, I.M., Yutzy, H. (1937), *Ind. Engng. Chem., Analyt. Edit.,* 9, 75.

Kremling, K. (1969), *Kieler Meeresforschung,* 25, 81.

Kremling, K. (1972), *Kieler Meeresforschung,* 28, 99.

Kremling, K., Wilhelm, G. (1997), *Mar. Poll. Bull.,* 3, 763.

Langmyhr, R.J., Klausen, K.S., Nouri-Nekoni, M.H. (1971), *Anal. Chim. Acta,* 57, 341.

Marquis, G., Lebel, J. (1981), *Anal. Lett.,* 14, 913.

Millero, F. (1982), *Ocean Sci. Eng.,* 7, 403.

Millero, F., Sohn, M.L. (1992), *Chemical Oceanography.* London: CRC Press.

Morris, A.W., Riley, J.P. (1966), *Deep-Sea Res.,* 13, 699.

Mucci, A. (1991), *Limnol. Oceanogr.,* 36, 409.

Noakes, J.E., Hood, D.W. (1961), *Deep-Sea Res.,* 8, 121.

Pate, J.B., Robinson, R.J. (1961), *J. Mar. Res.,* 19, 12.

Raff, P., Brotz, W. (1951), *Z. Anal. Chem.,* 133, 241.

Riley, J.P., Tongudai, M. (1967), *Chem. Geol.,* 2, 263.

Schwarzenbach, G., Kampitsch, E., Steiner, R. (1945), *Helv. Chim. Acta,* 28, 828.

Schwarzenbach, G., Senn, H. Anderegg, G. (1957), *Helv. Chim. Acta,* 40, 1886.

Thompson, R.G., Korpi, E. (1942), *J. Mar. Res.,* 5, 28.

Tsunogai, S., Nishimura, M., Nakaya, S. (1968), *Talanta,* 15, 385.

Uppström, L.R. (1968), *Anal. Chim. Acta,* 43, 475.

Voipio, A. (1959), *Suom. Kemistilehti,* B32, 61.

Wittig, G. (1950), *Angew. Chem.,* 62, 231.

12 Determination of trace elements

12.1–12.2.2 K. Kremling
12.2.3   M.O. Andreae
12.2.4   L. Brügmann
12.3   C.M.G. van den Berg
12.4   A. Prange and M. Schirmacher
12.5   F. Koroleff and K. Kremling
12.6   L. Brügmann and J. Kuss

12.1 Introduction

K. Kremling

12.1.1 Oceanic concentrations and distributions

Over the last two decades enormous progress has been made with respect to the knowledge of distributions and chemical behaviour of trace elements (TE) in seawater. Important factors initiating these advantages were the application of 'clean techniques' for collection and storage of samples, and major advances in analytical methods and instrumentation (*e.g., Wong et al., 1983; Bruland, 1983; Burton and Statham, 1990*). We now know the concentrations and distributions for most of the elements in the Periodic Table. We have gained much more insight into their biogeochemical behaviour in the oceans, and it is generally accepted that the interaction of dissolved TE with particles suspended in seawater is the predominant mechnism of the observed concentration and distribution patterns (*e.g., Clegg and Whitfield, 1990; Bruland et al., 1994*). Also, much attention has recently been focused on the nutritional role of selected TE in ocean surface waters. For discussion of the major advances we refer to *Kolber et al.* (1994), *Morel et al.* (1994), *Sunda and Huntsman* (1995) and *Coale et al.*, (1996).

For an overview, the North Pacific and North Atlantic surface and deep water concentrations are compiled in Table 12-1. We do not attempt here a detailed discussion of the chemistry of trace elements in the oceans. It seems, however, worthwhile to characterize briefly the three principal distribution types of elements as proposed by *Whitfield and Turner* (1987). Regarding their biogeochemical interaction with particles, the different groups can be distinguished as conservative (C), recycled (R) and scavenged (S) (see Table 12-1). Conservative trace elements (C) include monovalent cations such as Cs^+ or Tl^+ or negatively charged anions such as $UO_2(CO_3)_3^{4-}$ or MoO_4^{2-}. They interact weakly with particles and have long residence times ($> 10^5$ y) relative to the mixing time of the oceans. Recycled (R) or nutrient-type elements (*e.g.*, Cd and Zn) are involved in the internal cycles of biologically derived particulate matter; *i.e.*, their concentrations are depleted in surface waters and increase with depth (similar to phosphate, nitrate and silicate, see Chapter 10). The residence times are in the range of 10^3–10^5 y. Their fractionation between the deep oceans follows the deep water flow pattern; *i.e.*, Cd and Zn concentrations are 3 and 5 times higher, respectively, in the deep North Pacific than at comparable depths in the North Atlantic (Table 12-1).

Table 12-1. Concentrations and distribution types of the trace elements in ocean waters (according to *Donat* and *Bruland*, 1995).

Element	Concentration Units [a]	North Pacific Surface	North Pacific Deep	North Atlantic Surface	North Atlantic Deep	Distribution Type [b]
Be	pM	4–6	28–32	10	20	R+S
Al	nM	0.3–5	0.5–2	30–43	15–35	S
Sc	pM	8	18	14	20	R
Ti	pM	4–8	200–300	30–60	200	R+S
V	nM	32	36	23		R
Cr	nM	3	5	3.5	4.5	R
Mn	nM	0.5–3	0.08–0.5	1–3	0.25–0.5	S
Fe	nM	0.02–0.5	0.5–1	0.05–1	0.6–1	R+S
Co	pM	4–50	10–20	18–300	20–30	S
Ni	nM	2	11–12	2	6	R
Cu	nM	0.5–1.3	4.5	1.0–1.3	2	R+S
Zn	nM	0.1–0.2	8.2	0.1–0.2	1.6	R
Ga	pM	12	30	25–30	?	S
Ge	pM	5	100	1	20	R
As	nM	20	24	20	21	S
Se	nM	0.5	2.3	0.5	1.5	R
Y	nM				(0.15)	?
Zr	pM	12–95	275–325	100	(150)[e]	R+S
Nb	pM		(<50)			?
Ru	fM		(<50)			?
Rh			?			?
Pd	pM	0.18	0.66			?
Ag	pM	1–5	23			R
Cd	pM	1–10	1000	1–10	350	R
In	pM	<0.5–1.8		2.7	0.9	S?
Sn	pM	4	10–20	8		?
Sb	nM	(0.74–13)		1.7		?
Te	pM	1.2	1	1–1.5	0.4–1	S
Cs	nM	2.3				C
La	pM	20	50–70	12–15	80–84	S+R
Ce	pM	10	3	80	40–60	S
Pr	pM	3–4	7–9	3–4	10	S+R
Nd	pM	13–16	40–50	12.8	24.9	S+R
Sm	pM	2.6–2.8	8–9	3–4	7.6–8.0	S+R
Eu	pM	0.73	2.3	0.6–0.8	1.6–1.8	S+R
Gd	pM	3.8	12–13			S+R
Tb	pM	0.56	1.8–2.1	0.73	1.4–1.6	S+R
Dy	pM		6.1	4.8	5.1	?
Ho	pM	0.7–1.0	4–5	1.5–1.8	2.5–2.7	S+R
Er	pM		5.8	4.1	5.1	?
Tm	pM	0.3–0.5	2.0–2.5	0.7–1.0	1.1–1.3	S+R
Yb	pM	1.9–2.8	13–17	3.8-5.1	7.0-7.4	S+R
Lu	pM	0.3–0.4	2.3–3.1	0.7–0.8	1.5–1.6	S+R
Hf	pM	0.2–0.4	1–2	0.4	(0.8)[e]	?
Ta	pM		(<14)			?
W	nM		(0.6)			?

Element	Concentration Units [a]	North Pacific Surface	North Pacific Deep	North Atlantic Surface	North Atlantic Deep	Distribution Type [b]
Re	pM	28–82		32–43		?
Os				(?)		?
Ir	fM	(5–30)[c]				?
Pt	pM	0.4	0.3,1.2	0.2–0.4[d]	0.2–0.4[d]	?[f]
Au	fM	50–150		50–150		?
Hg	pM	0.5–10	2–10	1–7	1	S
Tl	pM	60–80	80	60–70	60	C
Pb	pM	14–50	3–6	100–150[g]	20	S
Bi	pM	0.2	0.02	0.25		S

[a] fM = 10^{-15} mol/L, pM = 10^{-12} mol/L, nM = 10^{-9} mol/L;
[b] R = recycled type, S = scavenged type, C = conservative type;
[c] SIO Pier sample;
[d] Indian Ocean concentrations; [e] depth = 1250 m;
[f] R, S, and C distributions have been reported;
[g] recent studies show declining concentrations (*i.e.*, 50–100 pM; *Kremling*, unpublished)

Scavenged (S) elements (*e.g.*, Al, Co, Pb or Mn) exhibit relatively short residence times ($< 10^3$ y) due to their strong interaction with particles. They show maximum concentrations near sources (*e.g.*, rivers, atmospheric deposition, hydrothermal vents, sediments) and decrease with distance from them. Deep-water concentrations in the North Pacific are generally lower than in the North Atlantic since scavenged elements decrease along the deep water flow with continuing particle scavenging (Table 12-1). A number of elements show hybrid distributions, *i.e.*, they are influenced by both recycling and scavenging processes (R + S types; Table 12-1). Such elements (*e.g.*, Cu and Fe) are often depleted in surface waters with high productivity and regenerated at depth. Both Cu and Fe concentrations, however, increase only gradually with depth due to the combined effects of regeneration and scavenging in deep waters. The relatively short oceanic residence times of about 1000 y for Cu and < 100 y for Fe reflect the difference in their scavenging efficiency, as compared with elements following true R or S modes. (For some further details on the chemistry of As, Sb, Ge, Hg, or Fe and Mn in seawater see Sections 12.2.3, 12.2.4 and 12.5, respectively).

In contrast to the oceans, the processes controlling the concentration and distribution of trace elements in dynamic shelf and coastal zones are still poorly understood. This is because the near steady state processes in offshore regions tend to be strongly altered close to the continents. Such disturbances are due to more intense inputs (from rivers, atmosphere or sediment) and removal of elements (by biological uptake or sorption onto sedimentary particles), or due to water exchange processes. Larger data bases are needed to establish and interpret the variations in these dynamic environments, particularly the evaluation of sources and sinks.

12.1.2   Analytical options

The analytical techniques most often used for the determination of dissolved and particulate trace elements in seawater are:
- electrothermal atomic absorption spectrometry (ETAAS);
- anodic or cathodic stripping voltammetry (ASV, CSV);
- mass spectrometric methods (*e.g.*, inductively coupled plasma mass spectrometry, ICP-MS);
- inductively coupled plasma atomic emission spectrometry (ICP-AES);
- X-ray fluorescence spectrometry (total reflection mode, TXRF);
- miscellaneous techniques such as chemiluminescence, gas chromatography or spectrophotometry.

Selected applications of the various approaches are presented in this chapter (Sections 12.2–12.6). Here the major advantages and/or disadvantages of the analytical techniques are briefly summarized.

Electrothermal atomic absorption spectrometry (ETAAS) has been the single most important technique in advancing our knowledge of the transition metal distribution in seawater. The graphite-furnace mode is used most frequently. It has the advantage of high sensitivity and therefore small sample volume (*e.g.*, 10–50 μL). Major disadvantages are the matrix interferences which usually necessitate a pre-concentration and/or a separation step (see Sections 12.2.1 and 12.2.2). Another application of ETAAS is the cold-vapour technique for the determination of mercury (Section 12.2.4).

Advantages of *voltammetric stripping methods (ASV, CSV)* are high sensitivity, multi-element capacity for a limited number of metals, small sample volume (*e.g.*, 5–50 mL), the possibility of performing real-time shipboard measurements and the capability of analysing chemical speciation directly (see Section 12.3). Another advantage is the relatively low capital investment. A disadvantage, in addition to the limited number of elements, is potential interference from organic substances.

Among *mass spectrometric methods (MS)*, the isotope dilution technique has been used by *Schaule and Patterson* (1981) to achieve the first Pb profile in ocean waters. The technique was also used for the simultaneous determination of transition metals *(Stukas and Wong,* 1983), and for the bulk of rare earth elements in seawater (*e.g.*, *Elderfield and Greaves,* 1982). Currently, the potential of inductively coupled plasma mass spectrometry (ICP-MS) is in the focus of interest. The power of this method lies in the fast, highly sensitive multi-element analyses (with detection limits of around 50 ng/L) and that ICP-MS can be applied in tracer studies using stable as well as radioactive isotopes (*e.g.*, *McLaren et al.,* 1985). Major disadvantages are interferences from the sample matrix, thus requiring a separation step before the analysis for most of the trace elements in seawater, and the high costs of the instrument.

Conventional *ICP atomic emission spectrometry (ICP-AES)* is not sensitive enough to allow the determination of dissolved trace elements in seawater. It might, however, majorly be applied in the analysis of collected particulate material. The technique combines fast multi-element measurements with somewhat lower investment costs when compared with ICP-MS (see Section 12.6).

Total-reflection X-ray fluorescence spectrometry (TXRF) is a powerful multi-element method, with detection limits in the pg/mL range, a dynamic concentration range of about 4–5 orders of magnitude and an easy quantification by means of internal standardization. The basic difference between classical X-ray fluorescence and TXRF is the improved beam geometry of the exciting and emitted radiation with respect to the sample. This has resulted in a drastic reduction of scattered radiation for TXRF compared with the conventional geometry and, therefore, in a substantial improvement of the peak/background ratio (see Section 12.4). Seawater analysis is limited to 12 trace elements and requires a matrix separation step. In particulate material, more than 30 elements can be determined simultaneously after digestion of the sample material.

In addition to the more popular techniques discussed, several other techniques have been developed or applied for various elements. Recently, some very sensitive *chemiluminescence procedures* for the determination of Co, Cu, Mn and Fe(*II*)/Fe(*III*), with detection limits in the picomolar and nanomolar ranges, have been developed (*e.g., Boyle et al., 1987; Sakamoto-Arnold and Johnson, 1987; Coale et al., 1992; Chapui et al., 1991; Elrod et al., 1991; Obato et al., 1993*). The workers coupled pre-concentration steps with chemiluminescence detection, or used flow-injection analysis with chemiluminescence detection. These methods are very promising because they are rapid, require only small sample volumes and are suitable for shipboard measurements. The measurement of trace element derivatives by *gas chromatography* is a method of limited use. The major reason is that the volatile organic metal chelates or derivatives are unstable at the temperatures required for sufficient volatility. Advantages of the technique include high sensitivity (with detection limits in the picomolar range), small sample volumes, relatively low instrumentation costs and the capability of providing real-time data. This has been proved, for example, with determinations of Se, Be and Al (*Measures and Burton, 1980; Measures and Edmond, 1986, 1989*). Finally, *spectrophotometric methods* can be applied directly for seawater with elevated trace element concentrations (in the μmol/L to mmol/L range). For example, Fe and Mn occur at such elevated concentrations in anoxic waters, such as in the Baltic or Black Seas, or in the deep ocean near active hydrothermal regions. They offer the advantages shared by all spectrophotometric methods, *i.e.*, they are direct, fast and very precise (see Section 12.5).

Most of the instrumental methods currently available do not provide the sensitivity or freedom from matrix interferences to determine trace elements in seawater at the picomolar and nanomolar levels directly. Therefore, in most cases a pre-concentration step (and separation from the matrix for some methods) is necessary before instrumental detection. These pre-concentration steps may be selected from a wide variety of selective techniques; for example, liquid–liquid extraction with chelating agents (such as dithiocarbamates), chelating cation exchange with resins (such as Chelex-100), electrochemical pre-concentration in ASV and CSV, hydride generation for the measurement of metalloids or co-precipitation with $Mg(OH)_2$ or selected metal chelates. Some of these techniques will be discussed in more detail in conjunction with the analytical procedures outlined in Sections 12.2–12.4.

12.1.3 Working environment and sample handling

In situ analyses are rarely available for marine trace element measurements. Therefore, once a sample has been collected, a number of operations have to be carried out almost immediately including subsampling, filtration, if required, and storage or pre-concentration. Because of the extremely low levels at which trace elements occur in most ocean waters (Table 12-1), these procedures must be carried out with as much care and attention against contamination as during subsequent analysis steps. Many of the systematic errors that have

Fig. 12-1. Clean bench module with analytical facilities for operations.

led to bias in reported trace element data from recent decades were probably introduced by improper preliminary sample handling.

Since sampling, filtration and storage techniques have already been discussed in greater detail in Chapters 1 and 2, only those aspects pertinent to potential trace element contamination will be mentioned here (see also Sections 12.2–12.6). It is general practice now that special conditions of cleanliness (*e.g.*, controlled atmosphere, clean room conditions) must be adhered to by laboratories performing trace element analysis in seawater. This is especially important on-board oceanographic vessels which present a very contamination-prone environment for this type of work. Since it is usually not practicable to reserve part of the ship's interior to 'clean room' conditions, special precautions must be taken. A good compromise is the installation of clean air benches in conventional laboratories. These benches (flushed with filtered air in horizontal or vertical laminar-flow mode) permit all stages of sample preparation under controlled conditions (Fig. 12-1). Sampling bottles might be placed outside the bench, but the seawater is transferred contamination-free *via* Teflon tubing to either the filtration apparatus or directly to storage containers. An alternative practised by several working groups (*e.g.*, *Wong et al.*, 1977; *Danielsson and Westerlund*, 1983; *Morley et al.*, 1988) is the use of portable sea-going modules equipped with complete cleanroom sections. This concept combines the flexibility dictated by multi-use research vessels with the requirements of a controlled environment for trace element sample handling. It also allows the trace element analysis or pre-concentration steps to be started soon after sampling, minimizing problems arising from long-term storage of aqueous solutions with extremely low analyte concentrations.

It is beyond the scope of this section to describe the lay-out of clean laboratory facilities in detail; see the excellent reviews by *Whyte* (1991) and *Lieberman* (1992). The basic concept behind the operation of such rooms is that they are under positive pressure and that the air is first drawn through one or two pre-filters, and then forced through a high efficiency particle air (HEPA) filter which removes typically > 99.97 % of particles $\geq 0.3\ \mu$m in diameter.

Standards for the major classes of clean room vary from one country to another. The US FS209D standard probably is the most common one. Usually it is defined in terms of the number of particles of a specified diameter in a certain volume of air. A designed class 100

Table 12-2. Guidelines for clothing protocols in clean environments (according to *Howard* and *Statham*, 1993).

Class of room/facility	Appropriate clothing
Unidirectional flow hood in conventional laboratory	Laboratory coat, polyethylene gloves; *i.e.*, body parts extending into the hood must be covered by non-contaminating clothing
Class 100 000	Minimal specialized clothing; smocks, street shoe covers, head covers
Class 10 000 or 1000	Complete clean room uniform with gloves, breathing zone cover; no cosmetics or rapid body movements
Class 100 or 10	Complete change of clothing, special gloves and foot covers, minimal contact with critical equipment
Below Class 10	Robotic systems, preferably no humans

room, for example, should contain no more than 100 particles greater than $0.5\,\mu m$ in about 28 L of its air. It should be emphasized that the laminar flow equipment and the clean rooms itself represent a potential source of contamination; *i.e.*, metallic screws or screens to protect the HEPA filter, the motor or fan assembly may corrode in the presence of aggressive chemicals. Therefore, metallic components in clean rooms should be either generally avoided, or coated with epoxy paint or similar corrosion resistant material.

Often the most serious source of potential contamination is the analyst himself or herself. Personnel therefore must be trained in contamination control measures. Obviously the most effective strategy is to isolate her or him from the samples by use of appropriate clothing. An important precaution is to wear disposable polyethylene gloves providing acceptable protection during all type of manipulation. For further information with respect to the selection of laboratory coats and other clothing see the excellent compilation by *Howard and Statham* (1993). For practical purposes Table 12-2 provides some guidance on clothing appropriate for different clean room classes.

It should be mentioned, however, that the standard of cleanliness may vary depending on the trace element and matrix under investigation. The determination of Fe and Zn in ocean-surface waters or Pb in deep ocean samples, for example, necessitates much stricter control than the determination of Mn in estuarine waters. However, the analyst will encounter fewer problems from the working environment the more he/she approaches the extreme cleanliness developed for 'severe elements' such as Fe or Zn.

12.1.4 Materials and cleaning procedures

Plastics have widely replaced metals and glassware in sampler construction, storage containers and sample processing, although most samplers still have external metal components (see Chapter 1). Materials that satisfy the high purity requirements for trace element studies are Teflon (PTFE), FEP, ETFE, polyethylenes, polypropylene, acrylic, quartz and Pyrex glass (for storage of samples for Hg determination only). Soft or soda-lime glass is not suitable because of its high ion-exchange capacity and high level of trace metal impurities.

Despite distinct advantages over glass, plastics may, as a result of processing, contain numerous fine foreign particles (and catalysts) just below the surface or surface-adsorbed, which provide potential sources of metal contamination (*e.g., Zief and Mitchell,* 1976; *Moody and Lindstrom,* 1977). In addition, a number of additives designed to stabilize the polymers may either contaminate the samples, react with metals in seawater solution to cause negative contamination or interfere in subsequent analyses. It is also strongly recommended to avoid all coloured plastics (*e.g.,* certain micropipette tips) as many of the pigments are potential sources of contamination. Besides storage and handling equipment, the sampling bottles in particular have to be cleaned very carefully. Representing the first foreign material in contact with seawater samples, contamination at this stage will negate all care taken in subsequent steps.

Numerous cleaning operations have been suggested and several have been shown to be effective. Although very time-consuming, such procedures are necessary to obtain reliable data. The following cleaning operations for *Teflon, high density polyethylene (PE), polypropylene (PP) or quartz* have been used successfully by the IfM Kiel over recent years for laboratory equipment such as pipettes, beakers and storage bottles *etc.* and will, therefore, be described here. The first step involves treatment with a detergent solution (*ca.* 2 %) for 2–3 days to de-grease the material. Then the equipment is thoroughly rinsed with warm tap water, followed by distilled water. Afterwards die materials are allowed to stand for about 1 week at room temperature with 1 : 1 diluted HCl (analytical-reagent grade, a.g.). After rinsing with distilled water cleaning is continued with 1 : 1 diluted HNO_3 (a.g.) for another week at room temperature. The materials are rinsed with the purest distilled water and allowed to dry under a clean bench before being wrapped in purified plastic bags. If the seawater samples are to be stored in acidic solution we propose storing the plastic sample bottles filled with 0.1–0.2 % solution of ultrapure HCl or HNO_3 for conditioning.

Pyrex and quartz glass bottles for storage of mercury samples are best cleaned using the following procedure: Bottles are filled with a 0.1 % solution of $KMnO_4$ and $K_2S_2O_8$ containing 2.5 % HNO_3 and heated in a water bath at 80 °C for 2 h. After cooling to room temperature, 2 mL of 12 % hydroxylamine hydrochloride ($NH_2OH \cdot HCl$) are added to reduce dissolved Cl_2 and particulate MnO_2. Then 10 mL of 10 % $SnCl_2$ solution is added to reduce mercury compounds to Hg^0, which is then removed by bubbling N_2 through the solution. Finally, the bottles are rinsed thoroughly with Hg-poor deionized water. To ensure a satisfactory blank for Pyrex bottles, the cleaning procedure must be repeated several times.

Plastic samplers are cleaned internally in a manner similar to that described for plastic bottles. *Go-Flo samplers* (Niskin type; see Chapter 1) are first thoroughly cleaned with a detergent, then rinsed with tap water and, after soaking in about 1 mol/L HCl (a.g.) for several days, flushed with a final rinse of purest distilled water. Afterwards, the samplers are wrapped in clean plastic bags for storage.

Nuclepore filters are cleaned by soaking in cold 6 mol/L HCl (a.g.) for 3 days, rinsed with distilled water and soaked for another 3 days at about 50 °C in 1 % ultrapure HCl. After a final rinse with purest water the filters are dried in open cleaned plastic petri dishes under a clean bench over several days, and then stored in closed plastic dishes if the amount of particulate matter has to be determined by weighing. Otherwise, the empty filters are stored in containers under purest distilled water. Prior to use the filters (placed in the filter holder) are rinsed with about 250 mL of purest water.

12.1.5   Purification of reagents

To take full advantage of the sensitivity and accuracy of an analytical method, the reagent blank should be no more than a few per cent of the amount being determined. Most analytical-reagent grade chemicals are not pure enough to be used for trace element analyses without further purification. Some commercial suppliers have focussed attention on this problem and have introduced special grades of ultrapure reagents. Most companies furnish maximum trace element values for these chemicals, which are of the order of 10^{-7}–10^{-6} % for available *suprapur* (Merck A.G.), *p.a.-plus, picopur* (Riedel de Haën) or *ultra* chemicals (Baker Chemical Co.). Although fulfilling many requirements in marine trace element analysis, ultrapure reagents often do not satisfy the standards required by extreme trace analysis (*e.g.*, analysis of open-ocean waters). Chemicals with considerable higher purity (trace elements with $\leq 10^{-9}$ %) are provided by Seastar Chemicals Inc. (PO Box 2219, Sydney, B.C., Canada V8L 3S8) as *baseline* products and include all the common mineral acids as well as acetic acid and ammonia.

Methods of preparing ultrapure reagents in the laboratory have been described in detail by *Zief and Mitchell* (1976) and *Howard and Statham* (1993). This section therefore presents only the basic techniques for easily purified reagents such as, in addition to water, hydrochloric and nitric acid, ammonia solution, and some organic solvents. Procedures for the purification of special reagents will be outlined in the chapters of the book where analytical methods are described in detail.

Water. Conventional distillation (with a quartz still) by boiling significantly reduces the trace element load of a feedstock. However, aerosol formation leads to droplets of unpurified material being carried over with the distillate. Smooth boiling, a long path length between the boiler and receiver, or the use of two stills in series may significantly improve the quality of the final distillate. The highest water quality can be obtained by utilizing a quartz sub-boiling distillation unit which may be fed with deionized water (Fig. 12-2). Here, radiative heating increases the vapour pressure of the liquid to be purified (without boiling) and this vapour then is recondensed. Water of similar quality can also be obtained in a multi-stage process (as practised by the author) involving a preliminary purification by cation-exchange and reverse osmosis followed by passage through a Millipore-Q system (acti-

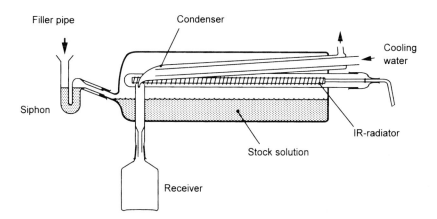

Fig. 12-2. Sub-boiling quartz still (module from *H. Kürner*, D-83022 Rosenheim, Germany).

vated charcoal and ion-exchange cartridges plus back-up membrane filters). This results in trace element concentrations of only about $10^{-12}-10^{-11}$ g/mL.

Acids. A review of purification operations suggests that sub-boiling distillation in quartz or Teflon equipment provides the purest acids (Fig. 12-2). *Kuehner et al.* (1972), who investigated the non-boiling procedure in great detail, reported Zn, Cd, Pb or Cu concentrations for HCl and HNO_3 of the order of $10^{-12}-10^{-11}$ g/g when feeding with a.g. acids.

Another rather simple sub-boiling distillation apparatus for the preparation of HCl, HNO_3 and HF as described by *Mattinson* (1972) is shown in Fig.12-3. Two FEP bottles are connected at right angles by a Teflon block, and heat is supplied by a 300 W IR lamp. The metal contents of the purified acids are comparably as low as those mentioned before. Special advantages of the system are minimum handling and operation in a closed system.

Isothermal (or isopiestic) distillation is another method for purifying very volatile acids such as HCl, HF or CH_3COOH (see Fig. 12-4). This procedure can be performed in a desiccator at room temperature. Two beakers, one filled with concentrated acid (a.g.), the other with ultrapure water, are placed close to each other. The final acid concentration depends on the time of exposure and on the volume ratio of acid to water and should be determined experimentally. In our laboratory ≈ 0.1 mol/L HCl was obtained when placing 25 mL of concentrated acid 5 cm apart from 25 mL of distilled water in a quartz beaker for 2 d. The purification of hydrofluoric acid by isothermal distillation requires inert materials such as PTFE.

Ultrapure HCl of different concentrations can also be produced by gassing distilled water with pure HCl, obtained from a tank or from concentrated HCl solution (by heating). Gas from commercially available cylinders should be filtered through 0.4 µm Nuclepore or Teflon filters to remove particles before entering the water.

Ammonia. Isothermal distillation (see Fig. 12-4) and gaseous saturation of ultrapure water are suggested for the preparation of high-purity ammonia solutions. When cylinder

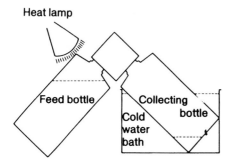

Fig. 12-3. Sub-boiling Teflon bottle still (after *Mattinson*, 1972).

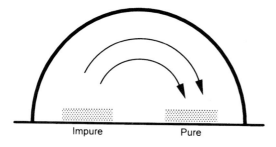

Fig. 12-4. Principle setup of isothermal distillation.

NH$_3$ is used (with an inline-filter), it should first be bubbled through an ammoniacal solution of EDTA (*ca.* 0.01 mol/L) to reduce copper and nickel impurities before passing into distilled water. Alternatively, gaseous NH$_3$ may be obtained from a 25 % ammonia solution (a.g.) by heating.

Solvents. Organic solvents such as Freon or chloroform, employed to extract and pre-concentrate trace elements from seawater solutions are commonly purified by smooth distillation in vitreous silica stills or better by sub-boiling distillation in quartz stills. Chloroform of reagent grade quality is best cleaned by first extracting with the purest available 2 mol/L HCl (to reduce the inorganic salt content) and subsequent distillation.

12.1.6 Reference materials

The use of carefully cleaned reagents is essential for low laboratory or method blanks (see section on 'Qualitiy control' in Chapter 1). In order to prove whether or not the laboratory analysis is correct, certified reference materials (CRMs) have been developed. The values certified for certain elements (with specified uncertainties) are commonly based on the results of analyses by at least two reliable independent methods.

Suitable reference material should match matrix composition and analyte concentration of the unknown samples as close as possible. Of the relatively large number of suppliers only very few offer reference materials for trace analysis in seawater (see *Howard and Statham, 1993*). Such material collected in the off-shore North Atlantic mixed layer (NASS) and Nova Scotia estuarine (SLEW) and nearshore coastal waters (CASS) respectively, is available from the National Research Council of Canada (Ottawa, Canada K1A OR6). These samples have been certified for 13 elements (As, Cd, Cr, Co, Cu, Fe, Pb, Mn, Mo, Ni, U, Se, Zn).

Analyses of certified reference samples are not only a great help in assessing the precision and accuracy of a newly developed analytical procedure, but such measurements also are essential for quality control programmemes in routine measurements (see Chapter 1). It should be emphasized, however, that accurate analyses of reference materials do not necessarily indicate accuracte results for collected field samples. If analytes have been lost, or were added through contamination during sampling, this bias will not be monitored, of course, by the use of CRMs.

12.2 Analysis by atomic absorption spectrometry

12.2.1 Cadmium, cobalt, copper, iron, lead, nickel and zinc by ETAAS

ETAAS is primarily used for samples that have been pre-concentrated and separated from major ions (see Section 12.1.2). In the literature, various techniques have been described by *Boyle and Edmond* (1977); *Smith and Windom* (1980); *Sturgeon et al.* (1981), *Bruland et al.* (1985); *Hartmann et al.* (1989) or *Van Geen and Boyle* (1990). The method outlined here is a procedure modified from *Danielsson et al.* (1978) and *Bruland et al.* (1979). It provides a liquid–liquid extraction of dithiocarbamates with Freon (or chloroform) and combines the convenience of a rather broad pH working range of extraction with quick separation of the organic solvent from the aqueous phase, and with low total blanks for the metals of interest.

12.2.1.1 Principle of the method

Solvent extraction

For effective extraction of a metal ion from the aqueous (M_{aq}) into the organic phase (M_{org}) it is necessary to form, usually by chelate reaction, non-polar complexes. If organic solvents, inmiscible with water, are mixed with seawater solution, then the distribution ratio (D) of the metal species is given by:

$$D = \frac{(M_{org})}{(M_{aq})} \tag{12-1}$$

The final distribution (D) is a function of the nature and concentration of the chelating reagent, the character of the organic solvent and, most importantly, the pH of the aqueous phase.

The most frequently applied chelating agents are pyrrolidine derivatives of dithiocarbamic acid ($NH_2-C \overset{S}{\underset{SH}{<}}$), mainly because of their non-selective complexing properties and a broad pH working range (optimum pH values for complexing of the elements under consideration: 3–7). The method described here involves chelation with a mixture of equal amounts of ammonium pyrrolidine dithiocarbamate (APDC) and diethylammonium diethyldithiocarbamate (DDDC) at pH 4–5, which is adjusted with an acetate buffer. Freon TF (1,1,2-trichloro-1,2,2-trifluoroethane) was chosen by *Danielsson et al.* (1978) as an organic solvent because of its low toxicity, its very low solubility in water (0.017 % m/m at 21.1 °C), its quick separation from the aqueous phase and its low metal blank even at technical grade quality. Alternatively, the metal chelates may also be extracted into chloroform (*Bruland et al.*, 1979).

The main disadvantage of metal carbamates is their poor stability in organic solvents limiting the time for instrumental analysis after extraction. This problem can be avoided by separating the organic phase soon after the extraction process and transferring the analyte into acidic aqueous solution. This can be performed by either back-extraction into dilute nitric acid or by smooth evaporation of the solvent and acidic redissolution of the metals. An important criterion also is the degree of extraction achieved. From spike experiments conducted in our laboratory a recovery of ≥ 98 % was observed with two-fold extraction. Filtered samples to be analysed for iron should be acidified (pH ≤ 2.5) prior to analysis. Otherwise, only part of the metal will be detected due to its colloidal Fe fraction usually present in natural waters. It must also be stressed that kinetically stable metal organic complexes that exist in seawaters (*e.g.*, humic substances) might not be affected by the chelating agents. For metal release, such compounds can be digested by chemical reaction or by UV irradiation before extraction. It is, however, our experience that samples stored for several weeks at pH values between 1.0 and 2.0 release most of the metals combined with such components.

Atomic absorption spectrometry (AAS)

This chapter only describes the principle of the technique; for more details see the survey by *Holcombe and Bass* (1988) or the text book by *Robinson* (1996).

Free atoms of any element are capable of absorbing a quantum of energy to pass into an excited state. When returning to their ground state, the excited atoms normally release the energy as radiation. Electrical or thermal excitation leads to energy release in the form of an emission spectrum. If, however, the atoms are excited by light energy, they can only take up defined amounts of energy, ε, *i.e.*, light of specific wavelength λ (frequency v). The relationship is given as

$$\varepsilon = hv = \frac{h \cdot c}{\lambda} \tag{12-2}$$

showing for each atomic species characteristic values of ε and v (or λ), where h is Planck's constant. After 10^{-9}–10^{-7} s, the atoms can re-emit this energy, thus returning to the ground state. It is essential for the process that the wavelength emitted is also absorbed. If, therefore, an emission spectrum of the element to be determined is passed through an 'absorption cell' (where atoms are produced from the sample by thermal dissociation) a portion of the incident light will be absorbed. What is in fact measured in AAS is the fraction of radiation absorbed, defined as the ratio I/I_o, where I_o is the intensity before and I the intensity after passage through the absorption cell. If the cell has a definite path length l, then Beer–Lambert's law may be applied for the absorption process in the form

$$A \equiv \log \frac{I_o}{I} = 2.303 \, \mathrm{k} \cdot N \cdot l \tag{12-3}$$

where the absorbance (A) is directly proportional to the total number of free atoms (N), or their concentration present in the cell, and k is the spectral atomic absorption coefficient. For the construction of an AA spectrometer see Fig. 12-5.

The most important process in AAS is the generation of gaseous atoms in the ground state from ions or molecules present in the sample. This process includes a number of dynamic equilibria each of which is easily influenced. In recent decades, atomization by electrothermal devices (*L'vov*, 1961) has gained widespread application and revolutionized the practical determination of trace elements in seawater. In particular, graphite furnaces (GFAAS) heated by resistance, show some striking advantages over conventional flame atomization. They offer very high absolute sensitivities (with detection limits down to 10^{-13} g for metals such as Cd and Zn), need only small amounts of sample and have a largely inert atmosphere. The sample is injected into the graphite atomizer, which is then heated at variable steps and temperatures to evaporate firstly the solvent, then secondly, to evaporate or to ash the matrix, and finally, to atomize the remaining residue.

In the early stages of development, the graphite atomizers suffered from relatively low analytical precision, severe chemical interferences and from the intense background signals. With the introduction of the L'vov platform and Zeeman-effect background correction considerable improvements have been made with respect to these deficiencies. The L'vov platform placed inside the graphite tube (first proposed by *L'vov* in 1978) is mainly heated by atomization from the tube wall as the temperature increases. Atomization of the sample from the platform delays atomization until near isothermal gas-phase conditions are present

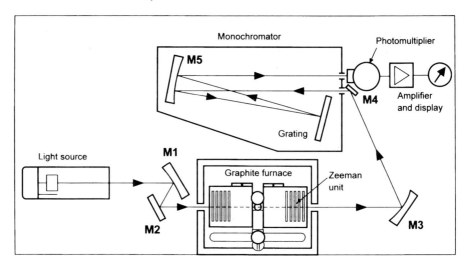

Fig. 12-5. Schematic construction of an atomic absorption spectrometer with graphite furnace and Zeeman correction mode (M1–M5 for mirrors). Modified after *Perkin-Elmer* (1991).

within the atomizer, unlike atomization into a rapidly changing temperature environment, which is the case with atomization from the tube wall. This leads to higher sensitivity especially with respect to measuring integrated absorbance (peak area) and often better precision and improved accuracy of data due to considerable reduction in interferences. Also, the time delay helps to reduce the often intense background peaks from tube wall atomizers. Such furnaces are now labelled stabilized temperature platform furnaces (STPF).

Technical improvements in furnace designs have provided enormous support to the STPF concept. Modern equipment (such as the Perkin-Elmer THGA furnace model) are fitted with a transversly heated graphite atomizer, THGA, which practically guarantees constant temperatures within the furnace's volume. This significantly reduces the risk of recombination and condensation of atoms, *i.e.*, reduces interferences, less tailing of specific element peaks and in reduction of non-specific light losses. An added advantage of an STPF is, that in contrast to the tube wall atomizers, it is made entirely of pyrolytic graphite coated electrographite providing better analytical signals and much longer lifetimes of tubes and platforms. Attempts to correct the non-specific molecular absorption have made considerably progress with application of the Zeeman effect (*Hadeishi and McLaughlin*, 1971). This technique, in contrast to the use of continuum sources, achieves background correction at the specific wavelength and can eliminate background levels of more than 2 absorbance units (for details see, *e.g.*, *Robinson*, 1996). The advantage of this technique also is that only one light source and only one detector are used; *i.e.*, in a sense the Zeeman-effect background correction acts as a double-beam system in measuring both sample and background signals with the same optical path (Fig. 12-5).

A chemical procedure more frequently used in earlier years to distinguish the analyte signal from the matrix absorption in GFAA technique is the addition of so-called chemical modifiers. They are chosen either to increase matrix volatility or decrease analyte volatility; for example, the addition of sufficient NH_4NO_3 to seawater samples causes the reaction $NaCl + NH_4NO_3 \rightarrow NaNO_3 + NH_4Cl$, which significantly increases the volatility of the matrix components. However, when measuring at the nanogramme and picogramme level, chemical modifiers may cause severe purity or blank problems. Furthermore, the sensitivity of the AAS technique restricts the direct measurement of trace elements in seawater to a few metals in coastal or brackish waters. For most seawaters, especially ocean waters, trace elements will still need a pre-concentration step such as described in detail in the following sections.

12.2.1.2 Reagents and equipment

1. *Ammonium pyrrolidine dithiocarbamate (APDC) and diethylammonium diethyldithiocarbamate (DDDC)*: Prepare an aqueous solution, 1 % (m/v) of each, and purify the reagents by extraction with Freon or chloroform (two extractions of 5 mL each) until metal blanks are negligible. Store the APDC/DDDC solution at about 4 °C. The solution is stable for 3 days.

2. *Ammonium acetate buffer*: Prepare a 20 % aqueous solution of CH_3COONH_4 (a.g.; to be stored at about 4 °C) and purify the buffer by repeated carbamate/Freon or chloroform extraction (*e.g.*, add 0.5 mL of 1 % APDC/DDDC for 100 mL of the buffer, and extract three times with 5 mL of Freon or chloroform).

3. *Freon TF*: 1,1,2-trichloro-1,2,2-trifluoroethane, density 1.565 g/mL (25 °C), b.p. 47.6 °C. It has to be re-distilled before use (see Section 12.1.5) and preferably stored in a quartz vessel. For purification of *chloroform* see Section 12.1.5. It has also to be re-distilled before use.

4. *Nitric acid:* Use concentrated HNO_3 of ultrapure quality (prepared by sub-boiling; see Section 12.1.5), prepare daily ≈ 0.1 mol/L HNO_3 solution by adding *ca.* 6 g of ultrapure concentrated HNO_3 to *ca.* 670 m/L of purest water.

5. *Standard metal solutions*: Prepare standard stock solutions (pH $\approx$ 2) of 1 g/L (*e.g.*, 'Titrisol', Merck). From this stock solution prepare composite working standards of 10^{-4}–10^{-6} g/L (pH $\approx$ 2) in 0.1 mol/L HNO_3 according to metal concentrations in the samples. Standard working solutions of $\geq 10^{-6}$ g/L can be used for around two weeks when stored in FEP bottles at $\approx$ 4 °C.

6. *Separatory funnels* (of Teflon or FEP) of 500 mL volume; 20 mL Teflon (PTFE) or quartz *beakers*; 100 mL polypropylene *volumetric flasks*.

7. *Infrared heater.*

8. *Ceramic hot plate.*

9. *pH meter.*

10. *Adjustable micropipettes*: Clean tips by shaking in 1 % APDC/DDDC solution for about 30 min (in separatory funnel) and extract twice with 10 mL of Freon or chloroform. Rinse thoroughly with the purest water available. Dry overnight in clean benches.

11. *AA spectrometer*: With graphite furnace cell and background correction.

12.2.1.3   Analysis of the sample

1. Extraction procedure

Transfer 400 g of ocean water (or less for shelf/coastal waters) to a 500 mL Teflon separatory funnel. Adjust the pH to 4.5 by adding acetate buffer. (The amount necessary should be determined in a pre-experiment; *e.g.*, with 50 g of the same sample.) Add 1 mL of the chelating reagent solution followed by 10 mL of Freon or chloroform. Shake the mixture vigorously for 5 min. Allow the phases to separate (typically 10 min), then drain the lower organic layer into a 20 mL Teflon beaker (performed in a clean bench). Then gradually evaporate the solvent to dryness, preferably on a ceramic hot plate and supported by means of a IR lamp, as shown in Fig. 12-1. In the meantime continue by adding another 10 mL of Freon or chloroform to the funnel and again shake for 5 min, allowing the phases to separate for 10 min. Repeat the draining and evaporation of the Freon or chloroform phase as described earlier. Evaporate to dryness, add 2x 100 μL of concentrated HNO$_3$ (to destroy metal carbamates) and evaporate to dryness again after each addition. The residue in the beaker is dissolved in 1 or 2 mL of warm 0.1 mol/L HNO$_3$ (depending on the enrichment factor needed for AAS analysis). When quantitatively transferred to suitable vials, the sample is ready for AAS measurement. The procedure is schematically outlined in the following diagramme:

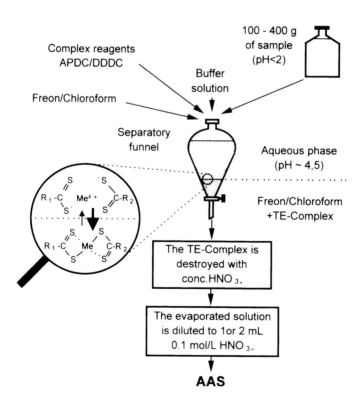

Complex reagents APDC/DDDC

Freon/Chloroform

Buffer solution

100 - 400 g of sample (pH<2)

Separatory funnel

Aqueous phase (pH ~ 4,5)

Freon/Chloroform +TE-Complex

R$_1$-C Me^{2+} C-R$_2$

R$_1$-C Me C-R$_2$

The TE-Complex is destroyed with conc. HNO$_3$.

The evaporated solution is diluted to 1 or 2 mL 0.1 mol/L HNO$_3$.

AAS

The procedure described here has a maximum concentration factor of ≈400 (volume ratio of sample to final extract). This is, for example, necessary for cobalt determinations in ocean samples. For other metals and sample origins (*e.g.*, Zn in estuarine waters), the final extracts may provide metal concentrations too high for direct ETAAS analysis. In this case, part of the acidic extract is diluted with 0.1 mol/L HNO_3 to bring the metal concentration into the proper working range.

2. Measurement by ETAAS

Generally, the manufacturer's recommended procedures for AA spectrometers (*e.g.*, lamp current, wavelength, slit width) should be followed. The operating conditions for the furnace, however, will be considered here in some detail.

Table 12-3. Example of graphite furnace heating conditions for selected metal solutions in 0.1 mol/L HNO_3 (Perkin-Elmer THGA furnace in conjunction with AA spectrometer 4100 ZL; protection gas is argon, 99.998 % with internal flow rate of 250 mL/min).

Element	Sample volume (μL)	Drying[a] steps		Charring steps		Atomization[b] step	Cleaning step
Cd	20	20 s (R) / 30 s (H) / 120 °C	30 s (R) / 30 s (H) / 130 °C	10 s (R) / 25 s (H) / 300 °C		0 s (R) / 5 s (H) / 1200 °C	1 s (R) / s (H) / 2400 °C
Co	60	30 s (R) / 50 s (H) / 120 °C	30 s (R) / 50 s (H) / 130 °C	30 s (R) / 30 s (H) / 1100 °C		0 s (R) / 5 s (H) / 2050 °C	3 s (R) / 5 s (H) / 2400 °C
Cu	20	20 s (R) / 40 s (H) / 110 °C	20 s (R) / 40 s (H) / 130 °C	20 s (R) / 20 s (H) / 700 °C	10 s (R) / 10 s (H) / 900 °C	0 s (R) / 5 s (H) / 1900 °C	1 s (R) / 5 s (H) / 2500 °C
Fe	20	10 s (R) / 40 s (H) / 110 °C	20 s (R) / 30 s (H) / 130 °C	30 s (R) / 20 s (H) / 1000 °C	10 s (R) / 5 s (H) / 1100 °C	0 s (R) / 5 s (H) / 1900 °C	3 s (R) / 8 s (H) / 2500 °C
Mn	20	20 s (R) / 30 s (H) / 110 °C	20 s (R) / 20 s (H) / 130 °C	20 s (R) / 40 s (H) / 1100 °C		0 s (R) / 6 s (H) / 1600 °C	4 s (R) / 5 s (H) / 2400 °C
Ni	20	20 s (R) / 30 s (H) / 110 °C	30 s (R) / 40 s (H) / 130 °C	30 s (R) / 10 s (H) / 700 °C	10 s (R) / 10 s (H) / 1000 °C	0 s (R) / 5 s (H) / 2200 °C	3 s (R) / 7s (H) / 2200 °C
Pb	40	30 s (R) / 40 s (H) / 110 °C	25 s (R) / 40 s (H) / 130 °C	20 s (R) / 25 s (H) / 550 °C		0 s (R) / 5 s (H) / 1350 °C	1 s (R) / 3 s (H) / 2400 °C
Zn	20	10 s (R) / 30 s (H) / 110 °C	15 s (R) / 30 s (H) / 120 °C	20 s (R) / 20 s (H) / 550 °C		0 s (R) / 5 s (H) / 1300 °C	3 s (R) / 2 s (H) / 2400 °C

[a] (R) and (H) are ramp and hold times of corresponding (nominal) temperatures;
[b] gas interrupt mode should be used during atomization step.

An example of ETAAS heating conditions for the determination of elements described in Sections 12.2.1 and 12.2.2 is presented in Table 12-3. Depending on the abundances of trace elements in ocean waters, their concentration factor during extraction and the sensitivity of the ETAAS instrumentation the sample volumes typically are between 20 and 60 μL. Temperatures and time were determined so that during the first step (drying) the solvent is evaporated quantitatively without spitting. In the second stage (charring) other components, such as acids or organic matter, are quantitatively removed. The pre-treatment temperature should be selected so that for the element under study no losses from the atomizer occur due to volatility.

To minimize potential chemical interferences during atomization (third stage) and to attain thermal equilibrium in the furnace the following procedure (in the STPF mode) is recommended:

(a) use graphite tubes with an platform installed;
(b) apply instantaneous atomization of the sample (*i.e.*, heating power at maximum with a 'Ramp time' of 0 s);
(c) keep the temperature difference between atomization and charring as small as possible;
(d) interrupt the gas flow ('Gas stop') during the atomization step; and
(e) evaluate the signal by peak integration ('peak area').

A cleaning step (fourth stage) should clear the furnace from residual amounts of the element and a final cooling step (not listed in Table 12-3) should prepare the furnace for the next analysis.

12.2.1.4   Calculation of results

Two procedures are most commonly used for evaluation of the samples. The routine method is based on a calibration curve, which can be prepared either from aqueous standards with the same HNO_3 concentration as in the procedure, or by spiking pre-extracted water samples (if the extraction efficiency is close to 100 %). The slopes of both curves, however, should agree with the slope of the standard addition method (see Fig. 12-6); only thus can it be guaranteed that all interferences have been eliminated. It is also good practice to check the sensitivity of the instrument from time to time (*i.e.*, to identify the mass of element per 0.004 absorbance) when analysing standard solutions, and to compare the determined 'characteristic mass' with the manufacturer's nominal values.

The sensitivity (slope) of ETAAS decreases with tube age. Therefore, if calibration curves are used, it is mandatory to re-check the slope of the calibration curve by repeated measurements of the standard sample.

Using calibration curves, the metal concentration, C, of the original sample can be calculated from:

$$C \ (g/kg) = \frac{a_{St} \cdot A_{Ex} \cdot v_{Ex} \cdot 1000}{A_{St} \cdot v_i \cdot m}$$

where

a_{St} =   amount of injected standard (g);
A_{St} =   absorbance of standard, corrected for blank;

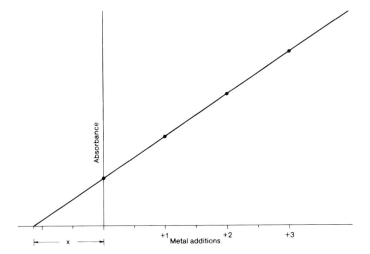

Fig. 12-6. Method of standard additions.

A_{Ex} =   absorbance of extract, corrected for blank;
v_i =   volume of injected extract (mL);
v_{Ex} =   total volume of extract (mL); and
m =   mass of seawater sample (g).

For the evaluation of samples showing interferences during atomization, the method of standard additions is recommended. Each sample is divided into four aliquots, to three of which are added different amounts of standard; they are extracted as described earlier. Select the standard so that the first addition raises the metal content by about 50 %, the next by another 50 %, and the third one by a further 100 %. The calculation of the metal concentration is illustrated in Fig. 12-6. The measured absorbance values and the amount of standard added to each aliquot of seawater are plotted as coordinates. The intersection of the calculated linear regression line with the abscissa (x) defines the element concentration (including total blank) in the unspiked aliquot of sample.

12.2.1.5   Precision, blank determination and detection limits

The precision of the method was determined by several sets of replicate measurements of NASS reference materials (see Section 12.1.6) and open North Atlantic surface waters. The coefficients of variation have been found to be between 5 % and 10 % for all elements discussed here.

The blanks should be determined by parallel extraction of previously processed seawater or purest distilled water following the same procedure. It is also worthwhile to distinguish between the reagent blank and the total procedural blank values to identify the factors contributing to the blank.

The evaluation of detection limits ($3s$ of the mean total blank values; see also Chapter 1) for the elements discussed here have been achieved under clean room conditions and been estimated as < 0.006 (nmol/L) for Cd, <0.01 (Co), < 0.05 (Cu), 0.03 (Fe), < 0.07 (Ni), < 0.005 (Pb), and 0.07 (Zn), respectively.

12.2.2   Determination of manganese

As Table 12-1 shows, seawater may contain Mn concentrations of <0.1 nmol/kg. There-fore, for most oceanic samples a pre-concentration step is necessary before analysis. (For the measurement of higher Mn concentrations in seawater by a spectrophotometric method see Section 12.5.) Here a relatively simple and widely used solvent extraction procedure is outlined. The method is based on the extraction of Mn oxinate into chloroform and was first described for seawater by *Slowery and Hood* (1971), but more extensively by *Klinkhammer* (1980) and *Landing and Bruland* (1980).

An alternative method has been described by *Hartmann et al.* (1989). It is based on pre-concentration by co-precipitation of manganese hydroxide with $Mg(OH)_2$ (from the sea-water magnesium) at a pH of ≈ 10. After separation from the bulk of the seasalts by centri-fugation and redissolution of the precipitate with nitric acid, Mn is measured by ETAAS. When applying a concentration factor of 40, the authors reported a detection limit for Mn of 0.02–0.04 nmol/kg.

12.2.2.1   Principle of the method

The first systematic studies to extract Mn(*II*) from aqueous solutions using 8-hydroxyqui-noline (quinolinol or oxin) in chloroform were presented by *Gentry and Sherrington* (1950) and *Stary* (1963), showing efficient extractions only at pH > 6.7. These findings were gener-ally confirmed for seawater solutions with an optimum pH range for the extraction of between 9.0 and 9.5.

Chloroform is used as the organic solvent with an excellent solubility for Mn oxinate. Analysis by ETAAS is performed after back-extraction into a nitric acid solution.

12.2.2.2   Reagents and equipment

1. *8-Hydroxyquinoline* (C_9H_7NO) *in chloroform:* Prepare a solution of 0.3 g 8-hydroxyqui-noline (oxin) in 500 mL of chloroform, redistilled. Store the reagent cool and in the dark. It should be prepared weekly.
2. *Ammonia solution, concentrated:* Prepare a ten-fold diluted solution of NH_4OH (in the purest water available).
3. *Nitric acid:* Prepare 7 mol/L HNO_3 of ultrapure quality (see Section 12.1.5).
4. *Standard solution:* Prepare working standards (pH 2) from a Mn stock solution (1 g/L). See also Section 12.2.1.2.
5. 100 and 250 mL *Teflon separatory funnels;* 10 mL Teflon or quartz *beakers.*
6. *Adjustable micropipettes:* Clean as described in Section 12.2.1.2.
7. *AA spectrometer* with graphite furnace and background correction.

12.2.2.3 Analysis of the sample

1. Extraction procedure (For general description see Section 12.2.1.1)
Transfer a subsample of ≥100 g of seawater to a 250 mL separatory funnel and add (1:10) diluted ammonia solution to adjust the pH between 9.0 and 9.5. (Determine the necessary ammonia volume using another subsample.) After pH adjustment, immediately add 10 mL of oxin reagent and extract by shaking the mixture for 4 min. After separation of the phases (*ca.* 10 min), drain the lower organic layer into a 100 mL Teflon separatory funnel. Add another 10 mL portion of the chloroform reagent to the sample and shake again for 4 min. After 10 min combine the two extracts and add 4.0 mL of 7 mol/L HNO$_3$. Then shake for 2 min to ensure back-extraction. After the phases have separated, the (lower) CHCl$_3$ phase is discarded.

The 4 mL acid phase then is drained into a 10 ml Teflon or quartz beaker, and the 100 mL separatory funnel is rinsed twice with 1 mL of 7 mol/L HNO$_3$. The total extract of about 6 mL is evaporated slowly to dryness on a ceramic hot-plate (see Fig. 12-1), and the residue is digested with two or three sequential volumes of 0.25 mL concentrated HNO$_3$. The residue then is dissolved in 1.000 mL of 0.1 mol/L HNO$_3$ and is ready for ETAAS analysis.

2. Measurements by ETAAS
The conditions for the ETAAS measurement have been described in Section 12.2.1.3 and Table 12-3.

12.2.2.4 Calculation of results

The overall extraction efficiency should be investigated by spiking pre-extracted water samples and repeating the extraction as outlined before. Investigations in our laboratory showed extraction efficiencies of around 96 %, which is used in calculating the final results (see also Section 12.2.1.4).

12.2.2.5 Precision and blank determination

The coefficient of variation estimated from samples with a Mn content of about 3 nmol/L by replicate measurements is between 5 and 10 %. This is in good agreement with the results of *Klinkhammer* (1980).

The total blank is determined by extracting 100 g of ultrapure water and has usually been found not to differ from the reagent blank (of the order of 0.01 nmol/L). Random sources of contamination during analysis under clean-room conditions therefore can be neglected.

12.2.3   Arsenic, antimony and germanium

M. O. Andreae

12.2.3.1   Arsenic

Arsenic occurs in seawater in a variety of chemical species, the most abundant of which are arsenate, arsenite, methylarsonate and dimethylarsinate (*Andreae*, 1979; *Cullen and Reimer*, 1989). The method described here permits the individual determination of these four species. It provides high sensitivity (limit of detection 4.0 pmol/L) and adequate precision (5–10 % relative standard deviation). In seawater and most natural waters, interferences are absent or minimal. This method was first published by *Andreae* (1977) and has been modified subsequently (*Andreae*, 1983). It has been used successfully by numerous workers, often with minor modifications. Some improvements have been proposed recently, such as the addition of cysteine to the reaction mixture (*Le et al.*, 1994; *Howard and Salou*, 1996) and the use of glass beads etched with hydrofluoric acid in the cold trap (*Howard and Comber*, 1992).The following description represents the procedure followed in the author's laboratory.

In addition to the well-defined species mentioned above, seawater also contains refractory arsenic species with poorly known molecular structures (Howard and Comber, 1989). These species are most important in coastal and estuarine regions, where they can account for as much as a quarter of the total dissolved arsenic. Techniques for their quantitative speciation are not available at this time. If the determination of total dissolved arsenic is required, we recommend that all arsenic species first be converted into inorganic As(V) by drying and ashing of the water sample (*Uthe et al.*, 1974; *Tam and Conacher*, 1977).

Principle of the method

The arsenic species in solution are reduced by sodium borohydride to the corresponding arsines: arsenate and arsenite to arsine (AsH_3), methylarsonate to methylarsine (CH_3AsH_2),

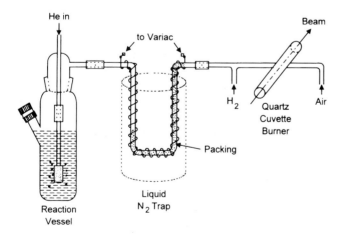

Fig. 12-7. Apparatus for the determination of arsenic species in water.

dimethylarsinate to dimethylarsine (($CH_3)_2AsH$). The arsines are purged from the solution by a helium carrier gas stream and collected on a cold trap filled with a gas chromatographic packing. After completion of this step, the trap is heated and the arsines elute in the sequence of their boiling points as chromatographic peaks. They are swept by the carrier gas stream into a quartz cuvette burner, where they are combusted in a hydrogen–air flame. The burner is mounted in the beam of an atomic absorption spectrometer, and the atom populations produced from the combustion of the arsines result in transient atomic absorption signals monitored on a chart recorder, reporting integrator/recorder or a chromatographic data acquisition system. Arsenate and arsenite are separated on the basis of the pH-dependency of the borohydride reduction: at near-neutral pH, only arsenite is reduced to the arsine, while at acid pH both arsenite and arsenate as well as the methylarsenic species are reduced. Arsenate is then obtained by difference. The analytical apparatus is shown in Fig. 12-7.

Sample collection, treatment and storage

Arsenic has a relatively high concentration in seawater (when compared with many other trace elements; see Table 12-1) and is not present at high levels in materials used in the construction of ships, marine equipment, and samplers. Therefore, the problem of sample contamination is much less severe for this element and its species than for many other trace elements. Clean room facilities are not required for arsenic species determination. Care should, however, be taken in the laboratory to avoid the possibility of contamination resulting from the preparation and handling of standards.

Most samplers used for seawater collection are adequate to obtain seawater samples for arsenic species determination: we have used Niskin and Go-Flo samplers (General Oceanics) as well as TPN and COC samplers (Hydrobios) for sample collection and have not seen evidence of contamination resulting from any of these samplers (see also Chapter 1).

Filtration of the samples can be done with any of the pressure or vacuum systems in common use (see Chapter 2). The filtering apparatus should be cleaned thoroughly by acid washing (see Section 12.1.4). Membrane filters release no detectable arsenic species, but should be rinsed by passing a small amount of the sample through them. Glass fibre filters may contain soluble arsenate or arsenite and should not be used. In general, filtration is not required, because oceanic particulates contain few soluble arsenic species that would be detectable with this method.

Samples should be stored in acid-washed polyethylene bottles (see Section 12.1.4). The arsenite/arsenate ratio can be preserved by 'quick-freezing' the sample. For this purpose, about 250 mL of sample in a polyethylene bottle are immersed in liquid nitrogen or in a dry ice–methanol slurry. (To prevent leakage, about 10–20 % of the bottle volume should be left as air space before freezing!) The sample should then be stored at a temperature below –30 °C. If samples are frozen simply by putting them into a freezer, arsenite is oxidized at unpredictable rates, often as fast as overnight. Storage at temperatures at which some residual brine coexists with ice also leads to arsenite oxidation. Arsenate and the methylarsenicals are stable in quick-frozen samples.

If the separation of arsenate and arsenite is not required, the samples can be acidified with 4 mL of concentrated HCl per 1000 mL of sample. Under these circumstances, arsenite will oxidize to arsenate, but total inorganic arsenic and the methylarsenic compounds will be stable for several months. The methylarsenicals undergo unpredictable changes in untreated samples.

Reagents and equipment

1. *Gas supply.* Helium, hydrogen and air are required. We have not observed any differences between the analytical performance of different purity grades of these gases; usually the standard quality gases are adequate. The tanks are fitted with dual-stage regulators. The gas flows are controlled by high precision control valves and monitored on rotameters. The gas flow rates are calibrated by soap film flow meters; the rotameters are used to monitor the constancy of the flow rates. Some improvement in analytical precision can be obtained by using mass flow regulators. Polyethylene or Nylon tubing is used for the gas lines, Swagelok fittings are used for the connections. To provide flexible connections, helium is brought to the reaction vessel, and hydrogen and air to the burner through 1/8 in (≈ 3 mm) tubing (Nylon or Teflon). The following flow rates should be used initially: 300 mL/min hydrogen, 150 mL/min air, 100 mL/min helium (measured at the **outlet** of the cold trap).

2. *Reduction vessel.* The reduction vessel is made from Pyrex glass. The head (Fig. 12-7) is blown from a 29/42 ground glass joint and 6 mm outer diameter tubing for the helium inlet and outlet. Fritted glass distribution tubes of different lengths are attached to the helium inlet tube by a short piece of Teflon tubing (1/4 in outer diameter, 5.8 mm inner diameter; *e.g.*, from Supelco Inc.). This tubing makes a gas-tight fit around the 6 mm glass tubing and is used for all glass-to-glass connections. Reaction vessels of different sizes can be attached to the head; for arsenic species determinations, 25–100 mL vessels are useful. A Teflon foil sleeve of the appropriate size is used to seal the ground glass connection. The reaction vessel is secured to the head with a steel spring. The reaction vessels have a side-arm (6 mm outer diameter, 2–3 cm long) that ends just below the liquid surface. It is closed off by a Teflon-coated silicone septum (as used in gas chromatography) held in a Teflon Swagelok union (1/4 in outer diameter).

3. *Cold trap/column.* The cold trap consists of a 6 mm outer diameter Pyrex U-tube; the length of the U is about 30 cm. The first 8 cm in the limb of the U adjacent to the reaction vessel is left unfilled, the rest is packed with a chromatographic packing (15 % OV-3 on Chromosorb W-AW-DMCS 60–80 mesh, Supelco Inc.), held by two plugs of glass wool. About 1 m of Chromel wire (*ca.* 3 Ω) is wound around the outside of the trap/column and connected to a variable transformer. This permits the trap to be heated at a controlled rate after the removal of the liquid nitrogen bath. A bridge clamp should be fitted between the upper ends of the two limbs of the U to prevent breakage during manipulation of the apparatus. The outlet of the cold trap/column is connected to the glass burner cuvette in the atomic absorption detection system.

   The cold trap/column is treated with a silylation reagent to deactivate the internal surface and the chromatographic packing. For this purpose, it is installed in a chromatographic oven (or a drying furnace), inert gas is supplied at a flow rate of 50 mL/min to the inlet side of the trap, and the temperature increased to 150–180 °C. It is conditioned at this temperature for several hours, and then two times 0.025 mL of Silyl-8 (Pierce Chemical Co.) or a similar silylation reagent are injected onto the trap, which is then left to condition at the same temperature overnight. This treatment prevents peak tailing and arsine losses through irreversible adsorption onto the trap materials. It should be repeated if a deterioration of the column performance is noted after prolonged use.

4. *Quartz cuvette burner.* The quartz cuvette burner consists of a quartz glass tube, 9 mm inner diameter and 7 cm long, which is mounted in the beam path of the atomic absorp-

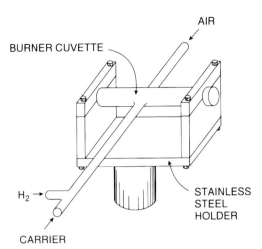

AIR

BURNER CUVETTE

H₂ →

CARRIER

STAINLESS
STEEL
HOLDER

Fig. 12-8. Quartz cuvette burner head.

tion spectrometer by a stainless steel bracket (Fig. 12-8). The design of this bracket depends on the AAS instrument used; a holder with dimensions similar to a standard flame AAS burner head has proved most convenient for our system (Perkin-Elmer AAS 5000). The socket of the mount has the same diameter as the base of the burner head and permits a rapid change-over. The quartz tube is held by stainless steel clamps; asbestos paper is put between the quartz tube and the steel surfaces to prevent breakage. The burner can be centered in the beam by the standard adjustment screws. The carrier gas from the trap/column is mixed with hydrogen, and the mixture enters the burner cuvette from the front. Air is introduced from the opposite side of the cuvette. The length and shape of the hydrogen and air inlets can be varied to suit the particular AAS used. The distance between the cold trap and the burner cuvette should be kept to a minimum to prevent diffusive peak broadening. The trap/column is attached to the burner glass-to-glass by a short sleeve of Teflon tubing. The air and hydrogen inlets are connected to the burner by $1/8$–$1/4$ in Nylon Swagelok reducing unions.

5. *Atomic absorption spectrometer.* As the arsenic line used (193.7 nm) is in the short wavelength UV region, a spectrometer with good performance in this region is required. As light source, an electrodeless discharge lamp should be used, which provides significant improvement in detection limits and linear range over hollow cathode lamps for arsenic. The use of a background corrector is normally not necessary; depending on the type of AAS, the use of a deuterium background corrector may lead to reduction of baseline noise. The atomic absorption signal is monitored on a chart recorder, reporting integrator/recorder, or chromatography data acquisition system. The latter permits the evaluation of both peak heights and peak areas, as well as internal calibration and direct output in units of arsenic concentration. See also Section 12.2.1.1.

6. *Reagents.*

   a) *Hydrochloric acid*: Use highly purified HCl (*e.g.*, G. Frederick Smith Double Distilled or Merck Suprapur) diluted 1:1 with water. (This dilution is not essential, but avoids the corrosive effects of HCl fumes on instrumentation, *etc.*)

   b) *TRIS–HCl*: A buffer solution is prepared which is 2.5 mol/L in TRIS (tris(hydroxymethyl)aminomethane) and 2.475 mol/L in HCl. This can be done either by using the

TRIS-HCl salt available from some manufacturers (*e.g.*, Sigma Chemical) and adding some TRIS base, or by neutralizing TRIS base with the required amount of concentrated HCl and bringing the solution up to volume. This buffer should have a pH between 6.0 and 6.5 after dilution to 0.05 mol/L; if necessary, the concentrated buffer can be adjusted by adding some more acid or base to give the desired pH upon dilution.

c) *Sodium borohydride (NaBH$_4$)*: A 4% solution of sodium borohydride is prepared in pure water and 1 mL of 2 mol/L NaOH solution is added per 100 mL of reagent. This solution is stable for a few days. Sodium borohydride is hygroscopic and the solid reagent hydrolyses when exposed to moist air. It should therefore be kept in a desiccator and the weighing performed quickly. The sodium borohydride is a major source of the arsenate reagent blank; arsenate concentrations between 1 and 30 ng/g have been found in this material. Its purity varies between manufacturers and even between different batches from the same manufacturer. We recommend testing a number of different lots and buying several years' supply once a satisfactory quality has been found.

7. *Standards.* The standard for arsenite is prepared from 'Primary Standard' grade arsenic trioxide. For this purpose, 989.2 mg of As$_2$O$_3$ are dissolved with a few NaOH pellets in about 100 mL of water, the solution is then neutralized using HCl and a glass electrode, and brought to a volume of 1000 mL to give a 0.01 mol/L As(*III*) solution. Fresh dilutions to the required concentrations are made daily. The stock solution is stable for several months. Care has to be taken that the water used for dilution is of high purity. It is especially important that no traces of free chlorine or hypochlorite are present (this sometimes happens with faulty deionizing systems). Otherwise oxidation of arsenic(*III*) to arsenic (*V*) may occur within a very short time.

Standards for arsenate, methylarsonate and dimethylarsinate (cacodylate) are prepared from the commercially available acids or salts. These are usually not completely pure and/or stoichiometric. For a 0.01 mol/L arsenate standard, 3.120 g of Na$_2$HAsO$_4 \cdot 7$H$_2$O are dissolved in 1000 mL of water. The arsenite content of this standard should be checked by running it through the procedure for arsenite described below (after dilution to 10^{-6} mol/L). Its arsenic content is verified by comparison with the arsenite standard using standard flame AAS. The stock solution is stable for some months, but should be tested periodically for arsenite. The methylarsonic acid standard is prepared from methylarsonic acid disodium salt (Pfaltz and Bauer Inc.); for a 0.01 mol/L standard, 2.920 g of this salt are dissolved in a few hundred mL of water, then 4 mL of concentrated HCl are added, and the solution made up to 1000 mL. The 0.01 mol/L dimethylarsinic acid standard is prepared in the same way, using 1.380 g of dimethylarsinic acid (hydroxydimethylarsine oxide, Baker). The commercially available methylarsenicals are usually not very pure, and the composition of the standards and their titer need to be checked using the methods discussed below and by standard flame AAS. Alternatively, the standard compounds can be purified using the method of *Dietz and Perez* (1976). In slightly acid solution (≈ 4 mL of concentrated HCl per 1000 mL), the methylarsenical standards are stable for many months.

Analysis of the sample

If only total inorganic arsenic and the methylarsenicals are to be determined, the analysis is performed in one step in acid medium. If the redox speciation of arsenic is also to be determined, a second analysis is run at near-neutral pH to determine arsenite separately.

The water volume to be analysed depends on the expected arsenic species concentration, the instrument sensitivity and its linear range. Total inorganic arsenic is so abundant in seawater that 10 mL would usually be adequate for analysis (see Table 12-1). However, the methylarsenicals, which can be determined in the same run as total inorganic arsenic, are usually present at concentrations of 20 % or less of the inorganic arsenic value. A compromise has to be found where the inorganic arsenic peak is still within the linear range of the instrument and where the methylarsine peaks are large enough to be quantifiable. Sometimes this is not possible, in which case separate runs are necessary for the determination of inorganic and methylated arsenic, with the sample volumes optimized independently. Arsenite is usually present at relatively low concentrations (except in anoxic waters); normally a sample size of 50–100 mL is optimal for this species. The different sample sizes should normally be analysed using reaction vessels of the corresponding volume; if this is inconvenient, smaller samples can be brought up to the volume of the reaction vessel by the addition of pure water.

For the determination of arsenite, 1 mL of TRIS-HCl buffer solution is added for a 50 mL sample volume. The reaction vessel is then attached to the head and purged with helium for 3 min to remove entrapped air. The liquid nitrogen bath is then moved up to immerse the cold trap/column; the liquid nitrogen is topped up as needed to maintain a constant level from analysis to analysis. Now 1 mL of sodium borohydride solution is injected through the side port (see Fig. 12-7) using a plastic syringe with a stainless steel hypodermic needle, and the reaction is allowed to proceed for 6 min. The liquid nitrogen bath is then removed, the Variac switched on to supply *ca.* 12 V to the heating coil on the cold trap/column, and the chart recorder or data acquisition device is started. After the peak has been recorded, the heating is allowed to continue until all water visible in the trap has evaporated.

To determine arsenate and the methylarsenicals, 2 mL of 6 mol/L HCl are added per 50 mL of sample (or less if the sample has been acidified previously). The reaction vessel is attached to the head and the system purged with helium for 3 min. Then the trap is cooled and 3.2 mL of sodium borohydride solution per 50 mL sample is injected slowly into the solution (the injection should take about 1 min). The reaction is allowed to continue for 6–10 more min and then the arsines are determined as described earlier.

Test procedures and sources of error

The system and procedure should be checked and optimised after being set up; sensitivity and blank checks should be done daily. In seawater there are normally no interferences; it is, however, advisable to test the recovery from time to time by the method of standard additions. The following operations should be performed initially and should be repeated when spurious results are suspected.

1. Prepare a 10 nmol/L working standard of arsenite and run it through the arsenite procedure. Repeat this, varying the flow rates of air and hydrogen (separately) to find the optimal signal-to-noise ratio.
2. Use the same reaction vessel as for the sample analyses and run the procedure on pure (distilled and/or deionized) water. Vary the amount of reagents used (HCl, TRIS–HCl, sodium borohydride) both together and independently (the latter within ±50 %). This will indicate the source of the blanks. There is normally no blank in the arsenite procedure. Most of the arsenate blank should be due to the sodium borohydride; it should be highly reproducible and not above 0.3 ng for a 50 mL sample size.

3. Prepare 10 nmol/L working standards of arsenite and arsenate in water and run them through the appropriate procedures. They should give the same peak areas (after subtraction of the arsenate blank).
4. Prepare a mixed standard containing about 10 nmol/L each of arsenate, methylarsonate and dimethylarsinate. Run this standard through the arsenate procedure and vary the heating voltage until peak sharpness is maximal at adequate separation. Record the retention times of the peaks; significant variations are usually an indication of trouble.

A few trouble-shooting hints:
1. Check the helium flow at the *outlet* of the cold trap/column: it should be 100 mL/min. Clogging of the trap or of the frit in the reactor, gas leaks, *etc.*, can often be diagnosed by this test.
2. Check the voltage and current through the heating wire and make sure there are no oxidized contacts.
3. Prepare fresh standards and sodium borohydride solution and repeat the experiments with the fresh solution.
4. Check the alignment of the quartz cuvette in the beam (with the background corrector *off*) as well as the settings of the monochromator.
5. If the sensitivity of the standard determination is seriously reduced and none of the above tests indicates any trouble, use pure water from a different source or run standard additions to a natural water. This may show if the water used for making standard solutions or reagents is contaminated with interferents.
6. If the sensitivity for arsenate is unusually low, try using a different batch of HCl. This acid sometimes contains free chlorine, which may interfere with the reaction.

Quantitation of the results

Even though there are usually no matrix interferences, we prefer to calibrate the system using standard addition to a seawater matrix. First, the reagent blank should be determined using the procedure described in the previous section by determining the value of the blank in terms of peak height or area and conversion into molar units using the distilled water sensitivity value. Then an appropriate volume of seawater is chosen and analysed. Additions of the arsenic species sought are then made to at least three times the value of the original sample concentration and the spiked samples run by the appropriate method. All determinations are made in duplicate or triplicate. From these data, the slope of the analytical curve can be calculated in units of peak height (or area) per nanomole by means of linear regression (see Fig. 12-6). A plot of the data verifies the linearity of the calibration curve over the desired range. The sample concentration is then obtained by dividing the sample peak absorbance by the analytical slope and subtracting the blank concentration. The following equations are used for these calculations:

$$Blank\ concentration = A_{wb} \times \frac{M_{ws}}{A_{ws} - A_{wb}}\ (mol/L)\ or\ (mol/kg)$$

where

M_{ws}	=	concentration of standard;
A_{ws}	=	absorbance peak area of standard in distilled or deionized water (*ca.* 10 nmol/L);
A_{wb}	=	absorbance peak area of unspiked water.

Sample concentration $= (A_s \times S)$ – blank concentration (mol/L) or (mol/kg)

where

A_s	=	absorbance peak area of sample; and
S	=	slope of analytical curve in absorbance peak area units per mole.

The absorbance can be expressed in terms of absorbance peak height or area (area was used in the expressions above). Each species has to be calibrated separately. Normally, there are no detectable blanks for arsenite and the methylarsenic species. The linearity of the calibration curve is dependent on the optical characteristics of the instrument and the intensity of the light source. It needs to be evaluated for each instrument and checked occasionally. The analytical curve should not be extrapolated beyond the highest standard concentration.

Precision and systematic errors

The precision of the method is dependent on the arsenic species, its concentration, the instrument used and the method of signal processing. Typically, at the nanomolar level, a precision of about 4 % can be expected with peak integration and about 8 % when peak heights are used. Somewhat poorer precision is usually obtained for the methylarsenicals.

In the absence of certified standards, the accuracy of the method cannot be rigorously assessed. Systematic errors can, however, be minimised by using the method of standard additions as described. The good agreement between arsenic data for deep ocean water obtained by different groups using this and other methods suggests that the accuracy is better than 10 %.

12.2.3.2 Antimony

Antimony is present in seawater as four different chemical species: antimony(*III*) as antimonous acid, antimony(*V*) as hexahydroxoantimonate(*V*) anion and two methylated forms, methylstibonic acid [$CH_3SbO(OH)_2$] and dimethylstibinic acid [$(CH_3)_2SbO(OH)$] (*Andreae et al.*, 1981; *Andreae and Froelich*, 1984; *Cutter and Cutter*, 1995). The dissociation constants of the methylated species are not known; by analogy with the arsenic homologues it is probable that they are present as anionic species. It is not known if refractory organic antimony species, such as have been described for arsenic, exist in natural waters.

The method described in this chapter permits the determination of all four species with a limit of detection of about 4 pmol/L and a precision of about 4 % relative standard deviation. It represents an adaptation of the procedure described in the preceding section (Arsenic) and was first published by (*Andreae et al.*, 1981). The instrumental requirements are identical to those for the arsenic procedure (Section 12.2.3.1).

Principle of the method

The dissolved antimony species discussed above react with sodium borohydride to form the corresponding stibines (antimony hydrides): stibine, methylstibine and dimethylstibine from inorganic antimony, methylstibonic acid, and dimethylstibinic acid respectively. The stibines are swept from the solution by a helium gas stream and collected on a cold trap filled with a chromatographic packing. Upon heating, this trap acts as a chromatographic column, separating the stibines in the sequence of their boiling points. They elute into the atomic absorption detector, which consists of a quartz cuvette burner mounted in the beam of an atomic absorption spectrometer. In the burner, the stibines are combusted in a hydrogen-air flame and the resulting atom population produces an absorption signal peak. The peaks are recorded on a chart recorder, reporting integrator/recorder or a chromatographic data acquisition system.

The differences in the reaction conditions required for the reduction of the different antimony species permit the separate determination of antimony(III) and antimony(V). The reduction efficiency for antimony(III) is independent of pH below a pH of about 8. Antimony(V), on the other hand, requires a pH below 1 and the presence of at least 0.12 mol/L iodide for complete reduction. Thus antimony(III) is determined at near-neutral pH, and total inorganic antimony at pH 0.8 in 0.15 mol/L KI. Antimony(V) is then obtained by difference. The methylantimony species are reduced optimally at mildly acidic pH (≈ 1.5). When they are reduced under the conditions specified for antimony(III), the yield for dimethylstibinic acid does not change significantly, while the monomethylstibonic acid peak decreases by about 30 %. Under many circumstances, the labour savings from combining the determination of antimony(III) and the methylantimonials will outweigh this minor loss in sensitivity. Calibration is by addition of antimony species standards to seawater samples and compensates for the yield variation.

Sample collection, treatment and storage

In our work with antimony determination in natural waters, we have not experienced serious contamination problems. We have used Niskin and Teflon-coated Go-Flo samplers (General Oceanics) and polycarbonate COC and TPN samplers (Hydrobios) for the collection of seawater samples without evidence of contamination resulting from any of these sampler types (see also Chapter 1).

The samples should be stored in acid-washed polyethylene bottles (see Section 12.1.4). Since antimony(III) oxidizes upon storage over several days, it should either be determined within two days of sample collection or preserved by 'quick-freezing' the sample. For this purpose, about 250 mL of sample in a polyethylene bottle are immersed in liquid nitrogen or in a dry ice–methanol slurry. To prevent leakage, about 10–20 % of the bottle volume should be left as air space before freezing. The sample should then be stored at a temperature below –30 °C. If samples are frozen by putting them into a freezer, oxidation may occur very rapidly in the brine formed during partial freezing. Antimony(V) and the methylantimony compounds are stable for weeks to months in quick-frozen samples.

Acidification of the samples with 4 mL concentrated HCl per 1000 mL of sample leads to relatively rapid oxidation of antimony(III), but total inorganic antimony and the organoantimony compounds are stable over at least a few months in acidified samples.

Reagents and equipment

The apparatus for the determination of antimony is identical to that used for the determination of arsenic (see Figs. 12-7 and 12-8).

1. *Gas supply.* See Section 'Reagents and equipment' under arsenic. The following flow rates should be used initially: 100 mL/min helium, 240 mL/min air and 300 mL/min hydrogen. These flow rates should be optimised as discussed for arsenic.

2. *Reduction vessel, cold trap/column, and quartz cuvette burner.* See Section 'Reagents and equipment' under arsenic.

3. *Atomic absorption spectrometer.* See also Section 'Reagents and equipment' under arsenic.

   The 217.6 nm antimony line is used at a slit setting of 0.2 nm. An electrodeless discharge lamp, due to its higher light output and consequently better signal-to-noise ratio and linearity, is preferable to a hollow cathode lamp. The deuterium background or Zeeman-effect correction mode is selected (see also Section 12.2.1.1). The atomic absorption signal is monitored on a chart recorder or a recording integrator/recorder or a chromatographic data acquisition system. The last makes possible the evaluation of peak heights and peak areas, as well as internal calibration and direct output in units of antimony species concentrations.

4. *Reagents.*

   a) *Hydrochloric acid*: Use high purity HCl (*e.g.*, G. Frederick Smith Double Distilled or Merck Suprapur) and dilute 1 : 1 with high purity water. (This dilution is not essential but does avoid the corrosive effects of HCl fumes on instrumentation, *etc.*).

   b) *TRIS–HCl*: The TRIS–HCl buffer solution is prepared by dissolving 230 g of TRIS (tris(hydroxymethyl)aminomethane) in *ca.* 750 mL of high purity water and adding 155 mL of high purity concentrated HCl to this solution. Adjust the solution until a pH of about 6 is indicated by a glass electrode, and dilute to 1000 mL.

   c) *Sodium borohydride* ($NaBH_4$): Sodium borohydride is dissolved in pure water to make a 4 % solution. After addition of 1 mL of 2 mol/L NaOH per 100 mL of reagent, this solution is stable for several days. Sodium borohydride is hygroscopic and the solid reagent hydrolyses if exposed to moist air. It should therefore be kept in a desiccator and the weighing performed quickly. The sodium borohydride is a major source of the antimony(V) blank, which can be expected to be around 0.6 ng of Sb per analysis depending on the quality of the borohydride. The purity of the borohydride varies between manufacturers and even between different batches from the same manufacturer. We recommend testing a number of different lots and buying several years' supply once satisfactory material has been found.

   d) *Potassium iodide*: Prepare a 5 mol/L solution of potassium iodide (830 g of KI per 1 L of solution) of at least analytical-reagent grade purity. Add 10 pellets of potassium hydroxide (KOH) per 1 L of reagent. The solution should be stored in the dark and discarded if significant discoloration forms during storage.

5. *Standards.* Antimony(*III*) potassium tartrate (3.3393 g) is dissolved in 1 L of pure water to make a 0.01 mol/L stock standard solution. This solution is stable for several months. The antimony(V) standard is prepared by dissolving 2.7190 g of potassium antimonate(V) ($KSb(OH)_6 \cdot \frac{1}{2} H_2O$) in dilute hydrochloric acid and diluting to a final volume of 1 L to make a solution of *ca.* 0.01 mol/L concentration. This solution should be calibrated (after further dilution) against the Sb(*III*) standard by standard flame atomic absorption tech-

niques to correct for the somewhat variable stoichiometry of the antimonate(V). If this correction is ignored, systematic errors of about 2 % may be encountered.

There are no standards currently available commercially for the methylantimony species. The standards used in our work had been prepared during work on the organometallic chemistry of antimony by *Meinema and Noltes* (1972). If organoantimony standards cannot be obtained, the use of the same calibration constant as used for Sb(*III*) will provide an estimate of the concentration of the methylantimonials. Depending on the condition of the trap and the efficiency of the reduction of the methyl compounds, this estimate may be below the true value by a factor of two or more. Therefore, if more than a semiquantitative estimate of the methylantimony species is required, the standard compounds will have to be synthesized according to the procedures of *Meinema and Noltes* (1972).

The standard solutions are stable at the 0.01 mol/L level for several months. The working standards should be prepared by serial dilution immediately before use. The stability of dilute antimony(*III*) standard solutions may be highly variable. In some instances we have found oxidation to antimony(V) to be almost complete after only a few hours, while some solutions showed little oxidation over several months. We suspect that trace amounts of oxidants present in the water used for dilution (*e.g.*, reactive chlorine not removed by the purification process) may be responsible for this problem. We suggest that standard dilutions be prepared immediately before use and that the stability of the standards in the particular water used in the laboratory be tested. If problems persist, the dilutions may be prepared using a natural water sample containing little or no antimony(*III*). The addition of *ca.* 1 g ascorbic acid per 1000 mL of solution also improves the stability of antimony(*III*) standards.

Analysis of the sample

The optimal reduction of the different species (antimony(*III*), antimony(V)), and the methylantimony acids) occurs under different conditions for each species. As discussed above, this effect is used for their selective determination. As the methylantimony species are chromatographically separated from inorganic antimony during the analysis, they can in principle be determined together with the inorganic species in one step. However, under the conditions required for the selective determination of antimony(*III*), the yield for methylstibonic acid reduction is decreased to about 70 %. If standards for the methylantimony acids are available, this can be compensated for by using the standard addition method of calibration. The determination of antimony(*III*) and the methylantimony compounds can then be performed in one step. If, on the other hand, maximum yield for the reduction of methylstibonic acid is desired, the reaction should be performed under the mildly acidic conditions outlined below; under these conditions some reduction of antimony(V) will occur together with that of antimony(*III*), so that the inorganic antimony peak from this step cannot be interpreted. In the following paragraphs the determination of total inorganic antimony, antimony(*III*), and the methylantimony acids is described using the conditions optimised for each determination. Reagent amounts are given for a sample size of 100 mL; for other sample sizes they should be adjusted proportionately.

1. *Total inorganic antimony.* For the determination of total inorganic antimony in seawater, sample volumes between 25 and 100 mL may be used, depending on the sensitivity and linear range of the instrument as well as the concentration of antimony in the sample

(typical values can be expected to lie near 1 nmol/L). The sample is measured gravimetrically or volumetrically into the reaction vessel and 3 mL each of 6 mol/L HCl and 5 mol/L potassium iodide solution are added. The reaction vessel is attached to the apparatus and purged with helium for 3 min to remove enclosed air. Then the trap is immersed in liquid nitrogen and 3 mL of the borohydride solution are injected slowly through the side port using a plastic syringe with a stainless steel hypodermic needle. The injection should take about 1 min. The reaction is allowed to proceed for 6–10 more min; then the liquid nitrogen bath is removed, the Variac is switched on to supply *ca.* 12 V to the heating coil in the cold trap/column, and the chart recorder or data system is started. After the peak has been recorded, the heating is allowed to continue until all water visible in the trap has evaporated.

2. *Antimony(III)*. For antimony(*III*), sample volumes of 100 or 250 mL are usually required. The sample is measured into the reaction vessel; 1 mL of 1.9 mol/L TRIS–HCl solution is added. The reaction vessel is attached to the system and purged with helium for 3 min. The trap is cooled and 3.1 mL of the borohydride solution are injected and the reaction allowed to continue for 6–12 min. The stibine is then determined by heating the trap and recording the signal as described above.

3. *Methylstibonic and dimethylstibinic acid*. The determination is performed as described for antimony(*III*), but instead of 1 mL of TRIS–HCl, 0.5 mL of 6 mol/L HCl is added to control the pH of the reaction.

Test procedures and sources of error

The same procedures as described in Section 12.2.3.1 for arsenic can be applied to the antimony determination, using the corresponding antimony species at concentrations of 1.0 nmol/L. As mentioned above, more problems were encountered with the stability of antimony(*III*) standards than with those for arsenic(*III*). If little or no peak is obtained using an antimony(*III*) solution, this problem can be isolated by running the same solution through the antimony(*V*) procedure.

Quantitation of the results

The interpretation of the results is performed as described in Section 12.2.3.1 for arsenic.

Precision and systematic errors

The precision of the determinations is dependent on the antimony species, its concentration, the instrument used and the method of signal processing. At the nanomolar level, about 4 % relative standard deviation can be expected for antimony(*III*) when peak height measurements are used. Peak integration may give significantly more precise results (as good as 1 % relative standard deviation), depending on the antimony concentration and the type of integrator used. Similar precision is obtained for total inorganic antimony. The determination of the methylantimonials is considerably less precise: 10–20 % relative standard deviation is typical for these compounds.

There are no certified standards for antimony species in aqueous solutions or natural waters currently available, making the direct evaluation of accuracy impossible at this time. We have tested for the possibility of systematic errors by checking the standards for the different species against each other and by standard additions to seawater and river water samples. This work showed no evidence of significant systematic errors. As a reference standard we use antimony potassium tartrate, which is available at high purity and with stoichiometric composition. The standards for the other species are referenced against this compound.

12.2.3.3 Germanium

The chemistry of inorganic germanium in seawater is closely linked to that of silicon: dissolved germanium occurs as germanic acid $(Ge(OH)_4^0)$ and its distribution follows closely that of dissolved silica. However, in anoxic basins and in the hydrothermal solutions emanating from ocean-floor hydrothermal systems, germanium anomalies have been observed that make the element an interesting geochemical tracer (*Froelich and Andreae*, 1981; *Froelich et al.*, 1985; *Mortlock et al.*, 1993). In seawater and many fresh water bodies, mono-, di- and trimethyl germanium species have been found to make up a significant fraction of the total dissolved germanium content (up to 99 % in some surface waters). These species are very unreactive in the ocean and are distributed in linear proportion to salinity (*Lewis et al.*, 1985; *Lewis et al.*, 1989).

In this section, a method for the determination of dissolved inorganic and methylated germanium species in seawater will be described. This procedure requires an atomic absorption instrument fitted with a graphite furnace, as well as some simple glassware and gas handling equipment. The method provides high sensitivity (limit of detection 4.0–10 pmol/L for the different species) and a precision of ± 1 to ± 4 %. In seawater and most natural waters, no significant interferences are encountered. This method was first published for inorganic germanium by *Andreae and Froelich* (1981), and subsequently extended to the methylgermanium species (*Hambrick et al.*, 1984). The procedures outlined here incorporate some further improvements and represent the method currently used.

Principle of the method

Dissolved germanic acid is reduced by sodium borohydride to germane (GeH_4), and the methylated germanium species are reduced to the corresponding methylgermanes $((CH_3)GeH_3, (CH_3)_2GeH_2,$ and $(CH_3)_3GeH))$. These compounds are volatile and are purged from the solution by a helium carrier gas stream and collected on a cold trap partially filled with a gas chromatographic packing and immersed in liquid nitrogen. After this step is completed, the trap is heated and the germanes elute into the atomic absorption detector. This detector consists of a graphite furnace with internal argon purge (Perkin-Elmer HGA 400 or similar) mounted in the beam of an atomic absorption spectrometer (see also Section 12.2.1.1; Fig. 12-5). The sample gas stream is introduced *via* the internal argon purge system. The heating cycle of the furnace is timed so that the atomization stage is reached when the germanes enter the furnace. The resulting transient atomic absorption signals are monitored on a chart recorder or on a chromatographic data system. Figure 12-9 shows the design of the analytical apparatus.

Sample collection, treatment and storage

Sample contamination with germanium and methylgermanium species is not a serious problem, as they do not occur at high concentrations in everyday materials (with the exception of semi-conductors, where germanium is usually well contained). Clean room facilities are not required for germanium species determination. Care should, however, be taken in the laboratory to avoid the possibility of contamination resulting from the preparation and handling of standards. This problem has been encountered particularly when solutions with high concentrations of the methylgermanium compounds were stored together with samples in the same refrigerator. Surface adsorption of methylgermanium species to glass surfaces is also a potential problem. This can result in memory effects when glassware is reused, as the adsorbed methylgermanium species can be released again when the glassware is heated (*e.g.*, during autoclaving) or filled with solutions of high pH.

Most samplers used for seawater collection are adequate to obtain seawater samples for the determination of germanium: we have used Niskin and Go-Flo samplers (General Oceanics) as well as TPN and COC samplers (Hydrobios) for sample collection and have not seen evidence of contamination resulting from any of these samplers (see also Chapter 1).

Filtration of the samples can be done with any of the plastic pressure or vacuum filtration systems in common use (see Chapter 2). The apparatus should be cleaned thoroughly by acid washing (see Section 12.1.4). Membrane filters release no detectable germanium and should be rinsed by passing a small amount of the sample through them. Glass fibre filters should not be used. In general, filtration is not required for open ocean seawater samples but is recommended for estuarine and coastal waters. When in doubt, follow the procedures established for the handling of samples for silica determination (see Chapter 10). River water samples should always be filtered immediately after collection.

Samples should be stored in acid-washed linear polyethylene bottles; 4 mL of concentrated hydrochloric acid (high purity grade) should be added per 1000 mL of sample. No changes in germanic acid concentration have been detected in samples stored for several years at pH ≈ 2.

Reagents and equipment

1. *Gas supply*. Helium is required as a carrier gas for the stripping and trapping system; argon is added to the gas stream before it enters the graphite furnace to increase the analytical sensitivity and prevent rapid deterioration of the graphite tubes. Standard quality gases are adequate. The helium gas flow is controlled by a high precision control valve and monitored on a rotameter. The analytical precision can be further improved by using a mass flow controller. The argon flow may be controlled using the adjustable low-flow purge system ('Miniflow' setting on the HGA 400) of the graphite furnace controller, or a system of the same type as used for helium. Polyethylene or Nylon tubing is used for the gas lines, Swagelok fittings are used for the connections. For flexibility, helium and argon are brought to the reaction vessel and the detector inlet by $1/8$ in Nylon or Teflon tubing from the rotameters or flow controllers. Unless mass flow controllers are used, the gas flow rates should be calibrated by soap film flow meters; the rotameters or the flow read-out on the furnace controller are then used to monitor the constancy of the flow rates. The helium flow rate, measured at the outlet of the cold trap after it has been immersed in liquid nitrogen, should be 40 mL/min. The argon flow rate is 100 mL/min; if one of the internal purge streams supplied by the HGA 400 is used for the argon supply, the flow indicator will read about 200 on the instrument. Nitrogen can be used instead of argon with comparable results, but due to the formation of cyanogen during the furnace operation effective ventilation must be ensured!

2. *Reduction vessel and water trap.* The reduction vessel design has been discussed in Section 12.2.3.1 for arsenic. Vessel sizes of up to 500 mL may be used. The water trap consists of a Pyrex *U*-tube (7 mm internal diameter, 30 cm long, reduced to 6 mm outer diameter at the ends (see Fig. 12- 9) immersed in a cold bath filled with isopropyl alcohol. This bath is cooled either with dry ice or with an immersed refrigeration probe and serves to remove most of the water vapour from the carrier gas stream. The type of refrigeration used is only a matter of convenience and does not influence the operation of the system.

If a large number of samples are to be analysed, it may be useful to connect two reaction vessels to the system so that one of them can purge (see below) while the other is reacting. For this purpose, two helium flow controllers are required to adjust the helium flow rate to 40 mL/min each through both reaction vessels (measured at the outlet of the cold trap). The flow through both channels must be identical to achieve the best precision. The helium is brought to each reaction vessel from the control valve by a length of $^1/_8$ in plastic tubing using Swagelok connectors ($^1/_4$–$^1/_8$ in reducing union, Nylon). Teflon tubing ($^1/_8$ in outer diameter) is attached to the outlet of the reaction vessels by the same type of connector. The lines from the reaction vessel are connected to a four-way valve (Teflon body sliding valve, ALTEX Co.) in such a way that one of the reaction vessels is always connected through the valve to the water trap, while the other is connected to a short end of tubing that exhausts into the air. In this fashion, one of the reaction vessels can be purged with helium (which is then simply exhausted into the room), while the other is attached to the cold trap to collect the germanium hydride. After the analysis is complete in the first vessel, the valve is switched and the second vessel is reacted while the first vessel is prepared for the next analysis and purged.

3. *Cold trap/column.* The cold trap/column consists of a 6 mm outer diameter Pyrex *U*-tube; the length of the *U is* about 30 cm. The first 8 cm in the limb of the *U* adjacent to the water trap is left unfilled, the rest is packed with a chromatographic packing (15 % OV-3 on Chromosorb W-AW-DMCS 60–80 mesh, Supelco Inc.), held in place by two 1 cm plugs of glass wool. About 1 m of Chromel wire (*ca.* 3 Ω) is wound around the outside of the trap/column and connected to a variable transformer initially set at 16 V. This permits the trap to be heated at a controlled rate after the removal of the liquid nitrogen bath. A

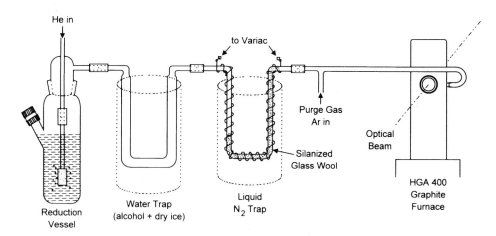

Fig. 12-9. Apparatus for the determination of germanium in water.

bridge clamp should be fitted between the upper ends of the two limbs of the *U* to prevent breakage during manipulations on the apparatus.

4. *Passivation of glass surfaces.* The passivation of all glass surfaces is critical for the determination of the methylgermanium species (*Lewis*, 1985). If only inorganic germanium is to be determined, passivation is less important and procedures such as those given for arsenic are adequate. For the accurate determination of the methylgermanium species, the following procedure must be repeated after each 40 h of use of the system.

All glassware is cleaned overnight in a hot ($\approx 80\,^\circ$C) bath of 6 mol/L HCl. It is then soaked in tapwater for over 8 h, rinsed well with deionized water (DIW) and placed in a muffle furnace overnight at 500 $^\circ$C. This is necessary to remove any organic compounds which may be sorbed on the glass surfaces. The glassware is then rinsed again with DIW and dried in an oven at $\approx 60\,^\circ$C. The head of the reduction vessel, the water trap and the transfer tube (between the column and graphite furnace) are then treated with 5 % dimethyl-dichlorosilane in toluene, rinsed twice with toluene, twice with methanol and dried by purging with inert gas. The reaction vessels and bubbler tubes (*i.e.*, the parts that come in contact with the liquid sample) should not be silanized.

The *U*-shaped cold trap/column is treated in a GC oven. After installation, it should be flushed with an inert carrier gas at a flow rate of about 100 mL/min, brought to a temperature of 200 $^\circ$C and kept at this temperature for 2 h. Then, the temperature is reduced to 150 $^\circ$C and two 25 μL aliquots of Silyl-8 (Pierce Chemical Company) or a similar reagent are injected 15 min apart. The column is then conditioned at the same temperature overnight.

5. *Atomic absorption spectrometer.* As the germanium line used (265.2 nm) is in the short wavelength UV region, a spectrometer with good performance in this region is required. As light source, an electrodeless discharge lamp should be used, which provides significant improvement in detection limits and linear range over hollow cathode lamps. The slit setting is 0.7 nm. The use of a background corrector is normally not necessary; depending on the type of AAS, the use of a deuterium or Zeemann-effect background correction may lead to reduction of baseline noise (see also Section 12.2.1.1). The atomic absorption signal is monitored on a chart recorder, reporting integrator/recorder or chromatography data acquisition system. The last permits the evaluation of both peak heights and peak areas, as well as internal calibration and direct output in units of germanium species concentration.

6. *Modification of the graphite furnace.* This system requires a graphite furnace with an internal argon purge system and quartz glass windows, *e.g.*, the Perkin-Elmer 400 or 500 models. During normal operation of these furnaces, argon gas enters the furnace through two internal purge inlets at either side of the graphite tube, flows from both sides towards the middle of the tube and is exhausted through the central injection hole. The quartz windows prevent the gas from escaping out through the sides of the furnace (see Fig. 12-5). To convert this furnace into a gas flow-through detector, the two 'internal purge' argon supply hoses are disconnected at the furnace housing and one of them is clamped off. The other is attached to the connector tube between the hydride generator and the furnace ('Ar in' in Fig. 12-9). This connection tube is then attached, using a short piece of Teflon or Tygon tubing, to one of the argon internal purge inlets. The tubing should be short enough so that the connector tube is attached glass-to-metal. As a result of these modifications, the helium gas from the hydride generator is now mixed with argon, the mixture is forced into the graphite tube at one end, flows through it and

escapes partly through the central injection hole of the graphite tube and partly through the second internal purge inlet. In principle it is possible to use graphite tubes without the central injection hole, so that all the gas flows through the full length of the tube; we have obtained such tubes from Ultracarbon (Bay City, MI, USA) but have not observed a significant increase in sensitivity when they are used. Graphite tubes without interior pyrolytic graphite coating give more consistent results than those with such a coating. The furnace is mounted and centred in the atomic absorption spectrometer in the usual manner.

The following programme is entered into the furnace controller (see also Table 12-3):

Step 1: 20 °C, 1 s Ramp, 5 s Hold, Full Purge Flow
(delay step before the actual atomisation step).

Step 2: 3000 °C, 1 s Ramp, 3 s Hold, Full Purge Flow
(during this step, any impurities are burnt out of the graphite tube).

Step 3: 2700 °C, 1 s Ramp, 80 s Hold, Miniflow
(Atomisation step. When the furnace argon controller is used to regulate the argon added to the helium carrier, the reduced flow mode ('Miniflow' on the HGA 400) should be selected during this step. If the methylgermanes, which elute after the inorganic germane peak, are not to be determined, the hold time during this step can be reduced to 12 s).

Step 4: 20 °C, 15 s Ramp, 5 sec Hold, Miniflow
(cool-down step).

7. *Reagents.*

a) *TRIS-HCl* (1.9 mol/L): Dissolve 230 g of TRIS (tris(hydroxymethyl)aminomethane) in *ca.* 750 mL of pure water and slowly add 155 mL of concentrated hydrochloric acid (high purity grade, *e.g.*, G. Frederick Smith Double Distilled or Merck Suprapur). After cooling to room temperature, adjust the pH by adding HCl or TRIS base until pH 6.0 is indicated by a glass electrode. Finally, bring the volume up to 1 L with pure water. Store in a polyethylene bottle in the refrigerator.

b) *Sodium borohydride* (NaBH$_4$): A 20 % solution of sodium borohydride is prepared in pure water and 1 mL of 2.5 mol/L NaOH solution is added per 100 mL of reagent. This should be prepared at least 8 h before use to allow it to stabilize. It is then stable for about 48 h and should be discarded after this time. Sodium borohydride is hygroscopic and the solid reagent hydrolyses if exposed to moist air. It should therefore be kept in a desiccator and the weighing performed quickly.

c) *EDTA* (0.2 mol/L): Dissolve 74.4 g of disodium EDTA dihydrate in 1 L of pure water. Keep refrigerated in a plastic bottle.

d) *NaCl*: Prepare a saturated solution in DIW ($\approx$ 350 g/L), store in plastic bottle.

e) *Decanol*: Use best available commercial grade.

8. *Standards.* To prepare a 0.01 mol/L stock standard solution, dissolve 1.0459 g of GeO$_2$ in 100–200 mL of pure water to which a few pellets of sodium hydroxide have been added. Titrate with hydrochloric acid to pH 5 and bring up to a volume of 1 L. Plastic laboratory ware and volumetric flasks are preferable, but not essential for these operations. The standards should be stored in linear polyethylene (LPE) bottles. Dilute standards are prepared from this stock solution by sequential dilution. As a working standard, a solution of 10 nmol/L of Ge is used. The standards are stable over several months, even at the 10 nmol/L concentration.

The methylgermanium standards are prepared from the commercially available mono-methyl, dimethyl and trimethyl chlorides (Alfa Johnson Matthey Corp.). Stock solutions at a concentration of 0.01 mol/L are made by dissolving the methylgermanium chlorides in DIW. They are then diluted sequentially to make working standards at the seawater concentration level (100–300 pmol/L). When seawater samples are to be analysed, the final working standard should be made in artificial seawater, or NaCl solution (see the following section) should be added to the reaction mixture.

The primary and secondary standards (0.01 mol/L and 10 μmol/L) are stable indefinitely. The working standards should be prepared fresh every few months. The standards should be stored in linear polyethylene bottles and may be kept at room temperature. The pure methylgermanium compounds and primary standards should not be stored in the same room as the dilute standards and water samples to avoid contamination.

Analysis of the sample

The water volume to be analysed depends on the expected germanium concentration; in seawater this concentration ranges between the limit of detection (4 pmol/L) and 500 pmol/L for inorganic germanium. The concentrations of mono- and dimethylgermanium in seawater are 330 and 100 pmol/L, respectively, and vary only slightly. Trimethylgermanium is below the detection limit in most natural waters. Reaction vessels to match the required sample size of between 50 and 500 mL are used. If possible, the amount of germanium per sample should be between 0.5 and 10 ng (7–140 pmol). In practice, a sample volume of 250 mL is most commonly used.

The sample size is determined either volumetrically or gravimetrically, and the sample is placed into the reaction vessel. Then 6 mL of the 1.9 mol/L TRIS–HCl solution, 10 mL of the saturated NaCl solution and 1 mL of EDTA solution are added per 100 mL of sample. The reaction vessel is attached to the apparatus and purged for 3, 5, 10 or 20 min for the 50, 100, 250 or 500 mL reaction vessels, respectively, to remove entrapped and dissolved air. The liquid nitrogen bath is then moved up to immerse the cold trap/column; the liquid nitrogen is topped up to maintain a constant level from analysis to analysis. Then 6 mL of sodium borohydride solution per 100 mL sample are injected through the side port using a plastic syringe with a stainless steel hypodermic needle. The germanes are stripped from the solution for a period of 10, 25, 45 or 90 min for the 50, 100, 250, or 500 mL reaction vessels, respectively. In rapid succession the following steps are then executed: (1) remove the liquid nitrogen bath from the cold trap, (2) switch on the Variac to supply the required voltage to the heating coil on the cold trap/column, (3) start the graphite furnace programme and (4) start the chart recorder or integrator. After the peaks have been recorded, the heating is allowed to continue until all water visible in the trap has evaporated.

The addition of NaCl solution is not required for samples of open ocean seawater, but should be made when estuarine waters are analysed. It facilitates the stripping of the germanes by the gas stream and eliminates a slight salt effect.

For strongly foaming samples, a drop of decanol can be added to the reaction mixture. This should only be done when absolutely unavoidable, however, because it results in a coating of the glass surfaces with decanol and a build up of pyrolytic graphite in the graphite furnace. Both effects reduce the sensitivity and precision of the germanium species determinations.

Test procedures and sources of error

The system and procedure should be checked and optimized after being set up; sensitivity and blank checks should be done daily. In seawater there are normally no interferences; it is, however, advisable to test the recovery from time to time by the method of standard additions. The following operations should be performed initially and should be repeated when spurious results are suspected.

1. Set up the apparatus and start the gas flows. Insert a graphite tube and initiate the programme cycle. Record the output on the chart recorder. With some graphite tubes, large blanks are seen at this stage. Repeat this process until no more blanks are observed. This may take as many as 20 burn cycles.

2. Use the 100 mL reaction vessel. Add 90 mL of water, the appropriate amounts of TRIS, EDTA and NaCl, and 10 mL of the 10 nmol/L standard (*ca.* 7 ng of Ge), and run it through the analytical procedure. A peak corresponding to *ca.* 0.07 absorbance units should be observed. If no peak appears, select a furnace programme consisting of a single step: 2600 °C, 1 s Ramp, 60 s Hold. This will help to locate the peak and determine its 'retention time'. Then adjust the programme parameters so that the peak lies well within the atomization interval. Repeat the determination on this type of standard, until consistent results are obtained. Then make small variations ($\pm 10 \%$ increments) in the flow rates of helium and argon (separately!) and check whether an increase in sensitivity can be obtained. Vary the atomization temperature in steps of 100 °C to see whether the temperature chosen lies on the plateau of sensitivity.

   Use a reaction mixture containing similar amounts of inorganic and methylgermanium species, *ca.* 0.1 nmol each. Vary the heating voltage until optimum peak sharpness is obtained, and the trimethylgermane peak elutes about 35 s after the inorganic germanium peak. Once an optimized set of conditions has been found, it should be carefully documented (flow rates, furnace programme, voltage and current through coil) and the sensitivity should be noted. This will help in future trouble-shooting.

3. Using the same settings as obtained in step 2, run the same procedure on pure water. The blank should be very small, if detectable at all. Usually with the 500 mL vessel there is a small but significant blank (*ca.* 0.3 ng). Record this blank and convert it to concentration units using the sensitivity constant from step 2.

A few trouble-shooting hints:

1. Check the helium flow at the *outlet* of the cold trap/column. It should be 40 mL/min here. If the two-reactor system is used, make sure both flows are balanced to better than $\pm 5 \%$. Clogging of the trap or of the frit in the reactor, gas leaks, *etc.*, can often be diagnosed with this test.

2. Check the voltage and current through the heating wire and make sure there are no oxidized contacts.

3. Prepare fresh standards and sodium borohydride solution and repeat the experiments with the fresh solution.

4. Check the alignment of the graphite furnace in the beam (with the background correction off) as well as the settings of the monochromator.

5. If erratic results are obtained and the graphite tube has been in use for a large number of analyses, it may be corroded at its ends. Gas can then escape before it gets into the furnace. Replace the tube.

6. A build up of pyrolytic graphite (*e.g.*, if decanol has been used) can be recognized by the presence of a large molecular absorbance peak during the pre-burn step of the furnace programme. The problem can usually be corrected by cleaning the graphite tube with a cotton swab. When this problem has occurred, it is usually also necessary to repeat the surface passivation procedure of the glass parts.

Quantitation of the results

Even though there are usually no interferences observed in the analysis of seawater samples, we prefer to calibrate the system using standard additions to a seawater or artificial seawater matrix. First the reagent blank should be determined using the procedure described in Section 12.2.3.1 for arsenic by determining the value of the blank in terms of peak height or area and conversion into molar units using the distilled water sensitivity value. Then an appropriate volume of seawater or artificial seawater is chosen and analysed. Additions of inorganic and methylgermanium standards are then made to cover the range 0–500 pmol/L and the spiked samples run by the appropriate method. All determinations are made in duplicate or triplicate. From these data, the slope of the analytical curve can be computed in units of peak height (or area) per nanomole by means of linear regression (see Fig. 12-6). A plot of the data verifies the linearity of the calibration curve over the desired range. The sample concentration is then obtained by dividing the sample peak absorbance by the analytical slope and subtracting the blank concentration.

$$Blank\ concentration = A_{wb} \times \frac{M_{ws}}{A_{ws} - A_{wb}}\ (mol/L)\ or\ (mol/kg)$$

where

M_{ws} = concentration of standard;

A_{ws} = absorbance peak area of standard in unspiked water (*ca.* 1 nmol/L); and

A_{wb} = absorbance peak area of unspiked water.

$$Sample\ concentration = (A_s \times S) - blank\ concentration\ \ (mol/L)\ or\ (mol/kg)$$

where

A_s = absorbance peak area of sample; and

S = slope of analytical curve in absorbance peak area units per mole.

The absorbance can be expressed in terms of peak absorbance or peak integrals. The linearity of the calibration curve is dependent on the optical characteristics of the instrument and the intensity of the light source. It needs to be evaluated for each instrument and checked occasionally. The analytical curve should not be extrapolated beyond the highest standard concentration.

Precision, detection limits and systematic errors

The precision of the method depends on the germanium concentration, the instrument used, the experimental parameters and the method of signal processing. Typically, at seawater concentrations a precision of *ca.* ±4 % relative standard deviation can be expected (using a 250 mL sample volume). Under optimal conditions, especially when mass flow controllers are used to regulate the gas flows, a precision of $\pm 1\%$ can be attained. The detection limits are 5 pmol/L for inorganic Ge, 6 pmol/L for monomethylgermanium, 9 pmol/L for dimethylgermanium and 4 pmol/L for trimethylgermanium. In the absence of certified standards for germanium species in a water matrix, the accuracy of the method cannot be properly assessed. Systematic errors can be minimized, however, by using the method of standard additions as suggested before. When an electrodeless discharge lamp is used, the system is linear from the detection limit up to 200 ng of Ge.

12.2.4   Mercury

L. Brügmann

12.2.4.1   Introduction

Mercury has been found in seawater in a variety of operationally characterized particulate (>0.4 or $>0.45 \mu m$) and 'dissolved' ('ionic', 'colloidal', 'gaseous', 'reactive', 'easily reducible', 'labile', 'organically associated', 'non-reactive') forms. In addition, chemically defined species have been determined, such as elemental (Hg^0), monomethyl- (MMHg) and dimethylmercury (DMHg) (*Bloom et al.*, 1995; *Coquery and Cossa*, 1995; *Iverfeldt*, 1988; *Leermakers et al.*, 1995; *Mason et al.*, 1995). The relationship between the different species of Hg and its total concentration (Hg_T) is highly variable in space (with sea area and water depth) and time (daily and seasonal changes). The procedure described here applies to the determination of 'total mercury'. A brief outline, however, is also given on the determination of some of the major species for which more or less 'standardized' methods are in common use.

Mercury is a unique metal. Its oceanic background concentration (Hg_T) is very low (≤ 1 ng/L; see also Table 12-1). The risks of rapid changes in speciation, losses and contamination of samples by contact with laboratory air, materials and reagents are high. In contrast to other trace elements, seawater reference materials certified for mercury concentration are not yet available. The quality assurance with regard to accuracy of data is further hampered by the fact that most laboratories, due to the lack of alternative methods, follow almost the same analytical approach.

In the past, mercury (Hg_T) in seawater was mainly determined by cold-vapour atomic absorption spectrometry (CVAAS) following oxidative pretreatment of the samples, reduction of the mercury ions, purging and collection of the Hg^0 on a trap, revaporisation and detection. During the last decade, the detection limit of the method has been lowered by a factor of >10, mainly due to further reduction of the blank values. In addition, replacing the AAS detection at 253.7 nm by more powerful fluorescence detectors allows detailed and reliable studies on different mercury species in relatively small sample volumes and even automation at ultratrace levels.

12.2.4.2   Principle of the method

Following oxidation of the sample, reduction of mercury ions to Hg^0, purging, collection and enrichment of the elemental mercury by amalgamation and its revaporisation, Hg_T concentrations in seawater are determined by cold-vapour atomic fluorescence spectrometric detection (CVAFS; *Bloom*, 1989). Compared with CVAAS, CVAFS has several advantages. These include, first of all, the higher sensitivity of commonly available instruments which are able to detect Hg^0 amounts of < 0.1 pg. Fairly short path length cells may thus be used whereby memory effects due to insufficient flushing and adsorption/desorption processes are reduced. Another advantage is the better immunity of CVAFS with regard to non-selective absorbances at 253.7 nm potentially caused by compounds other than Hg^0. In addition, the range of linearity has been extended up to five orders of magnitude. The detection limit of the analytical procedure, however, is still mainly controlled by the blanks contributed by the reagents and the analytical system, *i.e.*, memory effects, outgassing of Hg^0 from contacted materials and its penetration from ambient air.

The entire method involves the following steps:

1. To reduce losses by adsorption and evaporation of dissolved gaseous mercury (DGM) from solution during storage, the sample should be kept strongly acidified (pH ≤ 1), preferably with an oxidizing mineral acid such as nitric acid.
2. To convert the various species of mercury which occur even in strongly acidified seawater (*Cossa and Courau*, 1984) into a well-defined and more easily detectable chemical form, oxidation to labile Hg^{2+} ions is needed. This is carried out at room temperature with either BrCl (*Bloom and Crecelius*, 1983) or $KMnO_4$ (*Brügmann et al.*, 1985).
3. The excess oxidant has to be removed before the Hg^{2+} ions are reduced.. This is done by adding small aliquots of a concentrated solution of purified hydroxylaminehydrochloride ($NH_2OH \cdot HCl$) solution.
4. For reduction of ionic mercury to Hg^0, Hg^{2+} is reduced to elemental mercury with an excess of tin(*II*) solution in sulphuric acid.
5. The elemental mercury is purged from solution with argon and collected on gold foil by amalgamation.
6. Mercury is transferred from the collection trap into an analytical trap by electrothermal heating.
7. Heating the analytical trap drives Hg^0 as a cold vapour onto the AFS detector.

12.2.4.3 Reagents and equipment

1. Reagents

Standard solutions: A working standard solution of 10 µg/L of Hg is prepared in 1 mol/L HNO_3 by dilution of appropriate commercial stock solutions (1 g/L of Hg). When kept dark and cool ($\leq 8\,°C$) in a glass bottle, this working standard is stable for at least one year.

Gases: The purge (carrier) gas argon ($> 99.99\,\%$) taken from a pressure tank should be of sufficient quality regarding low impurities of oxygen and especially of mercury. It has to be cleaned properly by passing through both an inline dust filter and a gold trap. Mercury released from the analytical apparatus is scrubbed by a charcoal plug placed in a tube at the detector outlet.

Acids: The acids (HNO_3, HCl) added in relatively high amounts to the samples and/or reagents must be of superior quality, with blanks of ≤ 1 ng/L of Hg (*e.g.*, obtained by repeated sub-boiling distillation of reagent grade products; see Section 12.1.5). The acids are stored under clean-room conditions in screw-cap borosilicate glass bottles.

Oxidants: Bromine monochloride (BrCl) solution is prepared in a glass bottle by dissolving 11 g of $KBrO_3$ and 15 g of KBr, both of superior quality, in 200 mL of water freshly prepared by sub-boiling distillation (see Section 12.1.5). While swirling the bottle gently, the solution is made up to 1 L by slowly adding 800 mL of concentrated HCl. The solution is cooled and may be stored indefinitely at a cold ($\leq 8\,°C$) and dark place (*Bloom and Crecelius*, 1983).

(Alternatively, a 1 L saturated solution of potassium permanganate ($KMnO_4$) is prepared in a stoppered glass bottle and stored at least two weeks for self-purification; the bottle is shaken once a day. The solution is decanted and the decantate shaken and filtered making sure that a layer of precipitated MnO_2 rests on the glass fibre filter. After discarding the first 50 mL, the remaining solution is kept cool and dark in a well-stoppered borosilicate glass bottle until use.)

Solution to reduce oxidants: 100 mL aliquot of 50 % $NH_2OH \cdot HCl$ solution is prepared in a glass bottle. The solution is cleaned by passing it through a column (10×200 mm) containing acid-washed (activated) silver wool at a flow rate of <2 mL/min. The first 25 mL are discarded and the remaining portion is collected in a stoppered quartz bottle. For initial activation and regeneration, the column is treated with about 100 mL of 2 mol/L HNO_3 (*Brügmann et al.*, 1985). In addition to the cleaning with silver, the solution may be further purified by stirring overnight with about 0.5 g of cation exchange resin, *e.g.*, Chelex 100 (*Bloom and Crecelius*, 1983).

Reducing solution: Prepare a 10 % Sn^{2+} solution in 3 % H_2SO_4 in a glass bottle by dissolving the necessary amount of $SnCl_2$ (a.g.) in water, add 30 mL of concentrated sulphuric acid and make up to 1 L. The solution is purified by purging the bottle with argon or nitrogen for a minimum of 12 h at ≥ 200 mL/min (*Bloom and Crecelius*, 1983). (Alternatively, the reducing solution may be prepared in HCl.)

Working aliquots of reagent and standard solutions should be kept in small bottles for daily use. Contamination of stock solutions by repeated contacts with the laboratory environment is thus minimized.

2. Equipment

A flow chart for the analytical equipment used is given in Fig. 12-10. The purge gas flow (1) is measured with a flow meter (2). Between the flow meter and reaction vessel (5), the argon is purified by passing through a dust filter (3) and a trap with gold-coated quartz wool (4). All parts are connected with Teflon tubing of 2–3 mm inner diameter. Silicone rubber sleeves secure the connections between the pieces of tubing and equipment.

For oxidation, the samples are filled into glass-stoppered 500 mL borosilicate bottles which also serve as reaction vessels (5). Inserted into the neck of the bottle is a gas-washing bottle with a three-way cock (6) for short-cuts after completion of purging. The reaction vessel (5) sits on a magnetic stirrer. At its end the inlet tube of the gas washing head is equipped with a low porosity glass frit which produces bubbles of very small diameter. The outlet of the reaction vessel (5) is connected to the collection trap (8) consisting of a fused silica tube (about 10×100 mm) in which a plug of gold-coated quartz wool (of about 5 cm in length) is held in place with plug of uncoated quartz wool. A resistance wire wound around 70 % of the tube length heats the tube to a low temperature (hand-warm) during collection and to $\geq 800\,°C$ for vaporisation. The Hg^0 standards are injected between reaction The vessel and the collection trap through a septum (7) held in a PTFE screw cap.

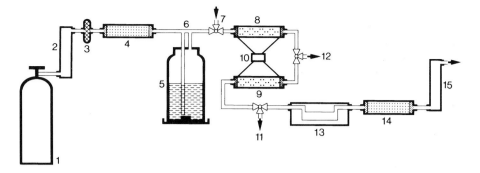

Fig. 12-10. Schematic view of the analytical system for mercury determination by CVAFS (CVAAS): 1 pressure bottle for inert gas (Ar, N$_2$,...); 2, 15 flow meters; 3 dust filter; 4 gas-cleaning trap (Au); 5 reaction vessel (500 mL) (including stirring bar, PTFE-encased), seawater sample (200 mL) and (magnetic) stirring device); 6 three-way cocks; 7 septum (Hg^0 injection); 8 collection trap (Au); 9 analytical trap (Au); 10 voltage control; 11, 12 gas outlets (optional via three-way cocks); 13 AFS (AAS) detector; and 14 mercury scrubber (charcoal trap).

The purge gas, after removal of Hg^0 may be vented through a three-way stop cock (12) between the collection trap (8) and the analytical trap (9). The detector (13) placed downstream from the analytical trap (8) usually measures the fluorescence (CVAFS). With Hg concentrations ≥ 100 pg/L, modern CVAAS monitors have almost identical sensitivity, accuracy and precision of measurements (*Bloom and Fitzgerald*, 1988; *Schmidt and Gerwinski*, personal communication). The resulting peaks may be recorded as analogue signals (*y/t* recorder) and/or they may be further processed (integrator, PC, *etc.*). Downstream from the detector, a charcoal trap (14) serves as a scrubber for released mercury; a second flow meter (15) allows detection of improper connections or other gas leaks within the system.

12.2.4.4 Analysis of the sample

In a 500 mL bottle, 200 mL of unfiltered seawater are acidified with 4 mL of HNO$_3$. Following addition of 1 mL of BrCl (saturated KMnO$_4$) solution, the sample is kept for oxidation ≥ 30 min (≥ 120 min with KMnO$_4$) at room temperature. Before analysis, a Teflon-coated stirrer bar and 0.25 mL of 50 % NH$_2$OH$\cdot$HCl solution are added to reduce the excess oxidants (chlorine and/or KMnO$_4$/MnO$_2$, respectively). Reduction of oxidants should be complete after about 30 s under stirring and can be detected by the change in smell and/or colour. Two mL of 10 % Sn^{2+} solution are added before the stirred solution is purged with Ar at 150 mL/min for 30 min.

During purging and collection, the argon leaves the system behind the collection trap (Fig. 12-10; (8)) *via* a three-way cock (12). After 30 min, the gas flow bypasses the solution for 1 min to dry the system. The argon flow is reduced to 75 mL/min. With current from a variable transformer (10) the resistance wire wound around the collection trap (8) is heated to a dull red (≤ 800 °C). The transfer of mercury from the collection (8) to the analytical trap (9) is complete within ≤ 20 s. To avoid interference from coatings of the surfaces in the optical system of the monitor, the gas flow during Hg transfer still bypasses the detector *via* (11). Following completion of the transfer, the gas flow is directed through the monitor (13) for 30 s to allow cooling of the collection trap (8), readjustment of the flow rate and stabilization of the detector baseline. Finally, the analytical trap (9) is heated to the pre-determined temperature maximum using a reproducible voltage ramp. Heating is continued until the peak has been recorded and the signal has returned to the baseline.

Additional remarks:

1. In total mercury determinations, filtration of seawater samples has been found to be a serious source of contamination (see also Chapter 2). Therefore, filtration should only be considered in cases where the suspended particulate matter (SPM) needs to be investigated separately and/or when the particulate fraction of Hg_T is $\geq 20\%$ and expected to fluctuate strongly as, *e.g.*, in estuaries and other coastal zones.
2. For the determination of particulate mercury, acid-cleaned Teflon and quartz fibre filters, the latter combusted at 500 °C (*Coquery and Cossa*, 1995), are recommended. A significant fraction of the mercury in seawater is present in colloidal forms and separated with the aid of acid-cleaned ultrafiltration cartridges (*Stordal et al.*, 1996).
3. If necessary, samples may be stored in the dark at 4 °C in acid-cleaned bottles made from Teflon (or FEP), quartz or borosilicate glass for weeks and even months without significant losses or contamination (*Bloom et al.*, 1995; *Schmidt and Gerwinski*, personal communication). Analyses for speciation studies, however, should be performed immediately after sample collection.
4. To acidify seawater samples for Hg determination, HCl has been shown to be a useful substitute for HNO_3. (High concentrations of H_2SO_4 in the samples, however, may cause problems due to higher blanks and/or interferences with the purging/amalgamation step.)
5. Photo-oxidation instead of wet-chemical oxidation is often considered an attractive method because of the very low blanks. However, irradiation with UV light does not seem to ensure complete recovery of mercury from organically associated and related species. Reduced signals have even been observed following extended irradiation times (*Gerwinski et al.*, 1996) which might indicate the formation of compounds with decomposition products from which Hg cannot be liberated by reduction with Sn^{2+}. Therefore, to determine true 'total' mercury concentrations, UV irradiation must be followed by wet-chemical oxidation. For samples with DOC concentrations of >25 mg/L, an initial pre-oxidation by UV-irradiation was found necessary to ensure total oxidation of DOC and accurate determination of Hg_T by the bromine monochloride method (*Olson et al.*, 1997).
6. Argon is the preferred inert gas, because it can be used for both sweeping the Hg^0 from solution *via* gold traps to the detector, and for purging the optical path. Nitrogen causes interferences in the detection step, and the oxygen in air is partly converted into ozone under UV irradiation which results in losses in sensitivity.
7. Often purge flow rates of between 400 and 500 mL/min are recommended. However, purging with lower flow rates but smaller gas bubbles in stirred solution was found to be equally efficient and also reduces problems regarding carry-over of acidic fumes and water vapour to the collection trap (*Schmidt and Gerwinski*, personal communication).
8. The gold traps may consist of plugs of either gold wire, gold foil, Au–Pt nets or different gold-coated materials placed in vitreous silica tubes. Traps containing the pure materials are more expensive and involve the risk that due to shrinking and clogging their collection efficiency is reduced. Therefore, Au-coated quartz wool or quartz sand should be given preference.

9. When establishing an analytical system for Hg determination, the efficiency of labora-
    tory-made traps should be scrutinized carefully. Initially, dual collection traps may be
    placed in a system to test for losses due to possible breakthrough.
10. The collection trap is kept hand-warm to avoid humidity accumulating. The use of
    additional inline drying tubes, prone to memory effects and lower reproducibilities, is
    thus avoided.
11. The use of only *one* Au-trap in analytical systems for Hg determination in seawater is
    not recommended. Revaporising the mercury directly from a collection trap into the
    detector may give rise to lower reproducibility. During the first collection step, mercury
    species other than Hg^0 will also accumulate. Their subsequent release may be con-
    trolled not only thermally but also by the reaction kinetic. In addition, the trap may
    become poisoned with time, necessitating repeated calibration and prolonged heating
    for regeneration. An additional integrated 'analytical trap' is decoupled from such
    interferences and therefore guarantees generation of reproducible signals over several
    months and even years.

This analytical procedure represents a typical example of how to carry out Hg_T determinations in seawater. Because of many procedural and equipment-related variables, users are
urged always to optimize the procedure with respect to the prevailing conditions. If, for
instance, the volume of seawater used for the analysis (200 mL) has to be increased due to
insufficient sensitivity of the available detector, it is necessary not only to change the volumes of applied reagents appropriately but also to optimize the analytical procedure under
the newly established conditions.

12.2.4.5 Calculation of results

For calibration, two different approaches should be used:

1. A calibration factor (*f*) is obtained by repeated analyses of a pre-purged ordinary sea-
    water sample which has been spiked with increasing amounts of mercury standard solu-
    tion. The concentrations of the spiked seawater, *e.g.*, 0.2, 0.5 and 1.0 ng/L of Hg, should
    cover the concentration range expected in the field. Between runs with spiked samples,
    the system blanks should be established, *i.e.*, by applying the analytical procedure without
    any addition of mercury.
2. For recovery tests and for performance checks of the system, small volumes of air satu-
    rated with mercury vapour are injected through a septum (see Fig. 12-10; (7)). The spike
    samples are taken with gas-tight GC microlitre syringes through a septum from a thermo-
    stated vessel with some Hg at the bottom. The temperature of the vessel has to be con-
    trolled to within $\pm 0.1\,°C$. The amount of Hg^0 injected into the system is controlled by the
    temperature and the selected volume (*ca.* 40–200 µL); it should cover the range expected
    from seawater analyses, *i.e.*, *ca.* 40–200 pg in a sample aliquot of 200 mL.

The Hg_T concentrations in samples and standards are determined using either the heights
or areas of recorded fluorescence (absorbance) peaks. The detector response of spiked samples is corrected for the 'system blank' and is used to obtain a calibration factor. Using this
calibration factor, the sample readings, corrected for both the system and reagent blanks,
allow calculation of the Hg_T concentration:

The calibration factor (f) refers to a sample volume of 200 mL and is obtained by

$$f = (F_{st} - F_{bls}) \, / \, pg \; Hg,$$

where F_{st} and F_{bls} denote the fluorescence (absorbance) of a pre-purged spiked seawater sample and of the 'system blank' (see Section 12.2.4.6), respectively. The calibration factor is the mean of at least triplicate measurements.

The Hg concentration in the seawater is then calculated according to

$$Hg \; (ng/L) = (F_{sw} - F_{blt})/f \cdot 200$$

where F_{sw} and F_{blt} denote the sample fluorescence (absorbance) of the seawater sample and of the total blank (see Section 12.2.4.6), respectively.

12.2.4.6   Precision and determination of blanks

Major problems might arise due to insufficient blank control. For instance, only materials with very low Hg concentrations, with a known history regarding previous contacts with higher Hg concentrations and with properties enabling thorough cleaning are usable. Blanks are reduced by (a) working in clean rooms and/or on laminar flow clean benches, (b) using only gases, chemicals and materials of controlled high purity with regard to Hg concentrations, and (c) by immediate preservation of seawater samples after collection.

The total blank is composed of the 'reagent blank' and the 'system blank'. Repeated purging and analysis of pre-purged seawater samples allows the determination of the 'system blank' and its standard deviation. The 'system blanks' should be ≤ 10 pg, otherwise the system and/or operating variables must be changed. Higher 'system blanks' may indicate memory effects and contamination caused by, *e.g.*, too large volumes of the system's compartments, improper performance of the Au-traps (deactivation, insufficient heating) or specific sites with risks of adsorption/contamination (septum, sleeves).

The reagent blank should not exceed 20 pg (100 pg/L); this is mainly due to the solutions described under 12.2.4.3. A major part of this blank originates in the oxidant solutions which must be cleaned carefully and protected against entrapment of Hg from the laboratory environment. The $NH_2OH \cdot HCl$ solution is less prone to later contamination but needs very thorough cleaning. Under clean laboratory conditions, the contribution of sub-boiling distilled HNO_3 (used to acidify the samples; see Section 12.1.5) can be reduced to about 2–5 pg (10–25 pg/L). Applying effective purging, Sn^{2+} solutions can be prepared almost free of mercury. To obtain the total blank value, a pre-purged sample is mixed twice with another set of reagents (4 mL HNO_3, 1 mL oxidant solution, 0.25 mL of 50 % $NH_2OH \cdot HCl$, 2 mL of 10 % Sn^{2+} solution) and analysed. The reagent blank is obtained after subtraction of the system blank.

The reagent and system blanks must be recorded at specified intervals. These intervals may be related to the number of samples processed or may be fixed periods of time. Changes in the analytical system, including replacement of parts of the detector and new sets of reagents, should always be followed by blank controls. The blank values should be monitored using control charts.

Detection limits ($3s$ of the blank values) of 0.05 ng/L have been reported for Hg_T (*Bloom et al.*, 1995). With concentrations ≥ 0.5 ng/L, relative standard deviations of between ± 5 and

± 10 % (*Bloom et al.*, 1995; *Iverfeldt*, 1988) may be obtained. At lower concentrations, this value can increase to ± 20 % (*Iverfeldt*, 1988).

12.2.4.7 Speciation studies

Many different methods are being used for speciation studies. Some of them are already well established and have produced comparable results in intercalibration studies (*Bloom et al.*, 1995; *Cossa and Courau*, 1984).

1. Concentrations of 'reactive' ('labile', 'easily reducible', 'ionic') Hg species are obtained by omitting the oxidation step from the above procedure proposed for Hg_T determinations in seawater. In order to obtain reproducible results, storage conditions (time, temperature, light, pH) must be kept constant. The use of $NaBH_4$ instead of Sn^{2+} as the reductant results in significantly higher concentrations representing 'reactive' plus 'non-reactive' mercury yet not true Hg_T (*Bloom et al.*, 1995; *Iverfeldt*, 1988). The reagent blanks in the procedure for 'reactive' mercury are lower in comparison with the Hg_T determination because only HNO_3 (HCl) and Sn^{2+} solutions are needed. The detection limits, however, are not lower. This is due to the lower precision of the measurements including the total blank determination, because the recorded values are more affected by analytical variables such as storage and purging conditions. To avoid the release of 'reactive' Hg species from otherwise recalcitrant compounds during storage of samples under acidified conditions, *Mason and Fitzgerald* (1990) preferred using untreated samples for the determination of 'active' Hg.

2. Dissolved 'gaseous mercury' (DGM), which is believed mainly to represent the sum of Hg^0 and DMHg, is determined by purging 1–2 L of untreated seawater for 1 h with 500 mL/min of Ar (*Coquery and Cossa*, 1995; *Leermakers et al.*, 1995). The mercury accumulates on the collection trap and is determined as described before for Hg_T. Owing to the absence of reagents the blanks are very low. The detection limit ($3s$ of the blank values) is approximately 5 pg/L (*Leermakers et al.*, 1995). (To ensure quantitative recovery of the non-Hg^0 fraction of DGM, the purged sample gas may first be passed through a reactor tube containing a plug of quartz wool pre-tested (900 °C) for thermal breakdown of volatile mercury compounds.)

3. For a more detailed operationally-defined speciation study of DGM, a trap containing graphitized carbon black (Carbotrap) can be used, which has been shown to collect DMHg almost exclusively. From the Carbotrap material, the organomercury compounds are thermally desorbed (250 °C) and transformed into Hg^0 for final AFS detection by passing through a reactor tube (900 °C) (*Bloom and Fitzgerald*, 1988). Using a different approach, DMHg may be separated by cryogenic gas chromatography and detected by AFS (*Bloom*, 1989). Detection limits of 3–5 pg/L ($3s$ of the blank values) have been reported for DMHg (*Cossa et al.*, 1994).

4. For the determination of dissolved monomethylmercury (MMHg) in seawater, the analyte is first separated from the interfering chloride matrix either by extraction with methylene chloride (*Mason and Fitzgerald*, 1990; *Mason et al.*, 1995; *Leermakers et al.*, 1995) or by distillation (*Bloom et al.*, 1995). This is followed by ethylation of the MMHg, cryogenic gas chromatography of the reaction products, and AFS detection (*Bloom*, 1989). The detection limit ($3s$ of the blank values) of this method is approximately 10 pg/L. Problems, however, have been observed with the recovery of MMHg from

anoxic waters (*Olson et al.,* 1997) and with respect to the formation of MMHg as an arte-fact from other mercury species during aqueous distillation (*Bloom et al.,* 1997; *Hintel-mann et al.,* 1997).

12.2.4.8   Automation

Some years ago, a microprocessor-controlled flow-stream system was introduced for the determination of ultratrace levels of 'easily reducible' mercury in water samples (*Stockwell et. al.,* 1989). The system is based on an autosampler, reduction unit, gas–liquid separator, gold trap and AFS monitor and allows processing of 8 samples per hour. For sample vol-umes of 45 mL, a detection limit ($3 s$ of the blank values) of 0.1 ng/L was found with an ana-lytical precision of 5 % (*Cossa et al.,* 1995).

More recently, a system has been proposed for the computer-controlled processing of dis-crete seawater samples involving photo-oxidation and dual amalgamation (*Gerwinski et al.,* 1996). Rather than by a peristaltic pump, the volume flow is exclusively driven by carrier gases.

Fully automated systems may become available in the future for reliable Hg_T determina-tions in seawater samples. Such miniaturized systems would probably allow high sample throughput rates and could be based on the following principles/parts: introduction of dis-crete samples, blanks and standards *via* pumps into the analytical system, injection of reagents with pneumatically driven micro-syringes, wet-chemical (BrCl) oxidation, purging and collection of Hg^0, dual amalgamation, and AFS detection. Critical steps controlling the sample rate are completeness of oxidation and the quantitative purging of elemental mer-cury from the reactor vessel.

12.3   Analysis by electrochemical methods

C.M.G. van den Berg

12.3.1   Introduction

Trace elements in seawater can be determined electrochemically using stripping voltam-metry, stripping potentiometry and possibly potentiometry. Stripping voltammetry has most applications for trace metal analysis, and the type of scans allow more control than stripping potentiometry. Potentiometry for trace metal analysis is still in its infancy. For this reason stripping voltammetry is recommended here for trace metal analysis in seawater generally. Notwithstanding this there may be specific applications in which one of the other methods may have advantages.

The name voltammetry originates (and is abbreviated) from voltamperometry. Here cur-rents are measured at a working electrode whilst the potential is being controlled by a potentiostat. Normally the voltage to the working electrode is gradually scanned whilst monitoring the currents due to oxidation or reduction reactions. Oxidation currents are pro-duced as a result of reduced species becoming oxidized by relinquishing one or more elec-trons at the electrode surface. Reduction currents are due to oxidized species accepting one or more electrons. Oxidation currents are called anodic, whereas reduction currents are called cathodic currents. Thus two types of voltammetry are distinguished depending on whether oxidation or reduction currents are measured: anodic stripping voltammetry (ASV) and cathodic stripping voltammetry (CSV). Oxidation currents are measured in ASV and reduction currents in CSV. Elements are oxidized during the ASV scan, whereas they are reduced during the CSV scan.

An important element of stripping voltammetry is the pre-concentration step which is carried out on, or in, the electrode prior to the voltammetric scan. In ASV this consists of plating of the metal onto the electrode: this means that the scan is preceded by a reductive step in which dissolved metal ions are reduced to the metallic state and plated on the electrode. During the ASV scan the plated metal is re-oxidized and goes back into solution. The change from plated metal to dissolved metal ion involves a change in the physical state in addition to the change in the redox state. The solid–solution change is kinetically slow and therefore results in electrochemical irreversibility and relatively poor sensitivity. This problem is alleviated by using a working electrode consisting of mercury in which metals can dissolve by forming an amalgam.

In CSV the element is pre-concentrated on the electrode in an oxidized form, either by adsorption as a surface-active complex or precipitation as an insoluble oxide or salt. The adsorption method has led to the successful development of procedures to detect some 20 elements (metals as well as metalloids) in seawater. The principle of this method will be set out in detail in this chapter, along with details of the determination of a few elements as examples.

12.3.2   Potential scanning methods

In voltammetry the potential is scanned whilst the current is monitored. The current has two components, one due to changes in the capacitance and the second due to changes in redox states (Faradaic currents). Usually the Faradaic currents are to be detected as they are specific to the element which is being determined, whereas the much less specific capacitance currents are eliminated as an interference (this is not always the case as changes in capacitance currents are used for instance to determine organic surfactants in natural waters (*Gasparovic and Cosovic*, 1994)).

The potential scans consist of a linear or stair-case sweep, or a sweep on which a modulation (usually a square-wave, sometimes a sinusoidal wave) is superimposed in which the potential region of interest is scanned rapidly. The effect of the modulation is two-fold. Firstly, the capacitance component of the current is largely eliminated by sampling the current a short time after each change in the potential. Secondly, when the reduction potential of the analyte is passed, the analyte is repeatedly being reduced and re-oxidized, as the square-wave height (typically 25–50 mV) is much greater than the step height (typically 2.5 mV) between each square-wave modulation. For this reason the same amount of analyte contributes repeatedly to the analysis, thus increasing the current associated with its reduction or oxidation.

Table 12-4. Analytical conditions for the direct determination of trace elements (metals) in seawater using CSV employing the reduction of the element in the adsorbed complex. The wave-form used for the voltammetric scan is indicated by DP (differential-pulse), SW (square-wave) and LS (linear-sweep). LD is the limit of detection standardized to an adsorption period of 60 s.

Element	Reagent [a]	pH range	Buffer (pH) [a]	LD(60 s) nmol/L	Scan type	Reference
As	copper		2 mol/L HCl	0.3	SW	*Sadana, 1983; Zima and van den Berg, 1994; Li and Smart, 1996*
Cd	oxine	7.5–8.5	HEPES (7.8)	0.1	DP	*van den Berg, 1986*
Co	nioxime/ nitrite	9.1	ammonia	0.003	DP	*Ostapczuk et al., 1986; Donat and Bruland, 1988; Vega and van den Berg, 1997*
Cr	DTPA	5.0–5.3	acetate (5.2)	0.2	SW	*Golimowski et al., 1985; Boussemart et al., 1992*
Cu	oxine	6–9	borate (8.5)	0.2	DP	*van den Berg, 1986;*
Cu	salicyl- aldoxime	5–9	borate (8.5)	0.1	DP	*Campos and van den Berg, 1994*
Fe	NN/bro- mate	7–8	PIPES or TRIS	0.5	LS	*van den Berg et al., 1991; Yokoi and van den Berg, 1992a; Rue and Bruland, 1995; Aldrich and van den Berg,1998*
Mo	mandelate/ chlorate	1.9	HCl	0.002	DP	*Yokoi and van den Berg, 1992b*
Ni	DMG	7–10	NH_3/NH_4Cl (9.2)	0.2	DP	*Pihlar et al., 1981, 1986; Donat and Bruland, 1988*
Pb	oxine	7.0–8.5	HEPES (7.8)	0.3	DP	*van den Berg, 1986; Wu and Batley, 1995*
Pt	formazone		0.5 mol/L H_2SO_4	0.0004	DP	*van den Berg and Jacinto, 1988*
Sb	catechol	5.8–6.8	MES (6.0)	0.6	DP	*Capodaglio et al., 1987*
Sn	tropolone	1.5–2.7	2.1	0.05	DP	*van den Berg et al., 1989*
Ti	mandelate/ chlorate	3	–	0.007	DP	*Yokoi and van den Berg, 1991, 1992b*
U	oxine	6.5–7.1	PIPES (6.8)	0.2	DP	*van den Berg and Nimmo, 1987; Sander et al., 1995*
U	TTA/TBP	2–3.6		1	LS	*van den Berg and Nimmo, 1987; Sander et al., 1995*
V	catechol/ bromate	6.6	PIPES (6.6)	0.14	SW	*Vega and van den Berg, 1994*
Zn	APDC	6.2–8.5	BES (7.3)	0.3	DP or SW	*van den Berg, 1984b*

[a] APDC = ammonium pyrrolidine dithiocarbamate, BES = N,N-bis(2-hydroxyethyl)-2-aminoethanesulphonic acid, DTPA = diethylenetriaminepentaacetic acid, DMG = dimethylglyoxime, HEPES = N-2-hydroxyethyl piperazine-N'-ethanesulphonic acid, MES = 2-(N-morpholino)ethanesulphonic acid, NN = 1-nitroso-2-naphtol, oxin = 8-hydroxyquinoline, PIPES = piperazine-N,N'-bis-2-ethanesulphonic acid, TBP = tri-n-butyl phosphate, TTA = 2-thenoyltrifluoroacetone.

Table 12-5. Electrochemical determination of other elements in seawater.

Element	Reagent	pH range	Buffer (pH)	LD (60 s) (nmol/L)	Scan type	Reference
NH_4^+	formaldehyde 13 %	3.8		4	DP	*Harbin and van den Berg, 1993*
I		8	borate	0.6	SW	*Luther III et al., 1988 Campos, 1997*
NO_2^-	sulphanilamide/ naphthylamine	varied	borate	0.4	SW	*van den Berg and Li, 1988*
Se(*IV*)	copper	4.5	NH_4SO_4	3	DP	*van den Berg and Khan, 1990*
Se(*IV*)	rhodium	1.6	0.1 mol/L H_2SO_4	0.03	LS	*Wang and Lu, 1993*
HS^-		8		0.5	SW	*Luther III and Tsamakis, 1989; Al-Farawati and van den Berg, 1997*

12.3.3 Cathodic stripping voltammetry

In CSV the element is reduced during the potential scan. The pre-concentration step typically involves adsorption of the element after complexation with a specific complexing ligand causing the formation of a complex with adsorptive properties. Occasionally adsorption of a complex with a metal ion is used to pre-concentrate and determine one of the anions in seawater. Several ligands are available for the pre-concentration step. A listing of the elements which can be determined in seawater, and the ligands used for their determination, is given in Table 12-4 (metals) and Table 12-5 (other elements).

The determination of five selected elements is discussed below in detail to illustrate the procedures which are common for all elements that can be determined by this technique.

12.3.4 Sample pre-treatment

Seawater samples should normally be analysed immediately upon sampling to prevent loss of analyte such as due to changes in the redox speciation or adsorption on the container (see also Chapter 1). Adsorption on the sample bottles can be mostly prevented in stored samples by sample acidification using nitric or hydrochloric acid. Addition of nitric acid can lead to interference with the voltammetric determination of chromium in seawater, so in that case hydrochloric acid should be used. The amount of acid should be minimized as it will have to be neutralized prior to the voltammetric determination of most elements. Reagent use should generally be minimized to avoid sample contamination.

Dissolved organic matter in seawater interferes with the voltammetric analysis as it reduces the sensitivity due to physical effects as a result of competitive adsorption on the electrode, and due to chemical effects by competitive complexation of metal ions. The interfering surface activity of the natural organic matter can be taken into account by calibration of the sensitivity by standard metal additions to each sample, but the competitive complexation causes a partial masking of several metals. Advantage has been taken of the competition between the added, analytical ligand and the natural organic matter to determine a 'reactive' metal concentration as a first step to determine the metal speciation in seawater (*van den Berg*, 1984a). However, it is necessary to either out-compete the natural organic complexing ligands (by adding a high concentration of analytical ligand), or remove the organic matter, to determine the total metal concentration. UV-digestion is a convenient method to destroy dissolved organic matter in seawater and usually requires no reagent additions (*Raspor et al.*, 1977; *Achterberg and van den Berg*, 1994).

12.3.5   Equipment

Voltammetric equipment is available from several sources: BAS (USA), Eco Chemie (Netherlands), EG&G (USA), Metrohm (Switzerland), Radiometer (Denmark/France), Sycopel Scientific (UK) and other companies such as in the Czech Republic and China. The equipment consists of a hanging mercury drop electrode (HMDE) and a potentiostat. Nowadays the electrodes have a valve to dispense a new mercury drop. These electrodes and potentiostats are computer controlled which is convenient as it automates the timing.

Interfacing with autosample-changers and burettes is possible with several of these instruments which leads to automation of the analysis, including calibration by standard additions to the sample in the voltammetric cell. Various flow cells for in-line analysis, essential for flow-injection analysis and flow-analysis, have been built (*Wang and Ariel*, 1978; *Fogg and Summan*, 1984; *Yarnitzky*, 1990; *Gutz et al.*, 1993; *Colombo et al.*, 1997), but their commercial availability is still pending at this moment due to difficulties with making a reliable flow-system for use with mercury drop electrodes. Nevertheless the use of flow-cells enables one of the major advantages of electrochemical techniques: their mobility and use on-board ship for monitoring of metals in coastal waters or the open sea.

12.3.5.1   Equipment pre-treatment

The voltammetric cell should be soaked in 2 mol/L acid prior to first use. During regular use it is sufficient to rinse the cell with pure water, unless the cell has become contaminated with very high metal levels when an acid rinse is recommended. Adsorption takes place which can cause the peak heights to decrease during measurements in freshly cleaned cells (*Cuculic and Branica*, 1996). All cell materials and components adsorb but the least problems are observed with quartz. It is therefore not advisable to rinse the cell with acid between measurements to allow the cell and electrode walls to become conditioned with seawater and the metal levels being determined.

The reference electrode normally has a slow leak into the solution at a rate of a few microlitre per hour. It is therefore advisable to ensure the filling solution (normally 3 mol/L KCl) is clean, either by using a high grade salt for its preparation, or by purification. Double-junction reference electrodes exist and these are useful for instance if silver or sulphide are to be determined due to the presence of ionic silver in the silver reference electrode cartridge.

The counter electrode consists normally of a platinum wire or a glassy carbon electrode. Either of these electrodes is suitable for trace analysis and requires no pre-treatment, unless platinum is being determined in seawater in which case a glassy carbon counter electrode should be used.

12.3.6   Reagents

12.3.6.1   Reagent purification

Purified water is used to prepare reagents. This is normally prepared using a preliminary clean-up stage using deionization, reverse osmosis or distillation of tap water, followed by a second, high-quality, deionization or distillation (see Section 12.1.5).

Acids and ammonia used for sample acidification and pH neutralization are purified by sub-boiling distillation using a quartz condenser, or isothermal distillation as described in Section 12.1.5.

Contaminating metal ions in pH buffers can be removed by equilibration with MnO_2 followed by filtration, or by adding a suitable chelating agent followed by passage through a C18-cartridge, or both. MnO_2 is normally prepared as a suspension by disproportionation of $Mn(II)$ and $Mn(VII)$ by mixing the appropriate amounts of manganese(II)chloride, potassium permanganate and sodium hydroxide in a molar ratio of $3:2:4$ (*van den Berg and Kramer*, 1979). Thereto 100 mL of 0.015 mol/L $MnCl_2$ are added to 100 mL of a solution containing 0.01 mol/L $KMnO_4$ and 0.02 mol/L NaOH. The mixture is stirred and immediately the MnO_2 will form as a brown precipitate. The pH of the suspension drops rapidly and is adjusted to neutral by addition of NaOH solution (by running 10–20 mL from a 0.1 mol/L solution from a pipette whilst measuring the pH). The MnO_2 is purified (2–3 x) by centrifugation and resuspension in water.

Borate pH buffer: A solution is prepared containing 1 mol/L boric acid and 0.35 mol/L ammonia. This buffer gives a pH of 8.35 when diluted 100-fold in seawater. Boric acid may precipitate if the pH of this solution is too low, in which case the ammonia concentration has to be increased. Contaminating metal ions can be removed from this buffer by adsorption on MnO_2 which is removed by filtration. Thereto MnO_2 suspension is added to a final concentration of 50–100 μmol/L; the mixture is equilibrated for several hours and then filtered over a 0.45 μm filter.

HEPES pH buffer: A solution is prepared of 1 mol/L *N*-(2-hydroxyethyl)piperazine-*N'*-ethanesulphonic acid in 0.5 mol/L ammonia solution. This buffer gives a pH of 7.6 when diluted 100-fold in seawater. Contaminating metal ions can be removed as for the borate pH buffer.

HEPPS pH buffer: A solution is prepared of 1 mol/L *N*-(2-hydroxyethyl)piperazine-*N'*-3-propanesulphonic acid in 0.5 mol/L ammonia solution. This buffer gives a pH of 7.8 when diluted 100-fold in seawater. Contaminating metal ions can be removed as for the borate pH buffer.

TRIS pH buffer: A solution is prepared of 1 mol/L tris(hydroxymethyl) aminomethane in 0.5 mol/L HCl. This buffer gives a pH of 8.07 when diluted 100-fold in seawater. Contaminating iron is removed from this buffer by addition of 20 μmol/L *N,N*-(1-nitroso-2-naphthol) and passing through a Sep-Pak C_{18} cartridge.

12.3.7   Analytical procedures

12.3.7.1   Copper

1. Principle of the method
Several complexing ligands can be employed to determine copper in seawater. However, the best results are obtained when copper is determined after adsorptive pre-concentration as a complex with salicylaldoxime (SA) (*Campos and van den Berg*, 1994). Analysis is carried out at neutral pH values (pH 6.5–8.5 is suitable) which means that the original sample pH can be maintained which is useful for speciation studies.

2. Reagent preparation
   Salicylaldoxime (SA): An aqueous solution containing 0.01 mol/L SA is prepared in 0.1 mol/L HCl. This solution is stable for at least several months.
   pH buffers: Several can be used (HEPES, HEPPS, TRIS or borate), prepared to their optimal pH values (in seawater 7.6, 7.9, 8.1 or 8.7, respectively) by dissolution of the weak acid to a final concentration of 1 mol/L in 0.5 mol/L ammonia solution. The pH buffers should be purified to remove traces of copper. This is done by addition of an aqueous suspension of MnO_2 to a final concentration of 50–100 μmol/L, with 2 h equilibration, followed by filtration. This does not work for TRIS which dissolves the MnO_2. This buffer is best purified by addition of SA (25 μmol/L) followed by passage through a C_{18} cartridge.

3. Voltammetric procedure
An aliquot of 10 mL of seawater is transferred into the voltammetric cell. The SA is added to the seawater to a concentration of 25 μmol/L and the pH of the solution is buffered to a value around 8 by addition of pH buffer to a final concentration of 0.01 mol/L. The pH of acidified seawater is approximately neutralized by addition of ammonia to a pH between 4 and 7 prior to the buffer addition. The amount of ammonia solution required to adjust the pH of acidified samples should be verified in a separate sample aliquot as it is not advisable to insert a pH electrode into a seawater sample prior to analysis as it is likely to cause metal contamination. Then the buffer is added.
   The seawater reagent mixture is purged for 6 min with water-saturated nitrogen to remove dissolved oxygen. The voltammetric parameters include a deposition potential of –1.2 V, a deposition time of 120 s, a reoxidation potential of –0.2 V, a re-oxidation time of 10 s, and then a CSV scan which is initiated from –0.2 V and terminated at –0.6 V. The copper peak appears at –0.35 V. A square-wave modulation at a frequency of 50 Hz is optimal for the scan, with a step height of 2.5 mV and a pulse height of 25 mV. However, the differential-pulse modulation can be also used too with as fast a pulse frequency as the instrument allows. Scans for copper in seawater are shown in Fig.12- 11. A peak occurs at –0.35 V; this peak is for the reduction of copper(*II*) in the adsorbed complex with SA to copper(0). The reduced copper(0) (metallic copper) dissolves in the mercury of the electrode during the scan.
   The sensitivity of the CSV method for copper is very good. The scans shown are for ≈3 nmol/L of copper in seawater, and were obtained using an adsorption time of 60 s. The sensitivity is enhanced by increasing the adsorption time. However, it is clear from Fig. 12-11 that copper levels much less than 3 nmol/L can be measured using a 60 s adsorption time.

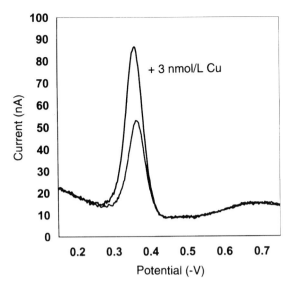

Fig.12-11. Scans for copper in UV digested seawater. Conditions: 25 μmol/L SA, HEPPS pH buffer, 60 s plating at –1 V, scan initiated from –0.15 V. The second scan is after a standard addition of 3 nmol/L Cu to the seawater.

4. Calibration, linear range, working range, limit of detection

The sensitivity is calibrated by a standard copper addition to the seawater in the voltammetric cell. The addition should be sufficient to at least cause a doubling of the original peak. The increase with the standard addition is used to calibrate the sensitivity of the measurement. The sensitivity, S (in nA/nmol/L), is calculated from the increase in the peak height at a given increase in the metal concentration:

$$S = (i_{ps} - i_{p0})/[Cu_{added}]$$

where

i_{ps} =   the peak height (in nA) for the sample including the standard addition;
i_{p0} =   the peak height (in nA) for the sample before the addition; and
$[Cu_{added}]$ is the concentration of copper (in nmol/L) added to the sample.

The response increases linearly with the metal concentration until the surface gets saturated with adsorbed metal complexes. The linear range is therefore a function of the metal concentration and the adsorption time. Using a 60 s adsorption time the response is linear until ≈ 35 nmol/L copper. At higher copper concentrations the linear range can be extended by reducing the sensitivity; this is achieved by using a shorter adsorption time or a slower stirring rate.

The limit of detection is reached at peak heights less than three times the standard deviation of the response at a low copper concentration (see also Chapter 1). A limit of detection of 0.1 nmol/L copper has been reported using a 60 s adsorption time (*Campos and van den Berg*, 1994).

5. Interferences

Organic surfactants lower the sensitivity due to their surface active effect: they adsorb competitively on the electrode. For this reason the sensitivity has to be calibrated by means of a standard metal addition to the sample which is being determined.

Organic complexing ligands in natural waters interfere by masking some of the metal as a result of competitive complexation. This is to a large extent alleviated by using a high concentration of SA. The stability of the copper complexes with SA is very high ($\alpha_{CuSA} = 10^{5.8}$ at 25 μmol/L SA) making it unlikely that complexes with natural complexing ligands can compete significantly. However, a higher SA concentration of 100 μmol/L could be used to raise the α-coefficient (α_{CuSA}) further to $10^{6.9}$.

A complication is the slow kinetics of reaction of the natural organic complexes of copper in seawater. The dissociation rate is fairly slow requiring up to an hour to reach equilibrium. Even if the added chelating ligand strips the copper from the complex with the natural chelating ligand this process is so gradual that the signal can be seen to increase until a new equilibrium has been established (Fig. 12-12). For this reason labile metal determinations have to be carried out with care. It is not essential that the measurement is from an equilibrium condition; it is essential that the measurement is reproducible and that the labile fraction is well-defined. It is more easy to define the labile fraction using equilibrium conditions so this is the preferred condition.

The interference by organic matter, including organic surfactants and ligands, is eliminated by UV irradiation (2–3 h) of the seawater. Peak instability can occur when freshly irradiated, acidified, seawater is analysed. This has been shown to be due to the formation of hypochlorite which produces a broad peak under that of copper (*Achterberg and van den Berg*, 1994). The hypochlorite is unstable at pH 8 and decays over a period of ≈ 30 min causing the apparent copper peak instability. Its removal is virtually instantaneous when a reductant (hydroxyl ammonium hydrochloride, final concentration 0.1 mmol/L) is added.

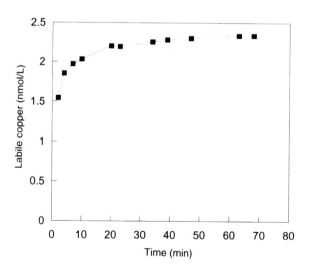

Fig.12-12. Slow increase of the peak for labile copper in seawater from the eastern Atlantic due to the gradual release of copper from organic complexes. Conditions: 25 μmol/L oxine, borate pH buffer (pH 8.3), 90 s adsorption at –0.15 V, scans initiated from –0.15 V. Data from the author (unpublished; Challenger cruise, March 1991).

6. Determination of blanks

Blank metal levels derive from carry-over between samples, and from reagents. The carry-over blank is minimized by using small standard metal additions (sufficient to double the original peak height) and by rinsing the cell with pure water between measurements. It is advisable to avoid using acid for rinsing as the cell is best used conditioned with metal concentrations similar to those being determined to minimize adsorption on the cell wall. The reagent blank is determined by determining copper in pure water using the standard reagent concentrations, and a comparative determination using much increased reagent concentrations (2x to 4x). A significant increase in the copper concentration would be due to a contribution from the reagents. If this contribution is large (a significant proportion of the expected copper concentration) then it is necessary to purify the reagents. A reagent contribution remaining after purification can be taken into account as it is a constant contribution. The reagent contribution of purified reagents to the copper blank typically is less than 0.1–0.2 nmol/L, so this is not likely to be a problem for copper determinations in seawater where the lowest copper concentrations are around 1 nmol/L.

12.3.7.2 Zinc

1. Principle of the method
Ammonium pyrrolidine dithiocarbamate (APDC) was used early on to determine zinc in natural waters by CSV (*van den Berg*, 1984b), and the procedure has not (yet) been improved. APDC is a general complexing agent and it is therefore perhaps surprising that there are no major interferences. Analysis is at pH values 7–8.5.

2. Reagent preparation
   Ammonium pyrrolidine dithiocarbamate (APDC): An aqueous solution containing 0.1 mol/L APDC is prepared by dissolution in pure water. Contaminating zinc is removed from this solution by extraction with a solvent such as trichlorotrifluoroethane, dichloromethane or formaldehyde. The solvent itself is first cleaned by shaking with acid.
   pH buffer: For example HEPES, HEPPS, TRIS or borate, is prepared by dissolution of the weak acid to a final concentration of 1 mol/L in 0.5 mol/L of ammonia solution. Traces of zinc in the buffer are removed by addition of MnO_2 (50 μmol/L) and filtration, or by addition of APDC (50 μmol/L) and extraction or passage through a C_{18} column.

3. Voltammetric procedure
The procedure involves adding APDC to a final concentration of 25 μmol/L, and a pH buffer (pH values between 7 and 8.5) to a final concentration of 0.01 mol/L, to seawater in the voltammetric cell. If the seawater had been acidified then its pH should be adjusted to a more neutral pH using ammonia prior to the reagent addition. Oxygen is removed by 7 min purging with water saturated nitrogen. A deposition potential of –1.3 V is used for a period of 3 min, followed by a 10 s equilibration time; the scan is initiated from –0.8 V and stopped at –1.3 V. The reaction is electrochemically reversible which means that a wave modulation can be used like the differential-pulse (10 Hz) or the square-wave (50 Hz or faster) modulation.

   Example scans for zinc in seawater are shown in Fig.12-13. These scans were obtained using a 60 s deposition time and illustrate the good sensitivity for zinc. The CSV method is

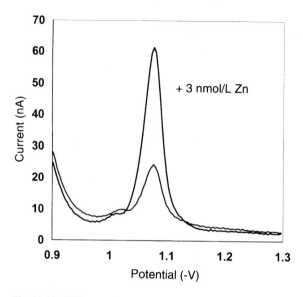

Fig. 12-13. CSV scans for zinc in seawater. Conditions: 25 μmol/L APDC, TRIS pH buffer (pH 8.1), 60 s deposition at –1.3 V. Lower scan for the original zinc in seawater, upper scan after an addition of 3 nmol/L zinc.

much more sensitive than that using ASV. The ASV method involves plating followed by re-oxidation during the scan. Drawbacks of that method are that the zinc is diluted in the mercury during the plating step so not all zinc is re-mobilized during the scan. The alternative is to use a mercury film electrode which has improved sensitivity but has a different drawback due to an interference with copper which may result in the formation of intermetallic compounds with the concomitant suppression of the zinc peak. These interferences are absent in the CSV method.

The response is linear up to about 150 nmol/L of zinc using 60 s adsorption and non-linear thereafter unless a shorter adsorption period is used. The sensitivity is improved by increasing the adsorption period with which the sensitivity is almost linearly related up to 10-min at low zinc concentrations. The limit of detection is 0.3 nmol/L using 60 s deposition, correspondingly less at longer deposition periods. Within the range of linear response the sensitivity is readily calibrated by means of a standard zinc addition to the sample. This addition should be sufficient approximately to double the peak height.

Labile zinc determinations. To evaluate the chemical speciation of zinc it may be of interest to determine the reactive or labile zinc concentration. This is best obtained by deposition also at a more positive deposition potential (*e.g.*, at –0.8 V) from where the scan is initiated too. This way a possible dissociation of natural organic complexed zinc during the deposition step is prevented. The sensitivity in seawater containing organic matter is generally somewhat less than in seawater subjected to UV-digestion and is possibly not constant. The sensitivity therefore has to be calibrated regularly, using a standard zinc addition to every sample until it is demonstrated that the sensitivity is constant.

4. Interferences

Peaks for nickel and cobalt are obtained at -1.08 and -1.2 V, respectively, compared with the zinc peak at -1.15 if deposition is carried out at an adsorption potential more positive than -1 V. However, the sensitivity for those elements is much less than that for zinc. The two potentially interfering peaks are readily eliminated by using a more negative deposition potential. This means that the metal is actually plated during the deposition step; the zinc is first plated at -1.3 V, then it is re-oxidized and re-adsorbed during the equilibration period at a more positive potential at -0.8 V (without stirring) prior to the potential scan which is in the usual CSV mode (in a negative potential direction). The nickel and cobalt are plated too but their re-oxidation is rather irreversible and their peaks do not show up.

For the labile zinc determination it is preferable to select a more positive adsorption potential. In this condition the sensitivity for zinc is still about ten times greater than that for nickel and cobalt, so these metals do not interfere unless their concentrations are much greater than usual.

More serious is the interference by dissolved oxygen which has to be removed diligently as the second oxygen wave is situated under the zinc peak and is the main cause of any irreproducibility of its determination. It is advisable to ensure that the solution is blanketed with a continuous stream of nitrogen during the entire analysis to prevent irreproducibility of the zinc analysis, and to repeat a brief purge between scans.

12.3.7.3 Nickel

1. Principle of the method

The accepted voltammetric method to determine nickel in seawater was almost the first one to be based on adsorptive CSV (*Pihlar et al.,* 1981). This successful method is based on adsorption of complexes with dimethylglyoxime (DMG); it was just preceded by a procedure using complexation by bipyridine *(Sawamoto,* 1980) but that method is not sufficiently sensitive to detect nickel in uncontaminated seawater. In addition to its high sensitivity the DMG method is very robust due to the high stability of the Ni–DMG complexes and the strong pH buffering provided by the recommended ammonium chloride–ammonia buffer. The same method can be used to determine cobalt but the sensitivity is not sufficient for oceanic cobalt levels. An alternative ligand is nioxime which can also be used simultaneously to determine very low levels of cobalt (*Vega and van den Berg,* 1997).

2. Reagent preparation

Stock solution of dimethylglyoxime: Final concentration 0.01 or 0.1 mol/L, is prepared either in methanol or in 0.5 mol/L sodium hydroxide solution. The methanolic solution can best be stored in a suitable glass *vial* as plastic containers, including polystyrene, tend to release plasticizers.

Ammonia–ammonium chloride pH buffer: Prepared containing 2 mol/L ammonia and 1 mol/L hydrochloric acid. By using purified ammonia and hydrochloric acid the nickel blank in this buffer is negligible. This buffer should give a pH of 9.2 when diluted 100-fold with seawater.

3. Voltammetric procedure

Optimal conditions include a DMG concentration of $1–2 \times 10^{-4}$ mol/L and a pH stabilized at 9.2 using 0.04 mol/L NH_4^+–NH_3. The solution is deaerated thoroughly to remove dissolved oxygen which tends to interfere as its second reduction wave is close to that of nickel. An adsorption potential of –0.7 V is selected for a 3 min adsorption period. The scan is carried out using either the linear sweep mode, sampled DC or with a fast square-wave mode. Good results are obtained with all these wave forms. The sensitivity is calibrated using a standard nickel addition to the seawater sufficient to at least double the peak height. Scans for nickel in seawater are shown in Fig. 12-14.

Dissolved organic matter causes minor interference due to competitive adsorption of surface active compounds and due to competitive complexation by natural chelating agents. Owing to the natural surfactants in seawater the sensitivity is slightly reduced, but this is taken into account by calibration of the sensitivity by a standard nickel addition to the sample. Natural organic complexing ligands tend to mask only a small fraction (typically 10 %) of the nickel from the DMG as the complex stability with DMG is very high. The organic matter can be removed by UV digestion of an acidified sample aliquot.

Labile nickel determination. Labile (reactive) nickel is obtained by CSV of seawater without UV treatment. Part of the nickel is masked by natural organic ligands of as yet unknown composition (*Nimmo et al.,* 1989; *Achterberg and van den Berg,* 1997). At the recommended concentration of DMG (2×10^{-4} mol/L) the α-coefficient (α_{NiDMG}) of complexation of nickel by DMG is very high ($\alpha_{NiDMG} = 10^{9.6}$), greater than that of natural organic complexing matter. To obtain a significantly lowered labile fraction it is necessary to use a much lower DMG concentration (2–10 μmol/L). The kinetics of complexation of DMG with nickel are then very slow in seawater requiring long equilibration times of several hours. Using this method it has been established that a small but significant fraction of nickel occurs to be complexed by organic matter in seawater (*Achterberg and van den Berg,* 1997). The different concentration profiles obtained with and without UV-digestion of the seawater are

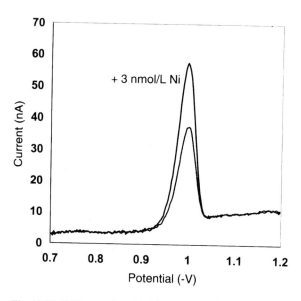

Fig. 12-14. CSV scans for nickel in seawater. Conditions: 2×10^{-4} mol/L DMG, 0.04 mol/L NH_4^+–NH_3 (pH 9.2), adsorption time 60 s. The lower scan is for the original nickel in the seawater, the upper scan is after a standard addition of 3 nmol/L nickel.

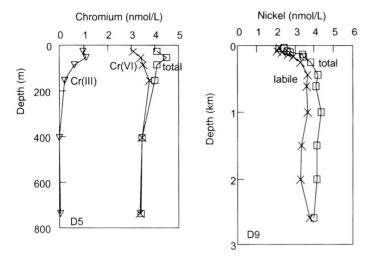

Fig. 12-15. Distribution of labile and total dissolved chromium and nickel in the water column of the western Mediterranean. Both metals were determined by CSV using the procedures described here. Data from *Achterberg and van den Berg* (1997).

shown in Fig. 12-15. The untreated seawater samples are clearly offset to lower nickel concentrations illustrating that the non-labile nature of metal complexes needs to be taken into account when total dissolved metal concentrations are determined.

12.3.7.4 Chromium

1. Principles of the method

The voltammetric method to determine chromium is sensitive in seawater and even more sensitive (10x more) in freshwater. It is an interesting method because of some unusual changes in the oxidation state of chromium during the analysis. The sensitivity is high due to a catalytic effect caused by the reduction product of chromium(*III*) (chromium(*II*)) on the electrochemical reduction of nitrate. Analysis is therefore carried out in the presence of a high concentration of nitrate which greatly enhances the sensitivity.

The determination of chromium is based on adsorptive collection of complexes with diethylenetriaminepentaacetic acid (DTPA) (*Golimowski et al.*, 1985; *Boussemart et al.*, 1992). The measurement of chromium in seawater is carried out at a pH between 5.0 and 5.3, whereas measurements in freshwater are carried out at a slightly higher pH of 6.2–6.4.

CSV responds to chromium(*VI*) as well as to (*III*) but the response to chromium(*III*) diminishes rapidly with time approaching zero after about 15 min due to chromium(*III*) adsorption on the cell wall. Thereafter the response is due to chromium(*VI*) only. For this reason it is recommended to convert all chromium into the *VI* state prior to the analysis. The only way to prevent the chromium(*III*) signal from diminishing is to use a flow cell instead of the standard, batch, voltammetric cell. This adsorption is not special to voltammetry: the chromium(*III*) adsorbs readily on almost anything, and its rapid removal from water samples is normally prevented by acidification. Acidification causes all chromium(*VI*) to become gradually reduced to chromium(*III*) by reaction with organic matter. The original chromium(*VI*) concentration in the sample is best preserved at the natural seawater pH.

2. Sample pretreatment

Total (combined) chromium is determined after oxidation of all chromium to chromium(*VI*). Samples are thus transferred into a silica tube immediately after sampling; storage can be maintained almost indefinitely if evaporation is prevented (capping and cold storage). Prior to analysis the samples are UV digested at neutral pH for a period of 1–2 h. The original pH of seawater (around 8) is optimal, but any pH between 7 and 8.5 is satisfactory. The pH of acidified samples has to be carefully adjusted to a neutral value with ammonia, and sodium hydrogen carbonate is added to a final concentration of 2–3 mmol/L which gives rise to a pH value of around 7.6. The pH of un-acidified seawater does not need to be altered and no sodium hydrogen carbonate is added.

Samples stored in plastic containers should be either acidified to retain all chromium in solution, or left at the original pH and stored cold or frozen to stabilize the original chromium(*VI*) concentration.

3. Reagent preparation

Diethylenetriaminepentaacetic acid (DTPA): An aqueous solution is prepared containing 0.25 mol/L DTPA.

Acetate pH buffer: A solution is prepared containing 2 mol/L sodium acetate and 0.8 mol/L ammonia. Addition of 100 μL of this buffer to 10 mL of seawater should give a pH of 5.2.

Nitrate stock solution: This solution is made close to the solubility of the nitrate salt (5 mol/L). Contaminating chromium in the concentrated nitrate solution is removed by coprecipitation with iron(III)hydroxide. Iron(*III*) chloride (final concentration 0.1 mmol/L) is thus added and allowed to become oxidized by dissolved oxygen. This is filtered off to produce the purified nitrate solution; this process reduces chromium(*VI*) to chromium(*III*) which adsorbs on the iron(III)hydroxide and is removed. Most conveniently the acetate buffer is premixed with the nitrate solution (to a final concentration of 0.2 mol/L acetate in 5 mol/L nitrate) and purified simultaneously. Thus the overall chromium reagent blank is typically reduced to less than 0.03 nmol/L.

Potassium chromate solution: Chromium(*VI*) is used for standardization; this is prepared by dissolution of potassium chromate in water to a final concentration of 1 mmol/L. This solution is stable for several months. Dilutions are made in pure water.

4. Voltammetric procedure

A 10 mL seawater sample is pipetted into a voltammetric cell. One mL of the nitrate solution is added to give a final concentration of 0.5 mol/L of nitrate. A higher nitrate concentration further increases the sensitivity (the sensitivity increases linearly up to 1 mol/L nitrate and continues to increase at higher levels) but this leads to sample dilution which begins to offset any signal gains.

The pH buffer is added to a final concentration of 0.02 mol/L; alternatively the buffer is added simultaneously with the nitrate using a pre-mixed solution. DTPA is added to a final concentration of 2.5 mmol/L.

Dissolved oxygen is removed by purging (5 min) with nitrogen. A new mercury drop is made and adsorption is carried out (1–3 min) at a deposition potential of –1.0 V. Then a quiescence period of 8 s is used, followed by a negative going potential scan. The square-wave modulation is used at an optimal frequency of 50 Hz; the differential-pulse mode can also be used. The chromium peak appears at –1.3 V. The scan is repeated to check for peak stability: a sig-

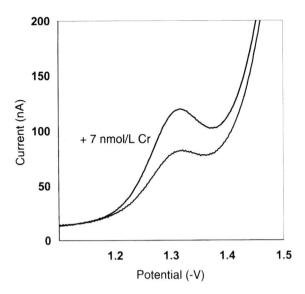

Fig. 12-16. CSV scans for chromium in seawater. Conditions: UV digested seawater, 2.5 mmol/L DTPA, 0.02 mol/L acetate pH buffer (pH 5.2), 0.5 mol/L nitrate, adsorption time 60 s. The lower scan is for the original chromium in the seawater, the upper scan is after a standard addition of 7 nmol/L chromium.

nificant decrease indicates the presence of chromium(*III*). Chromium(*VI*) is added to calibrate the sensitivity; the addition should be sufficient to double the peak height of the sample.

Scans for chromium in seawater are shown in Fig. 12-16. It can be seen that the chromium peak appears just before the hydrogen wave. The limit of detection is not so much determined by the electronic noise (which is very low due the large catalytic current) but by the slope of the background current. At low chromium concentration the peak becomes a shoulder on the hydrogen wave, and its determination becomes less and less accurate.

Redox speciation of chromium
The different chemistry of the chromium(*III*) and (*VI*) valences has been used to develop methods for their separation. Chromium(*III*) is much more reactive than chromium(*VI*) and is readily removed from pH 8 seawater by adsorption on fumed silica followed by filtration (*Boussemart and van den Berg*, 1994), or more easily by passing a seawater aliquot through a C_{18} cartridge (*Achterberg and van den Berg*, 1997). The redox speciation of chromium can then be determined from the total concentration of chromium (after UV digestion of the seawater) and that of chromium(*VI*) which is determined in the seawater from which chromium(*III*) has been removed. Data obtained by this method in the western Mediterranean show (Fig. 12-15) systematically increased levels of chromium(*III*) in the upper water column presumably as a result of photochemical effects (*Achterberg and van den Berg*, 1997).

12.3.7.5 Cobalt

1. Principle of the method
Cobalt is determined by catalytic CSV. The cobalt complex with nioxime is adsorbed on the mercury drop, and the reduction current due to the reduction of cobalt in this complex is recorded during the CSV scan. Even without catalysis the sensitivity of this method is very good and a detection limit of 6 pmol/L cobalt can be achieved using an adsorption time of 15 min (*Donat and Bruland*, 1988), but this cannot be achieved without UV-digestion to remove interfering organic matter.

The sensitivity is much improved in the presence of nitrite which re-oxidizes the reduction product causing the catalytic effect (*Vega and van den Berg*, 1997). The overall reaction is rather interesting as it involves oxidation of all cobalt(*II*) to cobalt(*III*) when nioxime, nitrite and ammonia are added to seawater, forming a mixed complex with cobalt(*III*). This mixed complex is adsorbed on the electrode and the cobalt(*III*) is reduced to cobalt(*II*) during the scan. The cobalt(*II*) catalyses the reduction of nitrite which in its turn re-oxidizes the cobalt(*II*) to cobalt(*III*) which is then reduced again at the electrode surface: this causes an enormous increase in the peak current for cobalt and very good sensitivity as well as selectivity. The catalytic effect is specific to cobalt so only the cobalt peak is enhanced whereas the nickel peak (which is also obtained with nioxime) is not changed upon the addition of nitrite. In this method the limit of detection is 3 pmol/L of cobalt with a 30 s adsorption time.

2. Sample pre-treatment
Experiments show that there is no significant adsorption of cobalt on silica UV digestion tubes at the natural seawater pH or from acidified samples (*Vega and van den Berg*, 1997). There is therefore no need to acidify seawater prior to UV digestion. UV digestion has to be used to ensure complete release of cobalt from natural organic complexes as a small fraction of the cobalt tends to be non-labile.

3. Reagent preparation
Aqueous stock solution of 0.1 mol/L *nioxime (1,2-cyclohexanedione dioxime)*: Prepared by dissolution in 0.2 mol/L sodium hydroxide (Aristar grade). This is diluted to 2 mmol/L nioxime by dilution with pure water.

Ammonia buffer solution, 4 mol/L: Prepared by addition of HCl (final concentration about 2.5 mol/L) to 4 mol/L NH_3 solution; addition of 100 μL of this buffer to 10 mL seawater should give a pH of 9.1.

Stock solution of nitrite, 5 mol/L: Prepared by dissolution of the sodium salt in 500 mL of water. This solution is the main source of the reagent cobalt contribution. Cobalt contamination in this solution is removed by electrolysis (24 h) above a mercury pool electrode at a potential of –1.35 V. Alternatively, hydrous manganese dioxide (MnO_2) can be added to a final concentration of 100 μmol/L followed by filtration (0.2 μm filter). The electrolytic purification lowers the cobalt contribution to below 4 pmol/L.

4. Voltammetric procedure
A 10 mL seawater sample is pipetted into the voltammetric cell, and 200 μL of the ammonia buffer are added (additional ammonia is added if the seawater is acidified) giving a final pH of 9.1. Then 20 μL of 2 mmol/L nioxime and 1 mL of 5 mol/L nitrite are added giving final concentrations of 4 μmol/L nioxime and 0.5 mol/L nitrite. Dissolved oxygen is removed by

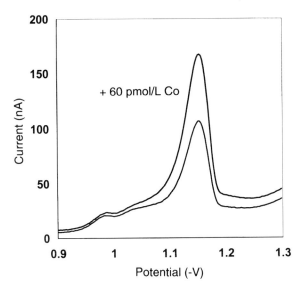

Fig.12-17. CSV scans for cobalt in seawater. Conditions: UV digested seawater, 4 μmol/L nioxime, 0.5 mol/L nitrite, 0.08 mol/L NH₃–NH₄Cl (pH 9.1); 30 s adsorption time. Lower scan for the original cobalt in the seawater, upper scan after an addition of 60 pmol/L cobalt.

purging with nitrogen gas (5 min). A new mercury drop is made, and the complex is adsorbed at –0.7 V whilst the solution is stirred for a period of 30 s. The CSV scan is initiated from –1.0 V after an 8 s equilibration time. The cobalt peak is at –1.12 V. The scan can also be initiated from –0.7 V and the cobalt is then preceded by that of nickel which can be determined under the same conditions. However, a low but broad peak due to reduction of free nioxime between the two peaks tends to deteriorate the general shape of the cobalt peak somewhat, which is eliminated by initiating the scan from –1.0 V. The sensitivity is calibrated by standard cobalt additions to the sample.

Scans for cobalt, and a standard cobalt addition, in seawater are shown in Fig. 12-17. The peak height increases linearly with the cobalt concentration up to least 3 nmol/L (much more than is present in uncontaminated seawater) using adsorption times of 15, 30 or 60 s, and increases with the adsorption time until a maximum is reached using 3–4 min. The limit of detection is 3 pmol/L using a 30 s adsorption time.

Acknowledgements

The voltammetric scans were kindly provided by Geraldine Sarthou. Some of the investigations of the author leading to the development of the voltammetric procedures were sponsored by grants from the NERC (PRIME, GST/02/1058 and SIDAL, SL2) and the EU (MERLIM, MAS3-CT95-0005).

12.4   Analysis by total-reflection X-ray fluorescence spectrometry (TXRF)

A. Prange and M. Schirmacher

12.4.1   Introduction

During the last ten years the TXRF technique has proved to be particularly suited to the analysis of seawater and has also become established for routine analysis and monitoring tasks. Relatively low instrumental detection limits in the pg or sub-ng/mL range, a dynamic range of about 4–5 orders of magnitude, easy quantification by means of internal standardization, small sample consumption and a large number of detectable elements are some of the main features of TXRF. Following a salt matrix elimination treatment, 12 elements (V, Mn, Fe, Co, Ni, Cu, Zn, Se, Mo, Cd, Pb and U) dissolved in seawater can be determined using TXRF. The number of elements is limited by the matrix elimination step, which is based on chelation with sodium dibenzyldithiocarbamate (Na-DBDTC). In other marine matrices such as suspended particulate material (SPM) and sediments, more than 30 elements can be determined simultaneously after digestion of the sample material.

Applications to seawater, estuarine water, suspended particulate matter and sediments using TXRF are numerous and have been described by many authors: Prange showed the applicability of TXRF in combination with Na-DBDTC chelation for the determination of trace elements in different seawaters and the use of the TXRF instrument on board research vessels in the Baltic Sea and the Pacific Ocean (*Prange, 1983, Prange et al., 1985*); the German Federal Maritime and Hydrographic Agency (BSH) routinely analyses North Sea water for monitoring tasks (*e.g., Freimann and Schmidt, 1989, Schirmacher and Schmidt, 1991, Haarich and Schmidt, 1993*); *Prange and Kremling* (1985) investigated V, Mo and U profiles in the Baltic Sea; *Kempe et al.* (1991) studied trace metal profiles in the particle layer of the Black Sea; *Schmidt et al.* (1993) investigated trace metal profiles at several deep water sites in the Atlantic Ocean within the framework of an IOC baseline study. *Hölemann et al.* (1997) studied seawater, sea ice, SPM and sediments of an Arctic shelf area (Laptev Sea and Lena River, Siberia) with the methods described here for a large number of elements, and *Koopmann and Prange* (1991) used TXRF for analyses in Wadden Sea sediments. Moreover, TXRF in combination with Na-DBDTC extraction has been employed successfully in the intercomparison and certification stage for the standard reference seawater CRM 403 (by Bureau of Reference of the Commission of the European Community) and in a feasibility study on estuarine water (*Freimann et al., 1993*). Finally, some work has also been published on investigations of estuarine and river water using TXRF (*Prange et al., 1990; Prange et al., 1993; Reus et al., 1993*).

This section presents a brief overview of the fundamentals of TXRF and the requirements for sample preparation of marine sample materials to be analysed by TXRF, followed by an extensive description of experimental work. Furthermore, we direct attention to the analytical capabilities and limitations of the technique.

12.4.2   Analytical method

With respect to performance and accuracy the most significant progress in energy-dispersive X-ray fluorescence analysis (EDXRF) was achieved when the phenomenon of total reflection, discovered in 1930 by Compton, was utilized in 1971 by *Yoneda and Horiuchi* for XRF analysis. In a non-traditional way, they directed the primary X-ray beam onto a polished quartz glass support, which served as a sample carrier, at glancing incidence angles smaller than the critical angle for total reflection. The sample to be analysed was deposited on the surface of the sample support by evaporating small amounts of solutions or fine-grained suspensions. In this mode of operation, amounts below 10^{-9} g became detectable for the first time using an energy-dispersive X-ray detector system. Three years later, European scientists followed up this work (*Aiginger and Wobrauschek*, 1974). Inspired by these publications, *Knoth and Schwenke* in the late 1970s designed a compact, stable and easily adjustable total reflection module and were able to increase the performance of TXRF stepwise to detection limits better than 5 pg for some metals (*Knoth and Schwenke*, 1978, 1982). This was the decisive step towards the introduction of the method into the analytical practice of chemical trace and ultra-trace element analysis and a few years later for the commercializa-

EDXRF

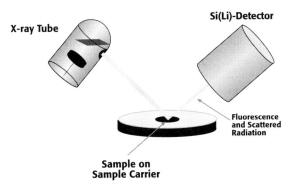

TXRF

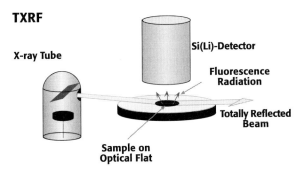

Fig. 12-18. Beam geometry of excitation/detection for conventional EDXRF compared with the geometry for TXRF.

tion of TXRF. In subsequent years, further improvements in TXRF spectrometers were achieved with instrumentation for extremely precise angle adjustment and the introduction of X-ray tubes with alloy anodes coupled with multilayer systems for monochromatisation of the incident radiation (*Knoth et al.*, 1997).

TXRF is a descendant of conventional energy-dispersive XRF, however, with detection limits improved by 4–5 orders of magnitude, to date below 10^{-12} g. It is important to point out that the TXRF technique differs fundamentally from classical X-ray fluorescence (EDXRF), in sample preparation, calibration, data analysis and detection performance, as well as in the objects under investigation.

12.4.2.1   Principle of the TXRF technique

From the physical point of view the basic difference between classical EDXRF and TXRF is the beam geometry of the excitation and the emitted radiation with respect to the sample (Fig. 12-18): whereas in EDXRF both angles are about 45°, TXRF uses a very small incidence angle of less than 1° and a detection angle of 90°. The incident radiation is then totally reflected in the forward direction and does not penetrate the reflecting surface by more than a few nanometres. Thus the primary X-ray beam virtually does not interact with the sample support, which results in a drastic reduction of scattered radiation compared with the conventional geometry and, therefore, a substantial improvement in the peak/background ratio. In addition, the fluorescence radiation is doubled in intensity because the sample is excited by both the incident *and* the reflected beam. TXRF is based on the development of standing X-ray waves above a sample support or in near surface layers, respectively, when using grazing incidence of the excitation radiation. For detailed information on its physics reference is made to *Klockenkämper et al.* (1992), *Prange* (1993), *Klockenkämper* (1997).

Figure 12-19 shows the basic design of a TXRF instrument. The primary beam is generated by an X-ray source with a line focus, either a fine structure tube (containing a Mo, W or

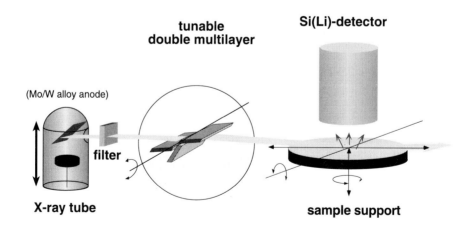

Fig. 12-19. Basic design of current TXRF instruments. Besides the total reflection conditions at the sample carrier, as shown in Fig. 12-18, the X-ray optical device in the form of a tunable band pass filter is the most important component of the new spectrometers.

Mo-W alloy anode) or a rotating anode tube. The beam is shaped like a strip of paper by means of precisely aligned slits; its vertical divergence is adjusted to less than 0.01°.

The X-ray optical component which processes the primary spectrum of the polychromatic radiation from X-ray tubes is the most important part of any TXRF equipment. A simple variant is a quartz glass mirror which acts as a low pass filter and merely cuts off the high energy bremsstrahlung. The effect can be intensified by a metallic foil filter. In the past, this combination found extensive use in trace analysis and microanalysis, since here the demands on angular divergence and spectral purity are comparatively low, while maximum intensity at the sample is of highest priority. In other applications, however, quantification of most of the effects of total reflection can only be performed for monochromatic X-rays of well defined energy. The incident beam must therefore be monochromatized, usually with multi-layer mirrors. These are characterized by artificial stacks consisting of thin (in the nm range), alternating layers (≈ 100) of high (*e.g.*, tungsten) and low density (*e.g.*, carbon) materials. Using such monochromators a certain small energy band can be selected, *e.g.*, the K_α line of the molybdenum or the L_β line of the tungsten spectrum by tuning the angle of incidence.

The new generation TXRF spectrometers are equipped with a double multilayer system: The two multilayer mirrors, which have identical specifications but different lengths, are arranged in parallel, at a distinct distance to each other, but staggered. The variation of the angle of incidence and hence the selection of the excitation energy is achieved simply by rotation of the double multilayer about a common axis. The length of the lower multilayer defines the lowest permitted angle and the highest energy which can be transmitted. In this TXRF set up the optimum conditions for total reflection are achieved by shifting the height of the X-ray tube and double multilayer monochromator together with respect to the sample carrier.

The fluorescence radiation is usually recorded with a Si(Li) detector and is registered by a multi-channel analyser as is the norm in energy-dispersive spectrometry. The detector is mounted directly above the sample, perpendicular to the sample support, at a distance of less than 1 mm in order to enlarge the angle of reception and to maximise the fluorescence intensity. Si(Li) detectors with active areas of 30 and 80 mm^2 and spectral resolutions of about 135 and 150 eV at 5.9 keV, respectively, are in common use. Normally measurements are performed in ambient air.

TXRF spectrometers equipped with a combination of a Mo-W alloy anode and a double multilayer monochromator to date show the best compromise with respect to detection limits and multielement capability and are state of the art (TXRF 8030 C, Multielement Tracea-nalyzer; Atomika Instruments, Oberschleißheim, Germany).

A further essential part of the instrumentation, aside from the excitation and monochromator is the reflecting medium, *i.e.*, the *sample carriers* for TXRF. Several sample carrier materials have been investigated and used for TXRF. There are some general requirements which sample carriers have to fulfil for optimum use in TXRF. The carrier material must be highly reflective, chemically inert, free of impurities, easy to clean, should not give rise to fluorescence peaks within the energy range used for measurements and, last but not least, must be cheap. Besides quartz glass carriers, Plexiglass (Perspex), glassy carbon and boron nitride have been used. Fig. 12-20 shows a TXRF spectrum of a clean quartz glass carrier. The Si peak originates from the carrier, the Ar peak from the air between the carrier and detector and the peaks between 16 and 20 keV originate from the Mo source. The residual Fe contamination, clearly visible in the inset spectrum, corresponds to a mass of 15 pg and gives an idea of the excellent sensitivity of the technique.

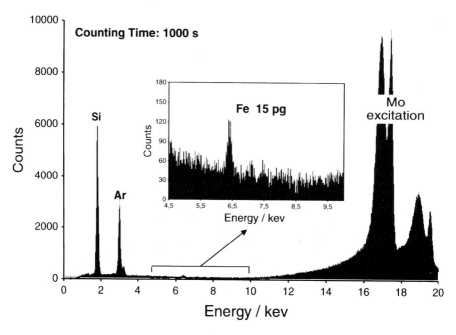

Fig. 12-20. TXRF spectrum of a clean quartz glass carrier.

12.4.2.2   Analytical features of the instrument

TXRF is characterised by multi-element determination, matrix independent calibration and single-element internal standardization, and low detection limits. In addition, the technique requires very low sample masses. Another significant feature of TXRF is its inherent surface sensitivity which makes it useful for surface analysis, a topic not discussed in this book.

1. Multi-element determination
Figure 12-21 shows two TXRF spectra of a multi-element standard solution (Merck No. 11355) using Mo and W excitation, respectively. The multi-element spectra show the registered pulses versus the energy of the X-ray photons. The upper spectrum was taken with Mo excitation. A 10 μL volume of the diluted standard solution was pipetted onto a pre-cleaned sample carrier and evaporated to dryness to give 1 ng each of 23 elements, of which 18 can be determined with TXRF. The 'organ pipe' structure of the signals from neighbouring elements (K series) nicely reflects the instrumental response function (see below). The total intensity for the L series is distributed over several peaks with similar intensities (Tl, Pb and Bi) as a result of the different relationships of the L peak intensities compared with the K series. For elements such as Ag, Cd, and In, Mo excitation is inefficient and the signals overlap with those due to K and Ca, both of which are abundant in many types of samples. Therefore, it is necessary to perform a supplementary measurement in which the sample is excited by continuous radiation of the W material of the tube, as shown in the lower spectrum. The energy range of the previous spectrum is now compressed into the first half of this spectrum with somewhat reduced signal intensities, but the elements

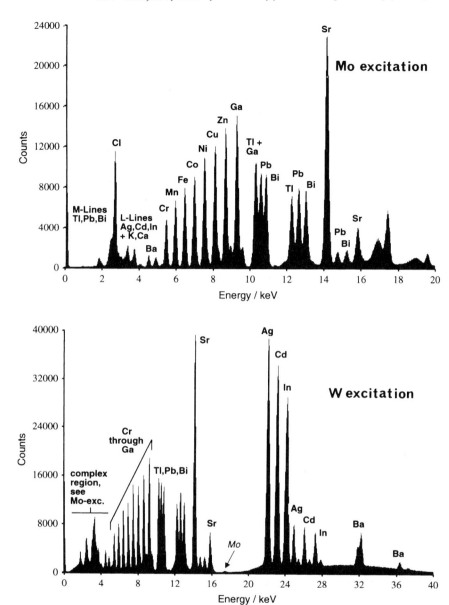

Fig. 12-21. Multi-element spectra from a multi-element standard solution using Mo and W excitation, respectively. The peaks arise from an amount of 1 ng per element. The different peak heights reflect different sensitivities.

from Mo at 17.5 keV to Ba, including Ag, Cd, In, can now be determined with higher efficiency and with no spectral overlap. In the region of Ba, the signal intensity becomes weak when using W excitation, so that at this point it is advisable to return to Mo excitation to obtain better results.

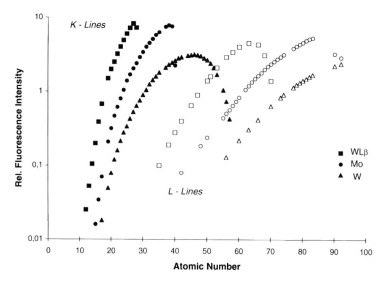

Fig. 12-22. Relative fluorescence sensitivities for different elements as function of their atomic number. The curves show three different excitation modes as described in the text.

X-ray spectra from real samples are complex. Their evaluation requires deconvolution of overlapping peaks and automatic background subtraction. Commercially available spectrum processing software fulfills these demands and, moreover, facilitates the operating procedures by use of a multi-channel analyser.

2. Calibration and quantification

Low sample masses avoid matrix absorption or enhancement effects which results in a constant relationship between the fluorescence yield and the atomic number. The calibration curve therefore is established only *once*, namely during the phase of installation, by repeated measurements of multi-element standard solutions. No alterations are necessary unless parts of the hardware or sensitive software are changed. It maintains its validity for all matrices, for the entire concentration range and for any element chosen as internal standard. Unlike conventional XRF, there is no need for matrix corrections and for external standards. The calibration for TXRF is independent of the properties of the sample matrix. Figure 12-22 shows calibration curves for three different excitation modes. Fluorescence may be excited by either tungsten L radiation at about 9 keV, by molybdenum K radiation at about 17 keV or by a tungsten white spectrum ranging from about 30 to 35 keV. For obvious physical reasons, each type of radiation is restricted to a certain range of atomic numbers where either K or L lines are excited. As can be seen, this combination results in similar signal intensities for all elements above $Z=20$ (Ca), and fades away for the lighter elements with decreasing Z.

Quantification is very simply performed by internal one-element standardization. A defined amount of internal standard is added prior to any sample pre-treatment; for quantification the standard signal is related to other signals occurring in the spectrum.

$$c_{\text{unknown}} = c_{\text{Standard}} \times cps_{\text{unknown}} / cps_{\text{Standard}} \times \varepsilon_{\text{Standard}} / \varepsilon_{\text{unknown}}$$

where

c = concentration, cps = counts per sec and ε = relative sensitivity.

With a single multi-element determination a dynamic range may be covered of up to 5 orders of magnitude.

3. Instrumental limits of detection (L_D)

The instrumental limits of detection (L_D) of course are directly correlated to the response function which also includes the influence of the spectral background. The instrumental L_D is derived from the peak to background ratio in the X-ray fluorescence spectra of pure standard solutions. Figure 12-23 shows the instrumental limits of detection for three different excitation modes. The limits of detection define the measured value or concentration above which the signal of an element can be recognized under stipulated borderline conditions (*e.g.*, amount of sample, measurement time, nature of the sample). The instrumental limits of detection are obtained by measuring isolated element peaks on matrix-free standard solutions and are derived, according to *Currie* (1968), by the following equation:

$$L_D = 4.65 \, (I_{\text{Background}} - I_{\text{Net pulse rate}})^{\frac{1}{2}} / I_{\text{Net pulse rate}} / t^{\frac{1}{2}} \, C$$

where

I = intensity, C = calibration factor and t = measurement time.

The limits of detection for the heavy elements are in the low or sub-pg-range. For a 1 ng Ni sample, a detection limit of 0.2 pg has been achieved. With a sample volume of 100 μL this corresponds to a concentration of 2 ng/L. For elements lighter than K, detection limits increase sharply with decreasing atomic number due to their low fluorescence yield.

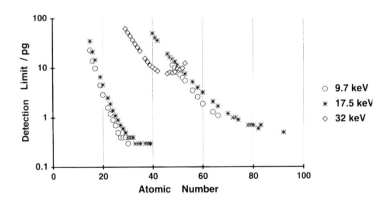

Fig. 12-23. Instrumental detection limits for the three different excitation modes. For more than 60 elements they are below 10 pg.

12.4.2.3   Sample pre-treatment

Although low instrumental detection limits are necessary for the majority of applications, in practice they alone are not sufficient. To take full advantage of the instrumental detection limits in a variety of applications samples must be pre-treated prior to the measurement. Much work has been invested therefore in the development of standardized sample preparation techniques for aqueous and solid marine samples including tools for sample handling and sample preparation as well as specially developed devices for various practical problems. Figure 12-24 gives an overview of the sample preparation procedures applied to marine matrices (seawater, SPM and sediments) which have been especially developed for TXRF analysis and are described in the following sections. Additional and/or alternative procedures for the pre-treatment and digestion of particulate material (mainly for the analysis by AAS and ICP-AES techniques) are presented in Section 12.6 of this chapter. It is mandatory that sample pre-treatments and analyses are carried out in clean-rooms and/or at least under clean bench conditions as described in Section 12.1.3 of this chapter.

1. Seawater

As mentioned, samples are usually produced by evaporation of μL volumes of the sample solution on the surface of a TXRF sample carrier resulting in a residue of small sample amounts. This is illustrated in Fig. 12-25. Application of this procedure directly to seawater would lead to the formation of a large amount of salt crystals on the surface of the sample carrier during the evaporation process. This would give rise to a strong increase in scattered radiation from the sample matrix. Moreover, the presence of the bulk elements Na, K, Mg

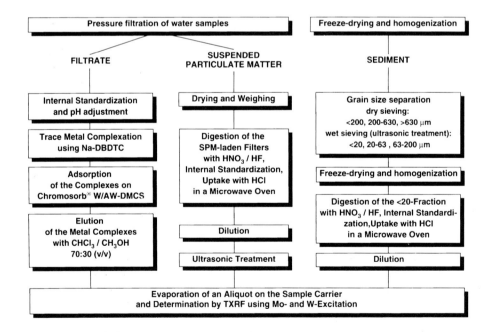

Fig. 12-24. Schematic runs for the sample preparation techniques applied to different marine matrices.

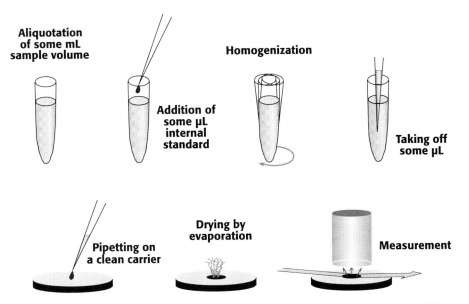

Fig. 12-25. Internal standardization and preparation of the measuring sample from aliquots of the sample solution.

and Ca would lead to peak overlapping with the trace elements. Owing to the very low concentration levels of some elements, especially in open ocean water, direct measurements would not be possible. It is therefore imperative to separate the trace elements from the salt matrix.

Separation and pre-concentration. A method first suggested by *Knapp et al.* (1975), developed further by *Prange* (1983) for seawater analysis using TXRF and later slightly modified by *Schmidt et al.* (1993) combines matrix removal with enrichment by a factor of about 50. The elements V, Mn, Fe, Co, Ni, Cu, Zn, Se, Mo, Cd, Pb and U are chelated using sodium-dibenzyldithiocarbamate (Na-DBDTC). The water-insoluble chelates are adsorbed on a column of a reversed-phase silica gel. Alkaline and alkaline earth elements elute unchelated with the aqueous phase and are discarded. The chelate complexes are removed from the dried column with a mixture of chloroform and methanol (methanol is needed for the quantitative elution of Mo). Some $100\,\mu L$ of this extract are evaporated on quartz sample carriers and measured for around 1000 s.

Limitations in the number of measurable elements result in the selectivity of the complexing agent. Using Na–DBDTC, *i.e.*, only Cr(*VI*), As(*III*) and Se(*IV*) are complexed. As a consequence of conservation with acids to a pH of about 2 after sampling and with increasing time of storage, Cr(*VI*) in seawater is mostly reduced to Cr(*III*) and therefore cannot be detected by this technique. A special method for the determination of chromium in seawater has been developed by *Geisler* (1992). *Haffer and Schmidt* (1995) described a method for detecting As together with V, Mn, Fe, Co, Ni, Cu, Zn and U with Na-DBDTC and TXRF. They applied an additional reduction step by means of sodium thiosulphate. Moreover, the level of complexation depends on the pH. *Prange* (1983) has shown that the formation of V and Se complexes, in particular, decreases with increasing pH values. Above pH 5, increasing losses of Se increase the complexation curve of V; it descends beyond pH 6. Thus accurate adjustment to a pH level of 4.8–5.3 is essential. pH values lower than 4.8 often lead to precipitation of the chelating agent.

With optimized reaction parameters the recovery of the metals in the eluate is around 80 %. Variations of 5–22 % of the analytical results determined by radio tracer experiments are higher than the 1–14 % achieved for the yields relative to the internal standard (95 %) obtained by TXRF measurements. If the relative yields are based on an internal standard for TXRF, variations in the recovery do not have any significant effect, as long as all elements display uniform behaviour during the sample preparation procedure. That this applies was shown by *Prange et al.* (1983). Thus, effects of incomplete precipitation, adsorption or elution are eliminated by the internal standard used for quantification.

For many applications the elements Co or Se are used for internal standardization, because they usually occur in very low concentrations. If all detectable elements are to be quantified, one has to perform the procedure twice with different standard elements. The concentration of the internal standard should be adjusted between the highest and lowest concentrations expected for all other elements in the sample. Details for the complete procedure are given in the Section 12.4.2.5.

2. Suspended particulate matter (SPM) and sediments

Prior to the determination of trace elements by TXRF in suspended particulate matter (SPM) and/or in sediments, it is imperative to define the separation procedure for SPM from the water phase (*e.g.*, filtration or centrifugation; for details see Chapter 2) and to separate the grain-size fractions of sediments in order to provide a good basis for meaningful comparisons of the element concentrations within individual areas or between areas. In addition, TXRF measurements require appropriate digestion procedures.

Grain-size separation. Sediment samples normally consist of materials of different grain-size. Suspended particulate material once collected on the filter, has to be analysed unsorted. Sediments, however, should be separated into grain-size classes. Applying the German classification system according to *Scheffer and Schachtschabel* (1989) they are usually defined as coarse and very coarse sand ($>600\,\mu$m), medium-coarse sand (200–600 μm), fine sand (60–200 μm), coarse silt (20–60 μm) and medium and fine silt, clay (<20 μm). Because the main fraction of heavy metals, *i.e.*, metals of environmental interest, has been found to occur in silt and clay, it has become a more or less accepted convention to carry out analyses of heavy metals in the $<20\,\mu$m class. The fine silt and clay fractions are preferred to avoid dilution effects induced by coarser grain-sizes with lower heavy metal contents but greater mass, which makes the results of sediment analyses incomparable.

To obtain the different grain-size fractions, a procedure is recommended which combines dry sieving for fractions $>200\,\mu$m and ultrasonication aided wet sieving for fractions $<200\,\mu$m. Details have been presented elsewhere (*Koopmann and Prange*, 1991, *Ackermann et al.*, 1983).

Digestion (see also Section 12.4.2.6). As outlined before, measurements with TXRF require the conversion of solid samples into fine-grained suspensions ($<1\,\mu$m), or even better, into homogeneous solutions. A widely used procedure is described in the German Industry Norm specification (DIN 38 414 - S7). It refers to the digestion of sludges and sediments with *aqua regia* for subsequent determination of the acid-soluble portion of metals. One disadvantage of the DIN specification is the incomplete digestion especially for the main compounds and elements bound to the clay mineral components. Therefore, this procedure is only useful for the determination of the acid-extractable fraction of an element which does not adequately represent its total abundance over the entire composition of the sediment. Only a quantitative digestion procedure which results in clear solutions guarantees a reliable determination of a large number of elements (*Krause et al.,* 1995).

For complete digestion, the procedure has to include the following 3 steps

– Oxidation (including the filter-material) with 65 % HNO_3 and simultaneous break down of the silicate matrix with 30 % HF.
– Evaporation to dryness to destroy highly insoluble fluorides of, *e.g.*, Al, by emanation of SiF_4.
– Dissolution in 30 % HCl *via* formation of soluble chlorides and chloro complexes.

Heat and pressure have little influence on the effectiveness of the digestion procedure. Good results have been attained using Teflon bombs in a microwave oven (*Knapp et al.,* 1997). The Teflon bombs can be cleaned to a high level of purity and are pressure-resistant up to some 10 bars (10^6 Pa). The decomposition of the filter material (polycarbonate) and, to a varying extent, of the organic sample material, leads to the release of CO_2. To attain a slow, controllable pressure increase, the temperature programmeme has to raise the microwave power stepwise.

Quantification is performed by spiking the sample with one element showing a low, natural concentration level in the sample. Cobalt or gallium have been used successfully for marine samples (*Schirmacher and Schmidt,* 1990; *Hölemann et al.,* in the press). The internal standard concentration should be adjusted between the highest and lowest concentrations expected for all other elements in the sample.

12.4.2.4 Apparatus and reagents

1. TXRF spectrometer

Much of the work described here and cited in the literature has been conducted using the TXRF spectrometers EXTRA II and EXTRA IIa distributed previously by Rich. Seifert & Co. (Ahrensburg, Germany) and more recently by Atomika Instruments. These instruments were equipped with double beam excitation (2 kW fine focus molybdenum and tungsten tubes), a Si(Li) detector (Link) and an automatic sample changer, suitable for 34 sample carriers. The electronic periphery equipment and peak deconvolution software, a computer-controlled multi-channel analyser system (QX 2000), was supplied by Link Analytical Ltd. (High Wycombe, UK). Today, successor instruments such as the TXRF 8030C, Multielement Traceanalyzer mentioned before, are also distributed by Atomika Instruments GmbH, Bruckmannring 40, D-85764 Oberschleißheim, Germany.

2. Apparatus and reagents for salt matrix separation and trace element enrichment

Apparatus
a) *Quartz columns*: According to Fig: 12-26 (made by Westdeutsche Quarzschmelze, D-21502 Geesthacht), equipped with a stopper and an adapter with vacuum connection.
b) *Erlenmeyer flask*: For mixing the seawater sample with all necessary reagents.
c) *Centrifuge tubes*: Glassy, conical, to collect the $CHCl_3$–CH_3OH extract.
d) *Vacuum pump*: For drying the column system.

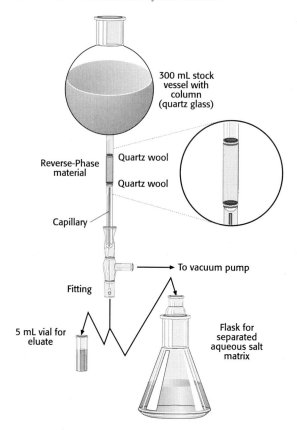

300 mL stock
vessel with
column
(quartz glass)

Reverse-Phase
material

Quartz wool

Quartz wool

Capillary

Fitting

To vacuum pump

5 mL vial for
eluate

Flask for
separated
aqueous salt
matrix

Fig. 12-26. Equipment for the separation and concentration procedure.

e) *Precision heater:* E.g., Gerhardt, Germany, for evaporation of the sample on a quartz carrier.

f) *Laboratory balance*: With 0.01 g precision for weighing the water sample.

g) *Wash bottles (FEP)*: For ultra-pure water, $CHCl_3$, CH_3OH and the $CHCl_3$–CH_3OH mixture.

h) *Metal-reduced microliter pipettes*: Several volumes (*e.g.*, Unipette by Eppendorf, Hamburg, Germany).

i) *pH meter and pH indicator sticks*: For several pH ranges.

j) *Quartz sample carriers*: 2–3 mm thick and with a diameter of 30 mm (Westdeutsche Quarzschmelze, D-21502 Geesthacht).

k) *Special device*: For preparing the sample to be measured from the eluate as shown in Fig. 12-27.

Reagents

Acids such as HNO_3, HCl and HF as well as organic solvents ($CHCl_3$ and CH_3OH) can be cleaned by sub-boiling distillation as described in Section 12.1.5.

Sodium acetate buffer solution: 3 g of 96 % CH_3COOH Suprapur, 4.1 g of CH_3COONa Suprapur, 497 mL of ultrapure water.

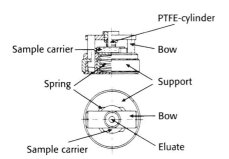

Fig. 12-27. Device for preparation of the sample to be measured from organic eluates.

30 % NaOH Suprapur.

Na-dibenzyldithiocarbamate: Purum (Fluka, Germany).

Standard metal solutions: *E.g.*, AAS or ICP standard solutions, Merck, Darmstadt, of the elements to be used as the internal standard (*e.g.*, Co).

Moreover, for the column material, silica gel (Chromosorb W/AW-DMCS 0.15–0.18, Merck, Darmstadt) and quartz wool are needed.

3. Apparatus and reagents for digestion of solid samples (see also Section 12.4.2.6)

Apparatus

a) *Microwave oven:* E.g., Intec Laborgeräte, Uhingen, Germany: MLS 1200 Mega-240, equipped with a suck off unit (e.g., Intec EM 30), a rotor (e.g., Intec HPR-1000/6 HP) and several Teflon bombs.

b) *Evaporation unit:* E.g., Intec MCR-6.

c) *Scrubber:* E.g., Büchi, Germany, for the collection and neutralization of acid-containing exhaust gas.

d) *Device for neutralization of electrostatic charge on filters and vessels:* E.g., Haug, Germany.

e) *Metal-reduced microlitre pipettes:* Several volumes (e.g,. Unipette by Eppendorf, Hamburg, Germany.

f) *Quartz sample carriers:* With 2–3 mm thickness and 30 mm diameter, (Westdeutsche Quarzschmelze, D-21502 Geesthacht).

g) *Special device:* To centre the sample drop adequately on the sample carrier.

h) *Analytical balance:* With 0.1 mg precision for weighing the SPM on filters.

Reagents

Acids: Such as 65 % HNO_3, 30 % HCl and 30 % HF are subboiled (see Section 12.1.5).

30 % H_2O_2 Suprapure.

Standard metal solution: *E.g.*, gallium AAS standard solution, Merck, Darmstadt, as internal standard.

Silicon oil: Serva, Heidelberg, Germany, to prepare a hydrophobic surface on the quartz glass carrier.

4. Cleaning procedures

All laboratory ware to come in contact with the sample has to be cleaned carefully. Standard cleaning procedures for quartz glass, Teflon and polyethylene are described in Section 12.1.4.

Only the special cleaning procedure for the TXRF sample carriers will be outlined here. Ultra-pure water, laboratory cleaning agent (*e.g.*, RBS detergent), and 30 % HCl are used as reagents. A Teflon stand for 25 quartz carriers is used for handling of the sample carriers (Atomika Instruments GmbH, Oberschleissheim, Germany).

The carriers are pre-cleaned mechanically with methanol- and/or acetone-drenched laboratory wipes. Then 25 sample carriers are placed in the carrier stand and transferred into a beaker. Approximately 0.5 L of ultra-pure water is added and approximately 15 mL of laboratory cleaning agent. The beaker is heated for 2 h at 60 °C. After cooling the stand and beaker are rinsed with ultra-pure water. The stand is placed into the beaker again. A 0.5 L aliquot of ultra-pure water are added and approximately 10 mL 30 % HCl. The acid bath is heated again for 2 h at 60 °C, the stand and beaker are again rinsed with water. The carriers are allowed to dry, *e.g.*, inside a clean bench. A blank is measured with the dried carriers to check if they have been cleaned successfully (approximately 300 s counting time). Carriers having residual impurities are singled out and the procedure is repeated until they are clean. What 'clean' means depends on the nature of the samples to be measured. In our experience, values of Co, Ni, Cu, Zn and Pb below 10 pg per carrier and Fe below 30 pg per carrier are sufficiently low for the vast majority of applications.

12.4.2.5   Analysis of seawater

1. Preliminary procedures

The buffer solution is prepared according to Section 12.4.2.4-Reagents. A solution is prepared of 400 mg of sodium dibenzyldithiocarbamate in 10 g of methanol. The solution can be stored, if cooled, for 1 or 2 days. A 2+1 (v : v) mixture is prepared of $CHCl_3$ and CH_3OH. A dilution of a metal standard solution (*e.g.*, 10 mg/L Co) is prepared.

A tuft of quartz wool is placed in a glass beaker and is leached for several hours in warm 10 % HCl. One layer of leached and rinsed quartz wool, approximately 5 mm, is placed in the undermost position of the augmented part of the capillary (Fig. 12-26). Above this layer, 80 % of the remaining space is filled with dry silica gel (approximately 1.5 mL of Chromosorb W/AW-DMCS). The column is closed with another 5 mm plug of quartz wool. The quartz column is connected to a vacuum pump. The column is rinsed slowly with approximately 300 mL of 3 % HCl and then with ultra-pure water until the obtained water passed is of neutral ph. The column is cleaned by performing repeated separation procedures as described in the following section until the blank values are constant.

2. Chelation and elution

Volumes of 100 to 200 mL of seawater (or ultra-pure water for blanks) are placed into a 300 mL pre-cleaned Erlenmeyer flask. Two mL of buffer solution and an appropriate volume of internal standard solution are added (*e.g.*,. 100 µL of 10 mg/L Co for each 200 mL, resulting in 5 µg/L). Concentrated and diluted NaOH solutions and HNO_3 are used to adjust the pH value to a value between 4.8 and 5.3 (seawater samples are normally conserved at pH 1–2). pH indicator sticks are used for rough and the pH meter for precise pH adjustment. It goes without saying that contact between sample and pH sticks or pH electrode has to be strictly avoided. Instead, a few µL are transferred with a pipette onto the stick. For exact pH measurements a few mL of sample solution are transferred into a special vessel large enough in diameter to accommodate the pH meter electrode. When the correct pH is

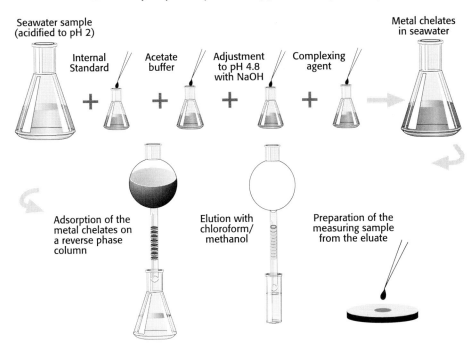

Fig. 12-28. Schematic run of the separation and concentration procedure for the analysis of seawater.

reached, 0.5–1 mL of methanolic sodium dibenzyldithiocarbamate is added and the flask shaken well. No persisting precipitation should occur. The sample is transferred to the sample reservoir above the silica gel column. A slight vacuum is applied to draw the sample through the capillary (2 drops per s) which is then rinsed twice with 2–3 mL of ultra-pure water. The column is dried by applying maximum vacuum for approximately 0.5 h. The column is dry if coolness due to evaporation can no longer be felt. The Erlenmeyer flask is replaced by precleaned conical centrifuge tubes. The complexes are eluted with 2–4 mL of the $CHCl_3$–CH_3OH mixture (the volume of eluent depends on the total element content of the sample; total element contents above 1.0 mg/L require eluent volumes greater than 4 mL (*Prange et al.* 1983). Sample carriers are prepared with 100–300 μl of the extract according to the next subsection. The separation and concentration procedure is schematically illustrated in Fig. 12-28.

The sample reservoir and the column are cleaned by rinsing without applying vacuum with approximately 4 mL of $CHCl_3$; subsequently they are conditioned with at least 2 mL of CH_3OH. The column is now ready for the next sample or blank.

An automated system for this separation and concentration procedure has been recently described by *Gerwinski and Schmidt* (1998).

3. Preparation for measurement of sample eluate

A 100–300 μL aliquot of of the organic eluate is transferred into the highly polished quartz sample carrier and evaporated to form a dry residue by means of a special device (see Fig. 12-27) designed to adjust the sample spot to meet the excitation beam.

Another way to prepare the sample is to place a cleaned quartz sample carrier on a heater that has been set to 100 °C. A 20 µL aliquot of the eluate is transferred slowly by means of a pipette onto the centre of the carrier. The procedure is repeated, with each subsequent delivery through the pipette tip adding to the solution already dried until an appropriate amount is evaporated (100–300 µL, depending on the type of sample). The spot should only appear within the central area (8 mm diameter) to be exposed to the X-ray beam. A pattern is drawn on the heater surface for correct positioning of the spot.

4. Measurement of samples

The sample carriers are inserted into the sample changer of the TXRF instrument. Seawater samples, especially those from open ocean waters, are usually measured by means of both molybdenum and tungsten excitation in order to ensure detection of Mo and Cd which cannot be determined with good efficiency using the molybdenum tube alone. The sample spot on the carrier should be prepared in such a way that maximum tube current can be applied. The duration of measurements is 3000–5000 s depending on the concentrations to be determined. The TXRF software will evaluate the data. Correct evaluation of the spectrum requires a complete setup of every peak occuring in the element list. If peaks can be seen in the spectrum but are not listed in the element list, the base-line of the background is calculated incorrectly, and this leads to a wrong evaluation. To be sure that all relevant peaks are listed, the spectra of unknown samples are scanned prior to measurement.

A typical TXRF spectrum of a seawater sample prepared according to the procedure described here is shown in Fig. 12-29 (obtained from the seawater reference material CASS-1; see Section 12.1.6).

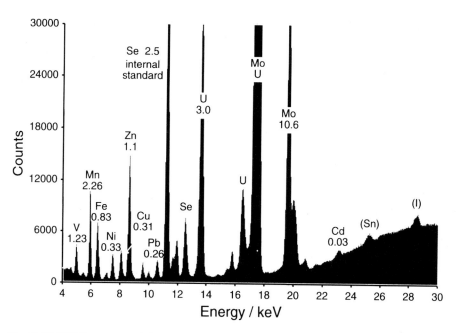

Fig. 12-29. TXRF spectrum of the seawater reference material CASS-1 using W-excitation. The values are given in µg/L. Selenium was used for internal standardization.

5. Some general remarks on the separation procedure

Sodium-dibenzyldithiocarbamate (Na-DBDTC) tends to precipitate as carbamic acid at pH levels less than 4.8. Introduced into the column it accumulates at the column on its top and later redissolves in the $CHCl_3^-CH_3OH$ mixture, resulting in poor sample spots on the carrier. This leads to a large background signal and deterioration of the detection limits. Low lead and cadmium concentrations in open ocean water can be difficult or impossible to detect under such conditions. Thus, precipitation of carbamic acid must be avoided.

If water samples with relatively low element concentrations follow samples with relatively high element concentrations (*e.g.*, 50 ng/L follows 50 µg/L) memory effects may occur. Whenever assessable, the order in a set of samples should be from lower to higher element concentrations. Memory effects are avoided when a blank is measured between any two samples or when the sample is separated twice using the same reverse-phase column.

12.4.2.6   Analysis of solid samples

1. Preliminary procedures

A stock of sub-boiled and blank-controlled acids is prepared and a dilution of a metal standard solution (*e.g.*, 100 mg/L of gallium). Several separation procedures are performed to clean the Teflon bombs (refer to next section), until the blank values are constant and low.

Filters are weighed to a precision of at least 0.1 mg. If any electrostatical charge of the filters occurs it must be neutralized (see Section 12.4.2.4-Apparatus.).

The sediment fractions to be analysed are dried and homogenized.

2. Digestion of suspended particulate matter (see also Section 12.6)

The weighed filter is transferred into the pre-cleaned Teflon bomb. Four mL of HNO_3 (65 %), 1 mL of HF (30 %), 0.5 mL of H_2O_2 (30 %) and an appropriate amount of internal standard (*e.g.*, 500 mg/kg final concentration) are added and well shaken to cover the filter with the acids. The bombs are closed and placed in the microwave oven, according to the instructions for the instrument. A typical temperature programmeme is, *e.g.*, 2 min 250 W, 3 min 500 W, 2 min 600 W, 2 min 700 W, 2 min 800 W, 5 min 250 W. The bombs are cooled and transferred into the evaporation device. The residue is heated (*e.g.*, 250 W) and pumped until dry (approximately 1 h). The steam is collected in two condensation traps, then led through the gas washing device (scrubber) and finally into the laboratory exhaust system. The bombs are allowed to cool. Two mL of HCl (30 %) are added to each residue and the bombs are closed again and put back into the microwave oven. They are heated again for 20 min with the power set to 400 W. The samples are then transferred to pre-cleaned, scaled PE vessels and diluted to an appropriate volume (*e.g.*, 20 mL). The bombs are rinsed twice with ultra-pure water. Before the next charge, a blank is performed using HNO_3, HCl and HF in combination and by applying the same temperature programmeme. Evaporation and the subsequent addition of HCl are omitted when performing a blank just for cleaning. When performing separation procedures for blank corrections, the procedure should be followed exactly.

3. Digestion of sediments

Amounts of 20 to 100 mg of a dried sediment sample are transferred into the Teflon bomb and the procedure given for samples of suspended material should be followed.

4. Preparation of SPM and sediment samples for measurement

A 20 μL aliquot of silicon oil (*e.g.*, Serva, Heidelberg, Germany) is placed onto a cleaned sample carrier and allowed to dry, in order to prepare a hydrophobic surface on the quartz carrier to prevent bleeding of the water (acid) drop. Approximately 10 μL of the sample solution are deposited at the centre of the carrier (the volume depends on the type of sediment, its decomposed mass and the dilution after digestion) and dried on a heater adjusted to 100 °C. The special device should be used for positioning the spot correctly in the central area (8 mm diameter) of the carrier.

5. Measurement of samples

The sample carriers are inserted into the sample changer of the TXRF instrument. SPM and sediment samples are usually measured with molybdenum anode excitation. The counting time normally lies between 1000 and 3000 s. The selected excitation depends on the combination of elements under investigation.

6. Some general remarks on the procedure

The formation and condition of the sample spot is of crucial importance to the measurement. In the case of SPM and sediment samples, evaporation of the acid solution sometimes leads to poor sample spots due to fractionated crystallization. Such samples may be improved by adding a small volume (a few μL) of a 0.2 % aqueous solution of poly(vinylalcohol) (PVA) to the sample volume, directly on the carrier.

Another important requirement is that the spot be placed exclusively inside the central area (8 mm diameter) of the carrier, so that no portion of any element can migrate out of the area to be irradiated with X-rays.

Thick spots produce a large background signal and thus give rise to deterioration of the detection limits. The sample should be diluted appropriately.

The extracted metal complexes should be measured quickly after elution. Adsorption processes alter the concentrations during storage, whereas the prepared sample spot can be stored for at least 6 months (*Prange,* 1983).

12.4.2.7   Blanks, detection limits, precision and accuracy

For general definitions see the section on 'Quality control' in Chapter 1.

1. Seawater

Blank values. The overall blank values of the analytical method, from weighing of the specimen to the preparation of the sample, are mainly caused by the added reagents, *i.e.*, the buffer solution, acids and bases, the dithiodicarbamate and, last but not least, the ultra-pure water if used instead of seawater for dilution or for the determination of the blank values. If no water of sufficient purity (ultra-pure water) is available, seawater extracted with Na-DBDTC is an acetable substitute. The use of reference seawater is another option (*e.g.*, NASS and CASS reference material; see Section 12.1.6). The blanks are derived from the difference between the certified and the measured values and should result in the same (similar) blank values from different reference materials. Typical blank values for Fe have been found to be between 100 and 150 ng/L, for Zn between 60 and 90 ng/L, and for Cd and Pb below 10 ng/L. Variations depend on the stocks of chemicals employed. In particular the blanks of the elements Fe and Zn pose a problem with the analyses of open-ocean waters in which Fe and Zn surface water concentrations in most regions have been reported to be much less than our blank values (see Table 12-1).

Detection limits. The detection limits of the analytical procedure depend on the sample volume, eluent volume, the volume used for the sample and the counting time. They may be derived from the peak to background ratio in X-ray spectra from real seawater samples. For a 200 mL sample volume, 2 mL eluate volume, 200 μL volume of sample to be measured and a measuring time of 1000 s, they are less than 10 ng/L for most of the heavy metals. However, this is only true for negligible blank values. In the case of higher blanks (in particular for Fe and Zn), the limits of detection have to be derived from the variations in the blank values (for definition see 'Quality control' section in Chapter 1). For example, in the case of Fe and Zn, relatively poor detection limits of about 50 ng/L were derived from the standard deviation of the blanks.

Precision. The precision of the method is between 5 and 15 %, whereas the values for Se, Cd and Pb at concentrations near the detection limits may reach 30 % or more. The precisions of independently prepared samples from the same separation procedure are distinctly better, relative standard deviations are always better than 5 %.

Accuracy. Analytical accuracy may be derived by analysing an appropriate standard reference material, if available. The results obtained by the procedure described here for the seawater standard reference materials NASS-1 and CASS-2 (see Section 12.1.6) were always in good agreement with the certified values. The deviations never exceeded 20 %. The TXRF results of trace metal intercalibration in estuarine water, organized by the International Council for the Exploration of the Sea (*ICES*, 1988) have also been accepted as 'true' values in the considered concentration range.

2. Suspended particulate matter and sediments

Blank values. Three sources contribute to the blanks and should be taken into account: The vessels, the acids used for digestion, and the filter material. The blanks of the vessels and acids can be obtained by performing the digestion procedure without the filter; the blank for the filter material can be derived from digestions of several pre-cleaned filters of each filter stock. Our experience indicates that the blanks for the elements Cr, Ni and Zn, in some cases also for Fe, are mainly caused by the filter material (see also Chapter 2). Obviously, the lower the content of SPM on the filter, the stronger the influence of the blank values for the corrections.

For sediments, blank values are usually negligible when working under clean analytical conditions as is the rule for marine investigations.

Detection limits. The detection limits mainly depend on the amount of SPM or sediment used for the element determination. For the elements which are influenced by blanks as mentioned before, the detection limits may be calculated from the standard deviation of the blank values. The total blank is constant for a given mixture of acids and filter stock but affects different amounts of SPM and sediments, depending on the amount of SPM on the filter or the employed sediment mass. Therefore, the detection limits may be 'normalized' for each 1 mg of SPM loaded on the filter. The detection limits are a factor of 2 better for 2 mg of SPM on a filter than for 1 mg. For other elements, detection limits are derived from the peak to background ratio taken from TXRF spectra of SPM or sediments, respectively. For SPM and sediments the detection limits vary between 3 μg/g (*e.g.*, for Cu, Ni or Zn) and and 25 μg/g (*e.g.*, for V).

Precision. The precision of the method for both sediments and suspended particulate matter varies between 5 % and 15 %.

Accuracy. Analytical accuracy has been assessed by analysing different standard reference materials (*i.e.*, MESS-1 from the National Research Council of Canada). The recovery lies between 90 and 110 % for Ca, V, Mn, Fe, Ni, Cu, Zn, As, Sr, Ba and Pb. Slightly lower recoveries between 80 and 90 % have been found for Ti, Cr, K and Rb (*Krause et al.,* 1995). The results of an ICES intercomparison exercise on the determination of trace elements in suspended particulate matter from the Baltic Sea (*Pohl*, 1994) show that the TXRF values fit nicely whithin the standard deviations of the intra-laboratory means with only Co being an exception.

12.5   Analysis by spectrophotometry

F. Koroleff and K. Kremling

12.5.1   Iron

The concentration and behaviour of iron in ocean waters have been briefly outlined in Section 12.1.1 and Table 12-1. (For overviews see also 'Marine Chemistry', Vol. 50 (1995; Nos. 1–4) and Vol. 57 (1997; p. 137–186)). For the determination of open-ocean Fe concentrations (*i.e.*, in the sub-nanomol range) sufficient care must be taken not to contaminate the samples. The methods, specialized procedures and equipment necessary to cope with these extremely low concentrations are described in detail in Section 12.1. and 12.2.1. Here, we outline a spectrophotometric procedure for dissolved Fe concentrations in the μmol/L range. Such amounts occur in the marine environment under anoxic conditions and pH values of around 7 (*e.g.*, in the Baltic or Black Sea). It can be explained by a steady diffusion of Fe(*II*) species from organic-rich sediments into stagnant bottom water resulting in the enrichment of rather soluble iron(*II*)sulphide.

12.5.1.1   Principle of the method

In weak acid solution, pH 3.5–5.8, iron(*II*) forms a violet complex with 2,4,6-tris(2-pyridyl)-1,3,5-triazine (TPTZ). The reagent was introduced by *Collins et al.* (1959) and modified by *Koroleff* (1974). In the procedure given here ascorbic acid is used to reduce Fe(*III*) to Fe(*II*). The molar absorptivity of the method (at 594–595 nm) is *ca.* 22 300 and the smallest amount of iron that can be determined directly is 20 nmol/L, which equals an absorbance of 0.004 with a 10 cm cell. Beer-Lambert's Law is valid to about 21 μmol/L. Interferences occur only from large amounts of Cu, Co or Ni (in mg/L range) in industrial waste waters.

12.5.1.2   Reagents and equipment

For the analysis use, unless otherwise stated, only reagents of recognized analytical-reagent grade (a.g.). All reagents should be stored in polyethylene or PVC bottles.

1. *Water*: Ordinary tap water distilled twice from an all-glass apparatus is satisfactory. De-ionized water can also be used (see Section 12.1.5).
2. *Standard solution*: Prepare standard stock solution (pH 2) of 1 g/L Fe (*e.g.*, 'Titrisol' from Merck, or 'Dilute-it' from Baker). For preparation of the calibration curve dilute this stock solution in two steps:
   a) Take 1000 μL, add 100 μL of HCl (to prevent adsorption of Fe on container wells) and dilute to 100 mL with distilled water (corresponds to 0.01 g/L or 10 μg/mL Fe).
   b) Continue by taking from the last solution (2a) 50, 100, 200, 500, 750 and 1000 μL, add 100 μL HCl and dilute to 100 mL with distilled water. These solutions should be prepared daily and correspond to concentrations of 5, 10, 20, 50, 75 and 100 μg/L Fe.
3. *Ascorbic acid solution*: Dissolve 7 g of ascorbic acid, $C_6H_8O_6$, in 100 mL of water. If stored in a refrigerator, the reagent is stable for several weeks. The reagent can be used as long as it remains colourless.
4. *Ammonium acetate solution*: Dissolve 5 g of ammonium acetate, CH_3COONH_4, in 100 mL of water.
5. *TPTZ solution*: To 0.08 g of TPTZ, 2,4,6-tris(2-pyridyl)-1,3,5-triazine, add 0.5 mL of concentrated HCl and dilute to 100 mL with distilled water. The reagent is stable for several months.
6. *Spectrophotometer*: With 5 or 10 cm cells.
7. *Erlenmeyer flasks*: 100 mL glass or plastic flasks.
8. *Glass or plastic pipettes of suitable volumes*: For cleaning, follow the instructions in Section 12.1.4.

12.5.1.3 Analysis of the sample

In unpreserved samples the analysis should be performed as soon as possible, preferably within half an hour after collection. To 50 mL of the sample add 1 mL of ascorbic acid (reagent 3), mix and wait for about 30 s. Then add 1 mL of buffer (reagent 4), mix and add 1 mL of TPTZ solution (reagent 5). Mix again and wait for 10 min. Then measure the absorbance at 595 nm in 5 or 10 cm cells. The colour of the blue-violet complex is stable for several hours; it is recommended, however, to measure within 1 h, using distilled water as the reference.

If the samples have bee preserved with acid (at pH 1–2), add 0.4 mL of NaOH solution (2 mol/L) to obtain the correct reaction pH.

12.5.1.4 Calibration

The method has no salt factor. Prepare the calibration curve by using the working standard solutions described above (see 2b in Section 12.5.1.2) and by following the procedure outlined for the analysis of sample.

12.5.1.5   Calculation of results

Establish the calibration curve (see 12.5.1.4) and calculate the sample concentration from the slope of calibration curve. Do not forget to correct for the reagent blank and, if necessary, for the turbidity of waters (see 12.5.1.6).

12.5.1.6   Precision and blank determination

The reproducibility (relative standard deviation) is about ±5 % except in the lowest range, < 0.1 μmol/L, where values can scatter about four times more.

For the determination of blank (A_b), add the reagents as described above (12.5.1.3) to 50 mL distilled water. Also determine A_{2b} by adding twice the amount of reagents, 2 mL each, to 47 mL of distilled water. The reagent blank $A_{rb} = A_{2b} - A_b$, and the numerical value is usually 0.005–0.010 with a 10 cm cell.

To correct for colour and/or turbidity add to a 50 mL portion of sample 1 mL of ascorbic acid only. Measure the absorbances as described earlier.

12.5.2   Manganese

The chemistry of manganese in seawater is complex. Dissolved Mn concentrations (< 0.4 μm) in the open ocean normally range from 0.2 to 3 nmol/kg (see Table 12-1). In oxygenated waters Mn(II) is thermodynamically unstable with respect to the oxidation to insoluble manganese oxides (probably in its Mn(III) or Mn(IV) oxidation states). However, owing to the relatively slow kinetics of oxidation of Mn(II) in seawater the low equilibrium concentrations are rarely attained. The oceanic distribution of the metal appears to be dominated by external input sources which lead to maxima in the surface waters (*e.g., Kremling,* 1985). An accumulation of dissolved Mn(II) does also occur in deep anoxic waters, such as in the Black Sea, and in the deep ocean near active hydrothermal regions, with concentrations in the μmol to mmol/kg range (*e.g., Sedwick et al.,* 1992; *Edmond et al.,* 1995). The concentration of dissolved Mn(II) in the anoxic waters is probably limited by its solubility with respect to MnCO$_3$.

Determination of the low levels of Mn in oceanic waters is outlined in Section 12.2.1. Here, we describe a fast and precise spectrophotometric method for dissolved Mn concentrations in the μg/L range which can be applied to samples containing several mg/L of hydrogen sulphide.

12.5.2.1   Principle of the method

The attractiion of the procedure is that manganese in alkaline medium gives a reddish-brown product with formaldoxime, $H_2C = NOH$. The compound is said to be $(CH_2NO)_3Mn$, with the metal in the trivalent state. The same complex is also obtained with Mn(II) and Mn(IV). The procedure given here was developed by *Koroleff* (1974), mainly based on the technique used by *Bradfield* (1957). The complex is formed at a pH of about 10.5 in the presence of citrate. The manganese formaldoxime is formed spontaneously before $Mg(OH)_2$ starts to precipitate. The sample is then heated to $80\,°C$, whereby other metal formaldoxime complexes are decomposed.

The molar absorptivity of the method (at 450 nm) is *ca.* 10 700 and the lowest Mn concentration that can be estimated directly with a 10 cm cuvette is about 100 nmol/L. The standard curve is a straight line up to several milligrams per litre. Variations in salinity have no effect and the method can be applied to samples containing several milligrams of H_2S per litre. Other metals forming coloured complexes with formaldoxime are Fe, Ni, Co and Cu. By heating to $80\,°C$, the chelate of Cu is completely destroyed, that of Fe disrupted to negligible amounts (*i.e.*, up to $20\,\mu mol/L$ of Fe, each $2\,\mu mol$ equals about 3 nmol of Mn, between 20 and about $200\,\mu mol/L$ of Fe, each $20\,\mu mol$ equals about 4 nmol of Mn). The complexes of Ni and Co are decolourized to 95 % and 98 %, respectively.

12.5.2.2 Reagents and equipment

For the analysis use, unless otherwise stated, only reagents of recognized analytical-reagent grade (a. g.). All the reagents should be stored in polyethylene or PVC bottles.
1. *Water*: As for iron (see 12.5.1.2).
2. *Standard solution*: Prepare standard stock solution (pH 2) of 1 g/L Mn (*e.g.*, 'Titrisol' from Merck, or 'Dilute-it' from Baker). For preparation of the calibration curve dilute this stock solution by two steps.
a) Take $1000\,\mu L$, add $100\,\mu L$ of HCl (to prevent adsorption of Mn on the container walls) and dilute to 100 mL with distilled water (corresponds to 0.01 g/L or $10\,\mu g/mL$ Mn).
b) Continue by taking from the last solution (2a) 100, 150, 200, 250, 300, 400, 500, 750 and $1000\,\mu L$, add $100\,\mu L$ of HCl and dilute to 100 mL with distilled water. These solutions should be prepared daily and correspond to concentration of 10, 15, 20, 25 $\mu g/L$ Mn.
3. *Sodium hydroxide*, 2.0 mol/L: Dissolve 80 g of NaOH in distilled water and dilute to 1 L.
4. *Saturated sodium citrate solution*: Dissolve trisodium citrate dihydrate, $C_6H_5Na_3O_7 \cdot 2H_2O$, in the proportions 24 g of salt to 40 mL of water. The solution is stable.
5. *Formaldoxime reagent*: Dissolve 10 g of hydroxylamine hydrochloride, $NH_2OH \cdot HCl$, in *ca.* 80 mL of water. Finally, dilute to 100 mL with water. The reagent is stored in the dark and is stable for several months.
6. *Spectrophotometer*: Preferably with 10 cm cells.
7. *Centrifuge*: With suitable glass tubes.
8. *Apparatus for an even heating of samples:* Up to maximum of $80\,°C$; preferably in small electrothermal mantles.
9. *Glass or plastic pipettes*: Suitable volumes (for cleaning, follow the instructions in Section 12.1.4).

12.5.2.3 Analysis of the sample

In unpreserved samples the analysis should be performed as soon as possible, preferably within an hour after collection. To 50 mL of the sample in a glass centrifuge tube, add 1 mL of citrate (reagent 4), mix and add 1 mL of the formaldoxime reagent (reagent 5). Mix again and wait for 1 min. Then add dropwise NaOH solution (reagent 3) until white magnesium hydroxide starts to precipitate. Note the amount of hydroxide solution added and adjust to a total volume of 60 mL with distilled water. Then cover the centrifuge tube and heat the sample to exactly 79–80 °C for 12±2 min (over 80 °C the manganese complex starts to decompose!). Directly after the heating, cool in running water and centrifuge hereafter for about 10 min with a speed of around 3500 rpm. Measure the absorbance of the clear solution within half an hour at 450 nm, using air or distilled water as the reference.

If samples were preserved with acid, add some more of the NaOH solution (reagent 3) to obtain the correct reaction pH.

12.5.2.4   Calibration

The method has no salt factor. Prepare the calibration curve by using the working standard solution described under 2b in Section 12.5.2.2 and by following the procedure outlined for analysis of the sample. However, since the working standard solutions do not contain any magnesium salt, the centrifugation step after the heating process is not necessary. The amount of NaOH solution necessary to establish the reaction pH of about 10.5 should be determined in a pre-treatment experiment.

12.5.2.5   Calculation of results

Establish the calibration curve (see Section 12.5.2.4) and calculate sample the concentration from the slope of the calibration curve. Do not forget to correct for the reagent blank and, if necessary, for the turbidity of the waters (see Section 12.5.2.6).

12.5.2.6   Precision and blank determination

The reproducibility (relative standard deviation) of the procedure is better than 10 %, except in the lowest range where measurements may scatter twice as much.

For the determination of reagent blank (A_{rb}) proceed exactly as described for iron (Section 12.5.1.6). The numerical value of A_{rb} is also about the same as that for Fe. To correct for colour and/or turbidity pour the sample from the cuvette back into the reaction flask and add 0.5 mL of about 5 mol/L sulphuric acid. The acid destroys the manganese complex. By reading the absorbance of this acidified sample, a measurement of colour/turbidity is obtained.

12.6   Analysis of marine particles

L. Brügmann and J. Kuss

12.6.1   Introduction

It is now well recognized that the production and decomposition of biogenic particles as well as their fractional removal to the deep sea control the distribution of most trace elements (TE) in the oceans (see also Section 12.1.1). Microbial decomposition, desorption and dissolution of suspended or sinking marine particles can release elements associated with labile (*e.g.*, organic) fractions back to the seawater. On the other hand, particles can scavenge TE from the dissolved phase and thereby transport them to sediments. An understanding of the behaviour and geochemical cycling of TE, therefore, requires the analysis of the composition and distribution of the particulate fractions in the oceans (see also Section 12.1).

In seawater, most TE occur primarily as 'dissolved' including colloidal species (< 0.45 or $< 0.4\,\mu m$; see Chapter 2). Therefore, in the analysis of TE filtration is often avoided, to minimize contamination or losses due to adsorption. Exceptions are made, however, when the suspended particulate matter (SPM) itself or selected elements are studied for which the particulate species constitute a significant fraction of the total concentration in seawater (*e.g.*, Fe, Pb). The SPM should also be separated when its concentration increases to $\gg 1\,mg/L$ and thus impacts the accuracy and precision of the methods used to determine the 'dissolved' TE fractions. Higher SPM concentrations are often observed in mixing zones of estuaries, in coastal waters, in the euphotic layer during plankton blooms or in intermediate turbidity layers and close to the bottom. Particle fluxes of TE in the ocean are best 'measured directly' with sediment traps deployed at different water depths (see Chapter 1).

12.6.2   Methods of sampling

The equipment for sampling marine particles is briefly described in Chapters 1 and 2. In this section we emphasize the specific requirements for sampling particulate trace elements. The major points are, first, to avoid contamination, and second, to collect enough material for the detection of as many TE as possible even at very low concentrations.

12.6.2.1   Sampling by pressure filtration

On-line pressure filtration from appropriate samplers (Fig. 12-30) and off-line pressure filtration *via* Nuclepore filters with 47 mm diameter and 0.4 μm pore size (Fig. 12-31) are adequate for the collection of SPM in TE studies of seawater. For initial cleaning, the filters are soaked 3 d in ultrapure 6 mol/L HCl, rinsed and leached overnight with high-purity distilled water, kept another 3 d in 0.5 mol/L HCl at 60 °C in a drying oven, rinsed again with high-purity distilled water, dried in a microwave oven and weighed on a microbalance. These and the following operations should be performed preferably under clean-room conditions. Special care has to be taken not to contaminate the purified filters during weighing. Electrostatic charging of the filters should be avoided and/or compensated for by using ionizing sources, such as 'ion guns' or radioactive emitters and antistatic scale pans. The cleaned and pre-weighed filters (with precision of ± 0.1 mg) are stored in polycarbonate Petri dishes of ≥ 50 mm diameter. For handling the filters, plastic tweezers with dull tips must be used.

Preferably, the filtration unit should provide a volume for ≥ 1 L of sample water. This will not only speed up the procedure, but also minimize contamination risks due to repeated opening and refilling. Inert gases (Ar, N_2) are preferred for pressure filtration of samples which are under-saturated with regard to oxygen. Otherwise, the use of pressurized air is possible. The pressure should be adjusted to allow a filtration rate of ≤ 100 mL/min but should not exceed 80 kPa (0.8 bar). An air-filter and a gas-washing bottle filled with acidified

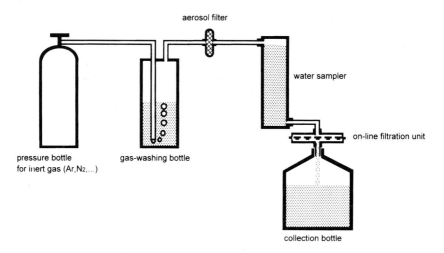

Fig. 12-30. Schematic diagramme for on-line sampling of SPM by pressure filtration.

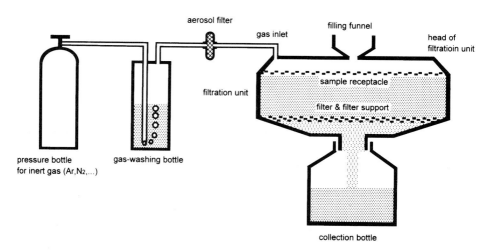

Fig. 12-31. Schematic diagramme for off-line sampling of SPM by pressure filtration.

high-purity distilled water are placed in the gas line to prevent particulate contamination. When performing 'dissolved' TE studies, the first 50 mL of the filtrate are usually discarded but must be included, if the total mass of particles per volume of sample is to be determined.

The filter should not run dry between refilling operations. After completion of the filtration, the loaded filter has to be washed to remove seasalt residues. This is best done with two portions of about 20 mL of high-purity distilled water. Then, after draining, the filter is pressurized for another 2 min to assist the following drying procedure and to prevent SPM losses from wet filters after dismantling the filter holder. To avoid memory effects between processing different samples, the filter holder should be rinsed with acidified water (1 mL conc. HCl diluted to 1 L with high-purity distilled water). The dry-weight of the collected SPM should not significantly exceed 2 mg. The salt-free filters are folded twice with the loaded surface inside, returned to the plastic Petri dishes, stored in a deep-freezer at ≤ -20 °C, dried and re-weighed in the land laboratory.

For waters with much less than 0.5 mg/L SPM, the amount of material obtainable from the typical 0.5–2 L sample volume is insufficient for subsequent precise and accurate TE analyses. Off-line processing of the much larger volumes of water then necessary becomes very laborious and should be replaced by on-line filtration from appropriate samplers (Fig. 12-30). To this end, the filter holder is attached to the drain spigot, and the necessary additional pressure is applied *via* the sampler's air-vent. The suspended particulate matter is collected by filtering all the water held by the sampler. To remove the seasalt from the loaded filters, the on-line filter holder is disconnected from the sampler and attached to a plastic syringe. Manually, two 20 mL portions of high-purity distilled water are drained through the filters. The further treatment of the loaded, salt-free filters is identical to the procedure applied to filters obtained by off-line filtration.

To speed up the filtration of large volumes of seawater and to collect SPM in excess of the recommended 1–2 mg loading capacity of 47 mm ($\varnothing$)/0.4 μm Nuclepore filters, the use of larger filters, *e.g.*, 142 mm ($\varnothing$), with the same physical and chemical characteristics is recommended. Loaded 142 mm ($\varnothing$)/0.4 μm Nuclepore filters must be washed twice with 50 mL of high-purity distilled water. These filters are initially cleaned as outlined before for their smaller 'relatives'. For storage and weighing, however, they might be kept rolled-up in cleaned plastic vials or PE tubing of appropriate size (see next section).

12.6.2.2 Sampling by *in situ* filtration

In situ filtration is reviewed briefly in Chapter 2 (Section 2.1.3.3). Here we outline the procedure for the collection of suspended particulate TE with the Kiel *in situ* Pump (KISP) introduced by *Petrick et al.* (1996) and also technically described in Chapter 2 (see Fig. 2-2). The system has been developed primarily for the determination of organic compounds at ultra-trace levels (see also Chapter 22).

For the determination of TE in SPM, sampling with the KISP is slightly modified: the water is drawn directly through the filter (142 mm ($\varnothing$) Nuclepore of 0.4 μm pore size) with the orifice of the inlet tubing about 1–1.5 m distance from the hydrographic wire. The components of the PTFE filter holder ($\varnothing$ 142 mm, Sartorius, Göttingen), the tube fittings and the PE tubings are thoroughly cleaned before use in accordance with the TE cleaning techniques described in Section 12.1.4. The cleaned and weighed filters are rolled up and stored in clean, weighed (with a precision of ±0.3 mg), and numbered PE tubing (16 mm i.d., 18 mm o.d., 160–180 mm length) and plugged with two PE stoppers.

Prior to deployment of the KISP, the filters are inserted into the filter holder under the hood of a clean bench. Then the inlet tubing is connected to a high-purity distilled water supply in order to calibrate the flow rate of the pump. After programming (*e.g.*, set-up of start time and end of pumping), the system is ready for deployment. We found an *in situ* pumping time in the deep ocean of about 4 h (*i.e.*, 300–400 L) as a reasonable compromise between shiptime costs and sample-mass requirement. After recovery of the KISP three portions (5 mL each) of pH ≈ 8 adjusted high-purity distilled water are drawn through the inlet tube *via* the filter. Under the clean bench, the filter holder is opened, the filter is folded twice, rolled up, placed in its original storage tube and stored at –20 °C. In this procedure we accept a certain residue of seasalt, because restricted washing limits the loss of labile components (*e.g.*, 30 % Cd loss was found by *Sherrell and Boyle*, 1992). Consequently, a measurement for sodium is necessary to check for seasalt residues in order to apply dry-mass corrections of the collected material.

12.6.2.3   Sampling by continuous-flow centrifugation

The amount of particles obtained by continuous-flow centrifugation and sediment traps is orders of magnitude greater than that obtained by common filtration; *i.e.*, gram *versus* milligram amounts. Continuous-flow centrifugation separates SPM from large volumes of seawater, typically sampled with pumps from the mixed surface layer (see Chapter 1). Water sampling may be performed from moving (steaming, drifting) or moored vessels. The continuous-flow centrifuge is fed either *via* a bypass directly from the sea or from previously filled plastic tanks (Fig. 12-32).

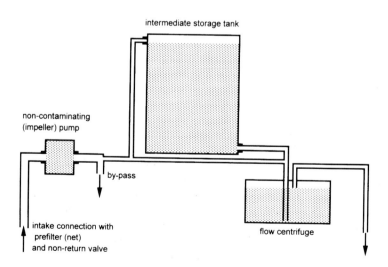

Fig. 12-32. Schematic diagramme for SPM sampling by flow centrifugation.
(For continuous pumping system with attached centrifuge see also Chapter 1, Fig. 1-2.)

For subsequent TE determinations in collected SPM, the following precautions must be taken:

– The sample water inlet must be positioned so that no particulate material originating from the ship (*e.g.*, paint, dust, wastes) is collected.

– The pumping system must not contaminate the samples. This may be achieved by the use of pre-cleaned Teflon or (HD)PE tubes and fittings, and of PTFE membrane or impeller pumps. Those parts of the centrifuge in contact with seawater or with the retained particulate matter must be made of chemically resistant and non-contaminating materials (*e.g.*, Ti, PTFE, Teflon, PP, PE).

– The efficiency of centrifugation must be checked against pressure filtration through $0.4\,\mu m$ Nuclepore filters, both with regard to the TE concentration of the collected SPM and the recovery of particulate elements from a certain volume of water. With lower rotational acceleration ($\leq 10\,000\,g$), the composition of the material collected by centrifugation may differ significantly as compared with filtration. With high-speed centrifuges ($\geq 18\,000\,g$; flow rates $\leq 1\,m^3/h$), however, comparability with filtration results was found to be satisfactory (*Brügmann*, 1993; *Schüßler and Kremling*, 1993).

– Recovery of the collected material from the rotating chamber of the centrifuge and its further treatment, *e.g.*, splitting into subsamples, washing for removal of seasalt or filtration of suspensions for sample transfer, should be performed in a dust-free environment (clean room or clean bench).

– Before further chemical treatment of the retained SPM, the material must be inspected for contamination by coarse particles of anthropogenic origin. For the inspection, the use of magnifying glasses and plastic-shielded magnets is recommmended.

Typically, between 0.5 and $20\,m^3$ of seawater are processed by continuous-flow centrifugation to obtain SPM samples with dry masses of about 1–10 g. Such relative large amounts of collected wet material are recovered with the help of plastic spoons and spatulas from the rotating chamber and transferred directly into PP vials for deep-frozen storage until analyses. Following (freeze-)drying, re-weighing and homogenization, the SPM sample is ready for chemical digestion and analysis.

12.6.2.4   Sampling by sediment traps

The downward flux of particles in the water column is best determined with sediment traps (*e.g.*, *Honjo and Doherty*, 1988; *Kremling et al.*, 1996; *Asper*, 1996). For general review, major design objectives and restrictions of this technique we refer to Chapter 1 (Section 1.3.5.2). Here we outline a somewhat routine procedure with special regard to the collection and handling of particulate TE (*e.g.*, *Kuss and Kremling*, submitted).

It is good practice to clean the cone, rotator and sampling bottles ($\approx 400\,mL$ PP bottles) of the sediment traps in the same manner as described in Section 12.1.4. Suspended, after deployment, from Keflar ropes (with a length of at least 50 m above the traps), a 2 week rinse under field conditions is performed before the sampling programme is started. This results in TE blank values generally $< 1\,\%$ of the TE amounts present in open-ocean trap material (evaluated from sampling periods of between about 1 and 6 weeks).

Several studies have shown that poisoning of the sampling cups (*e.g.*, by addition of formaldehyde or sodium azide) cannot entirely prevent the microbial and chemical degradation of biogenic material. Thus, large portions of certain constituents may be released from collected particles into solution. Based on the results of long-term flux studies in the northeastern Atlantic Ocean (*Kuss and Kremling*, 1999), significant dissolved portions were found in the supernatants for Cd ($>60\%$ of the total particulate flux), Mn ($\approx 37\%$), Ni ($\approx 22\%$), Cu ($\approx 19\%$) and Zn ($\approx 8\%$). Dissolved fractions of lithogenic elements (*e.g.*, Al, Co, Fe, V) contributed $<1\%$ to the total flux.

For reliable flux data both particles and supernatants, therefore, have to be analysed. For TE analyses a 1 % solution of sodium azide in seawater (*i.e.*, a 4:1 mixture of *in situ* seawater, collected by GoFlo-samplers, and 5 % sodium azide stock solution in high-purity distilled water) is recommended as a preservative.

Immediately after retrieval the sediment traps are covered with PE bags, and the collection cups are removed with great care to avoid contamination and stored at $\approx 4\,^\circ$C. In the home laboratory, zooplankton 'swimmers' are removed from samples by means of Teflon tweezers, because the inclusion of these organisms (entering the traps actively mainly at depth <1000 m) into the analysis may bias the results of sedimenting material. Afterwards, the samples are centrifuged twice (first in 500 mL centrifuge flasks and then in 100 mL centrifuge vials) for 30 min at 4000 rpm at 4 °C. The dissolved and particulate portion of each sample is stored separately (at 0–6 °C) for subsequent analysis.

For determination of dissolved fractions, the samples are acidified (pH 1–2) and purged for 2 h with 100 mL/min of filtered high-quality nitrogen to remove the interfering preservation reagent (hydrogen azide) from solution.

12.6.3 Sample pre-treatment

For the determination of TE concentrations in particulate material by wet chemical methods, total decomposition methods (digestion with strong acids), moderate extraction (with weak acids) or sequential material dissolution are commonly used. Often, simultaneously with the determination of TE, several of the major and minor constituents of the samples are determined to collect the necessary background information on sample origin, mineralogical composition, binding forms or potential 'bioavailability' of the TE.

Sample processing is best performed in clean rooms under positive pressure and/or under clean benches. For the cleaning procedures for the equipment see Section 12.1.4.

12.6.3.1 Weak acidic leaching

The fraction of the non-residual, easily leached elements is released from SPM by treating the loaded filters with diluted HCl in PP centrifuge tubes of appropriate size. The double-folded 47 mm ($\varnothing$) are commonly used filters are placed upright in the tubes, treated with 5 mL of 0.5 mol/L HCl, closed with a cap and kept under slow horizontal shaking for 12 h. The tubes are centrifuged for 10 min at 2500 rpm, and the leachate is decanted into purified PP vials.

For leaching of loaded 142 mm ($\varnothing$) filters, 25 mL of 0.5 mol/L HCl is used. Of the freeze-dried and homogenized material collected by continuous-flow centrifugation or with sediment traps, about 0.1 g amounts are weighed into PP centrifuge tubes and treated as the loaded 142 mm filters.

12.6.3.2   Total digestion

1. Loaded filters (sample amount < 10 mg)
Numerous procedures are described in the literature for digestion of SPM collected on filters (for an overview see *Knapp*, 1985). In the following a few of these digestion techniques are outlined which have been applied successfully by the authors of this chapter over several years.

Method A. In this method, hydrofluoric acid and *aqua regia* (*i.e.*, HCl–HNO₃ 3 : 1 by volume) are used to release the total TE content from the SPM or the residual fraction of a leaching procedure. The digestion is performed in sealed Teflon decomposition vessels ('bombs') under pressure. The pressure develops mainly by evaporation of the applied mixture of acids due to microwave heating and by decomposition products (mainly oxides of carbon and nitrogen) evolving from organic matter. For some TE, *e.g.*, Cr and Sn, the residual fraction may not entirely be released by this type of digestion.

For total digestion, either (a) the residues of the weak-acid leached or (b) previously untreated SPM samples are used. For the digestion of type (a) samples, the material first has to be washed twice with 5 mL of high-purity distilled water to remove quantitatively those TE portions which have been solubilised previously with 0.5 mol/L HCl but were not yet separated quantitatively. Following centrifugation, the supernatant wash solution is decanted. The residue is transferred into PTFE bombs of appropriate size, pushed down to the bottom of the vessels and allowed to dry under a clean bench.

Then each one of the 47 mm filters is treated with 0.9 mL of HCl (37 %) and 0.3 mL of HNO₃ (70 %) and left at room temperature overnight in loosely covered digestion vessels. This pre-oxidation under ambient conditions reduces pressure peaks occurring later in the sealed and heated vessels. Following addition of 0.3 mL of HF (49 %), the Teflon bomb is closed and heated in a microwave oven with at least a 750 W power rating for 90 s (Table 12-6). After cooling to room temperature, the digestion vessel is opened. Five mL of high-purity distilled water are added and mixed with the digest. After 5 min, the content is decanted into graduated 10 mL PP *vials*. The digestion vessel is rinsed twice with small amounts of high-purity distilled water. The solutions are combined and diluted to 10 mL.

Table 12-6. Digestion mixtures (volumes of added acid in mL) applicable either for 10–30 mg of dry material[a] or for prepared 142 mm (∅) polycarbonate filter sample in PTFE pressure bombs (in paranthesis are values for 47 mm (∅) filters; for details see text).

Method	HNO₃	HCl	HF	HClO₄	H₂O₂	Conditions
A	1 (0.3)	3 (0.9)	1 (0.3)	–	–	MW/90 s [b]
B1	1	–	3	3	–	165 °C/12 h [c]
B2	3	–	3	–	–	165 °C/12 h
C	4	–	1	–	0.5	MW [d]

[a] The powder is moistened with 0.5 mL of high-purity distilled water;
[b] microwave oven with at least 750 W;
[c] for resistant material (*e.g.*, lithogenic silcates), lower efficiencies for Pb, Cu, Zn have been observed;
[d] special temperature programme, followed by a digestion with 2 mL of concentrated HCl (see text).

In principle, the residues of 142 mm filters are treated in the same way as described for 47 mm filters. The differences only refer to the volume of applied acids (3 mL of HCl, 1 mL of HNO₃, 1 mL of HF) and of the final sample solution (25 mL instead of 10 mL).

Method B. The filters are dried to constant mass, transferred into the prescribed PTFE digestion bombs (Loftfield Analytische Lösungen, LAL, D-37249 Neu-Eichenberg, Germany) and the filter matrix is denaturated by treatment with distilled chloroform (2–3 digestions with 2 mL of CHCl₃, gently heating to dryness). The denaturated loaded filters are preferably digested with a 1:3:3 mixture of ultrapure HNO₃:HCl:HClO₄ at 165 °C for about 12 h (mixture B1; Table 12-6). The digests are evaporated to dryness and the residue is redissolved in 1 mL of HNO₃ plus 1 mL of high-purity distilled water by standing overnight. The solutions are diluted to 10 mL and are then ready for instrumental analysis.

Method C: (Used specifically for TE determination by TXRF; Section 12.4.) The filters are dried to constant mass and transferred to PTFE digestion bombs (Microwave digestion unit, *e.g.*, from Intec Laborgeräte, D-73066 Uhingen, Germany). Four mL of HNO₃ (65 %), 1 mL of HF (30 %), 0.5 mL of H₂O₂ (30 %) (mixture C; see Table 12-6) and an appropriate amount of internal gallium standard solution are added (*e.g.*, representing a Ga content in the dry material of 500 mg/kg) and mixed to cover the filters. The PTFE bombs are closed and placed in the microwave oven. A typical temperature programme involves, *e.g.*, 2 min heating at 250 W, 3 min at 500 W, 2 min at 600 W, 2 min at 700 W, 2 min at 800 W and, finally, 5 min at 250 W. The bombs are cooled and connected to the evaporation device (Intec MCR-6). The residues are heated (*e.g.*, at 250 W) under vacuum until dryness (*ca.* 1 h). After cooling and addition of 2 mL of HCl (30 %) the bombs are closed again and kept at 400 W for 20 min. Thereafter, the digests are transferred to PE *vials* and made up to 20 mL using high-purity distilled water.

2. Sediment trap and centrifuge material (sample amount > 10 mg)

The total wet particulate fraction is determined by differential weighing (*e.g.*, in PP vials or centrifuge cups). Afterwards the well mixed ('homogenized') sludgy material may be split quickly (under mass control) and transferred with a Teflon spatula into different vials for determinations of further compounds (*e.g.*, trace organic species; Chapter 22). The samples (TE splits) are freeze-dried to remove up to 90 % of the water content. Typically, dry material with salt contents of betweeen 20 and 50 % is obtained and then homogenized. The particulate dry mass has to be corrected for seasalt residues calculated from the water volume and its salinity. This can be confirmed by Na measurements. Some elements (*e.g.*, calcium) should be corrected for evaporated seawater *via* the element/salinity ratio or measurement of the dissolved fraction. Powdered particulate splits of 20–30 mg are digested with a total of a 7 mL 1:3:3 mixture of sub-boiled HNO₃, HF and HClO₄ (both Suprapur, Merck) at 165 °C in Teflon pressure bombs for about 12 h (B1; see Table 12-6). The digests are evaporated to dryness and redissolved in 10 mL of 1:10 diluted HNO₃. For the determination of Cu, Pb and Zn, a 1:1 mixture of HNO₃ and HF (mixture B2; see Table 12-6) is used to improve the recovery values.

12.6.3.3 Sequential extraction

The recommended five-step procedure for sequential extraction of TE from suspended and sedimenting particles is based on schemes proposed for sediment and soil samples (*Quevauviller and Maier*, 1994) and allows rough differentiation between (a) exchangeable (adsorbed), (b) extractable (by dilute acetic acid; *e.g.*, from carbonates), (c) reducible (*e.g.*, from Fe/Mn oxides/hydroxides), (d) oxidizable (*e.g.*, in organic compounds and sulphides), and (e) residual forms of TE (*e.g.*, in the silicate lattice). Taking into account the recommended maximum loading of 0.4 μm Nuclepore filters and the detection limits of the applied methods, the procedure is only applicable to SPM sampled by continuous-flow centrifugation or on 142 mm filters (with loadings of ≥ 50 mg) and to particles collected by sediment traps.

Reagents and materials
In addition to appropriate laboratory ware, including carefully cleaned beakers, volumetric cylinders, flasks, *etc.* (see Section 12.1.4), the following reagent solutions are used:
A   1 mol/L *Ammonium acetate, pH 6.0.*
B   0.1 mol/L *Acetic acid.*
C   0.1 mol/L *Hydroxylamine hydrochloride.*
D1 8.8 mol/L (30 %) *Hydrogen peroxide, pH 2–3.*
D2 1 mol/L *Ammonium acetate, pH 2.0 (adjusted with HNO$_3$).*
E   *Concentrated HCl (37 %), HNO$_3$ (70 %) and HF (49 %).*

Preferably, all chemicals and acids used for the preparation of the reagents should represent the highest analytical grade available with regard to TE impurities (see Section 12.1.5). Solutions A and B are stable indefinitely. Solution C should be prepared for the number of samples to be processed during one week. Solution D1 must be kept cool and dark and should be discarded when the volume in the bottle falls below 20 % of the total volume.

During the entire sequential extraction, except for step (e), the sample is kept in a centrifuge tube (volume 20–25 mL, preferred material PTFE). The sample either is a loaded 142 mm filter or 0.1 g of freeze-dried and homogenized SPM. A wrist action shaker, preferably of the rotary or end-over-end type is used running at moderate speed (about 30 rpm). If not specified otherwise, the extractions are performed at room temperature. There should be no delay between the addition of the extractant solution and the beginning of shaking. Generally, the samples are centrifuged at 2 500 rpm for 10 min. The supernatant solution is decanted (and filtered) into clean PP vials (with caps) of appropriate volume for later AAS analysis. To remove adhering extraction solution containing already released TE, the extraction residues are swirled with 5 mL of high-purity distilled water, shaken for 15 min and centrifuged. Those decantates are discarded. Heating (step d) may be performed on a steam bath or equivalent.

Extraction procedure
a) The sample is placed at the bottom of a centrifuge tube and 5 mL of solution A are added. The sample is shaken for 6 h, centrifuged, decanted and washed with high-purity distilled water.
b) The residue of step (a) is extracted overnight (16 h) by shaking with 5 mL of solution B, centrifuged, decanted and washed with high-purity distilled water.

c) The residue of step (b) is extracted overnight (16 h) by shaking with 5 mL of solution C, centrifuged, decanted and washed with high-purity distilled water.

d) Carefully, in small portions to avoid losses due to violent reaction, 2 mL of solution D1 are added to the residue of step (c). The centrifuge tube is covered with a watch glass and allowed to digest at room temperature for 1 h with occasional manual shaking. The digestion is continued for another hour at 85 °C, the watch glass is removed and the volume evaporated to ≤0.5 mL. Following addition of another 2 mL portion of solution D1, the covered sample is digested again for 1 h at 85 °C and finally evaporated uncovered to ≤0.5 mL. Four mL of solution D2 are added without further delay to cool the moist residue. Extraction is performed overnight (16 h) by shaking, followed by centrifugation, decantation and washing with high-purity distilled water.

e) The residue of step (d) is transferred into a Teflon bomb of appropriate size and allowed to dry under a clean bench. Following addition of 3 mL of HCl, 1 mL of HNO_3, and 1 mL of HF (see Table 12-6), the Teflon bomb is closed and heated in a microwave oven with at least a 750 W power rating for 90 s. After cooling to room temperature the digestion vessel is opened. Five mL of high-purity distilled water are added and mixed with the digest. After 5 min, the content is decanted into graduated 10 mL PP vials. The digestion vessel is rinsed twice with small volumes of high-purity distilled water. The rinsings are added to the PP vial, the volume is made up to the mark, and the contents are mixed.

Remarks

– The samples should be kept in suspension continuously during extraction. Cakes left in the centrifuge tubes after centrifugation and decantation must be broken up using plastic or silica bars before continuing with the next extraction step.

– Occasionally a filtration step (0.4 μm) may be necessary after centrifugation.

12.6.4   Instrumental detection

Frequently, the final determination of elements in digests, leachates and in supernatant solutions of samples from sediment traps is performed by either Inductively Coupled Plasma Atomic Emission Spectrometry (ICP-AES) or AAS. The decision as to which spectrometric method should be used mainly depends on the mass and composition of the particulate material and the range of elements to be determined. Measurements by AAS have considerably lower detection limits but are more time-consuming. ICP-AES is the method of choice if the concentrations are suitably high.

In Table 12-7, examples of the elemental composition of marine particulate matter of different origin are provided. The data are compared with the composition of surficial deep-sea sediments and the earth crust. In addition, sample-dependent detection limits of modern ICP-AES instruments are listed. These values (concentration ranges) are based on detection limits provided by producers and users of the instruments (*e.g., Pepellnik*, personal communication) for single element standard solutions which have been converted here into element concentrations in solids by assuming dissolution of 1 mg of particulate matter in 10 mL of digest (leachate) volume.

Table 12-7. Concentrations of selected elements in marine particles of different origin compared with detection limits of ICP-AES at preferred emission wavelengths (in nmol/g; Al, Ca, Fe, P and Si in μmol/g).

	Wave-length (nm)	SPM NA[a]	SPM NA[b]	SPM NA[c]	SPM BS[d]	Trap NA[e]	ICP DL[f]	Sed. NA[g]	Crust Ref.[h]
Al	396.1	9	35	17	456	95	0.4–4	460	3 050
Ca	393.3	1230	1515	801	–	–	0.02–0.2	8980	1040
Cd	228.8	123	84	3	14	23	30–900	1	2
Co	228.6/ 238.9	46	49	27	117	20	50–1500	93	424
Cu	324.7	95	926	166	299	303	80–200	447	865
Fe	238.2	1	13	4	269	16	0.2–2.0	140	1010
Mn	257.6	360	460	710	60 100	1600	10–200	9200	17 300
Ni	231.6	204	206	109	460	390	200–2000	189	1280
P	214.9	230	210	15	–	136	10–40	9	34
Pb	220.3	12	33	9	193	–	100–5000	–	–
Si	251.6	634	750	–	–	–	0.5–5	–	–
Ti	334.9	160	990	660	–	2300	100–2000	17 200	119 000
V	292.4	–	–	60	–	240	100–2000	522	2650
Zn	213.9	327	434	519	3105	1050	30–1500	288	1070

[a, b] Mainly biogenic (a) and 'ordinarily' composed SPM (b) in mixed layer of North Atlantic (*Kuss and Kremling*, unpublished);
[c] SPM taken at 1000 m depth of the North Atlantic (*Kuss*, unpublished);
[d] SPM taken in western parts of the Baltic Sea (*Kremling et al.*, 1997);
[e] sediment trap material, North Atlantic / 1000 m (*Kuss and Kremling*, 1999);
[f] ranges of estimated detection limits (D.L.) of modern ICP-AES instruments (assuming dissolution of 1 mg of dry material to 10 mL digest/leachate volume);
[g] sediment, North Atlantic (*Kuss and Kremling*, 1999);
[h] average crustal composition (*Taylor*, 1964).

12.6.4.1   Instrumental detection by AAS

Trace elements in leachates and digests of loaded filters, sediment traps and centrifuge materials can be detected using different atomic absorption spectrometric techniques. Depending on the amount of available sample solution and concentration of the respective elements either common flame, flame-injection or electrothermal AAS (ETAAS) are selected. The principle of atomic absorption spectrometry, its advantages and limitations have been summarized in Section 12.2 of this chapter. There are, however, a few peculiarities which have to be mentioned.

a) Because sample matrices of individual samples may differ significantly, the method of standard additions is strongly recommended (see Fig. 12-6).

b) For the analysis of leachates from sequential extraction by AAS (Section 12.6.3.3), the calibration solutions should be prepared with the appropriate extracting reagents (*i.e.*, with solutions A, B, C *etc.*, respectively).

12.6.4.2   Instrumental detection by ICP-AES

The application of ICP-AES has been pioneered since 1962 by the group of Albright and Wilson (*e.g.*, Greenfield) and by *Fassel* (1978). The electronic equipment, the aerosol injection and the optical systems have been steadily improved to reduce sample amounts and to lower detection limits. The recently developed semiconductor array detectors in ICP technology, in combination with a powerful data processing system, offer the advantage of simultaneous detection and calibration of 30 or more elements. For details of the physics and equipment we refer to *Atkins* (1987) or *Robinson* (1996).

This section focuses on ICP-AES, but it should be mentioned that ICP as an ion source coupled with a mass spectrometer (ICP-MS) is becoming more popular (see also Section 12.1.2). This more sophisticated multi-element technique has the advantages of very high sensitivity and a large linear dynamic range of 6–8 orders of magnitude. However, wide-spread application of ICP-MS is still hampered by the high price of the equipment and problems encountered with high concentrations of matrix elements (*e.g.*, *Allen et al.*, 1997).

Reagents and materials

Gas supply: Argon 4.8 (99.998 % purity); optionally N_2 (99.999 % purity) as purge gas.

Rinsing solution: For rinsing of the system between sample measurements, 0.5 mol/L nitric acid (*ca.* 42 g of concentrated HNO_3 are diluted to 1 L with high-purity distilled water).

Multi-element standard solutions: Certain volumes of single-element standards (1.000 g/L) of, *e.g.*, Al, Ba, Be, Ca, Cd, Co, Cr, Cu, Fe, K, Li, Mg, Mn, Na, Ni, P, Pb, Sc, Sr, Ti, V and Zn are mixed with 70 g of sub-boiled 65 % HNO_3 and diluted to 2 L with high-purity distilled water. The composition of the multiple standard should match as closely as possible the sample matrix (*e.g.*, see Table 12-7), except Ca and Na for samples with high proportions of, *e.g.*, biogenic calcite and/or seasalt residues. To avoid non-linearity of calibration curves, Ca and Na concentrations in the multiple standard should be about a factor of 10–100 lower than in the sample digests. (For measurement of these elements in the sample, the sample digest should be diluted to the concentration range of the standard, *e.g.*, 1 : 10 down to 1 : 100, if necessary). Single-element standards of either chloride or nitrate salts should be used, sulphates should be avoided due to formation of poorly soluble compounds (Ba, Ca, Sr).

Manganese 'warm-up solution' (10 mg/L Mn): 10 mL of Mn standard solution (1.000 g/L) and 5 g of HNO_3 are diluted to 1 L with high-purity distilled water. This solution may be used to monitor the performance of the instrument.

Sample vials and micropipettes.

Preferable emission wavelengths for elements in ICP-AES analysis are tabulated (in ICP manuals; Table 12-7) or are available in the literature (*e.g.*, *Zaidel et al.*, 1970; *Robinson*, 1996).

Analysis of the sample

How do you run a routine analysis using ICP-AES? The details depend on the manufacturer and, more importantly, on the age of the instrument. Modern instruments are run perfectly by computers with Windows or comparable software (done by mouse-clicks on icons). Older instruments require manual operations. However, the principles remain the same:

Turn on the spectrometer (at least one day before the measurement; it is recommended that the spectrometer be left on at all times).

Check the capillary tubings, the connections, the torch (deposits and indications of melting).

Turn on the gas.

Start the nebulizer by drawing in *rinsing solution.*

Turn the RF on, move the torch into the ignition position, ignite the plasma, examine the plasma and immediately move the torch back.

Start by measuring the *Manganese 'warm-up solution'* (the Mn spectral line may be recorded repeatedly until the counts reflect sufficiently constant instrumental output within 1 %).

Feed the autosampler with standards, blanks and sample solutions kept in appropriate vials.

Start with blank and standard, measure the blank and standard regularly after a set of 6–10 samples to check for instrumental drift.

Use certified samples to check the accuracy of analyses (*e.g.*, the marine sediment standard BCSS-1; Section 12.1.6).

Write an 'ID-WT file' (*i.e.*, identification of the sample, sample mass, dilution factor, conversion factor).

Select a list of elements (at one or two emission wavelengths each).

Calibrate the elements on the wavelength (*i.e.*, find the peak maximum, and the background, enter the element concentration).

Check for spectral overlap by comparing a 'one element' standard with a *multi-element standard solution.*

Write a method file; the number of replicates (commonly 3–5) depends on the available sample volume, on the instrument (sequential, simultaneous measurement) and the number of elements to be measured.

The elements should be grouped in appropriate dilution series to avoid extrapolation of the calibration curves. In sequential measurements, do not measure all selected elements in one run. The amount of sample solution can be reduced significantly by grouping elements exhibiting similar emission wavelengths together, because solution is drawn up during the time-consuming rotation of the grid or prism.

The data report should be saved as a data file and as a printed hard copy.

12.6.4.3   Calculation of results

The results of TE measurements on collected suspended particulate matter may be expressed in two different ways, *i.e.*, either in terms of concentrations of the respective element in a volume (mass) of water or in a certain amount of SPM. The concentrations of particulate TE in seawater (TE_{part} in nmol/L) and in the SPM (TE in nmol/g_{SPM}) are calculated according to

$$TE_{part} \text{ (nmol/L)} = (A_x - A_{bl}) \cdot c \cdot v_{dig} / (A_{std} \cdot v_{spl})$$

and

$$TE \text{ (nmol/}g_{SPM}) = (A_x - A_{bl}) \cdot c \cdot v_{dig} / (A_{std} \cdot m)$$

where

A_x =   average of measurements of the sample solution in AAS (absorbance) or ICP-AES (emission) units;

A_{bl} =   average of measurements of the total (reagent + 'system') blank in AAS (absorbance) or ICP-AES (emission) units, as contributing to the measurement of a single sample;

A_{std} =   average of measurements of the standard solution in AAS (absorbance) or ICP-AES (emission) units (alternatively, a certain c/A_{std} ratio may be used as 'calibra-ation factor');

c =   concentration of standard solution (nmol/L);

v_{dig} =   volume of the leachate or digest, respectively (L);

v_{spl} =   processed (filtered) water volume (L); and

m =   mass of SPM collected on the respective filter (g).

12.6.4.4   Accuracy, precision and blank values

Accuracy and precision. As a rule, data on the particulate element concentrations in a certain volume of water are more precise and accurate than those regarding the TE contents of the SPM. This is because errors of filtration and the applied wet-chemical analyses are additionally biased by the weighing and re-weighing procedure. Major uncertainties during weighing are insufficiently air-conditioned weighing rooms, disturbances by electrostatic charges and possible losses of the filter material during handling.

As shown by the results of intercomparison exercises, the TE content in samples comprising a few milligram of sediment reference material can be determined with an accuracy of 10–20 % (*Hovind and Skei*, 1992). The same range of analytical precision was obtained during intercalibration exercises on filters loaded with about 1 mg of SPM of natural marine origin (*Yeats and Dalziel*, 1987; *Pohl*, 1997). This value even included the variability introduced by filter cleaning, sampling and washing of the SPM. The precision of the weighing/ re-weighing procedure of 142 mm/0.4 μm Nuclepore filters loaded with around 30 mg SPM can be estimated to about 1 %.

Blank values. Preferably, all unloaded and loaded filters of a batch should be processed in parallel (leaching/cleaning, drying, weighing, storage, loading, digestion, *etc.*) using identical sets of chemicals and reagents. From each batch, at least three filters should be selected randomly for blank determinations. The results of the TE determinations on blank filters should be surveyed over longer period by using control charts.

The TE blanks in digests may differ significantly from batch to batch of filters from the same brand. In addition, despite careful cleaning of the filters, there are elements, such as Cr, which may be released into the total digests in rather high amounts. This is because the filters may contain, in addition to the plastic base material, inorganic additives for chemical and/or mechanical stabilization with TE impurities not leachable during cleaning.

In theory, the total contribution of the blanks should not exceed 10 % of the analytical signal. For studies on SPM collected by filtration, the blank values are determined by the quality of filters and reagents, and by contamination occurring during the analytical procedure ('system blank'). The contribution of the 'system blank' may be kept low by observing the relevant principles of 'good laboratory practice'. The reagents and acids are often commercially available in reasonable quality. If necessary, their blank contribution can be further reduced to ultratrace levels by repeated sub-boiling distillation as described in Section 12.1.5 of this chapter.

In contrast to reagent and system blanks, the filter blanks cannot be suppressed below detection limits by acid cleaning. Only by gentle leaching with weak, non-oxidizing acids are both the basic structure of the filters and the size of the holes left unaltered. This has some practical consequences. If filters are purified for several days with 6 mol/L HCl, the filter blanks for the determination of TE leached from the SPM (Section 12.6.3.3) are very low; however, filter blanks obtained after a *total* digestion of the SPM, may remain at significant levels. For a certain batch of 47 mm/0.4 μm Nuclepore filters, total filter blanks of 1–4 pmol Cd, 8–34 pmol Co, 5–14 pmol Pb, 50–160 pmol Cu, 120–170 pmol Ni and 90–180 pmol Zn per filter were observed during an intercalibration exercise (*Yeats and Dalziel*, 1987). Such blanks may be tolerated only when the filters have been loaded with SPM containing about ten times higher amounts of TE.

References to Chapter 12

Achterberg, E.P., van den Berg, C.M.G. (1994a), *Analyt. Chim. Acta*, 291, 213.
Achterberg, E.P., van den Berg, C.M.G. (1994b), *Deep-Sea Res.*, 44, 693.
Achterberg, E.P., van den Berg, C.M.G. (1997), *Deep-Sea Res. I*, 44, 693.
Ackermann, F., Bergmann, H., Schleichert, U. (1983), *Environ. Tech. Let.*, 4, 317.
Aiginger, H., Wobrauschek, P. (1974), *Nucl. Instr. Meth.*, 114, 157.
Aldrich, A.P., van den Berg, C.M.G. (1998), *Electroanalysis*, submitted.
AL-Farawati, R., van den Berg, C.M.G. (1997), *Mar. Chem.*, 57, 277.
Allen, L.A., Leach, J.J., Houk, R.S. (1997), *Anal. Chem.*, 69, 2384.
Andreae, M.O. (1977), *Anal. Chem.*, 49, 820.
Andreae, M.O. (1979), *Limnol. Oceanogr.*, 24, 440.
Andreae, M. O., Asmodé, J.F., Foster, P., Van't Dack, L. (1981), *Anal. Chem.*, 53, 1766.
Andreae, M. O., Froelich, P.N. (1981), *Anal. Chem.*, 53, 287.

Andreae, M. O. (1983), in: *Trace Metals in Sea Water*: Wong, C.S, Boyle, E., Bruland, K.W., Burton, J.D., Goldberg, E.D. (Eds.). New York: Plenum, 1983; pp. 1–19.

Andreae, M. O., Froelich, P.N. (1984), *Tellus*, 36B, 101.

Asper, V.L. (1996), in: *Particle Flux in the Ocean*: Ittekkot, V., Schäfer, P., Honjo, S., Depetris, P.J. (Eds.) Chichester: John Wiley & Sons Ltd, 1996; pp. 71–84.

Atkins, P.W. (1987), *Physical Chemistry*, Oxford: Oxford University Press.

Bloom, N.S., Crecelius, E.A. (1983), *Mar. Chem.*, 14, 49.

Bloom, N.S., Fitzgerald, W.F. (1988), *Anal. Chim. Acta*, 208, 151.

Bloom, N.S. (1989), *Can. J. Fish. Aquat. Sci.*, 46, 1131.

Bloom, N.S., Horvat, M., Watras, C.J. (1995), *Wat. Air Soil Pollut.*, 80, 1257.

Bloom, N.S., Colman, J.A., Barber, L. (1997), *Fresenius J. Anal. Chem.*, 358, 371.

Bothner, M.H., Robertson, D.E. (1975), *Anal. Chem.*, 47, 592.

Boussemart, M., van den Berg, C.M.G. (1994), *Analyst*, 119, 1349.

Boussemart, M., van den Berg, C.M.G., Ghaddaf, M. (1992), *Anal. Chim. Acta*, 262, 103.

Boyle, E.A., Edmond, J.M. (1977), *Anal. Chim. Acta*, 91, 189.

Boyle, E.A., Handy, B., van Geen, A. (1987), *Anal. Chem.*, 59, 1499.

Bradfield, E.G. (1957), *Analyst*, 82, 254.

Brügmann, L., Franz, P., Fröhlich, K., Gellermann, R., Hebert, D., Lange, D., Mohnke, M., Rohde, K.-H., Thiele, J., Weiß, D. (1985), *Geod. Geoph. Veröff.* R IV, 40, 110 pp.

Brügmann, L. (1993), *HERAEUS Sepatech Forum*, 5, 1-2.

Bruland, K.W. (1983), in: *Chemical Oceanography*: Riley, J.P., Chester, R. (Eds.). London: Academic Press, 1983; Vol. 8, pp. 157–221.

Bruland, K.W., Frank, R.P., Knauer, G.A., Martin, J.H. (1979), *Anal. Chim. Acta*, 105, 233.

Bruland, K.W., Coale, K.H., Mart, L. (1985), *Mar. Chem.*, 17, 285.

Bruland, K.W., Orians, K.J., Cowen, J.P. (1994), *Geochim. Cosmochim. Acta*, 58, 3171.

Burton, J.D., Statham, P.J. (1990), in: *Heavy Metals in the Marine Environment*: Furness, R. W., Rainbow, P.S. (Eds.). Boca Raton, FL: CRC Press, 1990; pp.5–25.

Campos, M.L.A.M. (1997), *Mar. Chem.*, 57, 107.

Campos, M.L.A.M., van den Berg, C.M.G. (1994), *Anal. Chim. Acta*, 284, 481.

Capodaglio, G., van den Berg, C.M.G., Scarponi, G. (1987), *J. Electroanal. Chem.*, 235, 275.

Chapiu, R.P., Johnson, K.S., Coale, K.H. (1991), *Anal. Chim. Acta*, 249, 469.

Clegg, S.L., Whitfield, M. (1990), *Deep-Sea Res.*, 37, 809.

Coale, K.H., stout, P.M., Johnson, K.S., Sakamoto, C.M. (1992), *Anal. Chim. Acta*, 266, 345.

Coale, K.H., Johnson, K.S., Fitzwater, S.E., Gordon, R.M., Tanner, S., Chavez, F.P., Ferioli, L., Sakamoto, C., Rogers, P., Millero, F., Steinberg, P., Nightingale, P., Cooper, D., Cochlan, W.P., Landry, M.R., Constantinou, J., Rollwagen, G., Trasvina, A., Kudela, R. (1996 b), *Nature (London)*, 383, 495

Collins, P.F., Diehl, H., Smith, G.F. (1959), *Anal. Chim. Acta*, 98, 47.

Colombo, C., Daniel, A., van den Berg, C.M.G. (1997), *Anal. Chim. Acta*, 346, 101.

Coquery, M., Cossa, D. (1995), *Neth. J. Sea Res.*, 34, 245.

Cossa, D., Courau, P. (1984), *ICES paper to MEQC*, 17 pp.

Cossa, D., Martin, J.-M., Sanjuan, J. (1994), *Mar. Poll. Bull.*, 28, 381.

Cossa, D., Sanjuan, S., Cloud, J., Stockwell, P.B., Corns, W.T. (1995), *Wat. Air Soil Pollut.*, 80, 1279.

Cuculic, V., Branica, M. (1996), *Analyst*, 121, 1127.

Cullen, W. R., Reimer, K.J. (1989), *Chem. Rev.*, 89(4), 713.

Currie, L.A. (1968), *Anal. Chem.*, 40, 586.

Cutter, G. A., Cutter, L.S. (1995), *Mar. Chem.*, 49, 295.

Danielsson, L.G., Magnusson, B., Westerlund, S. (1978), *Anal. Chim. Acta*, 98, 47.

Danielsson, L.G., Westerlund, S. (1983), *Ocean Sci. Eng.*, 8, 53.

Dietz, E. A., Jr., Perez, M.A. (1976), *Anal. Chem.*, 48, 1088.

Donat, J.R., Bruland, K.W. (1988), *Anal. Chem.*, 60, 240.

Donat, J.R., Bruland, K.W. (1995), in: *Trace Elements in Natural Waters*: Salbu, B., Steiness, E. (Eds.). Boca Raton, FL: CRC-Press, 1995; pp. 247–281.

Edmond, J.M., Campbell, A.C., Palmer, M.R., Klinkhammer, G.P., German, C.R., Edmonds, H.N., Elderfield, H., Thompson, G., Rona, P. (1995), in: *Hydrothermal Vents and Processes:* Parson, L.M., Walker C.L., Dixon, D.R. (Eds.). London: The Geological Society, 1995; pp. 77–86.

Elderfield, H., Greaves, M.J. (1982), *Nature (London)*,296, 214.

Elrod, V.A., Johnson, K.S., Coale, K.H. (1991), *Anal. Chem.*, 63, 893.

Fassel, V.A. (1978), *Science*, 202, 183.

Fogg, A.G., Summan, A.M. (1984), *Analyst*, 109, 1029.

Freimann, P., Schmidt, D. (1989), *Spectrochim. Acta*, Part B, 44, 505.

Freimann, P., Schmidt, D., Neubauer-Ziebarth, A. (1993), *Spectrochim. Acta*, 48B, 193.

Froelich, P. N., Andreae, M.O. (1981), *Science*, 213, 205.

Froelich, P. N., Hambrick, G.A., Andreae, M.O., Mortlock, R.A., Edmond, J.M. (1985), *J. Geophys. Res.*, 90, 1133.

Gasparovic, B., Cosovic, B. (1994), *Mar. Chem.*, 46, 179.

Geisler, C.D. (1992), Dissertation, University of Hamburg.

Geisler, C.D., Schmidt, D. (1991/92), *Dtsch. Hydrogr. Z.*, 44, H.4.

Gentry, C.H.R., Sherrington, L.G. (1950), *Analyt*, 75, 17.

Gerwinski, W., Michel, U., Schmidt, D. (1996), Proc. 4th Intl. Conf. on Mercury as a Global Pollutant, Hamburg, August 4–8, 1996.

Gerwinski, W., Schmidt, D. (1998), *Spectrochim. Acta*, 53B, 1355.

Golimowski, J., Valenta, P., Nürnberg, H.W. (1985), *Fresenius Z. Anal. Chem.*, 322, 315.

Gutz, I.G., Angnes, L., Pedrott, J.J. (1993), *Anal. Chem.*, 65, 500.

Haarich, M., Schmidt, D.(1993), *Dtsch. Hydrogr. Z.*, 45, H.6.

Hadeishi, T., McLaughlin, R.D. (1971a), *Science*, 174, 404.

Haffer, E., Schmidt, D. (1995), Berichte des Bundesamtes für Seeschiffahrt und Hydrographie Nr.6, Hamburg und Rostock, Germany.

Hambrick, G. A., Froelich, P.N., Andreae, M.O., Lewis, B.L. (1984), *Anal. Chem.*, 56, 421.

Harbin, A.-M., van den Berg, C.M.G. (1993), *Anal. Chem.*, 65, 3411.

Hartmann, M., Lars, H., Puteanus, D. (1989), in: *5. Colloquium Atomspektrometrische Spurenanalytik:* Welz, B. (Ed.). Überlingen: Bodenseewerk Perkin-Elmer GmbH, 1989; pp. 703-709.

Hintelmann, H., Falter, R., Ilgen, G., Evans, R.D. (1997) *Fresenius J. Anal. Chem*, 358, 363.

Hölemann, J.A., Schirmacher, M., Prange, A. (1997), in: *Land-ocean System in the Siberian Arctic: Dynamics and History*: Kassens et al. (Eds.); in the press.

Holcombe J.A., Bass, D.A. (1988), *Anal. Chem.*, 60, 226R.

Honjo, S., Doherty, K.W. (1988), *Deep-Sea Res.*, 35,133.

Hovind, H., Skei, J. (1992), *ICES Coop. Res. Rep.*, 184.

Howard, A. G., Comber, S.D.W. (1989), *Appl. Organomet. Chem.*, 3, 509.

Howard, A. G., Comber, S.D.W. (1992), *Mikrochim. Acta*, 109, 27.

Howard, A. G., Statham, P.J. (1993), *Inorganic Trace Analysis. Philosophy and Practice.* Chichester: John Wiley & Sons Ltd.

Howard, A. G., Salou, C. (1996), *Anal. Chim. Acta*, 333(1-2), 89.

ICES (1988), *ICES Coop. Res. Rep.*, 152.

Iverfeldt, Å. (1988), *Mar. Chem.*, 23, 441.

Jackson, L.L., Baedecker, P.A., Fries, T.L., Lamothe, P.J. (1995), *Anal. Chem.*, 67, 71R.

Kempe, S., Diercks, A.-R., Liebezeit, G., Prange, A. (1991), in: *Black Sea Oceanography*: Izdar, E., Murray, J.W. (Eds.). Dordrecht: Kluwer Academic Publishers; pp. 89–110.

Klinkhammer, G. (1980), *Anal.Chem.*, 52, 117.

Klockenkämper, R., Knoth, J., Prange, A., Schwenke, H. (1992), *Anal. Chem.*, 64, 1115.

Klockenkämper, R. (1997), *Total-reflection X-ray Fluorescence Analysis*. Weinheim: John Wiley & Sons, Inc.

Knapp, G., Schreiber, B., Frei, R.W. (1975), *Anal. Chim. Acta*, 77, 293.

Knapp, G. (1985), *Int. J. Environ. Anal. Chem.*, 22, 71.

Knapp, G., Panholzer, F., Schalk, A., Kettisch, P. (1997), in: *Microwave Enhanced Chemistry*: Kingston, H.M., Haswell, S. (Eds.). Washington D.C.: ACS Professional Reference Books, 1997; pp. 423–451.

Knoth, J., Schwenke, H. (1978), *Fresenius Z. Anal. Chem.*, 291, 200.

Knoth, J., Schwenke, H. (1980), *Fresenius Z. Anal. Chem.*, 301, 7.

Knoth, J., Prange, A., Schneider, H., Schwenke, H. (1997), *Spectrochim. Acta*, Part B, 52, 907.

Kolber, Z.S., Barber, R.T., Coale, K.H., Fitzwater, S.E. Greene, R.M., Johnson, K.S., Lindley, S., Falkowski, P.G. (1994), *Nature (London)*, 371, 145.

Koopmann, C., Prange, A. (1991), *Spectrochim. Acta*, Part B., 46, 1395.

Koroleff, F. (1974), in: *Report on Applied Methods for the Analysis of Selected Potential Pollutants in Baltic Laboratories*: Kremling, K., Slaczka, W. (Eds.). Kiel-Gdynia.

Krause, P., Erbslöh, B., Niedergesäß, R., Pepelnik, R., Prange, A. (1995), *Fresenius J Anal. Chem.*, 353, 3.

Kremling, K. (1985), *Deep-Sea Res.*, 32, 531.

Kremling, K., Lentz, U., Zeitzschel, B., Schulz-Bull, D.E., Duinker, J.C. (1996), *Rev. Sci. Instrum.*, 67, 4360.

Kremling, K., Schulz-Tokos, J.J., Brügmann, L., Hansen, H.-P. (1997), *Mar. Pollut. Bull.*, 34, 112.

Kuehner, E.C., Alvarez, R., Paulsen, P.J., Murphy, T.J. (1972), *Anal. Chem.*, 44, 2050.

Kuss, J., Kremling, K. (1999), *Deep-Sea Res.*, 46, 149.

Kuss, J., Kremling, K. (1998), *Mar. Chem.*, for publication.

Landing, W.M., Bruland, K.W. (1980), *Earth Planet. Sc. Lett.*, 49, 45.

Le, X. C., Cullen,W.R., Reimer, K.J. (1994), *Anal. Chim. Acta*, 285(3), 277.

Leermakers, M., Meuleman, C., Baeyens, W. (1995), *Wat. Air Soil Pollut.*, 80, 641.

Lewis, B. L. (1985), PhD Thesis, Florida State University.

Lewis, B. L., Froelich, P.N., Andreae, M.O. (1985), *Nature (Nature)*, 313, 303.

Lewis, B. L., Andreae, M.O., Froelich, P.N. (1989), *Mar. Chem.*, 27, 179.

Li, H., Smart, R.B. (1996), *Anal. Chim. Acta*, 325, 25.

Liebermann, A. (1992), *Contamination Control and Cleanrooms: Problems, Engineering Solutions, and Applications.* New York: Van Nostrand Reinhold.

Luther III, G.W., Tsamakis, E. (1989), *Mar. Chem.*, 27, 165.

Luther III, G.W., Branson Swartz, C., Ullmann, W.J. (1988), *Anal. Chem.*, 60, 1721.

L'vov, B.V. (1961), *Spectrochim. Acta*, 17, 761.

L'vov, B.V. (1978), *Spectrochim. Acta*, Part B, 33, 153.

Mason, R.P., Fitzgerald, W.F. (1990), *Nature (London)*, 347, 457.

Mason, R.P., Rolfhus, K.R., Fitzgerald, W.F. (1995), *Wat. Air Soil Pollut.*, 80, 665.

Mattinson, J.M. (1972), *Anal. Chem.*, 44, 1715.

McLaren, J.W., Mykytiuk, A.P., Willie, S.N., Berman, S.S. (1985), *Anal. Chem.*, 57, 2907.

Measures, C.I., Edmond, J.M. (1986), *Anal. Chem.*, 58, 2065.

Measures, C.I., Burton, J.D. (1980), *Anal. Chim. Acta*, 120, 177.

Measures, C.I., Edmond, J.M. (1989), *Anal. Chem.*, 61, 544.

Meinema, H. A., Noltes, J.G. (1972), *J. Organomet. Chem.*, 36, 313.

Moody, J.R., Lindstrom, R.M. (1977) *Anal. Chem.*, 49, 2264.

Morel, F.M.M., Reinfelder, J.R., Roberst, S.B., Chamberlain, C.P., Lee, J.G., Yee, D. (1994), *Nature (London)*, 3369, 740.

Morley, N.H., Fay, C.W., Statham, P.J. (1988), *Adv. Underwat. Technol., Ocean Sci. Offshore Eng.*, 16, 283.

Mortlock, R. A., Froelich, P.N., Feely, R.A., Massoth, G.J., Butterfield, D.A., Lupton, J.E (1993), *Earth Planet. Sci. Lett.*, 119(3), 365.

Nimmo, M., van den Berg, C.M.G., Brown, J. (1989), *Estuar. Coast. Shelf Sci.*, 29, 57.

Obato, H., Karatani, H., Nakayama, E. (1993), *Anal. Chem.*, 654, 1524.

Olson, M.L., Cleckner, L.B., Hurley, J.P., Krabbenhoft, D.P., Heelan, T.W. (1997), *Fresenius J. Anal. Chem.*, 358, 392.

Ostapczuk, P., Valenta, P. Nürnberg, H.W. (1986), *J. Electroanal. Chem.*, 214, 51.

Perkin-Elmer (1991), *Tecnical Documentation*, Publication B3110.01.

Petrick, G., Schulz-Bull, D.E., Martens, V., Scholz, K., Duinker, J.C. (1996) *Mar. Chem.*, 54, 97.

Pihlar, B., Valenta, P., Nürnberg, H.W. (1981), *Fresenius Z. Anal. Chem.*, 307, 337.

Pihlar, B., Valenta, P., Nürnberg, H.W. (1986), *J. Electroan. Chem.*, 214, 157.

Pohl, C. (1997), *Accred. Qual. Assur.*, 2/1, 2.

Prange, A. (1983), Dissertation, University of Hamburg.

Prange, A., Knöchel, A., Michaelis, W. (1985), *Anal. Chim. Acta*, 172, 79.

Prange, A., Kremling, K. (1985), *Mar. Chem.*, 16, 259.

Prange, A., Niedergesäß, R., Schnier, C. (1990) in: *Estuarine Water Quality Management, Coastal and Estuarine Studies*: Michaelis, W. (Ed.). Berlin: Springer-Verlag, 1990; pp. 429–436.

Prange, A., Schwenke, H, (1992), in: *Advances in X-ray Analysis*: Barrett et al. (Eds.). New York: Plenum Press, 1992; pp. 899–923.

Prange, A. (1993), *Nachr. Chem. Tech. Lab.*, 41, 40.

Prange, A. Böddeker, H., Kramer, K. (1993), *Spectrochim. Acta,* Part B, 48, 207.

Quevauviller, P, Maier, E.A. (1994), *European Commission, Report EUR 16000 EN*, Luxembourg, 1994, pp. 77–79.

Raspor, B., Valenta, P. Nürnberg, H.W., Branica, M. (1977), *Thalassia Jugoslavica*, 13, 79.

Reus, U., Markert, B., Hoffmeister, C., Spott, D., Guhr, H. (1993), *Fresenius J. Anal. Chem.*, 347, 430.

Robinson, J.W. (1996), *Atomic Spectroscopy*, New York: Marcel Dekker, Inc.

Rue, E.L., Bruland, K.W. (1995), *Mar. Chem.*, 50, 117.

Sadana, R.S. (1983), *Anal. Chem.*, 55, 304.

Sakamoto-Arnold, C.M., Johnson, K.S. (1987), *Anal. Chem.*, 59, 1789.

Sander, S., Wagner, W., Henze, G. (1995), *Anal. Chim. Acta*, 305, 154.

Sawamoto, H. (1980), *J. Electroanal. Chem.*, 113, 301.

Schaule, B.K., Patterson, C.C. (1981), *Earth Planet. Sci. Lett.*,54, 97.

Scheffer, F., Schachtschabel, P. (1989), *Lehrbuch der Bodenkunde.* Stuttgart: Verlag Ferdinand Enke.

Schirmacher, M., Schmidt, D. (1991), *Wissenschaftlich Technische Berichte*, Bundesamt für Seeschiffahrt und Hydrographie, Hamburg, Germany.

Schmidt, D., Gerwinski, W., Radke, I. (1993), *Spectrochim. Acta*, 48B, 171.

Schüßler, U., Kremling, K. (1993), *Deep-Sea Res.*, 40, 257.

Sedwick, P.N., McMurtry, G.M., MacDougall, J.D. (1992), *Geochim. Cosmochim. Acta*, 56, 3643.

Sherell, R.M., Boyle, E.A. (1992), *Earth Planet. Sci. Lett.*, 111, 155.

Smith, R.G., Windom, H.L. (1980), *Anal. Chim. Acta*, 113, 39.

Slowey, J.F., Hood, D.W. (1971), *Geochim. Cosmochim. Acta*, 35, 121.

Stary, J. (1963), *Anal. Chim. Acta*, 28, 132.

Stockwell, P.B., Thompson, K.C., Henson, A., Temmerman, E., Vandecasteele, C. (1989), *Int. Lab.*, 14, 45.

Stordal, M.C., Gill, G.A., Wen, L.-S., Santschi, P.H. (1996), *Limnol. Oceanogr.*, 41, 52

Stukas, V.J., Wong, C.S. (1983), in: *Trace Metals in Seawater*: Wong, C.S., Boyle, E., Bruland, K.W., Burton, J.D., Goldberg, E.D. (Eds.). New York: Plenum Press, 1983; pp. 513–536.

Sturgeon, R.F., Berman, S.S., Willie, S.N., Desaulniers, J.A.H. (1981), *Anal. Chem.*, 53, 2337.

Sunda, W.G., Huntsman, S.A. (1995), *Mar.Chem.*, 50, 189.

Tam, K. H., Conacher, H.B.S. (1977), *J. Environ. Sci. Health*, B12, 213.

Taylor, S.R. (1964), *Geochim. Cosmochim. Acta*, 28, 1273.

Uthe, J. F., Freeman, H.C., Johnston, J.R., Michalik, P. (1974), *J. Assoc. Off. Anal. Chem.*, 57, 1363.

Van den Berg, C.M.G. (1984a), *Mar. Chem.*, 15, 1.

Van den Berg, C.M.G. (1984b), *Talanta*, 31, 1069.

Van den Berg, C.M.G. (1986), *J. Electroanal. Chem.*, 215, 111.

Van den Berg, C.M.G., Jacinto, G.S. (1988), *Anal. Chim. Acta*, 211, 129.

Van den Berg, C.M.G., Khan, S.H. (1990), *Anal. Chim. Acta*, 231, 221.

Van den Berg, C.M.G., Kramer, J.R. (1979), *Anal. Chim. Acta*, 106, 113.

Van den Berg, C.M.G., Li, H. (1988), *Anal. Chim. Acta,* 212, 31.

Van den Berg, C.M.G., Nimmo, M. (1987), *Anal. Chem.,* 59, 924.

Van den Berg, C.M.G., Khan, S.H., Riley, J.P. (1989), *Anal. Chim. Acta,* 222, 43.

Van den Berg, C.M.G., Nimmo, M., Abollino, O., Mentassi, E. (1991), *Electroanalysis,* 3, 477.

Van Geen, A., Boyle, E.A. (1990), *Anal. Chem.,* 62, 1705.

Van Geen, A., Boyle, E.A., Moore, W.S. (1991), *Geochim. Cosmochim. Acta,* 55, 2173.

Vega, M., van den Berg, C.M.G. (1994), *Anal. Chim. Acta,* 293, 19.

Vega, M., van den Berg, C.M.G. (1997), *Anal. Chem.,* 69, 874.

Wang, J., Ariel, M. (1978), *Anal. Chim. Acta,* 99, 89.

Wang, J., Lu, J. (1993), *Anal. Chim. Acta,* 274, 219.

Whitfield, M., Turner, D.R. (1987), in: *Aquatic Surface Chemistry:* Stumm, W. (Ed.). New York: Wiley-Interscience, 1987; pp. 457–493.

Whyte, W. (1991), *Cleanroom Design.* New York: Wiley.

Wong, C.S., Boyle, E., Bruland, K.W., Burton, J.D., Goldberg, E.D. (1983), *Trace Metals in Sea Water.* New York: Plenum Press.

Wong, C.S., Cretney, W.J., Piuze, J., Christensen, P., Berrang, P.G. (1977), in: *Methods of Standards and Environmental Measurements:* NBS (Ed.). Washington, D.C.: NBS Special Publication, 1977, Vol. 464.

Wu, Q., Batley, G.E. (1995), *Anal. Chim. Acta,* 309, 95.

Yarnitzky, C.Y. (1990), *Electroanalysis,* 2, 581.

Yeats, P.A., Dalziel, J.A. (1987), *J. Cons. Int. Explor. Mer,* 43, 272.

Yoneda, Y., Horiouchi, T. (1971), *Rev. Sci. Instrum.,* 42, 1069.

Yokoi, K., van den Berg, C.M.G. (1991), *Anal. Chim. Acta,* 245, 167.

Yokoi, K., van den Berg, C.M.G. (1992a), *Electroanalysis,* 4, 65.

Yokoi, K., van den Berg, C.M.G. (1992b), *Anal. Chim. Acta,* 257, 293.

Zaidel, A.N., Prokov'ev, V.K., Raiski, S.M., Slavnyi, V.A., Shreider, E.Y. (1970), *Tables of Spectral Lines,* 3rd edn., New York: Plenum Press.

Zief, M., Mitchell, J.W. (1976), *Contamination Control in Trace Element Analysis.* New York: John Wiley&Sons.

Zima, J., van den Berg, C.M.G., (1994), *Anal. Chim. Acta,* 289, 291.

13 Determination of natural radioactive tracers

M. M. Rutgers van der Loeff and W. S. Moore

13.1 Introduction

Rates of many transport and reaction processes in the ocean can be evaluated with the natural clocks provided by the radioactive decay series of three primordial radionuclides present in the earth's crust: ^{232}Th, ^{238}U and ^{235}U (detailed review in *Ivanovich and Harmon,* 1992). In all three decay series, isotopes of relatively soluble elements such as U, Ra and Rn, decay to isotopes of highly particle-reactive elements (Th, Pa, Po, Pb) and *vice versa*. The resulting disequilibria in the decay chains can be interpreted as a measure of transport rates of the particulate and liquid phases relative to each other *(Ivanovich and Harmon,* 1992).

The highly different ocean chemistries of a large spectrum of parent–daughter pairs have been reviewed by *Cochran* (1992). The tracer pairs can be subdivided into three categories (*Rutgers van der Loeff and Boudreau,* 1997).

1. Pairs with a particle-reactive parent and a mobile daughter, suitable to study dispersion of material released from the seafloor into the bottom water. Applications include deep-sea mixing (^{222}Rn, ^{228}Ra) and the advective transport between shelves and the inner ocean (^{228}Ra).
2. Pairs with a readily-soluble parent and a particle-reactive daughter. The particle-reactive daughters are removed from solution and transported by settling particles to the sediment. The natural decay series supply these tracers on a wide variety of time scales. These tracers are widely used to study particle dynamics, ranging from export production (^{234}Th: *Coale and Bruland,* 1985; *Eppley,* 1989) through aggregation of colloidal material (all Th isotopes: *Honeyman and Santschi,* 1989; *Baskaran et al.,* 1992) and quantification of mass flux (Pa/Th ratios: *Anderson et al.,* 1983a,b; *Yu,* 1994) and water mass circulation (*Yu et al.,* 1996) to the calibration of sediment traps (^{234}Th, ^{230}Th: *Buesseler,* 1991; *Bacon et al.,* 1985; *Yu,* 1994; *Buesseler et al.,* 1994) and quantification of resuspension rates (*Bacon and Rutgers van der Loeff,* 1989).
3. Cosmogenic radionuclides such as ^{14}C, ^{7}Be, ^{10}Be, ^{26}Al.

In this chapter we will focus on the determination of those nuclides that are widely used as tracers, and where the analysis can be performed without access to highly specialized equipment such as AMS (accelerator mass spectrometry) or TIMS (thermal ionisation mass spectrometry). We describe detailed analytical methods for the nuclides ^{234}Th, ^{228}Th, ^{210}Po, ^{210}Pb, ^{7}Be, ^{228}Ra, ^{226}Ra, ^{224}Ra, ^{223}Ra and ^{222}Rn. For the determination of other nuclides key references have been listed in Tables 13-1 and 13-2.

A general review of chemical procedures for the separation of U-series nuclides has been given by *Lally* (1992) and for mass spectrometric determinations by *Chen et al.* (1992).

Other useful reviews and combined procedures are given in *Bhat et al.* (1969), *Baker* (1984), *Baskaran et al.* (1993) and *Luo et al.* (1995).

General remarks – The determination of radionuclides in seawater requires specialized training and equipment. These determinations are not to be undertaken without proper equipment and commitment. Because of the wide range of activities present in the ocean and the different uses that will be made of the data, each procedure should be researched adequately before its adoption. Ideally, a scientist seeking to begin making these measurements should spend time in the field and laboratory with a group that makes these measurements routinely. The procedures we report are not rigid, but are intended as a guide to the methods that are available. In most cases the procedure adopted may be somewhat modified from the specific procedures outlined here.

Table 13-1. Radioisotopes measured in seawater with reference to methods.

Isotopes	Sampling method	Typical sample volume (L)	Analysis	Reference	Section[a]
U/Th decay series					
Mobile elements: tracers for water mixing and transport					
^{238}U, ^{235}U, ^{234}U	$Fe(OH)_3$ prec		alpha counting	*Bhat et al.,* 1969	nm
	$Fe(OH)_3$ prec		TIMS	*Chen et al.,* 1986, 1992	nm
	no preconcentration		ICP-MS	*Klinkhammer and Palmer,* 1991	nm
^{228}Ra	MnO_2	50–1000	gamma	*Moore et al.,* 1985	13.8.5.3
	MnO_2	50–1000	ingrowth	*Moore et al.,* 1985	13.8.5.2
	$BaSO_4$	50–1000	gamma	*Reyss et al.,* 1995	13.6.3.3,13.8.5.3
	$BaSO_4$	50–1000	ingrowth	*Moore,* 1969	13.8.5.2
	$PbSO_4$	2–5	alpha spectrometry	*Hancock and Martin,* 1991	13.8.5.5
	review			*Orr,* 1988	
^{226}Ra	MnO_2	20	Rn-emanation	*Moore et al.,* 1985	13.8.5.1
	MnO_2	20	gamma	*Moore,* 1984	13.8.5.3
	$BaSO_4$	20	gamma	*Reyss et al.,* 1995	13.6.3.3, 13.8.5.3
	$PbSO_4$	2–5	alpha spectrometry	*Hancock and Martin,* 1991	13.8.5.5
		0.1	TIMS	*Volpe et al.,* 1991	nm
^{224}Ra	MnO_2	100–400	gamma	*Levy and Moore,* 1985	13.8.5
	MnO_2	100–400	alpha scintillation	*Rama et al.,* 1987	13.8.5
$^{223}Ra + {}^{224}Ra$	MnO_2	100–400	alpha scintillation	*Moore and Arnold,* 1996	13.8.5.4
	$PbSO_4$	2–5	alpha spectrometry	*Hancock and Martin,* 1991	13.8.5.5
^{227}Ac	$Fe(OH)_3/MnO_2$	500	alpha	*Nozaki,* 1993	nm
^{222}Rn	cool trap	20	alpha scintillation	*Key et al.,* 1979 *Mathieu et al.,* 1988	13.8.5.1
Particle-reactive elements: tracers for particle transport					
$^{234}Th (+ {}^{228}Th)$	$Fe(OH)_3$ prec	20	alpha and beta	*Coale and Bruland,* 1985	13.8.1,13.8.4
	MnO_2 cartridge	300	gamma	*Hartman and Buesseler,* 1994	13.6.3.1
	MnO_2	20	beta		13.6.3.2
	diatomite	20	liquid scintillation	*Kersten et al., Pates et al.,* 1992	nm
	$Fe(OH)_3$ prec		gamma counting	*Baskaran et al.,* 1992	13.6.3.4

Isotopes	Sampling method	Typical sample volume (L)	Analysis	Reference	Section[a]
$^{230}Th + {}^{232}Th + {}^{228}Th + {}^{231}Pa$	MnO_2	1000	alpha	*Cochran et al., 1987* *Buesseler et al., 1992* *Rutgers vd Loeff and Berger,* 1993	nm
	$Fe(OH)_3$ prec	200–500	alpha	*Nozaki and Nakanishi, 1985,* *Nozaki, 1993*	nm
$^{230}Th + {}^{232}Th$	$Fe(OH)_3$ prec	10	TIMS	*Chen et al., 1986* *Moran et al., 1995*	nm
			ICP-MS	*Shaw and Francois, 1991*	
^{231}Pa	Eichrom-TRU resin	30 10	ICP-MS TIMS	*Francois,* pers. comm. *Picket et al., 1994*	nm
$^{210}Po, {}^{210}Pb$	$Fe(OH)_3$ prec	20	autodeposition ingrowth	*Fleer and Bacon, 1984*	13.8.2, 13.8.4
	Co-APDC	20		*Fleer and Bacon, 1984*	nm
Cosmogenic nuclides					
^{14}C	acid, vacuum extraction	100	beta counting accelerator MS	*Schoch et al., 1980*	nm nm
^{10}Be	$Fe(OH)_3/$ $Mg(OH)_2$	20	accelerator MS	*Southon et al., 1982*	nm
^{7}Be	$Fe(OH)_3$	20	gamma counting	*Baskaran and Santschi, 1993*	13.8.3
	Fe-fibre		gamma counting	*Lee et al., 1991* *Kadko and Olsen, 1996*	nm

[a] nm: no method description in this chapter.

Table 13-2. Radioisotopes measured in particles with reference to methods.

Isotope	Method	Reference	Section[a]
Methods including acid digestion			
Uranium, thorium and protactinium isotopes	acid digestion ion exchange electroplating alpha + beta counting	*Anderson and Fleer, 1982*	13.7
^{210}Po	acid digestion autodeposition	*Anderson and Fleer, 1982* *Fleer and Bacon, 1984*	nm 13.7
^{210}Pb	ingrowth of ^{210}Po	*Fleer and Bacon, 1984*	13.7
Non-destructive methods to count filters			
^{210}Pb	gamma counting	*Yokoyama and Nguyen, 1980* *Cutshall et al., 1983* *Appleby et al., 1986*	13.6.2.1
Radium isotopes ^{234}Th	gamma counting beta counting	*Yokoyama and Nguyen, 1980*	13.8.5 13.6.2.2

[a] nm: no method description in this chapter.

13.2   Sampling

13.2.1   Large-volume sampling

The highest activity of all natural radionuclides dissolved in seawater (at a salinity of 35) is that of ^{238}U and ^{234}U with about 2.5 dpm/L (decays per minute; 60 dpm = 1 Bq). If all ^{238}U activity in a 1 L seawater sample could be counted for 1 d with 100 % efficiency, this would yield 3600 counts with a 1-sigma counting error of 1/3600 or 1.7 % (Section 13.9). Most activities are 2–4 orders of magnitude lower, and many collection and counting efficiencies are far below 100 %. All analytical procedures based on counting decay events therefore require large sample volumes (usually 20 L to several m^3) to obtain good counting statistics within an acceptable counting time. Substantially reduced sample volumes (1 L or less) can only be achieved with mass spectrometry, which is a powerful alternative for nuclides with half-lives well above 70 years (*Chen et al.,* 1992, compare with Section 13.3.1). For all isotopes with shorter half-lives there is no prospect for miniaturization, and there appears to be no way around large-volume sampling. We can select one of the following options (compare with Chapter 1):

1. *Rosette sampling* (up to 30 L) is most convenient, rapid and provides exact depth and knowledge of other parameters.
2. *Gerard sampling* (up to 400 L). The bottles have to be rinsed with a yoyo-technique in order to obtain water from the desired depth (*Roether,* 1971). There is a serious risk of exchange during recovery (*Broecker et al.,* 1986) due to non-tight closure of the lids, which should be checked with, *e.g.,* silicate measurements. Radionuclide sampling with 250 L PVC bottles has been described by *Nozaki* (1993).
3. *In situ* pumps (Fig. 13-1) combined with (MnO$_2$-coated) absorbers (Section 13.2.3.1) form the only alternative for subsurface sample volumes >400 L (see *Buesseler et al.,* 1995, *Rutgers van der Loeff and Berger,* 1993). The method is shiptime consuming, and a risk of loss of particles during recovery and of sticking of particles to the baffles in front of the filters (*Shimmield,* personal communication) have to be considered. We found a reassuring correspondence between particulate ^{234}Th profiles obtained with Rosette casts and with *in situ* pumps.

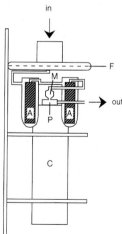

Fig. 13-1. COSS *in situ* stand-alone pump with 293 mm filter (F), two MnO$_2$ absorbers, (A), flowmeter (M), time-programmed pump (P), and control and power unit (C). Height approx. 1 m.

4. *Surface seawater supply* is a convenient underway sampling of surface water. However, there is a risk of artefacts, especially at insufficient throughput through long lines and clogging inputs, which can affect both the dissolved and particulate phase (*Buesseler*, personal communication). A hose attached to the CTD can provide detailed samples from the upper 100 m.

13.2.2 Particulate fraction

Filtration. Except for the *in situ* pumps, large-volume filtration has to be performed on board, for which a variety of pressure or suction techniques is available (compare with Chapter 2 and Chapter 12, Section 12.6). In order to prevent the loss of any particulate or dissolved activity, the risk of settling of particles and adsorption/adhesion to walls of the sampling bottles has to be considered. If a water pump is used before the filter, it should be carefully checked, *e.g.*, by microscopy, whether the pump leaves the particles intact. Centrifugal pumps are notoriously harmful to plankton. An elegant filtration method has been described by *Dehairs et al.* (1997). A 30 L calibrated Perspex cylinder with a conical bottom is filled with the sample obtained from a Rosette cast. The outlet is connected to a 142 mm Teflon filtration unit. Controlled (50–100 kPa) air pressure is applied to a series of these units from a simple small compressor.

Ultrafiltration. There is increasing interest in the measurement of radionuclides in colloids, especially since it has been suggested that particle-reactive nuclides such as Th can be used as *in situ* coagulometers (*Honeyman and Santschi*, 1989). These measurements are beyond the scope of this chapter; the reader is referred to *Baskaran et al.*, 1992, *Moran and Buesseler*, 1993, and to an intercalibration conducted by *Buesseler et al.* (1996).

13.2.3 Dissolved fraction

Radionuclides are usually extracted from filtrates of discrete samples (from the Rosette sampler, Gerard bottles or from the ship's seawater supply) by coprecipitation as $Fe(OH)_3$, $Mg(OH)_2$, MnO_2, $BaSO_4$, $PbSO_4$ or Co-APDC. If quantitative recovery cannot be guaranteed, yield tracers are used. The sample volume has to be known accurately. Large volumes can be metered with a water meter (approximately 1 % error). Samples of approximately 25 L can conveniently be weighed on board with a balance (precision approximately 50 g).

13.2.3.1 Extraction with MnO_2-coated fibres

A widely used method to extract radionuclides is to run the sample through columns filled with MnO_2-coated fibres. Such columns or cartridges can be made to remove Th, Pa, Ra, Ac (not U) with high efficiency from large volumes (several m^3) of seawater. As the efficiencies are seldom 100 %, two identical cartridges are used in line. The efficiency, E, can then be calculated from the ratio of activities A and B measured in the two cartridges:

$$E = 1 - B/A,$$

(the 'cartridge formula'). The assumption that both cartridges are identical and have equal efficiencies becomes especially critical at lower efficiencies. Cartridges should therefore always be used in pairs from the same batch. The formula cannot be used if there is a risk of transfer of MnO_2 with associated activity from the first to the second cartridge, either by physical entrainment of MnO_2 particles or by MnO_2 reduction in anoxic water masses.

Preparation of Mn-fibre and MnO₂-coated cartridges

Numerous procedures for preparing fibre or cartridges containing particles of MnO_2 have been published. Some earlier procedures prepared the MnO_2 and physically forced the particles through a fibre cartridge. Most procedures now in use chemically bond MnO_2 to acrylic or polypropylene fibre strands. Acrylic fibre is more reactive towards $KMnO_4$ and can be prepared with a higher absorption efficiency than polypropylene. Polypropylene in some cases may have a lower blank for Th isotopes (*Buesseler et al.,* 1992) and, due to the lower reactivity, it is less critical to keep the coating process under control. We describe here what we consider the best means of preparing Mn-fibres of each type.

Preparation of MnO₂-coated acrylic fibre (*Moore,* 1976)

Acrylic fibre (*e.g.,* Monsanto 'Acrilan', 3.0 denier, Type B-16) is treated with a hot solution of saturated $KMnO_4$ for ≈ 10 min. The $KMnO_4$ oxidizes specific sites on the acrylic molecule and deposits MnO_2 at these sites. The prepared fibre is washed and is ready for use. This process produces Mn-fibre having sub-micrometre sized particles of MnO_2 chemically bonded to the fibre. The MnO_2 constitutes 8–10 % by mass of the Mn-fibre. This procedure may be conducted in a beaker (100 mL) or vat (20 L) scale. At the larger scale it is important to recognize that the reaction is exothermic and proceeds very rapidly at $> 80 \,°C$. Unless excessive build-up of heat is prevented or dissipated quickly once it occurs, the entire fibre may be oxidized and spattering of hot $KMnO_4$ solution may injure the operators. It is recommended that small-scale preparations be made initially so the operators may recognize warning signs of rapid oxidation. Heavy aprons, rubber gloves and eye protection are essential for all operators.

Heat a volume of water to 80 °C and add enough $KMnO_4$ to produce a 0.5 mol/L solution. (*Note:* If a large volume is being prepared, the reaction vessel should be placed in a sand-bath and the solution heated with immersion heaters.) This solution will be approximately saturated at a temperature of 75 °C. Add acrylic fibre to the solution and monitor the temperature to keep it in the range 70–80 °C. (*Note*: If a large volume is being prepared, a plastic dipping basket is useful for transferring the prepared Mn-fibre to the wash-bath.) Gently mix the fibre with the solution. In a large vat, this is accomplished using pieces of Plexiglas as large stirring rods to rotate the mass of fibre through the bath. The reaction may cause the fibre mass to rise above the surface. If this occurs, use the stirring rods to force it into the solution. As the reaction proceeds the colour of the fibre will turn from white to orange to black. We refer to partially reacted fibre as 'orangutan hair' and prepared fibre as 'gorilla fur'. Check the colour of the fibre by rinsing a small amount with water to remove excess $KMnO_4$. As the reaction nears completion, you may notice steam escaping from the fibre, especially if the fibre/solution ratio is high. This is not a problem if the heat build-up in the solution is monitored closely. Mixing the steaming fibre into the solution should quell the steaming. When the fibre is black, remove it from the solution and allow it to drain as much as possible while squeezing excess solution back into the reaction vessel. As the fibre begins steaming, transfer it to the wash-bath immediately. Several washings will be necessary to remove excess $KMnO_4$ as well as MnO_2 that did not attach to the fibre. For large volume preparation, a laundry washing machine with a hand wringer is useful. The Mn-fibre should be partially dried and stored damp.

Several batches of Mn-fibre may be prepared from the initial solution. As the MnO_4^- activity of the solution is reduced, the reaction speed slows and each batch requires more time. To compensate, additional solid $KMnO_4$ may be added to the original solution. However, the common ion effect will ultimately render the solution ineffective. The decision when to discard the solution is a trade off between the rate of preparation and the cost of preparing a new solution.

A low-temperature method of preparing Mn-fibre from $KMnO_4$ is also effective. In this case the acrylic fibre is placed in a saturated $KMnO_4$ solution for several days at 30–35 °C. This method has been used to prepare cartridges of Mn-fibre from prewoven acrylic filter cartridges (*Michel et al.*, 1981). The solution may be circulated through the cartridges to distribute the MnO_2 more uniformly. A problem with this technique in some cases is contamination of commercial acrylic cartridges leading to unacceptable blanks (*Moore and Santschi*, 1986).

Preparation of MnO_2-coated polypropylene filter cartridges (*Buesseler et al.*, 1992; *Hartman and Buesseler*, 1994)

Polypropylene filter cartridges are produced by winding (CUNO Micro-Wynd) or thermally bonding (Hytrex) the fibre. For *in situ* pumps and large sample volumes (approximately 1 m³) 25 cm long cartridges are used. Smaller ones (approximately 10 cm length) are adequate for sample volumes up to several hundred litres and the Hytrex-type of this length can be squeezed to a small volume for non-destructive gamma counting (*Hartman and Buesseler*, 1994). The cartridges are wetted by flushing with detergent solution, rinsed with demineralized water, immersed overnight in a saturated $KMnO_4$ solution to produce the MnO_2 coating and rinsed with Milli-Q water. Each step is described here in more detail:

Wetting. This is essential to produce a uniform coating. Use a strong detergent solution (*e.g.*, 5 mL 'Decon' per litre of demineralized water). Flush the cartridge in the holder and apply a vacuum to remove air from the fibre. Repeat until production of air bubbles declines. Rinse with demineralized water, keeping the outlet tube at a higher level than the cartridge to prevent the production of new air bubbles. The saponification steps used in previous procedures can be left out.

Coating. Prepare a $KMnO_4$ solution saturated at 45 °C. (After each batch, $KMnO_4$ has to be added to make up for the chemicals removed.) Immerse the rinsed cartridges directly (do not allow water to drip off) in the $KMnO_4$ bath and leave them overnight.

Rinsing. Let the cartridges drip off, and rinse them in a series of holders with Milli-Q water. Turn them upside down and pass them on regularly ahead in the rinsing sequence to optimize the rinsing effect and to save water. After the final rinse the cartridges are allowed to drip dry and packed in plastic bags while still wet. Note the coating date and batch number. The cartridges can be stored for several months.

13.2.3.2   Removal of adsorbed elements from Mn-fibre

If non-destructive counting techniques (Section 13.6.3.1) cannot be used, the adsorbed elements have to be removed from the Mn-fibre. This may be accomplished by combustion or leaching. Combustion is rapid but produces a very fine residue that may prove difficult to recover quantitatively. Combustion of acrylic releases cyanide gas and should only be undertaken in a fume hood. For procedures to isolate radionuclides from the ash see *Cochran et al.* (1987) and *Buesseler et al.* (1992).

A number of reducing agents such as hydroxylamine hydrochloride and HCl will convert Mn^{4+} into Mn^{2+} and release adsorbed elements from the fibre, but the acid concentration should be sufficient (for Th at least 3 mol/L HCl) to keep the radioisotopes in solution and achieve isotopic equilibrium with added yield tracers. A convenient way to ensure the quantitative recovery of elements from Mn-fibre without producing large volumes of solution is to leach the fibre by refluxing HCl through it in a Soxhlet extraction apparatus. The reaction between MnO_2 and HCl produces chlorine gas and should only be undertaken in a fume hood. The Mn-fibre is placed in the Soxhlet glass thimble and covered with concentrated HCl for several hours. The HCl reduces Mn^{4+} into Mn^{2+} and releases the adsorbed nuclides. Dilute (6 mol/L) HCl is added to the extraction vessel to induce siphoning to the boiling flask and the system is refluxed until the fibre in the extraction vessel is clear (2–4 h). During the extraction the solution should stabilize at close to 6 mol/L HCl at 108 °C.

13.3   Analytical options

13.3.1   Radiometry *versus* mass spectrometry

In a theoretical comparison between alpha spectrometry and mass spectrometry, *Chen et al.* (1992) showed that alpha spectrometry gives better precision for isotopes with half-lives of up to 70 years, and similar arguments hold for beta and gamma counting. The practical break-even point is at considerably longer half-lives. Accelerator mass spectrometry (AMS) yields a precision which is already far superior to beta counting for ^{14}C (5730 y half-life). Thermal ionisation mass spectrometry (TIMS) is for ^{226}Ra (1600 y half-life), an element with a high ionisation efficiency, superior in precision to alpha spectrometry (*Volpe et al.,* 1991). The break-even point for established ICP-MS methods is close to the half-life of ^{230}Th (75 400 y, *Shaw and Francois,* 1991), but it is dropping to about 500 y as a result of technical improvements (*Francois*, personal communication).

13.3.2   Radioactivity measurements

Alpha spectrometry is characterized by good isotope separation, uniform (*i.e.,* energy- and isotope-independent) detector efficiency and a very low background (typically of the order of 0.001–0.003 cpm (counts per minute) for the 4–8 MeV energy range of a new surface barrier detector, increasing with time as a result of the accumulation of recoil products from measured samples). It allows precise measurements at low activities, and easy calibration with yield tracers. Alpha particles have a very short range. They are absorbed or slowed down by interaction with sample material or by impurities on the source, the flat planchet with the purified element plated on it. The technique is therefore not practicle for non-destructive measurements. The best results in terms of efficiency and energy resolution are obtained with clean sources prepared from well-purified samples.

Beta counting is not isotope specific. It requires relatively simple equipment. A gas-flow anti-coincidence counter with 10 cm lead shielding is frequently used, giving background values of 0.15–0.20 cpm for a source of 25 mm diameter (Risø National Laboratory, Roskilde, Denmark). It is the preferred method for ^{234}Th, can be used with non-destructive techniques (Sections 13.6.2.2 and 13.6.3.2), and can be used on-board ship.

Gamma spectrometry allows identification of a wide range of isotopes without chemical purification and is the ideal method for non-destructive techniques. Inherent (Compton) background and self-absorption by the sample, energy- and geometry-dependent detector efficiency, coupled for some elements with low gamma branching ratios, reduce the sensitivity and accuracy of the method. The overall efficiency is generally low compared with other techniques. The equipment can be used on board ship (*Buesseler et al.*, 1992).

Alpha scintillation is used for the detection of radioactive radon isotopes (Section 13.8.5.1).

Liquid scintillation has been used with success as an alternative method for ^{234}Th determination (*Pates et al.*, 1992; *Kersten et al.*, 1998).

Mass spectrometry

ICP-MS is being used with increasing success for nuclides with half-lives of > approximately 10^4 years: ^{238}U, ^{234}U (*Klinkhammer and Palmer*, 1991), ^{235}U and ^{230}Th (*Shaw and Francois*, 1991) and methods are being developed for ^{231}Pa (*Francois*, personal communication).

TIMS, AMS. For references to this more specialized equipment we refer to Table 13-1.

13.4 Special requirements for laboratory

1. *Permission to work with radioactive materials.* The analysis of natural radionuclides requires in many cases the addition of spikes of artificial nuclides as yield tracers. Radioactive material is also necessary for calibration purposes. Although very low activities, only equivalent to the natural activities expected in the samples to be analysed, are needed, permission to work with these nuclides has to be obtained in most countries from local authorities to comply with health and safety regulations governing the protection of employees at their working place in the individual country. For the analyses described in this chapter, Table 13-3 gives guidelines for the permissions to be applied for. Companies selling spikes may require a copy of the permission obtained.

2. *Fume hoods.* When suspended particles are to be analysed, and non-destructive techniques (Section 13.6) are not a satisfactory option, a fume hood is required which is designed for work with perchloric and hydrofluoric acid (*i.e.*, with water rinsing of the outlet channels).

3. *Clean-lab.* For the methods described in this chapter 'clean-lab' facilities are not required. However, if samples are prepared for analysis by TIMS or ICP-MS, 'clean-lab' facilities become essential (see Chapter 12).

Table 13-3. Guidelines for the amounts of radioactive material for which permission has to be obtained in a laboratory where ^{234}Th, ^{228}Th, ^{210}Po, ^{210}Pb and ^{226}Ra are determined in seawater.

To determine	You need: nuclide /spike	Function	Activity per sample (dpm)	Maximum activity in laboratory (kBq)
^{234}Th	^{230}Th	yield monitor	16 (0.27 Bq)	15
	natural U (U$_3$O$_8$)	calibration	60 (1 Bq)	15
	mid-depth sample*	standardization		
^{228}Th	^{230}Th	yield monitor	16 (0.27 Bq)	15
	^{229}Th	calibration	16 (0.27 Bq)	15
	UREM-11	calibration	60 (1 Bq)	15
^{210}Po, ^{210}Pb	^{209}Po (^{208}Po*)	yield monitor	6–60 dpm (0.1–1 Bq)	150
	^{210}Pb	calibration		100
	UREM-11/DL1A	calibration	60 (1 Bq)	15
^{226}Ra	^{226}Ra	calibration		15
	DL1A	calibration	60 (1 Bq)	15

* Alternative.

13.5   Reagents, spikes and standards

1. *FeCl$_3$ solution (50 g/L Fe)*: 121 g of FeCl$_3 \cdot$ 6H$_2$O are dissolved in 500 mL of 8 mol/L HCl. In a 1 L glass separating funnel, the Fe is extracted in three times 167 mL of isopropyl ether and the combined extracts are then back extracted with 2 times 250 mL of 0.1 mol/L HCl (separate overnight).
2. *Pb yield tracer (20 g/L Pb)*: Lead chemicals and standard materials contain ^{210}Pb. For sufficiently low background levels, carry out a digestion of a certified lead ore such as the sulphide ore galena (*Fleer*, personal communication).
3. *KMnO$_4$ solution (60 g/L)*: 15 g of KMnO$_4$ are dissolved and made up to 250 mL with distilled water. This concentration is just below saturation at 20 °C.
4. *Manganese(II)chloride*: 40 g of MnCl$_2 \cdot$5H$_2$O are dissolved and made up to 100 mL with distilled water. This solution is the same reagent as is used for the determination of oxygen (Winkler titration; Chapter 4).
5. *Barium chloride*: Dissolve 25 g of BaCO$_3$ in dilute HCl, dilute gravimetrically with demineralized water to 5 kg. Prepare weighed aliquots, each containing the equivalent of approximately 500 mg of BaCO$_3$, in tightly capped 100 mL polyethylene bottles.
6. *Uranium solution*: Dry U$_3$O$_8$ at 105–110 °C for 1 h. Dissolve a weighed amount (Note: Be very careful not to inhale any of this radioactive material) in 8 mol/L HNO$_3$ and dilute

gravimetrically with distilled water to approximately 4 mol/L in HNO_3. Determine U (μg/g) using the U content of U_3O_8 (714.09/842.09 g/g):

$$U \ (Bq/g) = U \ (dpm/g)/60 = U \ (\mu g/g) \cdot 0.7414/60$$

Note: In open ocean water below approximately 200 m and more than 100 m above the sea-floor (*Bacon and Rutgers van der Loeff*, 1989), total ^{234}Th can be assumed to be in equilibrium with ^{238}U, the activity of which, A_U, can be derived from the salinity; A_U (dpm/L) = 0.0704 · salinity, based on the average uranium (^{238}U + ^{235}U) concentration in seawater of 3.238 ng/g normalized to salinity 35 (*Chen et al.*, 1986). A deep water sample or an acidified aged ($\geq$6 months) sample of filtered seawater can therefore be used for calibration. An independent check is however preferable.

7. *Anion-exchange resin*: Make a slurry of Biorad AG 1-X8 resin (100–200 mesh) in distilled water. Pipette into the exchange column. When using cheaper 'Dowex' resin make up small amounts to prevent size fractionation. The slurry may be stored in closed bottle, but unacidified slurry may grow bacteria. It is common practice to discard the resin after use, but some laboratories develop cleaning procedures to recycle it.

8. ^{230}*Th spike 16 dpm/mL (267 Bq/L)*: Prepare a stock standard by diluting an uncalibrated spike in 8 mol/L HNO_3. Prepare the spike by further dilution in 8 mol/L HNO_3 to an activity of approximately 16 dpm/mL. If required, this spike can be calibrated against a ^{229}Th spike, which in turn is calibrated against a certified uranium/thorium ore such as UREM-11.

9. ^{209}*Po tracer:* Prepare a stock solution in 2 mol/L HNO_3. Add 200 mg/L of Bi carrier as $Bi(NO_3)_2$. Prepare a spike solution in 2 mol/L HNO_3 of approximately 20 dpm/mL (0.33 Bq/mL) (the half-life of ^{209}Po is 103 y). Add 200 mg/L of Bi carrier as $Bi(NO_3)_2$. ^{208}Po can be used instead of ^{209}Po as a yield tracer (note the 2.9 y half-life), but requires a more elaborate tailing correction (*Fleer and Bacon*, 1984).

   Calibration: Use a certified reference sediment (UREM-11, DL1A). Weigh 100–150 mg in a Teflon beaker, add ^{209}Po (or alternatively ^{208}Po) tracer (weigh) with an activity of about twice the ^{210}Po activity (in secular equilibrium with ^{226}Ra), and perform a total digestion (Section 13.7.2 or *Anderson and Fleer*, 1982). Plate Po, count and calculate (*Fleer and Bacon*, 1984) the tracer activity. A convenient alternative is the use of a certified ^{210}Pb standard solution (*Fleer and Bacon*, 1984).

10. *Certified uranium ore:* UREM-11 (*Hansem and Ring*, 1983) is a reference ore certified for ^{238}U with ^{230}Th in secular equilibrium, distributed by the Council for Mineral Technology, Private Bag X3015, Randburg 2125, South Africa. The homogeneity is satisfactory. DL1A is a reference ore certified for ^{232}Th and ^{238}U (with ^{226}Ra, ^{210}Pb and ^{210}Po in secular equilibrium), distributed by Mineral Sciences Laboratories, Canadian Centre for Mineral and Energy Technology, Energy Mines and Resources Canada, Ottawa K1A OG1, Canada. For small subsamples problems with inhomogeneity have been reported.

13.6   Instrumental techniques without radiochemical purification

13.6.1   Destructive *versus* non-destructive techniques

It is sometimes possible to measure the radioisotope composition of a sample without acid digestion and/or ion exchange separation. The self-absorption of radiation by the sample has to be accounted for, being lowest for high-energy gamma radiation, appreciable for low-energy gammas and betas and prohibitively high for alpha radiation. Most methods are based on gamma spectrometry, which has the disadvantage of the usually lower gamma sensitivity and accuracy (see Section 13.3.2) compared with radioanalytical methods followed by alpha and beta counting. Beta counting is not isotope-specific and self-absorption by the sample is always significant. The next sections describe methods that take advantage of non-destructive counting, especially for the short-lived isotopes that have to be measured on-board ship.

13.6.2   Particulate matter: direct counting of filters

13.6.2.1   Gamma spectrometry

Marine particulate matter contains relatively high activities of a number of particle-reactive radionuclides. This makes high-resolution gamma spectrometry an efficient way to analyse particulate samples. Samples are obtained by filtering large volumes of seawater or collecting settling particles using a sediment trap. Because the technique requires little sample pretreatment, it can be non-destructive, allowing additional measurements of the same samples.

Yokoyama and Nguyen (1980) described a method for the determination of ^{210}Pb, ^{234}Th and ^{226}Ra on filters without acid digestion. The filters are simply dried, folded and placed on the detector in a reproducible geometry. For highest sensitivity measurements, a germanium crystal with a well is recommended. *Moore and Dymond* (1988, 1991) used a germanium well detector to measure ^{210}Pb and ^{226}Ra in samples from deep-sea sediment traps.

The most useful energies for ^{234}Th are 63 keV (3.8 % intensity) and 92.4 and 92.8 keV (5.4 % intensity for the combined peaks). For ^{210}Pb a single peak at 46 keV with an intensity of 4.05 % is used. The measurement of ^{226}Ra and ^{228}Ra by gamma spectrometry is discussed in Section 13.8.5.3.

Because ^{226}Ra is usually determined using a daughter of ^{222}Rn, the sample must be sealed to prevent Rn escape and aged 2 weeks to allow ^{222}Rn to equilibrate with ^{226}Ra. *Moore and Dymond* (1991) covered sediment trap samples in a vial with a layer of epoxy cement to prevent Rn escaping. As will be discussed in Section 13.8.5.3, it is best to calibrate the detector using standards of known activity measured in the same geometry. For calibration and background of gamma detectors see *Moore* (1984) and *Buesseler et al.* (1992). For large or compressed samples, a self-absorption correction may be necessary. *Cutshall et al.* (1983) have described a method for making this correction. Samples from sediment traps loosely packed in a vial have low self-absorption coefficients; these may be approximated by standards dissolved in water.

13.6.2.2 Beta counting

^{234}Th can be counted by non-destructive beta counting. In the open ocean, ^{234}Th activity usually overwhelms other isotopes that might contribute to the beta signal from suspended particles, such as ^{40}K, ^{226}Ra or ^{210}Pb. This should be checked occasionally by following the decay over several months. If the activity does not decline with the half-life of ^{234}Th (24.1 d) this is an indication of contributions from other nuclides.

^{234}Th itself decays with relatively weak betas (maximum beta energy 0.20 MeV) to ^{234m}Pa, which is a high-energy beta emitter (maximum beta energy 2.29 MeV, half-life 1.17 min). Thus, if a filter is dried and carefully folded (possible with polycarbonate, *e.g.*, Nuclepore filters, not with membrane filters, which will break) to fit the beta detector holder and covered with a thin foil, the activity measured is due mainly to ^{234m}Pa, with some contribution from ^{234}Th.

Self-absorption. Beta particles have a continuous energy spectrum extending from zero energy up to a maximum energy E_{max}. If a flat beta source is measured with a thin absorber of thickness L between the source and detector, the count rate I is very closely represented by

$$I(L) = I(0) \, e^{-\alpha L} \tag{13-1}$$

where $I(0)$ is the count rate measured in the absence of the absorber and α is an attenuation factor which can be derived from E_{max} and the density of the absorber (*e.g.*, *Tsoulfanidis*, 1995). If the same activity is homogeneously distributed in the thin absorber, a good approximation in the case of a multiply-folded filter, the count rate measured with a detector on top of the absorber can be described just as in case of the self-absorption of gamma radiation (*Cutshall et al.*, 1983):

$$I(L) = I(0) \quad I(L) = I(0) \, \frac{1 - e^{-\alpha L}}{\alpha L} = I(0) \cdot E_t = I(0) \cdot E_t \tag{13-2}$$

where we call E_t the transmission efficiency.

If filters have to be measured with a suspended load that is at least comparable to the filter mass, E_t has to be calculated with Eq. (13-2). The parameter αL has to be measured for each individual sample from measurements of a (strong) beta source with and without the filter in-between, using Eq. (13-1). As the Eqs. (13-1) and (13-2) only hold for the energy distribution of a single beta decay, all beta measurements have to be made with a cover removing the weak betas of ^{234}Th. A cover of 30 mg/cm^2, approximately equivalent to two overhead sheets, transmits 82 % of ^{234m}Pa but only 4 % of ^{234}Th betas.

However, if the absorber has reproducible geometry and thickness, as in the case of a carefully folded filter with particle load << filter mass, or in the case of a filter with a reproducible load of MnO$_2$ precipitate (see Section 13.6.3.2), E_t will have a constant value, which can be determined with standard filters. Measurements can be made with a thin cover, improving counting statistics.

Procedure. 20–50 L are filtered over 142 mm polycarbonate filters with 1 μm pore size. The filters are drained by suction, folded twice in two, air dried and carefully folded four more times to produce a 18 x 18 mm, 64 sheet thick package which is wrapped in thin (*e.g.*, 0.01 mm) plastic (polyester or polyethylene) foil, and counted directly in a beta counter (count rate I_F).

Blanks. Polycarbonate (Nuclepore) filters contain some ^{137}Cs, which contributes to the blank. The count rate I_{bl} of blank filters (including instrument background) is typically around 0.5 cpm, but can vary widely between batches and between pore sizes and definitely has to be checked in advance.

Calibration. Standard filters can be prepared by making a Fe(OH)$_3$ precipitate of a U standard solution. In a small Teflon beaker add 50 µl of 50 mg/mL FeCl$_3$ solution to a weighed uranium spike containing an accurately known activity A_U of approximately 100 dpm ^{238}U (135 µg U), mix, add dilute (1 mol/L) NH$_3$ solution while mixing until the precipitate remains (pH is now 8–8.5). Put the filter on graph paper and pipette spiked slurry dropwise in a regular pattern onto the filter. Dry in a fume hood, fold as for the other filters and count (count rate I_S). This standard filter is sufficient for calibration of filters from subsurface ocean water, when suspended load << filter mass.

For filters with suspended load comparable to or larger than the filter mass, αL is determined from the measurement of a strong source (*e.g.*, a planchet prepared from a natural uranium spike) (Section 13.8.1), with (count rate I_2) and without (count rate I_1) the filter in-between, using a thick (30 mg/cm^2) cover. Analogous to Eq. (13-1) we have

$$I_2 = I_F + (I_1 - I_{bg})\, e^{-\alpha L} \tag{13-3}$$

where I_{bg} is the background count rate, giving

$$\alpha L = -\ln\left(\frac{I_2 - I_F}{I_1 - I_{bg}}\right)$$

which is used to determine the transmission efficiency of the filter E_F. The transmission efficiency of a blank filter (1 µm pore size polycarbonate), E_{bl}, is typically around 80 %.

Calculation. The activity A_F on a filter with count rate I_F is given by

$$A_F = \frac{I_F - I_{bl}}{I_S - I_{bl}}\, A_U\, \frac{E_{bl}}{E_F}\, e^{\lambda Th \Delta t}$$

where $\lambda Th = \dfrac{\ln(2)}{t_{1/2}} = 0.0287\,\mathrm{d}^{-1}$

and Δt is the time lapse between filtration and counting. For low filter loadings the self-absorption correction term E_{bl}/E_F is unity.

13.6.3 Dissolved fraction: counting of fibre or precipitate

13.6.3.1 MnO$_2$-coated cartridges

^{234}Th activity on cartridges can be determined by gamma spectrometry. Three methods have been described:

- pressing to obtain a well-defined geometry: *Hartman and Buesseler* (1994), *Buesseler et al.,* 1995;
- melting (*Buesseler et al.,* 1992); and

– ashing (*Cochran et al.*, 1987; *Buesseler et al.*, 1992; *Baskaran et al.*, 1993).
The measurement of radium isotopes collected on fibres is described in Section 13.8.5.

13.6.3.2   MnO$_2$ precipitate

To a weighed 20 L aliquot of filtered seawater add 6 drops of concentrated ammonia solution (25 % m/m NH$_3$) and 250 μL of concentrated KMnO$_4$ solution, followed after mixing by 100 μL of a concentrated MnCl$_2$ solution (use the Winkler I reagent, 400 g/L MnCl$_2$ · 5H$_2$O; (see Chapter 4). After mixing the purple colour disappears rapidly and a suspension of MnO$_2$ is formed. After 8 h, to allow for the MnO$_2$ particles to grow, the suspension is filtered over a 1 μm polycarbonate filter. Use vacuum filtration and rinse the container vigorously to bring the remaining adhering MnO$_2$ particles into the suspension and onto the filter. The filter is rinsed with distilled water, drained by suction and folded while wet in the same geometry as used for the filters containing suspended matter from seawater (Section 13.6.2.2). The folded filter is held together with a plastic paperclip and allowed to dry before it is wrapped in plastic foil. This filter is counted directly in the beta counter. As it is not necessary to stop the weak betas of ^{234}Th, all beta countings can be done with a thin cover.

Calibration: Prepare MnO$_2$ filter blanks by producing an MnO$_2$ precipitate in Milli-Q water. Count rates should be only slightly above the filter blanks (see Section 13.6.2.2). Extraction efficiency is better than 99 % (as determined from repeated extractions of the same sample), but some precipitate may stick on the walls, tubing or filter holder. The loss can be estimated by rinsing all equipment with a solution of 0.1 mol/L of H$_2$O$_2$ in 1 mol/L HCl and measuring Mn in the leach with atomic absorption spectrometry, using the spike solutions (250 μL KMnO$_4$ + 100 μL MnCl$_2$ + 1 mL dilute H$_2$O$_2$–HCl) made up with distilled water to 500 mL for calibration.

Self-absorption and overall counting efficiency of an MnO$_2$-coated filter can be determined as in Section 13.6.2.2. The transmission efficiency E_t is typically around 78 %. The overall efficiency of the entire procedure can be checked by determining ^{234}Th in a sample from mid-depth in the open ocean and comparing the result with the value expected from equilibrium with ^{238}U (*Chen et al.*, 1986).

As in the case of particles, ^{234}Th activity usually overwhelms other isotopes that might contribute to the beta signal from the MnO$_2$ precipitate. However, radium and lead are partially coprecipitated with MnO$_2$ and the contribution form ^{226}Ra and ^{210}Pb daughters (^{214}Bi and ^{210}Bi with maximum beta energies 3.27 and 1.16 MeV, respectively) has to be checked by following the decay over several months. If the activity does not decline with the half-life of ^{234}Th (24.1 d) this is an indication of contributions from other nuclides. The contribution of ^{226}Ra daughters is small as most of the intermediate ^{222}Rn escapes from the filter during storage and measurement before it decays further to beta emitting daughters. In the MnO$_2$ precipitate of a typical seawater sample containing approximately 2.5 dpm/L ^{234}Th, 0.15 dpm/L ^{226}Ra and 0.10 dpm/L ^{210}Pb, approximately 4 % of the initial count rate remains after complete decay of the initial ^{234}Th. This remaining count rate is due to the activity of ^{226}Ra and ^{210}Pb daughters and to traces of U which may have coprecipitated and produced new ^{234}Th.

Accuracy and precision: The procedure is reproducible to within 2 %, but accuracy is approximately 5 % as a result of uncertainties related to the contribution from other isotopes.

13.6.3.3   BaSO₄

To a 30–60 L sample add, while stirring, dropwise one aliquot of the barium chloride solution, using for this slow dispensing, *e.g.,* an empty ion exchange column. Transfer the Ba solution quantitatively into the sample. Mixing the sample for 1–2 h after adding the Ba facilitates the production of larger crystals of $BaSO_4$ and enhances recovery. Allow the precipitate to settle, decant and collect the precipitate by centrifugation. Rinse, dry and weigh into a tube that fits a well-type gamma detector. Determine the chemical yield gravimetrically using the $BaSO_4$: $BaCO_3$ mass ratio of 1.183. For calibration and interpretation of the gamma spectrum see Section 13.8.5.

13.6.3.4   Fe(OH)₃

Gamma counting of the $Fe(OH)_3$ precipitate after centrifugation has the disadvantage that the volume remains too large for usual well-type counters, whereas for other types the geometry of the source is not sufficiently well defined. The dried precipitate can however be measured with gamma spectrometry (*Baskaran and Santschi,* 1993).

13.7   Radiochemical methods for the separation of Th–Po–Pb

13.7.1   General analytical considerations

Silver can be used for plating of all elements mentioned here (U, Th, Po). Sheets of 0.1 mm thickness can be obtained with a protective plastic cover, which is especially convenient for Po plating.

Use Teflon beakers throughout. Clean them after use by soaking in warm 8 mol/L HNO_3, which is the best cleaning agent for Th.

The separation of isotopes by ion exchange requires repeated evaporation of acids. In order to protect fume hoods and the environment, this is preferably done by heating the beakers with an infrared lamp under a vented glass cover (Fig. 13-2).

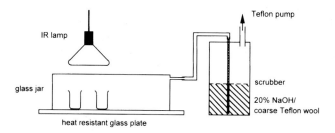

Fig. 13-2. Experimental setup for the evaporation of acids.

13.7.2   Total digestion

Filter digestion on-board is very inconvenient, and where possible non-destructive techniques (Section 13.6.2) should be considered for short-lived isotopes on long expeditions.
Acid leaching of filters is an option when it is not imperative to obtain the total latice-bound
concentrations of nuclides and their supporting parents. In all cases where non-destructive
techniques or leaching are not an acceptable option, the filter and filtered particles have to
be brought into solution after the addition of appropriate yield tracers. Procedures for the
total digestion in Teflon bombs (with or without microwave digestion) are described in
Chapter 12, Sections 12.4 and 12.6; see also *Anderson and Fleer*, 1982.

13.7.3 $Fe(OH)_3$ precipitation

Tare the balance with an empty canister. Weigh a 20 L filtrate sample. To 20 kg of filtered
seawater are added: 20 mL of concentrated HNO_3, appropriate yield tracers and 5 mL of
$FeCl_3$ solution (50 mg/mL Fe^{3+}).

Homogenize and leave to stand overnight for isotope equilibration. Neutralize with
NH_4OH solution to pH 8–8.5 (use pH paper). Allow sufficient time for the precipitate to
settle (usually 24 h), then syphon off the clear solution and take the volume of the $Fe(OH)_3$
precipitate to 10–15 mL by centrifuging the remaining suspension in a 50 cm centrifuge tube.
Wash the precipitate twice with distilled water and centrifuge again.

13.7.4   Polonium plating

Dissolve the $Fe(OH)_3$ precipitate in the minimum amount of 2 mol/L HCl. It may be
necessary to add a few (one or two) mL of 9 mol/L HCl. Then transfer quantitatively into
100 mL Teflon beaker and dilute with distilled water to approximately 40 mL. The solution
should be about 0.5 mol/L in HCl.

Heat in a sand bath on magnetic stirrer–heater at 85–95 °C (stirring balls are convenient).
Add ascorbic acid powder until the yellow colour (Fe) disappears.

Note: The amount required depends on the amount of Fe used. It takes a considerable
amount to bind 250 mg of Fe. In order to prevent adding too much, wait after each addition,
since the dissolution and complexation is slow.

Remove the protective cover from the hollow side of the silver disc, clean the disc with
ethanol and put it on the bottom of the beaker.

Note: If a disc is punched out of a sheet of silver, the borders remain a little curved.
When dropped into the sample, the disc will fall hollow side up on the bottom of the beaker.

Regularly check the colour: if slightly yellow, add some more ascorbic acid. If the colour
remains, there may be an excess of ascorbic acid being broken down. If the silver disc gets a
dark coating, it is wise to end the plating prematurely.

After 4 h, remove the disc with nylon forceps, rinse with 2 mol/L HCl into the sample,
rinse with Milli-Q water and then with ethanol. Note on the back of the sample the ID and
date of plating. Rinse the stirring ball into the sample with 2 mol/L HCl, then rinse with
Milli-Q water and store in 8 mol/L HNO_3.

The remaining solution can be used for the determination of ^{210}Pb (after ingrowth of ^{210}Po, *Fleer and Bacon,* 1984), or for the separation of Th and U (combined method, see Section 13.8.4). An alternative plating procedure, which works well when a subsequent determination of Th and U is not required, is given by *Flynn* (1968).

13.7.5   Chloride column

Take up the sample in 9 mol/L HCl. Fill a thick (15 mm ∅) column with anion-exchange resin up to 2–3 cm below the top plastic part. Drain, fill three times with 9 mol/L HCl. Discard the first eluate; collect the second and third eluates for cleaning purposes.

Note: This column serves to remove Fe (the dark brown colour) and U from the solution. The column diameter depends on the amount of Fe which has to be retained. A 15 mm ∅ column is required to separate the 250 mg of Fe used for coprecipitation from 20 kg of seawater.

Place a 100 mL Teflon beaker under the column. Add the sample. Rinse the sample beaker three times with a minimum volume (2 mL) of 9 mol/L HCl; drain each time. Elute the column three times with 9 mol/L HCl (filled to the top). Evaporate the eluate under an IR lamp.

13.7.6   Nitrate column

Dissolve the sample in 15–25 mL of 8 mol/L HNO$_3$. Allow several hours for complete dissolution, at best overnight. Heat will speed up the dissolution, but evaporation will increase the HNO$_3$ concentration to 16 mol/L. In this case add an equal volume of Milli-Q water to bring the HNO$_3$ back to 8 mol/L. Higher HNO$_3$ concentrations destroy the column, seen immediately by turning the resin into a dark red colour.

Fill a column (7 mm ∅) with resin up to 2–3 cm below the top part. Drain, fill three times with 8 mol/L HNO$_3$. Discard the first eluate and collect the second and third eluates for cleaning purposes. In the nitrate form, the anion-exchange resin retains Th, while most other elements including seasalts and Pb pass through. Pour the sample (which is now in 8 mol/L HNO$_3$) onto the column. Collect the eluate (Pb fraction) in a 50 mL PE bottle. Drain, rinse the beaker three times with a minimum volume (2–4 mL) of 8 mol/L HNO$_3$. Rinse the sides with 8 mol/L HNO$_3$, drain each time and elute the column three times with 8 mol/L HNO$_3$ (filled to the top). Ample elution is important.

The Th fraction is collected in a Teflon beaker by eluting the column three times with 9 mol/L HCl (filled to the top).

13.7.7   Thorium plating

Evaporate down the Th fraction to a drop or near dryness. Rinse down the walls with HNO$_3$ and evaporate down to near dryness with an IR lamp. This gentle drying avoids a hard cake which would be hard to dissolve in the next step. The resulting drop should be a tiny yellow or brown spot. Large spots with white or black material will result in poor plates; in this case the 8 mol/L HNO$_3$ column is repeated with a smaller column volume (about 2 mL). At this stage, the sample can be left until there is time for plating.

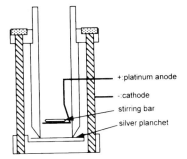

Fig. 13-3. Plating cell for the electrodeposition of U, Th and Pa on silver discs.

Preparation of plating cells: Cells for electroplating can be constructed according to Fig. 13-3, based on *Krishnaswami and Sarin* (1976). Teflon units with stirring bars laying on the platinum anode are stored immersed in 8 mol/L HNO_3. Rinse the Teflon units and put them on tissue. Clean a 23 mm diameter silver plate with ethanol, put it on stainless steel bottom plate, and put the cell together. Adjust the limiting values of the power supplies to 6 V and 0.8 A.

Take up the spot in 1 mL of 0.01 mol/L HNO_3. Be sure all material gets detached from the bottom of the beaker. Any spot that remains wetted may contain activity. Use a clean small pipette tip to detach all material and poor the solution into the plating cell. Rinse the beaker twice with 1 mL of NH_4Cl buffer (1 mol/L, brought to pH 2 with HCl) into the plating cell. Use the pipette tip again if required, and rinse this tip well. Rinse the beaker with 1 mL of saturated ammonium oxalate solution into the plating cell.

Put the plating cell on the magnetic stirrer in the fume hood. Connect the power supply (+ to the platinum anode) and switch the power on. The power supply should be current limited at 0.8 A. Strong gas evolution will tend to block the electrode surface and limit the current. This should be avoided by agitating the cell to remove large gas bubbles, and by maintaining the maximum possible stirring rate that the stirring bar can cope with.

After 20–30 min, the resistance increases: the power supply switches automatically to voltage limitation at 6 V. When the current has dropped to 0.2 A (after about 50–60 min; do not proceed to lower currents or longer plating time) the plating should be stopped in one quick procedure: add 1–2 mL of concentrated ammonia solution (25 %), immediately disconnect the plugs and add distilled water to the rim (while still in the fume hood: Cl_2 evolution) and rinse the cell with ample distilled water.

Take the cell apart. Remove the silver plate, rinse with distilled water and dry. Flame the plate gently in a gas flame to fix the Th to the silver. Heat until the colour changes in reflected light, but avoid overheating which would melt the silver. Mark the sample name and plating date on the back of the plate.

13.8 Analytical procedures of selected nuclides

13.8.1 ^{234}Th

This is the standard procedure for the determination of ^{234}Th in an $Fe(OH)_3$ precipitate from 20 L of seawater (*Fleer*, 1991; *Buesseler et al.*, 1994). High activities of ^{228}Th can also be measured with this procedure. The procedure is reliable but rather time consuming and alternative methods (Sections 13.3.2., 13.6.2.2 and 13.6.3.2) should be seriously considered.

Precipitate $Fe(OH)_3$, using 1 mL of ^{230}Th spike (16 dpm/mL) as the yield tracer (Section 13.7.3). Take up the precipitate in 9 mol/L HCl. Add some concentrated HCl to make up for reaction with 5 mmol of $Fe(OH)_3$. The solution should be 9 mol/L in HCl and have a volume of approximately 30 mL. Allow some time for complete dissolution.

Run a chloride column (Section 13.7.5), using a wide (15 mm $\varnothing$) column because a large amount of Fe has to be retained on the resin.

Evaporate the eluate from the chloride column under an IR lamp. Dissolve in 8 mol/L HNO_3 and run a nitrate column (Section 13.7.6). Collect the 8 mol/L HNO_3 eluate in a polypropylene bottle (Pb fraction). Repeat the nitrate column if the sample is not clean enough (Section 13.7.6).

Plate the Th (Section 13.7.7). Count ^{230}Th in the alpha spectrometer. Count the beta activity using a thin plastic cover (*e.g.,* overhead sheet, 0.1 mm polyester, 14 mg/cm^2) to absorb ^{230}Th alphas and most (78 %) of ^{234}Th betas. This cover lets most (91 %) of the ^{234m}Pa betas through. Calculate the ^{234}Th activity (see below for calibration factor). For decay correction, use the time of $Fe(OH)_3$ precipitation as the time of U and Th separation. This introduces some error, as some U is coprecipitated and only removed in the chloride column. The contribution of other beta emitters can be quantified by repeating the beta countings and fitting the results to the decay rate of ^{234}Th (*Buesseler et al.,* 1994).

Calibration: Add 1 mL (*ThVol*) of ^{230}Th spike to a weighed aliquot of the U stock solution (activity *Uact*, approximately 60 dpm) in a small Teflon beaker. Warm, evaporate to dryness under an IR lamp, take up in 9 mol/L HCl, separate Th from U by passing through a chloride column (Section 13.7.5; note time-of-separation), purify Th on a nitrate column (Section 13.7.6) and electroplate (Section 13.7.7). Count in alpha and beta counters. Calculate the calibration factor:

$$\text{Factor} = \frac{(countrate\ beta - bg\ beta)}{(countrate\ alpha - bg\ alpha)} \cdot e^{\lambda Th \Delta t} \cdot \frac{ThVol}{Uact}$$

where Δt is the time between U–Th separation (the chloride column) and counting. This factor includes the (relative) efficiencies of the alpha and beta counters and the activity of the ^{230}Th spike. These need not to be known individually.

Note 1: ^{230}Th is a clean alpha emitter with a narrow (double) peak, convenient for alpha counting. In laboratories where ^{230}Th is determined, a severe cross-contamination problem exists. Keep all Teflonware for natural and spike ^{230}Th separated. The use of ^{229}Th as an alternative tracer is not recommended. The slow ingrowth of beta emitting daughters complicates the calculation and contributes to the error. A further disadvantage is the build-up of ^{229}Th daughters in the alpha counter.

Note 2: The mixed ^{234}Th/ ^{230}Th planchet is worthless after the decay of ^{234}Th. A useful standardization can be made with a planchet with 60 dpm U electrodeposited on it (evaporate a weighed aliquot in a small Teflon beaker and follow electrodeposition procedure of Th; Section 13.7.7). After ingrowth of ^{234}Th to secular equilibrium with ^{238}U (several months), the planchet can be used for a check on the relative efficiencies of beta and alpha counters. Alpha detector efficiency determined this way for ^{238}U should also be applicable to ^{230}Th, but an independent calibration of the ^{230}Th spike is still required.

Notes on ^{228}Th: The ^{228}Th activity can be calculated from the ^{228}Th/^{230}Th ratio in the alpha spectrum. For accurate determinations of ^{228}Th, the ^{230}Th spike has to be calibrated.

^{228}Th has beta emitting daughters. Usually the contribution of ^{228}Th to beta activity is low, but in coastal waters, pore waters or resuspended particles it can be significant. The sensitivity of the beta counter for these betas can be determined by watching the change in beta count rate over time (see *Aller and Cochran*, 1976, who used a ^{228}Th spike).

13.8.2 ^{210}Pb and ^{210}Po

Po and Pb can be coprecipitated with Co/APDC (*Fleer and Bacon*, 1984) but here we describe the method using Fe(OH)$_3$ as it is more easily combined with other procedures and does not require an acid digestion.

Precipitate Fe(OH)$_3$, using 0.5 mL ^{209}Po spike (20 dpm/mL) and 0.5 mL Pb spike (20 mg/mL) as yield tracers (Section 13.7.3). Plate polonium (Section 13.7.4), count alphas of ^{209}Po and ^{210}Po and calculate the ^{210}Po activity according to *Fleer and Bacon* (1984).

Store the solution overnight with scrap silver to remove remaining traces of Po. Remove silver and note time. Add to sample approximately 3 mL of HNO$_3$ and evaporate to dryness. This procedure will oxidize the ascorbic acid. Rinse beaker walls with some 2 mol/L HCl, evaporate again to remove all traces of HNO$_3$, take up in 0.5 mol/L HCl and allow at least 6 months ingrowth of ^{210}Po. Add by weighing a new ^{209}Po spike (the spike should have approximately twice the expected activity of ^{210}Po and can be added anywhere during the ingrowth period). Plate Po (Section 13.7.4). Dilute plating solution to 500 mL and measure Pb with flame AAS. Calibrate with Pb spike diluted the same way to 500 mL, and calculate ^{210}Pb activity according to *Fleer and Bacon* (1984).

13.8.3 ^{7}Be

Prepare Fe(OH)$_3$ precipitate from a 20 L seawater sample following Section 13.7.3. *Dominik et al.* (1989) report for freshwater a chemical yield of 65 ± 5 %, determined with a stable (^{9}Be) spike, which is measured with AAS. According to *Baskaran et al.* (1992) and *Baskaran and Santschi* (1993) two consecutive precipitations secure quantitative recovery of Th and Be. Proceed according to *Baskaran and Santschi* (1993) to measure the dried Fe(OH)$_3$ precipitate by gamma spectrometry.

13.8.4   A combined procedure for measurement of dissolved ^{234}Th, ^{210}Pb and ^{210}Po in 20 L samples

Precipitate Fe(OH)$_3$, using a 1 mL ^{230}Th spike (16 dpm/mL), a 0.5 mL ^{209}Po spike (20 dpm/mL) and a 0.5 mL Pb spike (20 mg/mL) as yield tracers (Section 13.7.3). Plate Po (Section 13.7.4), add approximately 3 mL of HNO$_3$ (to oxidize the ascorbic acid) and evaporate to dryness.

Take up in 9 mol/L of HCl and proceed as for the ^{234}Th determination (Section 13.8.1) with the chloride and nitrate columns. Collect the eluate of the nitrate column in a polypropylene bottle (Pb fraction).

Plate the Th, count the alphas and betas and calculate ^{234}Th activity (Section 13.8.1).

Allow at least 6 months ingrowth of ^{210}Po in the lead fraction.

(*Note*: Even if the first Po plating was not 100 % effective, remaining traces of Po should be retained by the ion-exchange columns and a more rigorous separation appears not to be necessary.) Add by weighing a new ^{209}Po spike to the Pb fraction (the spike should have approximately twice the expected activity of ^{210}Po, and can be added anywhere during the ingrowth period). Evaporate the Pb fraction in a glass or Teflon beaker, rinse the beaker walls with some 2 mol/L HCl, evaporate again to remove all traces of HNO$_3$, take up in 0.5 mol/L HCl, plate the Po (Section 13.7.4) and proceed as in Section 13.8.2.

13.8.5   Radium and radon measurements in seawater

There is a fundamental trade-off in selecting a method for the determination of Ra in seawater, sample volume *versus* time. The larger the sample volume, the less time is required for an analysis. The procedure requiring the smallest volume (2–5 L) samples is alpha spectrometry, but considerable time for sample preparation and counting is required. Alpha scintillation counting of 20 L samples is the standard procedure for ^{226}Ra and ^{222}Rn measurement in seawater, but other Ra isotopes cannot be measured by this technique. Larger volume samples (100–400 L) and patience are required to measure ^{228}Ra in open-ocean samples *via* ^{228}Th ingrowth. For high activity estuarine or coastal samples, gamma spectrometry offers an easy method of measuring ^{226}Ra and ^{228}Ra and delayed coincidence scintillation counting can be used to measure ^{223}Ra and ^{224}Ra in the same sample. These five methods will be discussed in detail.

13.8.5.1   Alpha scintillation measurement of ^{226}Ra and ^{222}Rn

The most commonly used method for measuring ^{226}Ra and ^{222}Rn in seawater was first developed by *Broecker* (1965). This procedure begins with a 15–20 L sample collected in a 30 L Niskin bottle. If ^{222}Rn is to be measured, the water is sucked into an evacuated 20 L glass bottle (wrapped with tape or enclosed in an appropriate container in case of breakage). Containers made from 20 cm diameter plastic pipe are also used (*Key et al.*, 1979). Helium is used to transfer the Rn from the sample to a glass or stainless-steel trap cooled with liquid nitrogen or a charcoal-filled trap cooled with dry ice (*Broecker*, 1965; *Key et al.*, 1979; *Mathieu et al.*, 1988). The He may be repeatedly circulated through the sample and trap using a diaphragm pump, or passed through once and vented. Traps to remove water vapour and CO$_2$ are usually incorporated into the system. The Rn is transferred from the trap into a scintillation cell by warming the glass trap to room temperature or warming the charcoal-filled trap to 450 °C. Figure 13-4 shows a typical Rn extraction system.

The scintillation or Lucas cell (*Lucas*, 1957) is made by coating the inside of a Plexiglass, quartz or metal cell with silver-activated zinc sulphide (ZnS[Ag]). After transferring the Rn to the cell, it is stored for 1–2 h to allow ^{222}Rn daughters, ^{218}Po, ^{214}Pb, ^{214}Bi and ^{214}Po to equilibrate partially. Alpha decays from ^{222}Rn, ^{218}Po and ^{214}Po cause emissions of photons from the ZnS[Ag]. These are converted into electrical signals using a photomultiplier tube (PMT) attached to the cell and routed to a counter.

After the ^{222}Rn measurement, the water sample in the same container may be used for ^{226}Ra measurement by ^{222}Rn emanation. In this case the container is sealed for several days to several weeks to allow ^{226}Ra to generate a known activity of ^{222}Rn. Then ^{222}Rn is again

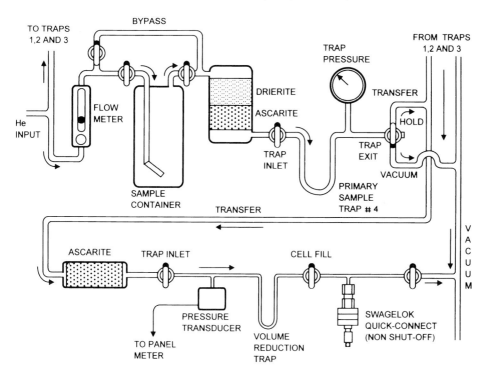

Fig. 13-4. Schematic of radon stripping and transfer system. The upper section is one of four channels used in this system; the lower (transfer) section is shared by all four. In this upper section and the three others like it, four samples are degassed, purified of H_2O and CO_2 and trapped. The samples are then moved individually, *via* the transfer manifold, to the sample reduction trap, then on to counting cells (after *Key et al.*, 1979).

stripped from the sample and measured using the procedure outlined above. In addition to the factors considered in the excess ^{222}Rn calculation, the fraction of equilibrium between ^{222}Rn and ^{226}Ra must be included to calculate the ^{226}Ra activity.

Schlosser et al. (1984) modified this technique to make high precision measurements of ^{226}Ra in seawater. They degassed the sample by boiling 14 L for 45 min and transferred the ^{222}Rn into an activated charcoal trap at –78 °C. The charcoal trap was warmed to 450 °C and the ^{222}Rn transferred into a proportional counter with a mixture of 90 % argon and 10 % methane. Details of the proportional counter and associated electronics are given in *Schlosser et al.* (1983).

The calculation of the excess Rn activity of the sample must include: (1) a decay correction from the time the sample was collected until the mid-point of the counting time, (2) the fraction of equilibrium attained with the Rn daughters (^{218}Po, ^{214}Pb, ^{214}Bi) before counting, (3) the efficiency of the detector, (4) the background of the detector and (5) the blank associated with the sample container and extraction system. These calculations and the errors associated with the measurements have been discussed by *Lucas and Woodward* (1964), *Sarmiento et al.* (1976) and *Key et al.* (1979). The best precision (2σ) obtained for the scintillation counting procedures is approximately 13 %. *Schlosser et al.* (1984) claimed a precision of 11 % for the proportional counting technique.

In some cases it is more practical to concentrate ^{226}Ra from the sample at sea to reduce the blank and avoid the problem of shipping large samples of water. In this case ^{226}Ra may be quantitatively removed using a small column (2 cm diameter × 10 cm long) containing a few grams of Mn-fibre (*Moore*, 1976). If the pH of the sample has been lowered for other purposes, *e.g.*, ^{14}C extraction, it must first be readjusted to ≈ 7. The sample is passed through the fibre at a flow-rate of 0.1–0.3 L/m and discarded after the volume is recorded. In the laboratory the ^{226}Ra may be removed from the Mn-fibre using HCl, or the ^{222}Rn may be determined by direct emanation from the Mn-fibre. In either case a gas system is used to transfer the Rn to a scintillation cell as described above. *Moore et al.* (1985) determined that the precision of the Mn-fibre extraction technique followed by alpha scintillation counting of ^{222}Rn is 13 % (2σ).

A variation on the scintillation technique for ^{226}Ra measurement was suggested by *Butts et al.* (1988). After concentrating the ^{226}Ra on Mn-fibre, the fibre was partially dried, placed in a glass equilibrator, flushed with nitrogen and sealed to allow partial equilibration of ^{222}Rn. The equilibrator was connected directly to an evacuated Lucas cell to transfer a fraction of the ^{222}Rn to the cell. The fraction of ^{222}Rn transferred was calculated by measuring the volumes of the equilibrator and Lucas cell and applying the gas law. *Butts et al.* (1988) demonstrated that this passive technique was much simpler and faster than quantitatively transferring the ^{222}Rn, and gave comparable results for samples containing 8–75 dpm ^{226}Ra.

Obviously, great care must be taken to assess the blank associated with any Ra measurement. Glass containers are a source of Rn contamination that can be difficult to assess accurately when low levels of ^{226}Ra are being determined by ^{222}Rn ingrowth. Barium salts used to precipitate Ra from solution (discussed later) can contribute significant ^{226}Ra and ^{228}Ra blanks. We suggest screening 100 g lots of Ba salts by gamma-ray spectrometry to help select the ones with lowest Ra contamination.

13.8.5.2   Measurement of ^{228}Ra *via* ^{228}Th ingrowth

Open-ocean waters have low activities of ^{228}Ra (< 2 dpm per 100 L). To measure ^{228}Ra in these waters, large volume samples and sensitive counting techniques are required. Most measurements are made by concentrating the Ra from 100–200 L samples onto Mn-fibre, separating and purifying the Ra, allowing ^{228}Th to equilibrate partially with ^{228}Ra, extracting the ^{228}Th and measuring its activity in an alpha spectrometer using ^{230}Th as a yield tracer. A separate sample of the same water is measured for ^{226}Ra activity using the ^{222}Rn emanation technique. Fig. 13-5 illustrates a procedure used to measure ^{228}Ra and ^{226}Ra on the same samples that were being measured for ^{14}C and ^{85}Kr.

Water samples are obtained from a large volume collector such as a 270 L Gerard barrel, by pumping the sample into a processing tank on the ship, or by concentrating Ra *in situ* on Mn-fibre. The *in situ* extraction may utilize a submersible pumping system to force water through an extraction column containing the Mn-fibre (Section 13.2.1; Fig. 13-1), or by sealing the Mn-fibre in a mesh bag and exposing it to water at a certain depth. This large volume sample is used to determine the ^{228}Ra/^{226}Ra activity ratio (AR) of the water.

Radium is removed from Mn-fibre by leaching with a mixture of hot hydroxylaminehydrochloride and HCl or HCl alone. Leaching may be easily accomplished in a Soxhlet extraction apparatus (Section 13.2.3.2). The extract is filtered and mixed with 10 mL of saturated Ba(NO$_3$)$_2$ solution followed by 25 mL of 7 mol/L H$_2$SO$_4$ to coprecipitate Ra with

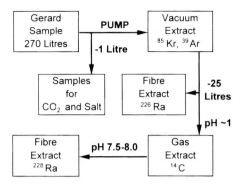

Fig. 13-5. Schematic diagramme for sample extraction.

BaSO$_4$. Warming the extract to near boiling produces larger particles of the precipitate and facilitates its separation.

After precipitating Ba(Ra)SO$_4$, the precipitant is washed with 3 mol/L HCl and water to remove all remaining Mn and dried. The Ba(Ra)SO$_4$ is converted into Ba(Ra)CO$_3$ by fusing it with a mixture of K$_2$CO$_3$ and Na$_2$CO$_3$ (*Moore et al.*, 1985). The solid is washed with water to remove all traces of sulphate and dissolved in HCl. An Fe carrier is added and precipitated with ammonia to remove Th. After removing all traces of Fe(OH)$_3$ from the solution, Ba and Ra are coprecipitated with K$_2$CO$_3$ solution and the precipitate stored for 5–20 months to allow partial equilibration of ^{228}Th. Approximately 30 % equilibration is attained in 1 y. The Ba(Ra)CO$_3$ precipitate is dissolved in HCl and the solution is spiked with ^{230}Th. After adjusting the pH to 1.5, Th is extracted into a trifluoroacetone (TTA)–benzene solution (*Moore et al.*, 1985) and this solution is mounted on a stainless-steel disc. The ^{228}Th/^{230}Th AR is determined by alpha spectrometry and ^{228}Th is calculated from the activity of the spike. The initial ^{228}Ra activity of the sample is calculated by multiplying the measured ^{228}Th activity by the reciprocal of the fraction of ^{228}Th/^{228}Ra equilibrium and this result is decay corrected for the time elapsed from sample collection to the initial purification and precipitation of Ba(Ra)CO$_3$. The solution containing the Ra is measured for ^{226}Ra using the ^{222}Rn scintillation technique to calculate the ^{228}Ra/^{226}Ra AR of the water sample. The activity of ^{228}Ra in the water is obtained by multiplying this AR by the ^{226}Ra activity determined from a separate sample of the same water. The overall precision of this technique, which includes a 13 % error on the ^{226}Ra measurement is 15 %.

Orr (1988) evaluated various methods of measuring ^{228}Ra in open-ocean samples and concluded that results could probably be obtained more quickly and with equal precision using beta–gamma coincidence spectrometry (*McCurdy and Mellor*, 1981) or liquid scintillation alpha spectrometry (*McKlveen and McDowell*, 1984). However, these techniques have not been applied to open ocean samples.

13.8.5.3   Gamma spectrometry measurement of ^{226}Ra and ^{228}Ra

This technique is applicable to samples containing relatively high activities of ^{226}Ra and ^{228}Ra (> 10 dpm) due to the low detection efficiency of most germanium detectors (*Moore*, 1984). Generally, 100 L samples are required for ^{226}Ra measurements. However, recent advancements in the production of large, high efficiency detectors have extended the technique to 20 L open-ocean samples (*Reyss et al.*, 1995; *Schmidt and Reyss*, 1996). The ^{228}Ra in estuarine, coastal and large volume surface ocean samples is also measured using this technique; however, it is not applicable to ^{228}Ra measurements in the ocean interior unless a high efficiency detector is available.

The Ra may be quantitatively extracted from a known sample volume on Mn-fibre or simply concentrated on Mn-fibre from an unknown volume. In the latter case the gamma technique is used to establish the ^{228}Ra/^{226}Ra AR and a separate small volume sample is processed to measure ^{226}Ra quantitatively. Alternatively, the Ra may be coprecipitated with BaSO$_4$. In this case the recovery may be determined gravimetrically (*Reyss et al.*, 1995).

If the Mn-fibre sample is to be used to determine Ra activity quantitatively, all extractions and purification must be quantitative. This can be accomplished by extracting the Ra on a column of Mn-fibre at a flow rate of 1 L/min followed by the Soxhlet extraction apparatus described in Section 13.2.3.2. This procedure ensures the complete removal of the Ra from the fibre into a relatively small volume of acid. After precipitating Ba(Ra)SO$_4$, the precipitant is washed and concentrated into a small vial. The vial is stored for 3–4 weeks to allow ^{228}Ac to equilibrate with ^{228}Ra and ^{222}Rn and daughters to equilibrate with ^{226}Ra.

The ^{226}Ra and ^{228}Ra activities of the sample are measured using a germanium detector. The detector actually measures gamma ray emissions that accompany the decay of ^{214}Bi and ^{214}Pb (^{226}Ra daughters) and ^{228}Ac (^{228}Ra daughter). There are three prominent gamma emissions commonly used for each Ra isotope. For ^{214}Pb, emissions occur at 295 and 352 keV; ^{214}Bi has an emission at 609 keV. For ^{228}Ac, emissions at 338, 911 and 968 keV are commonly used. These are not the only peaks that can be used for measurement of these isotopes, but they are the most prominent for most detectors. However, if a planar or low energy detector is being used, the 209 keV peak from ^{228}Ac and the 186 keV emission from ^{226}Ra may be more useful than the higher energy peaks.

To quantify the signal from the gamma detector, the detector must be calibrated with respect to its efficiency (*E*) for detecting each gamma emission and the intensity (*I*) or probability of gamma emission for each decay must be known. In laboratories that measure a variety of gamma emitting radionuclides, detectors are usually calibrated for detection efficiency with respect to energy using a set of standards of known activity. This *E versus* energy calibration curve can be used to determine the *E* at each energy of interest. The intensity of gamma emission for each peak can be ascertained from the literature. However, there are problems with this method for Ra measurements. The literature values for *I* may include a component derived from coincidence summations. The fraction of the summation component measured by the detector is a function of the counting geometry. Differences are observed when the sample is placed near or far from the detector. When germanium crystals with wells are used to measure samples, the literature values for some emission intensities are considerably different from measured values (*Moore*, 1984). Also, the lower energy gamma rays are preferentially absorbed by the sample matrix. BaSO$_4$ is a strong gamma ray absorber. Therefore, the best way to calibrate a germanium detector for Ra measurement is to prepare standards containing ^{228}Ra and ^{226}Ra in the same matrix and geometry as will be used for samples. For each gamma emission peak that will be used to calculate the Ra activity, determine a factor that converts counts per minute (cpm) into decays per minute (dpm) or Bq (60 dpm = 1 Bq). This factor is the reciprocal of $E \times I$ for each peak of interest.

Peaks of interest in the signal from the germanium detector must be separated from (1) other peaks in the spectrum, (2) emissions due to impurities in the detector housing and shielding and (3) scattering of higher energy emissions (Compton scattering, commonly called background). The final activity (A) is calculated from the following equation:

$$A = \left[\frac{(counts - background)}{time} - \frac{(counts - background)}{time} \right] F$$

where $F = \dfrac{1}{E \cdot I}$

and the term in italics (due to impurities) represents a count made without a sample.

There are a number of computer programmes that perform these functions, but they are often not flexible enough to allow the operator to enter individual factors for each peak. For Ra measurement it is best to use two programmes, one that only identifies and quantifies the peaks by separating them from other peaks and Compton scattering and another that converts the peaks into Ra activities using the factors and detector backgrounds (impurities) for each peak. If activities are determined for each of three peaks, a weighted means assessment can be used to obtain a final result. An excellent programme for resolving low-activity peaks is HYPERMET (*Phillips and Marlow, 1976*).

13.8.5.4 Delayed coincidence measurement of ^{223}Ra and ^{224}Ra

Giffin et al. (1963) reported a highly sensitive and specific system for the measurement of ^{219}Rn ($t_{1/2} = 4$ s) and ^{220}Rn ($t_{1/2} = 55$ s). This system was based on measuring delayed coincidence signals generated by the decay of each of these Rn isotopes to a short-lived Po isotope. ^{219}Rn decays to ^{215}Po ($t_{1/2} = 1.8$ ms) and ^{220}Rn decays to ^{216}Po ($t_{1/2} = 150$ ms). By transferring the Ra rapidly to a counting cell, the differences in the decay of each Rn–Po pair could be used to identify the radon isotope uniquely.

Moore and Arnold (1996) married the delayed coincidence system of *Giffin et al.* (1963) with a gas flow system described by *Rama et al.* (1987) for the measurement of ^{223}Ra and ^{224}Ra in seawater. In this procedure Ra is quantitatively extracted from a known volume of seawater onto a column of Mn-fibre. The measurement is based on the observation that Rn produced by Ra decay is quantitatively ejected from the Mn-fibre (*Butts et al., 1988; Rama et al., 1987*). The partially dried Mn-fibre is placed in an air circulation system and helium is circulated over the Mn-fibre and through a scintillation cell where alpha particles from the decay of Rn and daughters are recorded. Fig. 13-6 is a schematic diagramme of the gas flow system.

The scintillation detector is made from a 1 L Plexiglass cell or Pyrex flask coated internally with silver activated ZnS and mounted on a photomultiplier tube (PMT). With the exception of the window to the PMT, the cell is coated externally with reflective paint and wrapped in black tape. Inlet and outlet tubes are connected to an air circulation system consisting of a small diaphragm pump (Cole Parmer 'air cadet') and flow meter. The system is purged with He and the pump connected to circulate He through the Mn-fibre and carry the Rn to the counting cell at a flow rate of 5–7 L/min. The flow rate is controlled by a valve. At a flow rate of 6 L/min, most of the ^{219}Rn decays take place during the first pass of the gas through the detector. Because the gas is recirculated through the system, ^{220}Rn reaches steady state within the system after 5 min of initiating circulation.

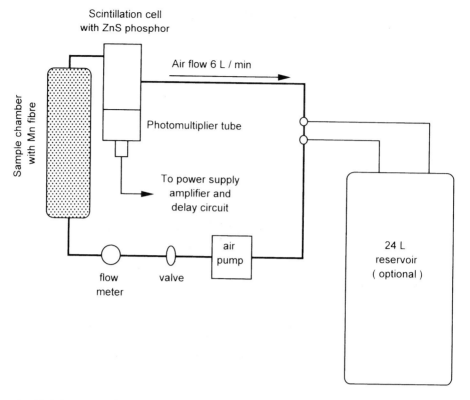

Fig. 13-6. Schematic diagramme of the circulation/counting system (after *Moore and Arnold*, 1996). With permission of Am. Geophys. Union.

Any alpha particle detected in the scintillation cell produces a signal which is routed to a total count register and also to the 219 and 220 circuits shown in Fig. 13-7.

In the 219 circuit the signal is delayed for $10 \, \mu s$ to allow the circuit to stabilize. The signal then opens a gate which remains open for 5.6 ms, about three half-lives of ^{215}Po. Any second count detected in this time interval is recorded in the 219 channel. The count itself is most likely due to ^{215}Po decay; but it would not have been recorded if a decay of ^{219}Rn had not opened the gate within the prior 5.6 ms. All signals are also fed to the 220 circuit. Here they are delayed for 10 ms to allow any ^{215}Po produced from ^{219}Rn to decay. Then the 220 circuit opens for 600 ms, four half-lives of ^{216}Po. If a signal occurs while this gate is open, it is recorded in the 220 channel.

The delay on the 220 circuit effectively prevents signals from ^{219}Rn – ^{215}Po from registering in this circuit. Some events due to ^{220}Rn – ^{216}Po fall into the 219 circuit because the 5.6 ms time constant is long enough for 2.5 % of the ^{220}Rn – ^{216}Po decays to occur in this window. These are subtracted along with chance coincidence counts. To measure low activities of ^{219}Rn in the presence of high activities of ^{220}Rn, the efficiency of the system for measuring ^{220}Rn can be reduced by installing a gas reservoir in the line. This lowers the overall count rate and hence the chance coincidence and cross talk rate for ^{219}Rn.

Delayed Coincidence Circuit

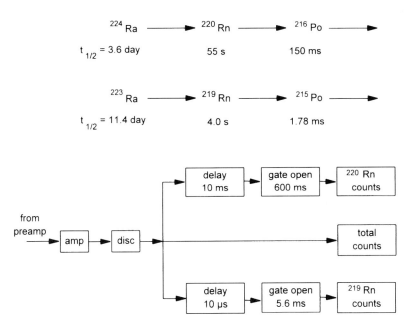

Fig. 13-7. Delayed coincidence circuit (after *Moore and Arnold*, 1996). With permission of Am. Geophys. Union.

13.8.5.5   Alpha spectrometry measurement of ^{226}Ra, ^{228}Ra, ^{223}Ra and ^{224}Ra

All Ra isotopes may be measured in the same sample by chemically separating and purifying Ra, mounting it on a planchet and discriminating among the isotopes by alpha spectrometry. A spike of ^{223}Ra or ^{225}Ra is used as an internal tracer. Often several counting periods are necessary to distinguish between the Ra isotopes and their daughters.

Koide and Bruland (1975) measured ^{226}Ra and ^{228}Ra using lead carrier with ^{223}Ra as a spike. They precipitated Ra and Pb from 20 L seawater samples with aluminium phosphate, dissolved the precipitant in nitric acid and purified the Ra by adding 75 % cold nitric acid to the solution to coprecipitate Ra with lead nitrate. The lead was separated from the Ra by adsorbing it on an anion-exchange column. The Ra was further purified using a cation-exchange column and electrodeposited on a platinum planchet from dilute HCl. ^{226}Ra and ^{223}Ra were measured in an alpha spectrometer immediately after plating. ^{228}Ra was measured by beta counting and by storing the planchet for several months to allow some ^{228}Th to equilibrate followed by alpha spectrometry. The ^{222}Rn was removed from the planchet by flaming prior to measuring the ingrowth of ^{228}Th.

Hancock and Martin (1991) modified the above procedure to measure all four Ra isotopes in a single sample. They used ^{225}Ra as a tracer and separated Ra from the sample by coprecipitation with $PbSO_4$. The precipitate was dissolved in a few mL of alkaline (pH = 10) 0.1 mol/L EDTA solution, mixed with a few drops of a slurry of anion-exchange resin (Bio-

Rad AG1-X8) and added to a small anion-exchange column. The Ra was removed from the column using 0.005 mol/L EDTA–0.1 mol/L ammonium acetate solution at pH = 8 while Pb remained on the column. The Ra was then passed through a cation-exchange column (Bio-Rad AG50W-X8) in 3 mol/L HNO_3 to remove residual Th. The purified Ra was electro-deposited on a stainless-steel planchet and after 15–20 min, placed in an alpha spectrometer. The initial count was used to measure ^{225}Ac, the daughter of ^{225}Ra, to determine the recovery of the ^{225}Ra yield tracer. Additional counts were required to determine the activities of the other Ra isotopes.

13.9   Propagation of errors

For the estimation of the errors associated with counting and the calculation of activities we refer to the literature on the measurement of radiation like *Tsoulfanidis* (1995) and *Ivanovich and Harmon* (1992). Here we will only mention the main principles:

Counting error

The standard deviation associated with the measurement of N disintegrations is:

$$\sigma\,(N) = \pm \sqrt{N}$$

Propagation of errors

If a physical magnitude Y is to be obtained by summation or difference of independent observations y_i with errors r_i the error R in Y is found by:

when $Y \pm R = y_1 \pm r_1 + y_2 \pm r_2$

then $R^2 = r_1^2 + r_2^2$

If a physical magnitude Y is to be obtained by multiplication or division of independent observations y_i with errors r_i the error R in Y is found by:

when $Y \pm R = \dfrac{y_1 \pm r_1}{y_2 \pm r_2}$

then $\dfrac{R^2}{Y^2} = \dfrac{r_1^2}{y_1^2} + \dfrac{r_2^2}{y_2^2}$

Error due to background

An observed count rate consists of a contribution from the background and a contribution from the source. The error in the net count rate (s) is given by

$$S = s \pm \sqrt{s/T_s + \mathrm{b}/T_s + \mathrm{b}/T_b}$$

where

S =   best estimate of net count rate
b =   background count rate during time T_b
$s+b$ =   total count rate during time T_s

Note that the background (b) appears twice in this calculation.

Acknowledgments

MRvdL is indebted to Mike Bacon, Alan Fleer and Rebecca Belastock for their kindness sharing their knowledge and experience and for teaching him many of the procedures described here. Many procedures were further refined in cooperation with Heike Höltzen and Gijs Berger. We thank Alan Fleer, Ken Buesseler, Roger Francois and Hans-Jürgen Walter for helpful comments on the manuscript. WSM acknowledges support from the US National Science Foundation.

References to Chapter 13

Aller, R.C., Cochran, J.K. (1976), Earth Planet. Sci. Lett., 29, 37.
Anderson, R.F., Fleer, A.P. (1982), Anal. Chem., 54, 1142.
Anderson, R.F., Bacon, M.P., Brewer, P.G. (1983a), Earth Planet. Sci. Lett., 62, 7.
Anderson, R.F., Bacon, M.P., Brewer, P.G. (1983b), Earth Planet. Sci. Lett., 66, 73.
Appleby, P.G., Nolan, P.J., Gifford, D.W., Godfrey, M.J., Oldfield, F., Anderson, N.J., Batterbee, R.W. (1986), Hydrobiologia, 143, 21.
Bacon, M.P., Rutgers van der Loeff, M.M. (1989), Earth Planet. Sci. Lett., 92, 157.
Bacon, M.P., Huh, C.-A., Fleer, A.P., Deuser, W.G. (1985), Deep-Sea Res., 32, 273.
Baker, C.W. (1984), Nucl. Instrum. Meth. Phys. Res., 223, 218.
Baskaran, M., Santschi, P.H. (1993), Mar. Chem., 43, 95.
Baskaran, M., Santschi, P.H, Benoit, G., Honeyman, B.D. (1992), Geochim. Cosmochim. Acta, 56, 3375.
Baskaran, M., Murphy, D.J., Santschi, P.H., Orr, J.C., Schink, D.R. (1993), Deep-Sea Res., 40, 849.
Bhat, S.G., Krishnaswami, S., Lal, D., Rama, Moore, W.S. (1969), Earth Planet. Sci. Lett., 5, 483.
Broecker, W.S. (1965), in: Symposium on Diffusion in Oceans and Fresh Waters: Ichiye, T. (Ed.). New York: Lamont-Doherty Geological Observatory, 1982; pp. 116–145.
Broecker, W.S., Patzert, W.C., Toggweiler, J.R., Stuiver, M. (1986), J. Geophys. Res., 91, 14345.
Buesseler, K.O. (1991), Nature (London), 353, 420.
Buesseler, K.O., Cochran, J.K., Bacon, M.P., Livingston, H.D., Casso, S.A., Hirschberg, D., Hartman, M.C., Fleer A.P. (1992), Deep-Sea Res., 39, 1103.
Buesseler, K.O., Michaels, A.F., Siegel, D.A., Knap, A.H (1994), Global Biogeochem. Cycles, 8, 179.
Buesseler, K.O., Andrews, J.A., Hartman, M.C., Belastock, R., Chai, F. (1995), Deep-Sea Res. II, 42, 777.
Buesseler, K.O., Bauer, J.E., Chen, R.F., Eglington, T.I., Gustafsson, O., Landing, W., Mopper, K., Moran, S.B., Santschi, P.H., VernonClark, R., Wells, M.L. (1996), Mar. Chem., 55, 1.
Butts, J., Todd, J.F., Lerche, I., Moore, W.S., Moore, D.G. (1988), Mar. Chem., 25, 349.
Chen, J.H., Edwards, L.R., Wasserburg, G.J. (1986), Earth Planet. Sci. Lett., 80, 241.

Chen, J.H., Edwards, L.R., Wasserburg G.J. (1992), in: Uranium-series Disequilibrium, 2nd. edn: Ivanovich, M., Harmon, R.S. (Eds.). Oxford: Clarendon, 1992; pp. 174–206.

Coale, K.H., Bruland, K.W. (1985), Limnol. Oceanogr., 30, 22.

Cochran, J.K. (1992), in: Uranium-series Disequilibrium, 2nd. edn: Ivanovich, M., Harmon, R.S. (Eds.). Oxford: Clarendon, 1992; pp. 334–395.

Cochran, J.K., Livingston, H.D., Hirschberg, D.J., Surprenant, L.D. (1987), Earth Planet. Sci. Lett., 84, 135.

Cutshall, N.H., Larsen, I.L., Olsen, C.R (1983), Nucl. Instrum. Meth., 206, 309.

Dehairs, F., Shopova, D., Ober, S., Veth, C., Goeyens, L. (1997), Deep-Sea Res. II, 44, 497.

Dominik, J., Schuler, Ch., Santschi, P.H. (1989), Earth Planet. Sc. Lett., 93, 345.

Eppley, R.W. (1989), in: Dahlem Workshop on 'Productivity of the Ocean: Present and Past': Berger, W.H., Smetacek, V., Wefer, G. (Eds.). New York: Wiley, 1989; pp. 85–97.

Fleer, A.P. (1991), in: Marine Particles: Analysis and Characterization. Geophys. Monogr. Ser. Vol. 63. Hurd, D.C., Spencer, D.W. (Eds.). Washington, D.C.,1991; pp. 227–228.

Fleer, A.P., Bacon, M.P. (1984), Nucl. Instrum. Meth. Phys. Res., 223, 243.

Flynn, W.W. (1968), Anal. Chim. Acta, 43, 221.

Giffin, C., Kaufman, A., Broecker, W.S. (1963), J. Geophys. Res., 68, 1749.

Hancock, G.J., Martin, P. (1991), Appl. Radiat. Isotop., 42, 63.

Hansem, R.G., Ring, E.J. (1983), MINTEK Report No. M 84, Council for Mineral Technology. Randburg, South Africa.

Hartman, M.C., Buesseler, K.O. (1994), WHOI Technical report WHOI-94-15. Woods Hole, MA, USA.

Honeyman, B.D., Santschi, P.H. (1989), J. Mar. Res., 47, 951.

Ivanovich, M., Harmon, R.S. (1992), Uranium-series Disequilibrium, Applications to Earth, Marine, and Environmental Sciences, 2nd. edn., Oxford: Clarendon Press.

Kadko, D., Olson, D. (1996), Deep-Sea Res. I, 2, 89.

Kersten, M., Thomsen, S., Priebsch, W., Garbe-Schönberg, C.–D. (1998), Appl. Geochem., 13, 339.

Key, R.M., Brewer, R.L. Stockwell, J. H., Guinasso, N. L., Schink, D.R. (1979), Mar. Chem., 7, 251.

Klinkhammer, G.P., Palmer, M.R. (1991), Geochim. Cosmochim. Acta, 55, 1799.

Koide, M., Bruland, K.W. (1975), Anal. Chim. Acta, 75, 1.

Krishnaswami, S., Sarin, M.M. (1976), Anal. Chim. Acta, 83, 143.

Lally, A.E. (1992), in: Uranium-series Disequilibrium, 2nd edn.: Ivanovich, M., Harmon, R.S. (Eds.). Oxford: Clarendon Press 1992; pp. 95–126.

Lee, T., Barg, E., Lal, D. (1991), Limnol. Oceanogr., 36, 1044.

Levy, D.M., Moore, W.S. (1985), Earth Planet. Sci. Lett., 73, 226.

Lucas, H.F. (1957), Rev. Sci. Instrum., 28, 680.

Lucas, H.F., Woodward, D.A. (1964), J. Appl. Phys., 35, 452.

Luo, S., Kusakabe, T.-L., Ku, M., Bishop, J.K.B., Yang, Y.-L. (1995), Deep-Sea Res.II, 42, 805.

Mathieu, G.G., Biscaye, P.E., Lupton, R.A., Hammond, D.E. (1988), Health Physics, 55, 989.

McCurdy, D.E., Mellor, R.A. (1981), Anal. Chem., 53, 2212.

McKlveen, J.W., McDowell, W.J. (1984), Nucl. Instrum. Meth. Phys. Res., 223, 372.

Michel, J., Moore, W.S., King, P.T. (1981), Anal. Chem., 53, 1885.

Moore, W.S. (1969), J. Geophys. Res., 74, 694.

Moore, W.S. (1976), Deep-Sea Res., 23, 647.

Moore, W.S. (1984), Nucl. Instrum. Meth. Phys. Res., 223, 407.

Moore, W.S., Arnold, R. (1996), J. Geophys. Res., 101, 1321.

Moore, W.S., Dymond, J. (1988), Nature (London), 331, 339.

Moore, W.S., Dymond, J. (1991), Earth. Planet. Sci. Lett., 107, 55.

Moore, W.S., Santschi, P.H. (1986), Deep-Sea Res., 33, 107.

Moore, W.S., Key, R.M., Sarmiento, J.L. (1985), J. Geophys. Res., 90, 6983.

Moran, S.B., Buesseler, K.O. (1993), J. Mar. Res., 51, 893.

Moran, S.B., Hoff, J.A., Buesseler, K.O., Edwards, R.L. (1995), Geophys. Res. Lett., 22, 2589.

Nozaki, Y. (1993), in: Deep Ocean Circulation, Physical and Chemical Aspects: Teramoto, T. (Ed.). Amsterdam: Elsevier, 1993; pp. 139–146.

Nozaki, Y., Nakanishi, T. (1985), Deep-Sea Res., 32, 1209.

Orr, J.C. (1988), J. Geophys. Res., 93, 8265.

Pates, J.M., Anderson, R., Cook, G.T., Mackenzie, A.B. (1992), in: Advances in Liquid Scintillation Spectrometry (Vienna, Sept. 14–18).

Phillips, G.W., Marlow, K.W. (1976), Nucl. Instrum. Meth. Phys. Res., 137, 525.

Picket, D.A., Murrell, M.T., Williams, R.W. (1994), Anal. Chem., 66, 1044.

Rama, J., Todd, F., Butts, J.L., Moore, W.S. (1987), Mar. Chem., 22, 43.

Reyss, J.-L., Schmidt, S., Legeleux, F., Bonte, P. (1995), Nucl. Instrum. Meth. Phys. Res., 357, 391.

Roether, W. (1971), J. Geophys. Res., 76, 5910.

Rutgers van der Loeff, M.M., Berger, G.W. (1993), Deep-Sea Res. I, 40, 339.

Rutgers van der Loeff, M.M., Boudreau, B.P. (1997), J. Mar. Syst., 11, 305.

Sarmiento, J.L., Hammond, D.E., Broecker, W.S. (1976), Earth Planet. Sci. Lett., 32, 351.

Schlosser, P., Kromer, B., Roether, W. (1983), Nucl. Instrum. Meth. Phys. Res., 216, 155.

Schlosser, P., Rhein, M., Roether, W., Kromer, B. (1984), Mar. Chem., 15, 203.

Schmidt, S., Reyss, J.-L (1996), J. Geophys. Res., 101, 3589.

Schoch, H., Bruns, M., Münnich, K.O., Münnich, M. (1980), Radiocarbon, 22, 442.

Shaw, T.J., Francois, R. (1991), Geochim. Cosmochim. Acta, 55, 2075.

Southon, J.R., Vogel, J.S., Nowikow, I., Nelson, D.E., Korteling, R.G., Ku, T.L., Kusakabe, M., Huh, C.A. (1982), Nucl. Instrum. Meth. Phys. Res., 205, 251.

Tsoulfanidis, N. (1995), Measurement and Detection of Radiation, 2nd ed., Washington, DC: Taylor & Francis.

Volpe, A.M., Olivares, J.A., Murrelli, M.T. (1991), Anal. Chem., 63, 913.

Yokoyama, Y., Nguyen, H.V. (1980), in: Isotope Marine Chemistry: Goldberg, E., Horibe, Y., Saruhashi, K. (Eds.). Tokyo: Ochida Rokakuho, 1980; pp. 259-289.

Yu, Ein-Fen (1994), PhD Thesis, Woods Hole, USA.

Yu, Ein-Fen, Francois, R., Bacon, M.P. (1996), Nature (London), 379, 689.

14   *In situ* determination of pH and oxygen

H. P. Hansen

14.1   Introduction

The initial motivation to development of *in situ* determinations as opposed to discrete sampling and subsequent analysis has been the difficulty to obtain samples from sites such as sediments, biota or small scale water layers and interfaces.

Today the much higher spatial and temporal resolution of measurements is a powerful incentive for *in situ* determinations rather than time-consuming sampling, sample transfer and batch determinations in the laboratory.

While laboratory determinations are performed under constant or controlled conditions, *in situ* determinations have to eliminate or compensate for wide ranges of environmental changes. The rule of matching calibration and measuring conditions cannot be obeyed during *in situ* determinations which, consequently, very often suffer from reduced accuracies. The choice of either laboratory or *in situ* determination should be carefully considered. The effort required to achieve acceptable accuracies of *in situ* determinations may well outweigh the benefits of their seemingly fast and simple application.

In situ determinations of pH and oxygen in seawater involve two different types of registration: registration of temporal variations at fixed locations, *e.g.*, monitoring applications, and registration of spatial variations, *i.e.*, by vertically and/or horizontally moving probes.

For the stationary registration, long-term stability of the sensor or analytical unit is the dominant criterion, while moving sensors have to be optimized for minimum response times. For example, the fast response time of membrane covered amperometric oxygen sensors and long-term stability are mutually exclusive. The sensors can only be optimized for one or the other criterion.

All environmental variables which may affect *in situ* registrations (temperature, pressure, conductivity/salinity) have to be recorded simultaneously with the variable under study.

14.2   *In situ* determination of pH

The potentiometric determination of pH by means of glass electrodes as described in Chapter 7 and the corresponding calibration procedures can be used for *in situ* determinations in surface water (to a depth of about 30 m) without any modification other than using a submersible electronic unit. There is no pressure dependency of the electrical electrode signal according to the Nernst equation (see Chapter 7) but the mechanical construction of the electrode has to consider the ambient pressure. High pressure affects the asymmetry potential. *Ben-Yaakow and Ruth* (1974) observed a shift of the asymmetry potential of a high pressure glass electrode of 2 mV at 6700 dbar ($67 \cdot 10^6$ Pa). Deploying electrodes to greater depths, *i.e.*, varying ambient pressure, may conflict with the electrolyte junction between the outer medium (seawater) and the reference electrode (commonly an Ag/AgCl electrode). One-rod pH electrodes may be combined with standard CTD (conductivity temperature depth) probes used for *in situ* registrations within the limits of their specified pressure stabilities.

We have tested several types of commercial electrodes to depths of 200 m and calculated operational accuracies from offsets between repeated calibrations before and after 200 m vertical profiles (about 20 min duration). The standard deviations were ±0.03 pH (liquid electrolyte) to ±0.05 pH (gel electrolyte). The best results were obtained with liquid electrolyte electrodes having a small Tygon bag filled with electrolyte attached to the refill opening. This flexible reservoir provides a gravity driven and pressure neutral electrolyte flow through the diaphragm (*Hansen*, unpublished). If no electrolyte flow is provided, the diaphragm is easily polluted by particles and seawater is pressed into the electrolyte and contacts the reference electrode resulting in a signal shift, noisy signal or total electrode failure.

Gel electrodes have no diaphragm, however during operation in seawater the gel behind the junction hole is partly consumed. In the resulting excavation, small air bubbles or particles are readily trapped when the probe is lowered through the sea surface which block the electrolyte junction and cause a malfunction of the electrode.

The response times of common glass electrodes (usually > 3 s) in *in situ* applications are generally sufficient with respect to the pH changes. However, in cases of rapid environmental temperature changes, the rather high heat capacities of standard pH electrodes with considerable electrode masses cause a temperature gradient between environmental seawater and inner electrode buffer. This results in undefined temperatures of the electrode surfaces and errors in the electrode potential.

Precision measurements of the *in situ* pH below a depth of about 400 m require separate glass and reference electrodes with electrolyte junction and pressure compensation. A pH electrode withstanding 15 000 dbar has been described by *Distèche* (1959, 1962). The first practicable deep-sea pH electrode was introduced by *Ben-Yaakow and Kaplan* (1968) and improved by *Ben-Yaakow and Ruth* (1974). The electrode can stand 6700 dbar hydrostatic pressure and has been tested to a depth of 5000 m. The pressure is compensated for by replacing a part of the glass electrode housing by Tygon tube. The accuracy is ±0.02 pH units. A correction for a shift of the asymmetry potential (2 mV for 6700 dbar) was included.

Meanwhile a variety of high-pressure pH electrodes using different modes of pressure compensation have been reported, see, *e.g.*, *Dexter and Perkins-Rohrbach* (1986) or *Queck and Auras* (1985). Some commercial CTD probes (*e.g.*, 'Aquatrace' by Chelsea Instruments, 55 Central Avenue, West Molesey, Surrey KT8 2QZ, UK) are provided with pH sensors which can operate down to 4000 m and provide accuracies of < 0.05 pH units.

14.3   *In situ* determination of oxygen

Nearly all sensors used for *in situ* registration of dissolved oxygen in seawater are of the amperometric membrane covered electrode type introduced by *Clark et al.* (1953). A de-

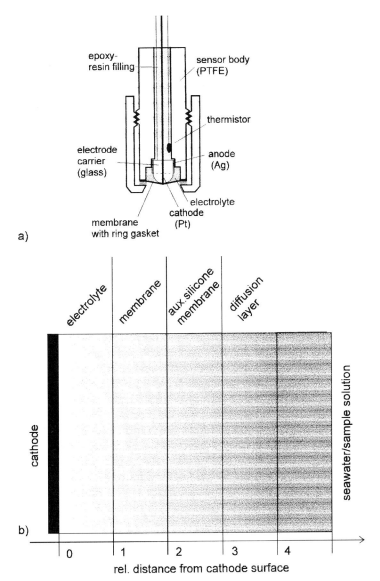

Fig. 14-1. Schematic diagramme of (a) a membrane covered dissolved oxygen sensor and (b) diffusion ranges between cathode and outer solution.

tailed description of the theory of this sensor has been given by *Grasshoff* (1981). Some valuable comments have been published by *Short and Shell* (1984). The following brief description is restricted to properties of the sensor important for application and calibration.

The amperometric membrane covered oxygen sensor (Fig. 14-1a) consists of a noble metal cathode (usually platinum or gold) and a silver anode. The electrodes are situated in an inner electrode compartment filled with electrolyte and separated from the environment by a gas-permeable membrane close to the cathode.

Oxygen from the outside can diffuse through the membrane into the electrolyte in the inner compartment. Applying a voltage of about 0.8 V to the cell, oxygen is reduced at the cathode according to

$$O_2 \text{ (gas)} + 2H_2O + 4e^- \quad \leftrightarrow \quad 4OH^-$$

With a neutral or acidic sodium chloride electrolyte, the oxidation process at the anode is

$$Ag + Cl^- \quad \leftrightarrow \quad AgCl + e^-$$

As the cathodic reduction of oxygen produces hydroxyl ions, the electrolyte becomes alkaline and the anodic oxidation changes to

$$2Ag + 2OH^- \quad \leftrightarrow \quad Ag_2O + H_2O + 2e^-$$

To avoid a shift of the anodic processes during operation, most oxygen electrodes use alkaline electrolytes and thus Ag/Ag_2O instead of $Ag/AgCl$ as the reference electrode ($+0.35$ instead of $+0.222$ V *versus* the Normal Hydrogen Electrode).

If the applied sensor voltage is in the polarization plateau range, *i.e.*, each oxygen molecule arriving at the cathode is spontaneously reduced and the resulting oxygen concentration at the cathode surface is zero, the sensor current depends only on the diffusion of oxygen through the membrane and the electrolyte layer and the oxygen concentration (partial pressure) immediately outside the membrane.

The formation of a polarization potential at the silver anode is suppressed (or at least minimized) by designing the anode surface much larger (about 100 times) than the cathode.

The cathodic consumption of oxygen leads to a decreasing oxygen concentration at the outer surface of the membrane. This reduced oxygen concentration represents a poorly defined diffusion layer and has to be removed by forcing strong turbulence in front of the membrane, *e.g.*, motion of the probe or addition of a stirrer.

Modern oxygen sensors use very small cathode surfaces and correspondingly low sensor currents ($\ll 1\,\mu A$) and oxygen consumptions. A spherical shape of the probe surface (Fig. 14-1a) also enhances the front mixing.

If the sensor has to be operated without front turbulence or if the oxygen consumption of the probe cannot be tolerated as, *e.g.*, in some respiration experiments, the sensor may be operated in pulsed mode (*Langdon*, 1984). In this mode the polarization voltage is pulsed after a period of delay. The oxygen concentration in front of the membrane is allowed to equilibrate and the total oxygen consumption is reduced. The plateau current is determined from the current-time registration during the pulse.

The time constants of membrane covered polarographic oxygen electrodes are determined by the shape and diameter of the electrode and by the diffusive fluxes through the diffusion layers (polarization layer and electrolyte, membrane and front layer (Fig. 14-1b)).

Minimum time constants are achieved with spherical electrodes of small diameters, thin diffusion layers and high permeability of the membrane material, *i.e.*, high diffusive flux (*Grasshoff*, 1981). A review of data sheets of 25 commercial dissolved oxygen sensors revealed response times from 200 ms to 30 s (according to the manufacturers specification).

An electrode of a given geometry may be equipped with different types of membranes and thus optimized for applications requiring either maximum mechanical and long-term stability or fast response time. Thin Teflon membranes provide fast response but low mechanical stability and *vice versa*.

The oxygen permeability of silicone rubber is about 60 times higher than that of Teflon. The diffusive flux through a 10 µm Teflon membrane may be increased by a factor of about four by covering the environment (outer) side of the membrane with a 25 µm silicone rubber layer (*Ben-Yaakov and Ruth*, 1980).

An oxygen electrode of extreme mechanical and long–term stability is produced by Interocean Sytems Inc. (3540 Aerocourt, San Diego, CA, USA). The Teflon membrane of this electrode is covered with a stainless-steel web and fixed with a small amount of liquid silicone rubber. After consolidation of the silicone rubber the membrane is tightly fixed and covered with a thin silicone rubber layer (imbedding the steel web). The 90 % response time of this electrode is about 22 s. This electrode has been deployed in a field experiment for 8 weeks with less than 2 % deviation from the initial 100 % value (*Hansen*, unpublished).

Standard oxygen electrodes in oceanic applications as, *e.g.*, the oxygen sensor by Meerestechnik Elektonik GmbH (24610 Trappenkamp, Germany) have response times of from 3 to 10 s (depending on the choice of membrane).

Calibration of oxygen electrodes

Independent of the brand of polarographic oxygen sensor, a general calibration procedure has to be followed. The geometry and the membrane of the sensor define, within limits, the diffusion layers and consequently the temperature dependence of the oxygen probe.

It has to be noted that the calibration of dissolved oxygen sensors of the Clark type requires homogeneous temperatures from the cathode to the seawater environment. Whenever the probe-body temperature differs from the environment, the diffusive fluxes are not defined and the calibration is not valid. Most sensors require thermal equilibration times of about 20 s or more due to their considerable heat capacities, which is considerably longer than the response times.

Initial temperature and linearity calibration

The initial temperature and linearity calibration is generally provided by the manufacturer of the sensor unit but may be carried out using a calibration tank and the procedure as described by *Grasshoff* (1981).

The oxygen probe is run in oxygen-free water (addition of 35 g/L sodium sulphite) until the output signal is stable close to zero, indicating that the polarization is at a steady state and oxygen contained in the electrolyte is consumed. The sensor signal for zero oxygen should be close to 'true zero', *i.e.*, the signal of the sensor without membrane and electrolyte and a totally clean and dry surface between the electrodes. A minor deviation may be caused by impurities in the electrolyte or a contamination of the surface between the electrodes with electrically conducting material.

The full range calibration is performed in a temperature controlled water-bath. Reduced and elevated oxygen concentrations are established by ventilation with nitrogen or air, respectively. After ventilation the water should be stirred modestly and allowed to homogenize.

Data pairs of sensor signals (voltage or equivalent) and oxygen saturations determined by Winkler-titration are collected for various temperatures (commonly 5–30 °C in steps of 5 K).

A second order polynomial is calculated for the oxygen saturation in % *versus* sensor signal S_s :

$$O_2 = 100 \cdot (S_s - S_0) / \sum A_{(i)} \cdot T^i$$

where S_0 is the temperature independent sensor signal for zero oxygen and $A_{(i)}$ are the polynomial coefficients for $i = 0, 1$ and 2.

Field calibration

Mechanical stress during operation modifies the electrode membrane and consequently the sensor signal as does a membrane change. These changes only affect the slope of the sensor (changes in diffusive flux) and not the zero reading.

The oxygen saturation calculated with the sensors polynomial response equation only has to be corrected by a simple factor close to 1. The interval between field calibrations very much depends on the type of electrode, the mechanical stress during operation and the desired precision.

Some electrodes maintain the calibration for weeks within <2 % while others require a calibration daily or even immediately before every use (profile).

The field calibration generally is a two point calibration which assumes constant zero reading. If required, the zero value may be checked with a freshly prepared sodium sulphite solution (about 35 g/L). The second calibration point is a defined oxygen saturation value close to 100 % saturation. The preparation of a calibration sample with a well defined oxygen content is rather difficult and time consuming and not practicable at sea. The only recommended method is the calibration of the sensor against a Winkler-titration (see Chapter 4) of sample(s) from the same location (at sea) or from the above mentioned calibration stand (in a landbased laboratory). The depth of the calibration sample(s) should be selected in a layer of homogeneous oxygen concentrations with no vertical gradients. The calibration can be performed as post measurement calibration based on the raw oxygen sensor data, the titration results of the Winkler samples and the recorded temperature, salinity and pressure data. High pressure affects the sensor mebrane and thus the diffusive flux. For many applications the pressure dependency of the dissolved oxygen sensor may be neglected. The precision and accuracy of the calibration thus obtained is about 2 %.

For oxygen determinations within the framework of WOCE (World Ocean Circulation Experiment) an algorithm to convert the CTD oxygen sensor measurements into oxygen profiles based on the documented sensor physics and *in situ* oxygen data of discrete samples is described in the *WOCE* Operations Manual (WHP, 1994). The achievable accuracy and precision is <1 % and ±0.1 %, respectively.

References to Chapter 14

Ben-Yaakow, S., Kaplan, I.R. (1968), *Rev. Sci. Instrum*. 39, 1133.
Ben-Yaakow, S., Ruth, E. (1974), *Limnol. Oceanogr.*, 19, 144.
Ben-Yaakow, S., Ruth, E. (1980), *Talanta*, 27, 391.
Clark, L.C., Wolf, R., Granger, D., Taylor, Z. (1953), *J. Appl. Physiol.*, 6, 189.
Dexter, R.I., Perkins-Rohrbach, A. (1986), *Corrosion* '86, 57.
Distèche, A. (1959), *Rev. Sci. Instrum.*, 30, 474.
Distèche, A. (1962), *J .Electrochem. Soc.*, 109, 1084.
Langdon, C. (1984), *Deep-Sea Res.*, 31, 1357.
Grasshoff, K. (1981), in: *Marine Electrochemistry*: Whitfield, M., Jagner, D. (Eds.). New York: John Wiley & Sons, 1981; pp. 327–420.
Queck, C.H., Auras, S. (1985), *Informationen aus dem Forschungsinstitut 'Kurt Schwabe', Meinsberg*, issue 3.
Short, D.L., Shell, G.S.G. (1984), *J. Phys. E. Sci. Instrum.*, 17, 1085.
WHP (1994), WOCE Report No. 68/91, Revision 1. Woods Hole, MA, USA.

15 Determination of dissolved organic carbon and nitrogen by high temperature combustion

G. Cauwet

15.1 Introduction

15.1.1 Organic matter or organic carbon analysis

The estimation of organic matter concentrations in most natural environments was always considered to be a crucial parameter for many different reasons. Almost all methods were based on the total oxidation of organic matter by combustion or chemical oxidation and the estimation of the loss in mass (ignition loss) or the consumption of the oxidant (potassium permanganate or potassium dichromate). These rough and sometimes rather imprecise methods were easily applicable to solids (soils, sediments) but with more difficulty to water samples. Methods appeared in the literature somewhat later which are based on the same reactions (oxidation of organic matter), but these techniques are combined with precise measurements of the carbon dioxide produced in the oxidation. What used to be called organic matter analysis then became 'carbon analysis' instead. The precision of the measurement itself increased and analysts began to know exactly what they measured.

15.1.2 DOC and DON in seawater

The biological importance of the ocean, owing to its richness as well as its size, is recognised universally. Knowledge of the different carbon reservoirs and their evaluation are considered key parameters in the description of the marine environment. Among them, dissolved organic matter (DOM) has the greatest mass, representing about 1000×10^{15} g of carbon, and not least because of its importance for the global climate there is a need to obtain accurate and comparable data of dissolved organic carbon (DOC) concentrations. The same importance may be attributed to the dissolved organic nitrogen (DON), the role of which in biological processes still has to be understood. This organic pool is subject, more intensely than we believed in the past, to dynamic processes such as excretion and consumption by marine organisms at very variable time scales, giving rise seasonally to accumulation (*Williams*, 1995). Methods for the determination of DOC developed at a rather slow pace due to difficulties related to the composition of seawater. While DOC concentrations are around 1 mg/L, seawater also usually contains more than 35 g/L of salt and more than 25 mg/L of inorganic carbon as CO_2, HCO_3^-, and CO_3^{2-}. It is still a demanding task to process samples with such a huge excess of dissolved inorganic salts and almost 25 times more inorganic than organic carbon. Being less concentrated by an order of magnitude and accompanied by variable concentrations of the inorganic forms (NO_3^- and NH_4^+), DON needs even more attention.

15.1.3   Wet *versus* dry methods

Until now, nobody has invented a method different from but as universal as the production of CO_2 by oxidation and its determination by chemical or physical methods. Various methods were used, successfully in most cases, for the determination of CO_2 (*e.g.*, manometry, titration, gas chromatography, non-dispersive infrared analysis), but the most difficult and controversial step in DOC and DON determinations was always the oxidation. The challenge is to devise a method which quantitatively transforms the carbon and the nitrogen bound in very complex mixtures of organic molecules into carbon dioxide and nitrate, without formation of artefacts (*Wangersky*, 1975). Despite some more or less successful attempts at dry combustion, most of the earlier techniques employed wet-chemical oxidation near the boiling point of water, with oxidants such as peroxodisulphuric acid, chromic acid or activated oxygen generated by UV irradiation. These methods are not affected by the presence of sea salt. They have been demonstrated to be largely effective ($> 95\,\%$) with many organic molecules, including some biopolymers. The few compounds exhibiting low oxidation yields such as urea, are the exception, but the conversion efficiency for natural DOM is unknown (*Hedges and Farrington*, 1993). Another approach, more attractive to many analysts, is dry oxidation at a high temperature ($600–900\,°C$) in the presence of oxygen. The large amount of salt present in seawater for some time limited the use of such methods, and wet chemical methods were preferred until the last decade (*Menzel and Vaccaro*, 1964; *Collins and Williams*, 1977; *Cauwet*, 1984). It seems that the analysis of very small volumes, made possible by improvements in sample handling and increased sensitivity of detection, gave a fresh boost to dry combustion techniques. The most momentous event of the last decade, however, was the report of elevated DOC and DON levels measured with a dry oxidation technique (*Suzuki et al.*, 1985; *Sugimura and Suzuki*, 1988). Their data not only induced a strong effort on the part of the DOC community to improve the methodology and to try and understand why different methods gave different results, but it also increased interest in the fate of organic carbon in the ocean.

15.2   Methods and instruments

15.2.1   Dry combustion or high temperature catalytic oxidation (HTCO)

The dry combustion methods may actually be divided into two groups: high temperature combustion of DOM with elemental oxygen with and without a catalyst. The first dry combustion methods derived from measurements of particulate organic carbon (POC): the water samples first were dried in an oven or freeze-dried and then analysed as solids. Although this method was used with some success (*McKinnon*, 1978), it was often plagued by high blanks and variable results. More recently, equipment has been developed employing on-line combustion tubes which allow direct injection of liquid samples into furnaces. At the very high temperatures ($900\,°C$) needed for total combustion without catalysis, the sublimation of sodium chloride is a detriment; it tends to damage the system and reduces the precision of measurements. The utilisation of an oxidation catalyst was a key improvement for obtaining equivalent oxidation yields at lower temperatures ($680–700\,°C$). In both cases the presence of oxygen (air or pure oxygen) is needed. The methods using a catalyst, are identified as high temperature catalytic oxidation (HTCO).

15.2.2 Some instruments

For HTCO analysis, manufacturers offer various combinations of instruments. Since it is difficult if not impossible to draw up an exhaustive list of all commercially available and laboratory-made apparatus, we only mention, with one example each, a few of the most frequently used configurations. Some instruments (*e.g.*, Ionics) use a combination of high temperature (900 °C) and pure platinum as catalyst. Others (*e.g.*, Shimadzu), use platinised alumina catalysts at lower temperature (680 °C). Analytik Jena sells an instrument working with a CuO catalyst at 950 °C. Antek produces a total nitrogen analyser based on the same principle (900 °C, on silica beads). The combination used by *Sugimura and Suzuki* (1988) which triggered the controversy on DOC concentrations was close to the Shimadzu design. The sample was injected vertically into a furnace filled with 3 % Pt on aluminium oxide, at 680 °C.

Most of these instruments give satisfactory results, each having its advantages and its disadvantages. Not all of these instruments were designed for analyses of seawater; some were constructed for a freshwater medium. For instance, high temperature exacerbates the volatilisation of sodium chloride which deposits on tubes and tubing walls, rapidly destroying quartz parts. It also produces more chlorine, dangerous for the infrared cell. Sodium chloride sublimation starts to be substantial at temperatures higher than 500 °C, and a compromise, therefore, must be reached between the efficiency of oxidation and the sublimation of sodium chloride. Catalytic oxidation at 680–700 °C seems to be a good combination; the method described in this chapter is based on these conditions.

The following data and information mostly derive from personal experience (field and laboratory) with a Shimadzu TOC 5000 apparatus for carbon and a Sievers NOA 270 B for total nitrogen.

15.3 Preparation of samples

15.3.1 Contamination problems

Precise and sensitive as the method may be, one only can measure what is present in the sample at the moment it is analysed. The main problem with carbon determinations is that this element is ubiquitous, in solid, liquid, dissolved and volatile forms; contamination, therefore, lurks at every step.

For the first step, chronologically, *i.e*, sampling and conditioning of samples, plastic materials should be avoided as far as possible. Their properties vary considerably and we do not always know what exactly they are. The chemical nature of the plastic is a crucial parameter, some usually are non-contaminating (polyethylene, polypropylene, Teflon, silicone); some others are slightly (PVC) or highly so (Bakelite, rubber). Generally, plastics with a strong odour are highly contaminating. However, the release of organics by polymers also depends on the state of the surface in contact with the sample and on the time span during which the sample is in contact with it. For precise determinations, plastics must be avoided systematically as material for bottles, vials and tubes in which the samples are stored for long periods of time (days to months). The time of contact with tubing used for transfer usually is too short to contaminate a sample; or contamination is undetectable, if clean water and sample are flushed through it for some time before a sample is collected. All plastic polymers release small amounts of organics; we could therefore use an equation to express the degree of contamination as proportional to the plastic surface and the time of contact and inversely proportional to the sample volume. High sample flow rates in tubing dilute possible contamination so that it becomes undetectable.

Another important source of contamination is dirty surfaces in contact with samples. It is self-evident that any grease or oily material must be absent from the environment where samples are handled. However, the most universal source of contamination with carbon remains the hands. The outside of all flasks touched with bare fingers is covered with a film of organic compounds released by the skin. When pouring a sample from one flask into another container, one frequently sees a drop running along the outside wall. Even if it does not seem to enter the tube or vial, the drop is contaminated so rapidly that most of the time a slight contamination occurs, probably due to molecular diffusion of small molecules such as amino acids. Contamination from atmospheric volatile organic compounds is difficult to trace, but it seems that water in any tube or flask, even if acidified (thus unable to fix CO_2), left in contact with the atmosphere for more than a few minutes, can be contaminated. An increase of DOC concentrations of the order of 100 μg/L C in seawater samples is sometimes observed after they were opened to the atmosphere for handling or analysis in the laboratory. Sample handling in the controlled atmosphere of a glove box would probably improve the data quality, but it is rather cumbersome. A clean bench with filtered laminar flow is a more convenient option, at least it offers a protection against dust. Putting samples into a small box under clean gas flow during bubbling and sampling would also certainly improve cleanliness and avoid atmospheric contamination.

15.3.2   Sampling

In chemical oceanography the sampling methodology is fairly variable, due to local constraints, field conditions, depth and volumes to be sampled, and availability of sampling devices. One may cite, from large to small systems: stainless-steel samplers (50–200 L), oceanographic samplers (*e.g.*, Niskin or Go-Flo samplers, 2–30 L), direct pumping from the boat (centrifuge or peristaltic pumps, vacuum pumping), special devices such as glass bottles, Teflon bottles or syringe systems. The sample volume depends on the scientific objectives. For multiparameter studies a rather large volume is needed, to be shared among participants. All these aspects have thoroughly been discussed in Chapter 1 of this book. For specific studies, it can be more convenient to have a personal, ultra-clean system, optimized for the purpose. This practical aspect of measuring DOC has been well studied by *Peltzer and Brewer* (1993).

For surface samples (down to 5 m), vacuum pumping into a glass bottle seems to be the most convenient method. A lead weighted Teflon tube with the intake opening at the required depth is connected to the bottle which is also connected to a vacuum pump *via* a second bottle. The second bottle guards against water accidentally entering and thus damaging the vacuum pump. This method is perfectly clean and fast. For deep casts, Teflon coated Go-Flo samplers, carefully cleaned before use, are unsurpassed.

15.3.3 Filtration

The determination of DOC implies that the samples are filtered. In many cases (especially in the open sea), samples may not have to be filtered considering that particulate organic carbon (POC) is only a few percent of the total organic carbon (TOC); that the limit between particulate and dissolved matter is arbitrary; and that filtration is a possible source of contamination (see also Chapter 2). The results are then given as TOC. If POC is determined separately, it can be subtracted to calculate DOC, with reduced precision, however. The limit between dissolved and particulate organic carbon is determined by the filter porosity. Since Whatman GF/F glass fibre filters with an average pore size of $0.7\,\mu$m are most commonly used for this purpose, most bacteria and colloids are included in the 'dissolved' fraction. Inorganic alumina membranes of lower porosity ($0.2\,\mu$m) recently appeared on the market; their use could be more rational, even though any size limit is discretionary. Filtration through $0.2\,\mu$m pores at least would exclude most living particles from the dissolved phase. Viruses, however, will pass these filters which might raise the question: are viruses living organisms?

For obvious reasons, glass is the preferred material for the filtration systems. All glassware is kiln-fired at $500\,°$C or cleaned with sulphochromic acid. Filters are combusted at $450\,°$C for a minimum of 4 h. Even with such precautions, contamination in DOC determinations is not uncommon. For handling large (100–1000 mL) samples voluminous flasks are needed with large and potentially contaminated surfaces. Subsampling also sometimes is critical. At the opposite end, handling small volumes needs more care, because there is no dilution effect on the contamination when it occurs. The cleanest method apparently is to avoid any transfer in open air. Samples are drawn directly into a syringe and filtered on-line. All glass and stainless-steel syringes, with a Teflon tipped piston, equipped with a three ways connection with anti flushback system and a stainless-steel filter holder, provide clean samples for DOC determinations. We recommend flushing the filtration unit 4–5 times with Milli-Q water and twice with the sample water. Very low DOC blanks are obtained with this method.

15.3.4 Sample storage

After filtration, samples for DOC measurement may be stored for a few minutes or for a few months before analysis. The exclusive use of glass is the only option for this purpose. Storage in a sealed ampoule probably is cleanest, but some experience is needed for proper flame-sealing, especially during field work. Glass tubes (Pyrex) with Teflon-lined screw caps are a very convenient alternative. Their disadvantage is the Bakelite of the cap which is highly contaminating; great care is needed to avoid any contact of it with the sample water. This sometimes occurs during transportation when vibrations slightly unscrew the caps. It also occurs when tubes are stored in cold place due to the different coefficients of thermal expansion of glass and Bakelite.

Storage also means preservation. Experience has shown that addition of a few ppm mercury(*II*)chloride ($HgCl_2$) is the most suitable and efficient method. Acidification suffices for samples to be processed within a few hours, but not for long term storage. Freezing is not suitable except if done fast and immediately after sampling, and even then only when rapid thawing is possible.

15.4   Blanks

15.4.1   Water blank

To prepare standards and to estimate the procedural blank of the method, water is needed with a zero or very low carbon content. Several methods have been suggested by different workers to prepare carbon-free water, including multiple distillation with oxidants and UV treatment (see also Chapter 16). Several manufacturers now offer equipment producing water with very low carbon concentration (3–5 $\mu g/L$). The difficulty then is to maintain such purity during transport from the water producing unit to the analyser. Carbon blanks of 20 $\mu g/L$ are low enough for most purposes.

15.4.2   Instrument blank

The total procedural blank measured after injection of supposedly pure water includes the water blank and the blank originating in the instrumention, comprising all materials that the sample comes in contact with, *e.g.*, filters, filter holder, injection system, tubing, catalyst. With most instruments it is almost impossible to separate these two sources of blanks, and therefore carbon-free water must be analysed. With the Shimadzu TOC 5000, however, it is possible to estimate the instrument blank. 'Pure' water is injected into the furnace; all traces of carbon are completely oxidised by catalytic high temperature combustion, the water vapour is condensed and the condensate collected in a quartz flask. This water then is reinjected into the furnace for analysis. The value obtained represents the instrument blank alone. Providing that the circuit is clean, most of this blank originates in the catalyst. This blank is not constant; it is high with a new catalyst and decreases to stable values after thorough washing by injections of pure water. The final blank value depends upon the type of catalyst. Shimadzu offers two types of catalysts: Pt-coated alumina and Pt-coated quartz wool. Blank determinations with different catalysts have shown that alumina always exhibits higher blanks than quartz wool based catalysts (Figs. 15-1; 15-2). This observation by *Benner and Strom*, (1993) was confirmed (*Cauwet*, 1994), and the chemical nature of the support was proposed to be responsible. Alumina is an amphoteric oxide and adsorbs acid as well as alkaline molecules. However, CO_2 is an acid, and thus alumina adsorbs it in non-negligible amounts, depending on its partial pressure in the combustion tube. Silica, on the other hand, is an acid oxide, and its capacity to adsorb CO_2 is low. The consequence is that when a high carbon content sample is injected, alumina has a tendency to adsorb some of the CO_2, which is slowly released when samples are injected with low DOC concentrations. This mechanism is much less important with silica. Former suggestions that some organic carbon could remain on the catalyst do not seem to be realistic (*Skoog et al.*, 1997). With the TOC 5000, using platinum coated quartz wool or silica as the catalyst, blanks are about 5–6 μmol/L C, while with alumina blanks are hard to reduce to < 15–22 μmol/L C), and they are more variable. The instrumental blank must be taken into account when calculating concentrations. The most convenient method is to shift the calibration curve to the origin and to subtract the instrument blank from the calculated concentration. All quoted values are based on injection volumes of 100 μL.

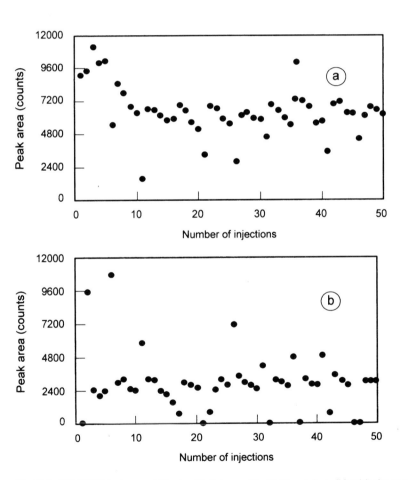

Fig. 15-1. TOC 5000 instrument blank with (a) new Pt–Al$_2$O$_3$ catalyst, (b) with the same catalyst after 5 d of washing with Milli-Q water.

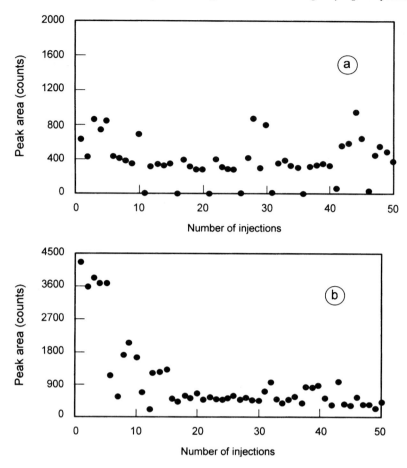

Fig. 15-2. TOC 5000 instrument blank with (a) Pt-quartz wool catalyst, (b) with a Pt-silica beads catalyst.

15.5   Calibration

15.5.1   Carbon calibration

Reliable determinations of DOC concentrations need precise calibration. The software of the TOC 5000 instrument allows computation of a calibration curve through four data points. For seawater, standards generally range between 50 and 500 μmol/L. There are many substances that may be used as reference material to prepare standards, but the selected material must exhibit the following characteristics:
– it must be a widely distributed commercial products of high purity;
– it must not be hygroscopic, so that it can be handled and weighed easily and
– in a reliable fashion;

– it must be soluble in water;
– it must be stable in time to temperature and light; and
– if possible, it should resist microbial degradation.

Potassium hydrogenophthalate ($C_8H_5O_4K$, KHP) is probably the most commonly used standard substance. It is water-soluble and stable, and its aromatic structure guards against easy degradation by bacteria. Glucose is used sometimes, but standard solutions cannot be preserved, even in a refrigerator, for more than a few hours. Among other substances, acetanilide can also be used, especially if nitrogen calibration is needed. Its C/N atomic ratio of 8 makes it a convenient standard for both elements.

A stock solution is prepared 500–2000 times more concentrated than the working standards which are prepared by dilution in pure water or in stabilised (old) seawater. No significant difference has ever been demonstrated between these two methods, and pure water is generally easier to obtain on a permanent basis and in constant good quality. Although most calibration curves cover the range between 50 and 300 μmol/L C, calibration curves at lower concentrations (20–80 μmol/L C), also show good linearity (Fig. 15-3). The curve is memorised in the 'Calibration curve file list' from where it can be recalled at any time. The curve generally does not pass through the origin because the total blank (water + instrument) is not zero. As mentioned above, it is best to shift the curve to the origin and to subtract the instrument blank from the calculated values. In the standards the blank is composite (water + instrument) while in samples there is no water blank.

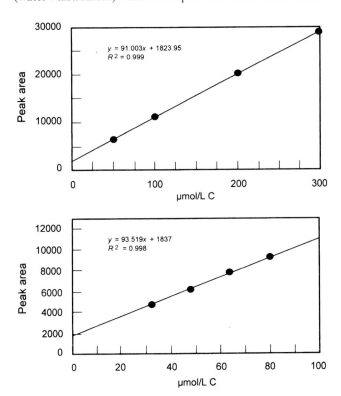

Fig. 15-3. DOC calibration curves obtained with high (50–300 μmo/L) and low (30–80 μmo/L) doses of acetanilide .

15.5.2 Nitrogen calibration

For simultaneous determinations of carbon and nitrogen, a calibration curve must also be computed for nitrogen, using acetanilide as the standard substance (Fig. 15-4).

15.6 Samples processing

15.6.1 Removal of inorganic carbon

Before DOC can be determined in a sample, all inorganic forms of carbon must be eliminated. This is done by acidifying the sample to an optimum pH of 2–3 and bubbling pure gas through it until complete removal of CO_2. Acidification can be carried out using 2 mol/L hydrochloric or orthophosphoric acid. HCl is very efficient, but it adds some more chlorine to the combustion gases. Phosphoric acid is slightly slower to react, but it is not volatile and

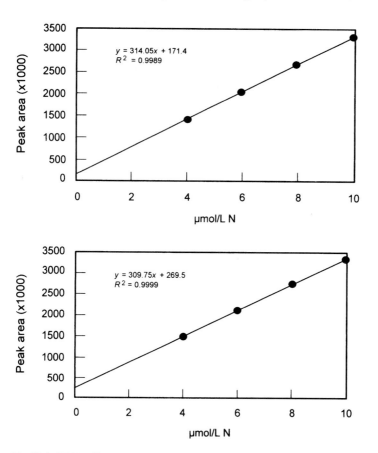

Fig. 15-4. TDN calibration curves obtained with nitrate and acetanilide standards (4–10 μmol/L).

does not contaminate the circuit to the infrared analyser. It was sometimes thought to poison the catalyst, but at the concentration we used no strong effect was seen.

For samples of approximately 10 mL in 16 mm diameter glass vials, 50 μL of 2 mol/L acid are sufficient for acidification. The sample must then be bubbled for about 10 min to eliminate the inorganic carbon completely. The exact conditions depend on the volume of the sample, the shape of the tube or flask used and the size of bubbles. Each configuration needs some individual adjustment.

Without an autosampler, the TOC 5000 is designed to allow bubbling in the same tube as is used for sampling. Only after sparging the sample is it taken up for injection. To save time, the purge gas can be directed so as to sparge the next sample, while the preceding sample is analysed.

15.6.2   Determination of DOC

Figure 15-5 is a schematic representation of the Shimadzu TOC 5000 instrument. Samples are injected directly onto the catalyst which sits in a vertical furnace. The combustion gases are swept by the carrier gas (pure oxygen or synthetic air) first into a cooling coil, then through a plastic vial containing 25 % phosphoric acid (also designed to inject samples for dissolved inorganic carbon (DIC) determinations). The water vapour is then condensed in a cooler (Peltier effect), and the reaction gases are completely dried and purified by passing through two traps (magnesium perchlorate and pure copper to remove water and halogen gases). The purified gases are then transferred to the infrared cell (Fig. 15-5).

The injected volume may vary between not too small for better precision and not too large to reduce accumulation of salt in the combustion tube. Most of the time, 100 μL are a good compromise. Smaller volumes (50 or 25 μL) are sometimes used without significant lack of reproducibility or precision. With the TOC 5000, two or three injections are generally enough to obtain reliable data; the coefficient of variation (CV, the ratio of the standard deviation to the mean) usually is below 2 %. The peaks are integrated and directly compared with the calibration curve. Up to three calibration curves can be programmed, when the concentration range of the samples is unknown.

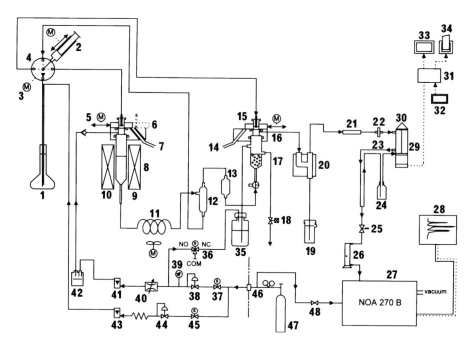

Fig. 15-5. Gas flow diagrams of TOC 5000 and NOA 270 B analysers. (1) sample vessel; (2) microliter syringe; (3) syringe pump type sample injector; (4) 4-port valve; (5) check valve; (6) TC injection port, slide type; (7) drain; (8) TC furnace, 680 °C; (9) combustion tube; (10) catalyst; (11) cooling tube; (12) ultra-pure water trap for checking blanks; (13) non-return trap; (14) drain; (15) IC injection port, slide type; (16) IC rection vessel; (17) IC solution; (18) pinch valve for draining; (19) drain receiver; (20) electronic dehumidifier; (21) halogen scrubber; (22) membrane filter; (23) gas cuvette; (24) CO_2 absorber; (25) needle valve; (26) flow controler; (27) NOA 270 B; (28) integrator; (29) NIDR optical system; (39) light source; (31) PC; (32) key board; (33) screen; (34) printer; (35) IC reagent container; (36) valve; (37) solenoid valve for carrier gas; (38) pressure controler; (39) pressure gauge; (40) mass flow controler; (41) carrier gas flow meter; (42) humidifier; (43) sparge gas flow meter; (44) pressure controler; (45) solenoid valve for sparge gas (46) carrier gas inlet, with filter; (47) purified air; (48) needle valve.

15.6.3   Determination of total dissolved nitrogen (TDN)

The oxidising conditions are such that the carbon and nitrogen in organic matter are transformed into carbon dioxide and nitrogen monoxide (NO), respectivley. Attaching a nitrogen oxide analyser on-line after the infrared cell thus allows the measurement of total nitrogen. Calibration curves obtained in experiments with standards made from organic compounds or the inorganic ions nitrate and ammonium were found to be identical (Fig.15-4), suggesting that the yield of transformation of any form of nitrogen to NO is the same (and probably 100 %). The Sievers NO 270 B analyser is designed to determine NO by chemoluminescence. In a low pressure cell NO is partially transformed into NO_2^*, which, when reverting to the ground state, emits a photon that is determined with a photomultiplier. The signal is recorded and integrated with an integrator. The reproducibility of peak areas is comparable to that of the TOC analyser (CV < 2 %). This configuration allows the measurement of DOC and TDN on the same injection. The DON can be calculated by subtracting the concentration of inorganic nitrogen ($NO_3^- + NO_2^- + NH_4^+$). The same device has already been used successfully by *Hansell*, (1993).

15.6.4   Personal comments

With years of experience, any analyst will be enticed to modify details in the design of the instrument he or she uses and will be able to recognise the origin of small troubles. Here is a list of points to verify in case of non-satisfactory results and of small changes in the analytical instrument.

In the first generation of TOC 5000 instruments, the cooling coil situated immediately after the combustion tube is a frequent source of problems. The glass-to-metal connection is always difficult to seal perfectly, and often some liquid enters between the metal and the glass coating of the tubing. Acids in the combustion gases rapidly corrode the metal, leading to very small holes (like a needle hole), even several centimetres away from the connection. This results in a non–permanent leak flow occurring when the pressure increases during injections. The problem can be solved easily by replacing the glass cooling tube with Teflon tubing. Teflon resists the high temperature, if the injected volume if not too large; 100–200 μL are acceptable. With higher volumes, Teflon will probably soften.

Another source of leaks that is hard to find is the inorganic carbon injection system. When this channel is not used frequently, the tubing draining the acid flask could become somewhat pinched and the evacuation of excess liquid is not efficient. The level rises and bubbling of gas injects droplets in the injection unit. The red O-ring then becomes white and crumbly, and it leaks; in that case drain the acid and change the O-ring. A very sensitive point for reproducibility is the perfect retreat of the last drop in the injection needle. It is not always perfect and depends on the zero calibration of the syringe. Check from time to time, and frequently check (and wash) the injection system.

With the original design, the pressure in the circuit is sometimes high, resulting in leakage through the drain pot. This becomes even more critical, when additional traps (magnesium perchlorate) are installed. Increasing the diameter of tubing in the gas circuit is an effective countermeasure.

15.7   Conclusions

The HTCO method, amended with improvements originating in a decade of experience of many analysts, now seems to be a convenient and reliable way to determine DOC in seawater. Vertical profiles published recently for different parts of the world's ocean (Fig. 15-6) exhibit comparable trends and similar concentration ranges (*Carlson et al.*, 1994; *Carlson and Ducklow*, 1995; *Peltzer and Hayward*, 1996).

Most of the differences observed in results have their origin in analytical conditions, not in the method itself (*Cauwet et al.*, 1997). Supposed very high values could mostly be traced to high blanks that were not well estimated. After several recent comparisons between HTCO and wet methods, it is unlikely that these two techniques give very different results. The methods are not the sources of the discrepancies, it is more likely that the analysts are. Most differences between old and new data were due to poor blank estimations and inferior stability of instruments response (*Sharp*, 1997). The care to be taken and the difficulty associated with these parameters depend upon the samples and the objectives. The lower the concentration to be measured, the more careful we have to be. Studies devoted to estimating concentrations and their distribution on a wide scale do not require the same precision as measurements of DOC variation during biological experiments, where only a small percentage of the total concentration is concerned.

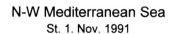

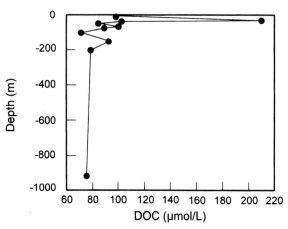

Fig. 15-6. Vertical distribution of DOC determined with the HTCO method at a station in the western Mediterranean.

References to Chapter 15

Benner, R., Strom, M. (1993), *Mar. Chem.*, 41, 153.
Carlson, C.A., Ducklow, H.W., Michaels, A.F. (1994), *Nature (London)*, 371, 405.
Carlson, C.A., Ducklow, H.W. (1995), *Deep-Sea Res.*, 42, 639.
Cauwet, G. (1984), *Mar. Chem.*, 14, 297.
Cauwet, G. (1994), *Mar. Chem.*, 47, 55.
Cauwet, G., Miller, A., Brasse, S., Fengler, G., Mantoura, R.F.C., Spitzy, A. (1997), *Deep-Sea Res.*, 44, 769.
Collins, K.J., Williams, P.J.LeB. (1977), *Mar. Chem.*, 5, 123.
Hansell, D.A. (1993), *Mar. Chem.*, 41, 195.
Hedges, J.I., Farrington, J. (1993), *Mar. Chem.*, 41, 5.
MacKinnon, M.D. (1978), *Mar. Chem.*, 7, 17.
Menzel, D.W., Vaccaro, R.F. (1964), *Limnol. Oceanogr.*, 9, 138.
Peltzer, E.T., Brewer, P.G. (1993), *Mar. Chem.*, 41, 243.
Peltzer, E.T., Hayward, N.A. (1996), *Deep-Sea Res.*, 43, 1155.
Sharp, J.H., (1997), *Mar, Chem.*, 56, 265.
Skoog, A., Thomas, D., Lara, R., Richter, K.U., (1997), *Mar. Chem.*, 56, 39.
Sugimura, Y., Suzuki, Y. (1988), *Mar. Chem.*, 24, 105.
Suzuki, Y., Sugimura, Y., Ito, T. (1985), *Mar. Chem.*, 16, 83.
Wangersky, P.J. (1975), *Adv. Chem. Ser.*, 147, 148.
Williams, P.J.LeB. (1995), *Mar. Chem.*, 51, 17.

16 The automated determination of dissolved organic carbon by ultraviolet photooxidation

P.J. Statham and P.J. le B. Williams

16.1 Introduction

The dissolved organic material (DOM) contained in ocean water represents a major component in the global organic carbon budget. Ironically, despite the enormous amount of organic material present in the world's oceans, its accurate determination has in the past presented the marine chemist with considerable analytical problems. The DOM content of deep ocean water appears to be relatively constant whilst in the productive zone of the ocean the levels are generally higher and more variable. The more our understanding of the planktonic cycle of production and decomposition improves, the more it is apparent that a great deal of metabolism and interaction between organisms is occurring through the extracellular organic pool (see, *e.g.*, *Aubert and Gauthier*, 1977; *Williams*, 1981; *Cherrier et al.*, 1996). Organic compounds dissolved in seawater can react with and alter the chemical state of many inorganic species, sometimes with important biological consequences (see, *e.g.*, *Barber et al.*, 1971) and a complex interaction is set up between the chemistry and biology through the external organic environment. Although these interactions could be mediated by minor constituents of the dissolved organic pool, bulk analysis of dissolved organic material is needed to determine the overall state of the production-decomposition cycle, and its spatial and temporal variability (*e.g.*, *Carlson et al.*, 1994).

Physical methods, such as ultraviolet (UV) absorption, are attractive in that they are usually simple, readily automated and may provide real-time data. However, no suitable single physical property is common to all organic compounds. Furthermore, UV absorption may suffer interference from dissolved inorganics (bromide and nitrate; see *Ogura and Hanya*, 1966) and particulate material. As a consequence it is necessary to resort to chemical methods. The oxidation of the organic material and subsequent determination of the produced carbon dioxide thus seems the most logical approach to DOM analysis and has been generally adopted. The oxidation procedure itself becomes the critical issue in the consideration of the available alternative methods.

There was a major revival in interest in DOM following the claims (*Suzuki et al.*, 1985; *Sugimura and Suzuki*, 1988) that the earlier methods for dissolved organic nitrogen (DON) and dissolved organic carbon (DOC) were seriously in error. Both claims turned out to be false (*Suzuki*, 1993). The debate as regards techniques for the measurement of DOC has almost exclusively become directed at the persulphate and the high temperature catalytic oxidation (HTCO) methods (see Chapter 15). Sadly in this debate the third method, UV photooxidation, was given little consideration despite some special virtues. *Sharp* (1997) has correctly observed that the outcome of the flurry of activity following the original papers in the mid-to late 1980s was instruments of greater convenience and consistency. However, a survey of a number of papers (*Peltzer et al.*, 1996; *Fry et al.*, 1996; *Chen and Wangersky*, 1993a; *Ridal and Moore*, 1993) and the contributions to the 1993 Special Issue of *Marine Chemistry*) leads one to a view that we are still some way from being able to make unequivocal statements about our methods for DOC determinations.

There has been a tacit assumption that the HTCO procedures should give complete oxidation and, providing the blank is properly assessed, an accurate analysis for DOC. The blank associated with the HTCO method was the root of the earlier erroneous claims of substantially higher DOC values for seawater. The careful analysis of a major inter-comparison (*Sharp et al.*, 1995) and the meticulous examination of the instrument blank of the high temperature catalytic oxidation (HTCO) method by *Cauwet* (1994) brought the community back to the previously held view (see *Gershey et al.*, 1979), that the discrepancy between HTCO and the persulphate method was of the order of 15 % at most rather than a factor of two. The expectation that the HTCO procedure is accurate in that it converts all the marine DOC into carbon dioxide is at odds with the observation by *Fry et al.* (1996) that sealed tube combustions (whose blank can be accurately assessed) gave deep water DOC values some 15–20 % greater than HTCO determinations. The situation is somewhat different in the case of the UV-photooxidation procedures. The reagent blank may be much more readily assessed by altering the dosing of the reagents and by switching off the lamp. It is also (see *Collins and Williams*, 1977) a simple matter to determine the first order rate constant for the reaction which may be used to help assess accuracy. Without supplementary information this would not be seen as a definitive test. The conclusion of *Gershey et al.* (1979) that the difference between persulphate methods and high temperature dry combustion was about 15 %, is consistent with present views and, accordingly, their associated conclusion that the difference between the dry combustion and the photooxidation techniques was less than 5 % must logically also stand.

This view of the comparability of data within *ca.* 5 % between UV and HTCO methods apparently runs in the face of observations by *Chen and Wangersky* (1993b), *Miller et al.* (1993) and *Ridal and Moore* (1993). There are a number of inhibitions to accepting these observations without question. Firstly, all three comparisons were made at the time prior to when there was widespread recognition that the HTCO blanks were seriously underestimated. There is also a question mark over the performance of the photooxidation procedure used by *Chen and Wangersky* and *Ridal and Moore*. *Chen and Wangersky* reported low recoveries (76 %) for glycine for their photooxidation analysis, whereas *Collins and Williams* found complete oxidation, suggesting that the photooxidation conditions were substandard in the former case. *Ridal* and *Moore* noted incomplete oxidation (*ca.* 70–90 %) with their continuous photooxidation procedure yet 95 % oxidation in the case of a batchwise method. Significantly they reported using a 550 W lamp for the continuous oxidation but a 1200 W lamp for the batchwise procedure. Experience is that the reduction in performance in the case of UV lamps is more than proportional to the reduction in wattage (*Armstrong and Tibbits*; 1968), *e.g.,* a 380 W lamp gave rates five-fold faster than a 240 W source. One would suppose that *Chen and Wangersky's* work was also done with a 550 W rather than a 1200 W source, and the performance will have suffered accordingly. Circumstantial evidence for the effectiveness of photooxidation procedures comes from the previously mentioned report by *Peltzer et al.* (1996) that seawater with very low organic carbon content ($<3\,\mu\text{mol/L}$ C) can be prepared by exposing it to sunlight in the presence of hydrogen peroxide. If these relatively mild UV photon doses in large volumes can be effective in decomposing DOC it may *a priori* be expected that the intense doses emitted from the medium-pressure lamps and the small thickness of water (1–2 mm) exposed would be as, if not much more, effective.

The conclusion from the foregoing arguments must be that given our current state of knowledge, and when carefully used, the photooxidation technique provides a viable and effective technique for the determination of DOC in seawater. As with other procedures, its accuracy will always need to be established empirically; the technique however has the overwhelming advantage that its blank can be established systematically and does not rely on the availability of presumed organic free seawater. The position will be further clarified when reliable certified reference materials for dissolved organic carbon in seawater become available. The UV photooxidation method as described in this section has been in regular use at the Southampton laboratory for over 15 years, and arguably during that period has provided the most consistently accurate analyses of DOC available.

16.2   Analytical strategy

Some early wet chemical techniques for measuring DOC in seawater related the amount of a chemical oxidant consumed to the amount of DOC present in the sample. See *Riley* (1965) for a review of these methods. However, more resistant organic matter will not be destroyed, and the techniques are prone to interferences from inorganic chemical reductants in samples. A more rigorous approach is to convert all organic carbon present in the sample completely into carbon dioxide and then to determine the amount of this gas produced. Therefore, contemporary determinations of DOC in seawater typically consist of three stages: (1) initial removal of inorganic carbon species, (2) oxidation of the organic material into carbon dioxide and (3) quantification of the carbon dioxide produced.

16.2.1   Removal of inorganic carbon

The inorganic carbon content of seawater is typically between 2000 and 2400 μmol/L C, whilst the organic carbon concentration is much lower and typically lies between 40 and 200 μmol/L C. The inorganic component must be totally removed before the determination of DOC, and this is done by acidifying the sample to reduce the pH below 4, under which conditions it is converted into carbon dioxide, which can be purged from solution by a suitable gas. Such gas purging of the sample can remove volatile organic carbon (VOC) compounds which may be present in the sample. For most seawater samples remote from active biological activity, the VOC appears to be < 5 % of the total carbon content of the sample (*MacKinnon*, 1981). However, in reducing environments the VOC fraction may be significant and *Barcelona* (1980) reported that in some of the oceanic sediment porewaters studied, > 50 % of the organic carbon was removed by gas purging. Some commercial systems (*e.g.*, Dohrman DC-50) have used an initial purge and analysis cycle for VOC, and specialised systems exist for trapping and quantification of individual VOC compounds in marine samples (*e.g., Bianchi and Varney*, 1989).

16.2.2   Oxidation of organic carbon to carbon dioxide

Methods for the oxidation of DOC fall into two basic categories, high temperature catalytic oxidation (HTCO) and wet chemical techniques including UV photooxidation. The use of high temperature catalytic oxidation for the conversion of organic carbon into carbon dioxide is described in Chapter 15 of this volume, and the use of a system based on UV photooxidation of dissolved organic carbon is described in the following sections.

A widely used wet oxidation method is the persulphate technique of *Menzel and Vaccaro* (1964), in which a known volume of seawater is purged of inorganic carbon, autoclaved in a sealed ampoule in the presence of peroxydisulphuric acid to convert DOC into carbon dioxide, and this gas is then measured by an infrared gas analyser (IRGA). The method has the advantages of using relatively inexpensive and unsophisticated equipment (with the exception of the IRGA), and it being possible to use the technique at sea. However, several replicates are normally required to ensure good precision, making the technique relatively slow and labour intensive, and the efficiency of the persulphate oxidation process for oceanic dissolved organic matter as used in earlier studies appears to be slightly less than that of the HTCO and UV methods (*Gershey et al.*, 1979). The persulphate technique has been used in a refined form in a recent intercalibration exercise (*Sharp et al.*, 1995), and the data show a closer agreement with HTCO measurements than those observed with the older persulphate technique.

The use of high intensity UV radiation for the decomposition of inorganic material in fresh and saline waters is well established (*e.g., Armstrong and Tibbets*, 1968; *Henriksen*, 1970). Automated systems for seawater DOC determination have been developed by *Ehrhardt* (1969), *Collins and Williams* (1977) and *Mantoura and Woodward* (1983) using medium pressure lamps, and by *Schreurs* (1978) using a low pressure lamp. These automated photooxidation procedures have many advantages: high sampling frequency, reduced human error, discreet and continuous shipboard analysis and the self-cleaning action of the equipment under the UV irradiation gives low and consistent blanks. The precise nature of the photooxidation process is still poorly understood, and much of the development of existing techniques has been done on an empirical basis. Whilst low pressure UV lamps are appealing as they do not have ozone generating problems and the greater bulk associated with the medium pressure lamps, it is essential (*Schreurs*, 1978) to add persulphate and a buffer to the reagent stream to ensure complete oxidation of organic matter; this addition of reagents has an associated potential blank problem. Earlier techniques using medium pressure lamps also used persulphate additions, but when optimized, as described here, the low concentrations typical of open ocean and coastal waters (*i.e.*, $< 400\ \mu$mol/L DOC) the persulphate additions can be omitted. If samples with high DOC concentrations are to be analysed, the persulphate and buffer can be added, or the sample diluted to bring the concentration of DOC below *ca.* $400\ \mu$mol /L DOC.

There have been extensive intercalibrations (*e.g., Gershey et al.*, 1979; *Sharp et al.*, 1995) and comparisons of data collected by different techniques for stable oceanic water masses (*Sharp*, 1997). The current view is that there is little difference between the wet chemical techniques used and the careful application of the HTCO methods, which infers consistent and accurate data are produced by these techniques (see also comments in the introduction).

16.2.3 Carbon dioxide detection

Several methods exist for the determination of carbon dioxide, but only a limited number have been applied in seawater DOC determinations.

With the conductivity cell as described by *Ehrhardt* (1976), the change in conductivity of a stream of dilute sodium hydroxide solution is related to the mass of carbon dioxide absorbed. Disadvantages to this method include the temperature dependence of electrical conductivity and the potential for interference from compounds other than carbon dioxide, which must be removed from the gas stream. Flame ionisation detector (FID) systems measure methane, which results from the reduction of the carbon dioxide in the gas stream from the DOC analyser by an in-line reductor. Whilst FID systems have a wide dynamic range, and good sensitivity, additional gases are required in the analysis, and the hydrogen used for the flame is explosive. A colorimetric technique described by *Schreurs* (1978) relies on the partial neutralisation of an acid solution by the carbon dioxide. The neutralization is followed by the change in absorbance of a coloured phenolphthalein solution. The technique is compatible with autoanalyser systems; it is not as sensitive as the other techniques mentioned here.

The instrument almost universally used in current techniques for determining carbon dioxide in DOC determinations is the non-dispersive IRGA, in which the absorption of long wavelength radiation by carbon dioxide in an analysis cell is measured relative to the absorption of gas free from carbon dioxide in a reference cell. As the technique measures concentration, the flow rate of the gas through the cell must be constant. With the solid state detectors used, the working concentration range is wide, and the technique is non-destructive. These modern instruments are stable, very reliable, self-contained and suitable for making measurements at sea.

16.2.4 Sampling, filtration and storage

Great care is required if truly representative samples of seawater are to be collected for DOC determinations (see general comments about sampling elsewhere in this volume, *e.g.,* Chapters 1, 2 and 15). A major potential problem is contamination. Sources of contamination can be very variable, and clearly materials which come into direct contact with samples must be free of soluble organic materials; specific approaches to cleaning are described below. A particular problem for DOC samples is contamination by volatile water soluble compounds such as ketones and alcohols, which may be in common use in laboratories, and can be introduced through the vapour phase from the atmosphere into samples. It is sensible to limit exposure of the sample to the laboratory atmosphere and to have dedicated areas for this type of work away from potential contamination sources.

16.3   Sampling

Research ships potentially are significant sources of a wide variety of oils and other organic compounds, which may be introduced into samples, and constant vigilance is required to ensure that the collected samples are not contaminated. Modern water samplers are frequently made of plastic, but it appears that if well conditioned in seawater, any leaching of organics is small (see, for example, use of Niskin type samplers by *Sharp*, 1973; *MacKinnon*, 1978; *Sharp et al.*, 1995). It is also possible for hydrophobic organic materials to be adsorbed onto the inner walls of sampling bottles. This may be a particular problem when open bottles pass through the organic rich surface micro-layer, and use of Go-Flo type samplers (General Oceanics), which pass through the surface layer closed, may be advantageous. Samplers suitable for use in hydrocarbon studies should in general also be suitable for DOC sampling.

16.4   Filtration

Filtration from the stand point of particulate organic carbon (POC) is considered elsewhere in this volume (Chapters 15 and 17). Natural waters contain a spectrum of particle sizes from atomic species to large organisms, and the conventional filtration cut-off (about $0.4 \mu m$) to differentiate between 'dissolved' and 'particulate' forms of organic material is thus arbitrary, and an increasing number of studies are examining the nature of DOC in the colloidal size range. Filtration using the conventional size cut-off has, however, several advantages: organisms that may alter the DOC content even in the short time prior to storage and analysis can be removed and the separated biologically derived particulate fraction can provide information on the biological processes. Temporal and spatial variability may be missed using total organic carbon (TOC), *i.e.*, dissolved plus particulate determinations, and for meaningful TOC determinations, the homogeneity of the sample must be ensured. HTCO is the only procedure that can be used reliably for direct TOC determinations, as quantitative destruction of 'particulate' material cannot be assured in the chemical oxidation, and particularly photooxidation, techniques (*e.g.*, see *Gershey et al.*, 1979).

Whatman glass fibre filters with smallest effective pore size (GF/F; $0.7 \mu m$) are most commonly used for filtration, after cleaning by heating at $500 \,°C$ for 4 hours or longer, in combination with all glass filtration equipment. Only a minimal vacuum, or preferably low positive pressure filtration, should be used with samples containing significant plankton biomass, as cell lysis may lead to elevated DOC levels. Glass fibre filters with their large surface area do appear to adsorb some organics from solution (*Banoub and Williams*, 1973), but conditioning by rinsing the filter with the sample leads to a small loss from the filtrate. If a smaller particle size cut-off is required, polysulphone filters ($0.2 \mu m$) have been recommended (*Norrman*, 1993). Filtration should always be done as rapidly as possible after sample collection to minimise changes in the DOC content of the sample.

16.5   Sample storage

Ideally, analysis of the sample should be done immediately after collection and filtration. However, this is not always possible and an effective storage protocol, which includes a means of stopping microbial activity which can change the DOC concentration, is needed. Samples in our laboratory have been frozen immediately after filtration, in pre-heated (500 °C for at least 4 h) soda glass bottles fitted with caps having prewashed Teflon liners. The bottles are supported at an angle of approximately 45° during freezing to allow expansion without breaking the bottles; low salinity and freshwater samples may still break the bottles on freezing. In such cases the addition of 2.5 mL of 1.6 % (m/v) mercury(*II*)chloride solution per litre of sample and storage in the dark at about 4 °C has been shown to be an effective preservation method (*Parker*, 1981). *Tupas et al.* (1994) working in the central North Pacific compared shipboard measurements of DOC with measurements on samples stored frozen and subsequently analysed in a shore laboratory, and found no significant difference between the two data sets.

16.6   Analysis

16.6.1   Equipment

The equipment is based on the system described by *Collins and Williams* (1977) and is essentially the same as that in *Statham and Williams* (1983). A flow diagram of the analysis is shown in Fig. 16-1.

After acidification and removal of inorganic carbon, oxidant and buffer are added to the sample if required, and the liquid stream passes through a silica coil surrounding a high intensity UV light source. Organic material is oxidised and the resulting carbon dioxide is collected in a stream of oxygen and its concentration measured with an IRGA. The general organisation of the DOC analyser is shown in Fig. 16-2. An important modification of the standard autosampler is the replacement of the tray with one capable of holding 40 of the 15 ml glass tubes used to contain the seawater samples during the analysis. The DOC analyser has been used in continuous mode on-board ship; modifications to allow this application are given in *Statham and Williams* (1983).

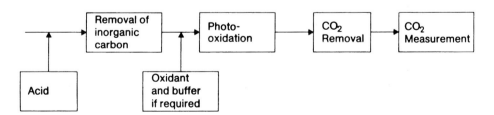

Fig. 16-1. Flow diagram of analytical procedure.

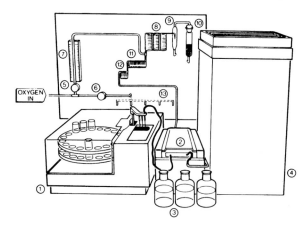

Fig. 16-2. Configuration of major components in the DOC analyser. (1) Auto sampler with arm in wash position; (2) peristaltic pump; (3) wash water and reagent containers; (4) irradiator assembly housing; (5), (6) flow regulators; (7) flow meter; (8) CO_2 stripping coils; (9) gas–water separator; (10) drying column; (11) hydroxylamine hydrochloride addition and mixing coils; (12) hydrochloric acid addition and mixing coils; (13) oxygen outlet, for degassing and reference cell of IRGA.

16.6.2   Initial removal of inorganic carbon

Below a pH of approximately 4 all inorganic carbon in seawater is present as carbon dioxide. Addition of dilute hydrochloric acid is used to reduce the pH, and the carbon dioxide is removed by purging with oxygen. The arm of the AutoAnalyser AA2 sampler has been modified to hold three narrow stainless steel tubes in addition to the sample removal tube; these additional tubes allow purging of the sample being sampled, and the two following samples. Oxygen at approximately 150 mL/min is passed through the additional tubes to remove carbon dioxide from solution. Thus the vial being sampled and the two preceding it are bubbled simultaneously. The normal sampling time is 5.5 min, giving each sample a minimum of 11 min bubbling which reduces the inorganic carbon content of the sample to less than 0.1 % of its original concentration. If shorter sampling, and therefore bubbling, times are to be used, the quantitative removal of the carbon dioxide must be checked.

16.6.3   The irradiator assembly

The silica coils and UV lamp are housed in a cylindrical aluminium light shield above a cooling fan (Fig. 16-3). The sample irradiation coil is constructed from 30 m of high purity quartz tubing (2 mm internal, 4 mm outside diameter), wound on an octagonal frame of silica (mean diameter 120 mm; the octagonal geometry simplifies the coil construction). In the present design a further 30 m silica tube helix surrounds the inner one and is used to supply irradiated wash water to the sampler. Attenuation of radiation by the sample coil and a reduced UV flux because of the shorter coil height leads to a slightly reduced photooxidation efficiency. This causes no problem here, as the second irradiation is a polishing step for the water, which is already very low in DOC. Positioning wash and sample coils above one another may have some advantages (as used by *Mantoura and Woodward*, 1983), but it is important to ensure the decomposition of DOC during irradiation is optimised and quantitative.

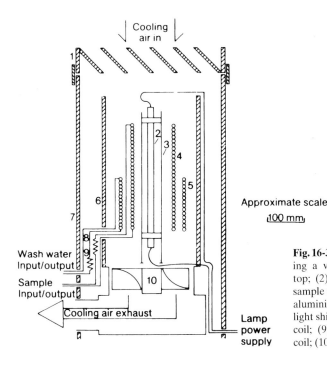

Cooling
air in

Wash water
Input/output

Sample
Input/output

Cooling air exhaust

Approximate scale
⌊100 mm⌋

Lamp
power
supply

Fig. 16-3. The radiator assembly, show-ing a vertical cross section. (1) louvre top; (2) UV lamp; (3) quartz tube; (4) sample coil; (5) wash water coil; (6) inner aluminium light shield; (7) outer wooden light shield; (8) sample output air-cooling coil; (9) wash water output air-cooling coil; (10) fan.

The axially mounted UV lamp (1 kW medium pressure mercury arc, Hanovia tube type 6751; Primarc, 753 Deal Avenue, Slough, UK) must be jacketed by a quartz tube (55 mm outside diameter) to prevent over-cooling of the electrodes and a consequent reduction in lamp life. The correct functioning of the lamp is checked by measuring the current flow with an in-series ammeter, and a timer monitors lamp usage. A down-draft fan (air flow *ca.* 20 m³/min) is used to remove ozone and provide cooling, as the irradiator design is simpli-fied without the gas tight seals that would be required for a reversed air flow. Further shields may be required to prevent UV radiation leakage, particularly at the top of the assembly.

The UV radiation output of the lamp will deteriorate after a time, which is typically in excess of 500 h in our experience. It is essential that the performance of the lamp is monitor-ed routinely, and empirically this can be done by comparing the decomposition of thiourea standards relative to equivalent DOC concentration standards of the photo-chemically labile potassium oxalate.

16.6.4   Removal of carbon dioxide from the irradiated sample, and drying of the gas stream

After leaving the irradiator coil, the liquid stream is mixed with 2 mol/L hydrochloric acid to ensure a pH < 2. Hydroxylamine hydrochloride is then added to the fluid stream, to reduce any chlorine gas produced during the irradiation back to chloride. The carbon diox-ide is stripped from the liquid stream into oxygen which is added before passage of the gas–liquid mixture through a series of coils. The liquid is separated from the gas stream and dried (see Fig. 16-4). The oxygen gas flow to this unit is smoothed and controlled by an auto-

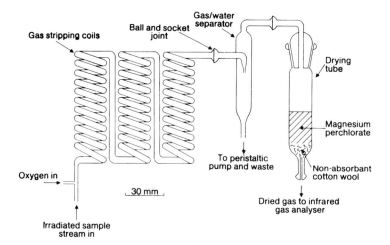

Fig. 16-4. Gas stripping coils for removal of carbon dioxide form the liquid stream, gas–water separator and drying tube assembly.

matic mass flow regulator (*e.g.,* Brooks Model 8774), and is monitored using a gas flow meter. Magnesium perchlorate (14–20 mesh) is used as desiccant; a column 18 mm diameter × 25 mm long is adequate for several hours of operation. The drying agent is held in a tube with ball and socket connectors at either end to allow easy changing of the unit.

16.6.5   Determination of carbon dioxide in the gas stream by an IRGA

Suitable IRGA instruments include the ADC Model 2250 (Analytical Development Company Ltd., Hoddesdon, Hertfordshire, EN11 0AQ, UK) and the Li-Cor Model 6252 (Li-Cor Inc., PO Box 4425, Lincoln, NE 68504, USA). Both have solid state detectors, which provide stability and a wide working concentration range, and data can be sent to integrators or a computer with appropriate data processing software; in the examples given here output is to a chart recorder. The gain of the instrument is adjusted so that a 100 ppm carbon dioxide standard gas mixture gives a full-scale response. This range is suitable for samples containing 0–300 μmol/L DOC. The reference cell is continually flushed with oxygen. Some small baseline differences can arise from flow rate differences and the carbon dioxide purity of the gases if different analysis and reference gas streams are used; it is advisable, therefore, to set the instrument zero with oxygen carrier gas flowing through both cells.

16.6.6   The autoanalyser manifold

The manifold and flow diagram are given in Fig. 16-5.

To reduce the potential for contamination for the low concentrations of DOC typical of coastal and oceanic systems, it is important to minimize contact with plastic materials in the analytical system. All linking tubing should be butt-jointed glass (ideally with ground flat ends) connected with a sleeve of silicone, or PTFE heat-shrink, tubing. Careful butt-joining prevents breakdown of the segmenting bubble pattern and consequent mixing of samples.

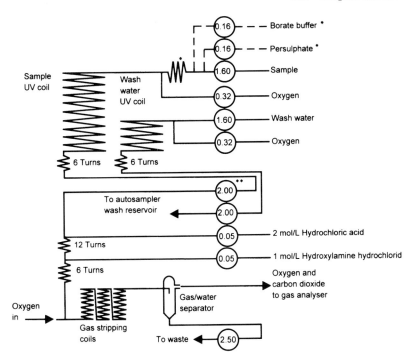

Fig. 16-5. The autoanalyser manifold and pump assembly. * High DOC samples only; ** increase to 2.5 ml/min if persulphate and borate are added.

Where flexible connections are required, as with the sampler and between reagents and the pump, narrow bore PTFE tubing is suitable. No detectable difference has been found in the blank level when either conventional PVC or silicone rubber pump tubing has been used, although the latter may be more resistant to the chlorine produced during the photooxidation process. Contamination from airborne particles, which may contain soluble carbon, is minimised by an acrylic cover over the sample tray of the system. Volatile water soluble organic solvents, cigarette smoke and other sources of spurious organic contamination must be kept away from the instrument, and all reagent solutions kept stoppered between analytical runs.

16.7   Reagents and standards

16.7.1   Low organic carbon content (LOCC) water

Low organic carbon content water can be readily prepared by bulk UV irradiation of distilled water using an irradiator similar in design to Fig. 16-3, but with the coil replaced by 200 mL capacity (35 mm diameter, main body 255 mm long) silica tubes positioned around the lamp. Free radical formation is aided by the addition of *ca.* 0.25 mL of 30 % (w/v) hydrogen peroxide per full tube, and the sample is normally irradiated for at least 4 h. Ideally, this water should be used immediately after preparation and cooling, on order to minimize potential for contamination during storage.

16.7.2   Analytical reagents

Hydrochloric acid, 2 mol/L: 178 mL of analytical reagent grade (a.g.) concentrated hydrochloric acid is diluted to 1 L with LOCC water

Hydrochloric acid, 0.1 mol/L: LOCC is used to give a 20–fold dilution of the 2 mol/L hydrochloric acid.

Hydroxylamine hydrochloride, 1 mol/L: 17.4 g of reagent (a.g.) is dissolved in 250 mL of LOCC water.

Oxygen: High purity grade.

Potassium persulphate (if required): 25 g reagent (a.g.) is dissolved in 1 L of LOCC water.

Sodium tetraborate (if required): 70 g reagent (a.g.) is dissolved in 1 L of LOCC water.

The blank obtained from different batches of persulphate and the tetraborate may vary significantly from batch to batch and from manufacturer to manufacturer. The blank from the persulphate solution can be substantially reduced, with little destruction of the reagent, by heating at 70 °C for 2 h (*Parker*, 1981).

16.7.3   Standards

A stock standard solution of 83.3 μmol/mL DOC (1 mg/mL C) is prepared by dissolving 0.7669 g potassium oxalate monohydrate (a.g.) in 100 mL of LOCC water. This stock solution, stored in the dark at room temperature, is stable for at least 8 months in our experience. Oxalate has the advantage as a standard that it is resistant to microbiological destruction and is photochemically labile. Standards for calibration of the instrument are prepared by dilution of the stock solution with LOCC water immediately before the analysis. No detectable variation in analytical signal has been noted when LOCC distilled water is replaced by LOCC seawater for the preparation of standards.

The issue of the accuracy of measurements of DOC in seawater is still open, and it is clear that development of certified reference materials for DOC in seawater will be necessary to improve the inter-comparability of data between laboratories and the different techniques in use.

16.7.4   Oxygen supply

As oxygen gas is used to purge the sample of carbon dioxide prior to UV irradiation it is essential that the oxygen gas stream does not contain any potentially contaminating organic compounds. The oxygen stream from the cylinder is therefore passed through a gas purification tube containing sequentially aliquots of soda lime (for carbon dioxide removal), 5 Å molecular sieve and dessicant (granular calcium sulphate) to remove contaminants prior to the use of the gas in the equipment and IRGA instrument.

16.8 Analytical procedure

16.8.1 Pre-analysis steps

Normally a 10 mL sample is used, and acidification with 250 μL of 0.1 mol/L HCl reduces the pH of the seawater to below 4. Samples are contained in 15 mL glass specimen tubes (diameter 18 mm, height 76 mm), and 40 of these are held in the modified sampler tray. Sample tubes are cleaned by heating overnight at 500 °C and covered with aluminium foil, also decontaminated by heating, until use. Screw cap versions of the 15 mL tubes, fitted with PTFE liners can be used to store samples, and the tubes can then be directly put onto the sample tray thus reducing handling steps and potential contamination.

Although the precise volume taken from the sample tube is determined by the pumping rate and sampling time, the sample should be dispensed with an accuracy of better than 5 %, as addition of the acid does give a small dilution factor. Before starting a series of analyses, the reagents and first sample should be purged with oxygen to remove carbon dioxide.

16.8.2 Analysis

After tensioning the pump tubes and fitting the pressure plate, the pump can be turned on. The oxygen flow to the analyser should be started immediately to prevent build-up of liquid in the gas stripping coils, which otherwise could lead to flooding of the drying column. The irradiator fan and then the UV lamp are turned on. The lamp current normally takes a few minutes to stabilise. For the first few minutes of operation, whilst the tubing is being flushed, the sample input can be temporarily attached to the wash water reservoir. During the analysis the oxygen gas flow through the stripping columns should be monitored carefully and maintained at 150 mL/min.

16.8.3 Sampling frequency

The sample to wash ratio on the AutoAnalyser AA sampler is controlled by a cam system which revolves once every 6 min; more recent autosamplers use electronic or other mechanical methods to control the sampling frequency. A sample time of 5.5 min and a wash time of 0.5 min, *i.e.*, 10 samples per hour, has been found to give a good peak plateau and adequate separation between traces. With the system described here, there is about 45 min time lag between sampling and the corresponding gas analyser recorder trace appearing. It is good practice to send a series of standards through the apparatus to ensure the system is functioning correctly, before any samples are analysed.

16.8.4   Blanks and calibration

LOCC water blanks are run between sets of standards and samples to give an analysis baseline. If samples containing dissolved organic carbon in excess of 400 μmol/L DOC are to be analysed, a significant amount of sample carryover may be encountered, and two LOCC water samples in succession may be required to establish a true baseline. Standards are usually run after every ten samples and an example of a typical trace from an analysis is shown in Fig. 16-6.

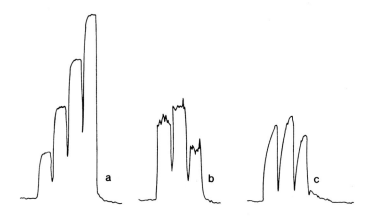

Fig. 16-6. Chart recorder output for infrared gas analyser during DOC analysis runs. (a) Good traces; 83, 167, 250 and 333 mol DOC/L; (b) traces affected by particles in the liquid stream and (c) traces affected by degraded or excessive amount of drying agent.

16.8.5   System shut-down

When the final vial has been sampled, the sampler arm can be stopped in the wash position until the last trace on the gas analyser record has ended. The pump, and then without delay, the oxygen flow, irradiator fan and UV lamp are all turned off. Before removing the pressure plate from the autoanalyser pump, it is advisable to clamp the transmission tubing leading to the irradiator coils, thus keeping them filled with irradiated water until the next analyses.

16.9   Data analysis

Peak heights of calibration standards, measured from the LOCC baseline, are used for calibration; the best fit line is found by linear regression analysis. Peak measurement using an integrator and/or suitable computer software is an alternative method for calibrating the instrument.

16.10 Method performance

An estimate of the blank for filtration and storage on-board ship and shore laboratory analytical steps has been determined by passing replicate photo-oxidised seawater samples through the procedure. The mean overall blank was 6 μmol/L DOC with a coefficient of variation of 10 % (n=5), giving a detection limit (taken as 3σ of the blank) of 2 μmol DOC/L. The blank of the analytical procedure alone (with no persulphate or blank additions) is about 2 μmol/L DOC (*Gershey et al.*, 1979). Replicate analyses of two stored seawaters gave mean DOC concentrations of 57 μmol/L DOC (coefficient of variation 6.0 %, n=30) and 92 μmol/L DOC (coefficient of variation 2.4 %, n=30).

16.11 Identification and solution of common problems

This section describes problems that have been encountered using the analyser, and their remedies; general problems associated with autoanalyser systems are covered in Chapter 10 of this volume.

If the bubble pattern breaks down, segments of sample solution will begin to mix with a consequent degradation of peak shape. An irregular bubble pattern is usually caused by poor joints between sections of tube, or the build up of a film of hydrophobic material on the inner walls of the tubes. The latter problem can be solved by passing a warm 10 % m/v solution of sodium triphosphate through the system. All tubing must be thoroughly washed before and after cleaning with this solution to prevent formation of a precipitate on contact with seawater samples, which may clog the apparatus.

Irregular peak shapes may be caused by several factors. Spikes on the peak plateau, as in Fig. 16-6 b, often indicate the presence of particulate organic matter in the sample stream; each pulse of carbon dioxide monitored by the gas analyser represents the partial destruction of a particle. The poorly defined peaks shown in Fig. 16-6 c have been associated with the use of excessive amounts of drying agent. It appears that there can be an exchange of carbon dioxide between the gas phase and the magnesium perchlorate giving rise to smearing of the signal.

A greatly reduced lamp life (*i.e.*, less than 100 h) can be caused by differential cooling of the electrodes. The aluminium tubing supporting the silica shielding as shown in Fig. 16-3 is intended to prevent cooling of the lower end of the lamp.

References to Chapter 16

Armstrong, F.A.J., Tibbits, S. (1968), *J. Mar. Biol. Assoc. U.K.*, 48, 895

Aubert, M., Gauthier, M.J. (1977), *Mar. Chem.*, 5, 553

Banoub, M.W., Williams, P.J. le B. (1973), *J. Mar. Biol. Assoc. U.K.*, 53, 695.

Barber, R.T., Dougdale, R.C., MacIsaac, J.J., Smith, R.L.(1971), *Inv. Pesq.*, 35, 171.

Barcelona, M.J. (1980), *Geochim. Cosmochim. Acta.*, 44, 1977.

Bianchi, A., Varney, M.S. (1989), *Analyst*, 114, 47.

Carlson, C.A., Ducklow, H.W., Michaels, A.F. (1994), *Nature, London*, 371, 405.

Cauwet, G. (1994) *Mar. Chem.*, 47, 55.

Chen, W., Wangersky, P.J. (1993a), *Mar. Chem.*, 41, 167.

Chen, W., Wangersky, P.J. (1993b), *Mar. Chem.*, 42, 95.

Cherrier, J., Bauer, J.E., Druffel, E.R.M. (1996), *Mar. Ecol. – Progr. Ser.*, 139, 267.

Collins, K.J., Williams, P.J. le B. (1977), *Mar. Chem.*, 5, 123.

Ehrhardt, M. (1969), *Deep-Sea Res.*, 16, 393.

Ehrhardt, M. (1976), in: *Methods of Seawater Analysis* (1st edn.): Grasshoff, K. (Ed.). Weinheim: Verlag-Chemie, 1976, pp. 289–297.

Fry, B., Peltzer, E.T., Hopkinson, C.S., Nolin, A., Redmond, L. (1996), *Mar. Chem.*, 54, 191.

Gershey, K.M., MacKinnon, M.D., Williams, P.J. le B., Moore, R.M. (1979), *Mar. Chem.*, 7, 289.

Henriksen, A. (1970), *Analyst*, 95, 660.

MacKinnon, M.D. (1978), *Mar. Chem.*, 7, 17.

MacKinnon, M.D. (1981), in: *Marine Organic Chemistry*: Duursma, E.K., Dawson, R. (Eds). Amsterdam: Elsevier, 1981, pp. 415–443.

Mantoura, R.F.C., Woodward, E.M.S. (1983), *Geochim. Cosmochim. Acta*, 47, 1293.

Menzel, D.W., Vaccaro, R.F. (1964), *Limnol. Oceanogr.*, 9, 138.

Miller, A.E.J., Mantoura, R.F.C., Suzuki, Y., Preston, M.R. (1993), *Mar. Chem.*, 41, 223.

Norrman, B. (1993), *Mar. Chem.*, 41, 239.

Ogura, N., Hanya, T. (1966), *Nature, London,* 212, 758.

Parker, D.A. (1981), Ph.D. Thesis, University of Liverpool, U.K.

Peltzer, E.T., Fry, B., Doering, P.H., Mckenna, J.H., Norrman, B., Zweifel, U.L. (1996), *Mar. Chem.*, 54, 85.

Ridal, J.J., Moore, R.M. (1993), *Mar. Chem.*, 42, 167.

Riley, J.P. (1965), in: *Chemical Oceanography*, Vol. 2: Riley, J.P., Skirrow, G. (Eds.). London: 1965; Academic Press,, pp 295–424.

Schreurs W. (1978), *Hydrobiol. Bull.* (*Amsterdam*), 12, 137.

Sharp, J.H. (1973), *Mar. Chem.*, 1, 211.

Sharp, J.H. (1997), *Mar. Chem.*, 56, 265.

Sharp, J.H., Benner, R., Bennett, L., Carlson, C.A., Fitzwater, S.E., Peltzer, E.T., Tupas, L.M. (1995), *Mar. Chem.*, 48, 91.

Statham, P.J., Williams, P.J. le B. (1983), in: *Methods of Seawater Analysis*, 2nd edn.: Grasshoff, K., Ehrhardt, M., Kremling, K. (Eds.). Weinheim: Verlag–Chemie, 1983, pp. 380–395.

Suzuki, Y., Sugimura, Y., Itoh, T. (1985), *Mar. Chem.*, 16, 83.

Sugimura, Y., Suzuki, Y. (1988), *Mar. Chem.*, 24, 105.

Suzuki, Y. (1993), *Mar. Chem.*, 41, 287.

Tupas, L.M., Popp, B.N., Karl, D.M. (1994), *Mar. Chem.*, 45, 207.

Williams, P.J. le B. (1981), *Kieler Meeresf. Sonderh.*, 5, 1.

17 Determination of particulate organic carbon and nitrogen

M. Ehrhardt and W. Koeve

17.1 Introduction

Seawater contains particulate organic material (POM) with a size spectrum ranging from colloids to organisms as large as a whale. This chapter describes a method to determine organically bound carbon (POC) and nitrogen (PON) in particles that typically fall into the lower range of this spectrum, *i.e.*, between approximately 0.45 and 300 μm. Phytoplankton, yeasts, most bacteria and other small heterotrophic organisms such as microzooplankton are found in this category as well as detrital particles and aggregates originating in life processes of larger zooplankton (fecal pellets, chitinous shells, larvacean houses). POM may also form abiotically as 'marine snow' from suspended small particles and dissolved or collodial organic matter *(Alldredge and Silver,* 1988), and it is usually sampled by filtration. With modifications mainly to ensure statistically adequate sampling the method may be used to analyse larger, aggregated particles and even individual mesozooplankton organisms. It thus covers nearly all important functional compartments of the biological pump that drives the oceanic carbon cycle *(Longhurst and Harrison,* 1989). Methods to measure POC and PON therefore are important tools in the quest to quantify and understand carbon fluxes in the ocean.

The intensity of primary particle formation of autotrophic organisms is the prominent factor controlling the distribution of particulate organic carbon and nitrogen. High concentrations of POC and PON are found during phytoplankton blooms near the ocean surface, under upwelling conditions, and in coastal waters *(e.g., Lochte et al.,* 1993; *Smetacek et al.* 1984). Lower concentrations are observed, *e.g.,* in oligotrophic subtropical gyres or during intervals between blooms. Very low concentrations of suspended POM are usually encountered in the interior of the ocean *(Gordon,* 1977). However, significant amounts of POM are frequently exported to the deep ocean after spring blooms and other events of mass accumulations of biogenic matter at the surface *(Wefer,* 1989; *Honjo,* 1996). In a review paper *Riley* (1970) offered a thorough discussion of experimental results pertaining to the distribution, chemical composition and formation of particulate organic matter in seawater.

The patchiness of phytoplankton blooms may account for sudden changes in the concentrations of POC and PON in the surface ocean both in time and space. However, significant variations in POC and PON concentrations have been reported even in deep waters. Based on the results of a systematic study *Wangersky* (1974) concluded that concentrations of POM in seawater appear to be characterized by a relatively uniform background at any given depth upon which local clouds of particles are superimposed. The observation of *Wangersky* presumably relates to large aggregates of particulate organic matter, nowadays well known as 'marine snow' *(Alldredge and Silver,* 1988), that are found in the ocean at highly varying abundances. These observations should be taken into account when designing a sampling strategy (see also *Gordon et al.,* 1979).

17.2   The elemental ratio of carbon and nitrogen

Elemental ratios of carbon and nitrogen in particulate organic matter were first determined by *Redfield* (1934). The average ratio of 6.6–7 may vary in sympathy with the proportion of living plankton cells, because proteins are remineralized more readily than the refractory structural polysaccharides. Thus, *Ehrhardt* (1969) found C/N ratios of 7.5 at depths of between 10 and 20 m in the central Baltic Sea increasing to 10.5 near the bottom. *Gordon* (1977) reported C/N ratios ranging from 6.8 to 8.3 in surface water, and from 11.8 to 20.6 at depth beyond 100 m at several stations on a transect from Halifax to Bermuda. *Solorzano* (1977) found C/N ratios in particulate material from Loch Etive, Scotland, as high as 20.0 during the winter months compared with 6.5 in summer.

17.3   Sampling and filtration

Water for the analysis of suspended particles may be collected with any clean water sampler used for oceanographic work, *e.g.,* Niskin bottles, Go-Flo bottles, Nansen bottles, Hydrobios water samplers (see also Chapter 1). The sample volume will depend on the expected concentration of POM. In nearshore and/or biologically productive water, 0.5–2 L usually is an adequate sample volume. Ten litres may be required in particle-poor open ocean waters such as the Sargasso Sea. For sampling large volumes of water, 30 L Niskin bottles are recommended. Even larger volumes may be sampled with *in situ* pumps. (See Chapters 1, 2 and 13).

Since the abundance of large aggregates is highly variable and aggregates tend to be fragile (*Alldredge and Silver,* 1988), their sampling appears to be very difficult with standard waters samplers in a statistically reliable fashion. These particles may be sampled with Plexiglass (Perspex) cones by divers (*Trent et al.,* 1978) or with particle interceptor traps (*Kremling et al.,* 1996) either tethered to drifting surface floats or moored on the seafloor. Samples from sediment traps require specific treatment (picking of swimmers, splitting of subsamples etc.) described elsewhere (*e.g., Michaels et al.,* 1990; *v. Bodungen et al.,* 1991).

17.3.1   Filtration of suspended particles

Immediately after recovery of a sampler a suitable volumetric flask is rinsed twice with about 50–100 mL of sample seawater. The flask is then filled to the mark to subsample a precisely known water volume. Volumetric flasks or Stohmann bottles with ground glass stoppers were found to be fairly handy for the purpose.

For POC and PON determinations suspended particles are collected on filters. Occasionally, it may be advantageous to remove coarse particles by straining the water through a stainless steel screen before it is passed through the filter. Since organic carbon and nitrogen are to be measured, filters must be made of inorganic material, *e.g.*, glass fibre or metal foil. *Gordon and Sutcliffe* (1974) recommended gravity filtration through 0.8 μm silver filters (47 mm diameter precombusted for 2 h at 450 °C) as the most reliable method for oceanic waters. *Altabet* (1990) suggested the use of aluminium oxide filters, which appeared to show good retention of submicrometre particles (*Koike et al.*, 1990), but are fairly expensive and difficult to handle. Glass fibre filters are commonly used to collect particles for POM determinations and analyses (*e.g.*, UNESCO, 1994; *Karl et al.*, 1991). Because of their better sampling efficiency Whatman GF/F filters with a nominal pore size of 0.8 μm are recommended rather than the coarser GF/C filters (pore size 1.3 μm).

The subsample is filtered immediately through clean glass fibre filters under slight vacuum (0.027 MPa or 200 Torr). Higher pressure gradients may cause cell rupture on the filters resulting in loss of body fluid. When the water volume above the filter has been reduced to a few mL, the filtration cone should be rinsed twice with a small amount of filtered seawater (not distilled water!). The filter must not run dry before rinsing.

Most commercial elemental analysers have rather small combustion chambers, and fitted autosamplers therefore usually do not accommodate large samples which restricts the size of filters. A diameter of 2.5 cm is usually the upper limit for glass fibre filters. When large glass fibre filters are used, *e.g.*, in conjunction with large volume *in situ* filtration or to sample greater amounts of particulate material for detailed analyses of constituent compounds, a cork borer is a convenient tool for cutting out small discs that may then be used to determine POC and PON in aliquots. If the particles on large filters show a patchy distribution, at least duplicate or triplicate subsamples should be processed for POC and/or PON determinations.

Samples from sediment traps subsampled by wet splitting techniques (*e.g.*, *v. Bodungen et al.*, 1991), should be filtered as described above. Flushing at the end of filtration is important.

17.4   Sample preservation

After filtration and flushing some air is drawn through the filters to remove most of the water. The moist filters are then dislodged from the filter holder with a pair of stainless steel tweezers and placed into a shallow cavity in a clean Plexiglass (Perspex) transport box (Fig. 17-1) which is easily made by any competent workshop.

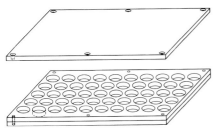

Fig. 17-1. Perspex (Plexiglass) filter transport box.

The transport box with filters is temporarily stored in a freezer until each cavity (usually 50) has been filled. The transport box without cover is then put into a drying oven and the filters are dried at 60 °C for 30 min. Subsequently, the cover of the transport box is screwed into place and the box is stored frozen until immediately prior to analysis. Some 20 filters of each shipment should be retained unused so that their carbon and nitrogen blank values can be determined.

17.5   Separation of particulate inorganic carbon from particulate organic carbon

The global mean production of particulate inorganic carbon (PIC) in the sea is approximately 25 % of the production of particulate organic carbon (*Broecker and Peng*, 1982). Nevertheless, it has often been found that PIC concentrations are very low (almost undetectable) compared with POC concentrations (*e.g., Gordon*, 1969) so that usually no precautions are necessary to distinguish between PIC and POC (*Ehrhardt*, 1983). Well known exceptions, however, are blooms of the calcite forming coccolithophorids that have been observed over large areas (*Holligan et al.*, 1983) in the North Atlantic during late spring and summer. Molar calcite concentrations during such blooms might be as high as 25–186 % of the associated POC concentrations (*Fernandez et al.*, 1993). Samples from sediment traps are also usually enriched in calcium carbonate, with typical PIC:POC molar ratios of approximately 1 : 1. The median of selected data at globally distributed deep open-ocean stations presented by *Honjo* (1996; their Table 7.2) was 0.8 with 25 and 75 % percentiles of 0.6 and 1.6, respectively. Sediment trap samples definitely will need separation of PIC and POC prior to POC determination.

Perhaps because it is seldom necessary for samples of suspended particles, no generally accepted method has yet been established to separate PIC and POC. For particles collected on filters (suspended particles, aggregates, wet splits from sediment traps) fuming with concentrated HCl in a desiccator for 48 h has been recommended (*Hedges and Stern*, 1984; *v. Bodungen et al.* 1991; *Knauer* 1991). Filters with samples rich in large foraminifera may have to be examined under a dissecting microscope after fuming (*Sellmer*, 1995) to determine if the calcareous shells have disappeared. If not, fuming must be repeated until shells are no longer detected (24–48 h). Acid rinsing of filters is not recommended, since significant losses of PON and POC have been reported (*Karl et al.*, 1991). Differentiation of PIC and POC by different combustion temperatures (*e.g., Hirota and Szyper*, 1975) is also not endorsed.

For samples collected in sediment traps and split after freeze drying (dry splitting), either the protocol of *Verardo et al.* (1990) or a similar method published by *Fischer and Wefer* (1991) may be applied. Briefly, the method includes the following steps: after weighing a homogenized aliquot into an aluminium boat, a drop of distilled water is carefully added to wet the sample. Subsequently 10 µl of sulphurous acid is added to the sample and the sample is re-dried in a drying oven at 60 °C (15 min). This procedure is repeated with increasing amounts of acid (30, 50, 100 µL) until a total volume of 400 µL has been added and/or no effervescence is noticed in the sample cup (*Verardo et al.*, 1990). Alternatively, samples may be decalcified with 6 mol/L HCl in silver boats (*Fischer and Wefer*, 1991).

17.6 Analysis

Gordon (1969), who used a Hewlett–Packard 185 CHN Analyzer, and *Kerambrun and Szekielda* (1969), who adopted a Perkin Elmer 240 Elemental Analyzer for the purpose were among the first to describe methods for simultaneous measurements of organic carbon and nitrogen in particles recovered from seawater. Modifications of these early attempts, some optimized for particular sample composition (*e.g.*, carbonate poor suspended particles *versus* carbonate rich sediment samples), have since been published (*Sharp*, 1974; *Ehrhardt*, 1983; *Hedges and Stern*, 1984; *Verardo et al.*, 1990).

The following paragraphs describe a method for determining POC and PON with the commercial Heraeus CHN-O-Rapid analyser. A variety of similar instruments is currently on the market. In particular Carlo Erba and Hewlett–Packard CHN analysers have been frequently used for this purpose. However, the method is largely independent of the type of CHN-analyser and is modified easily to accommodate the technical features of a different instrument. The method described here has been applied for approximately ten years at the Institut für Meereskunde, Kiel, Germany, to analyse samples of suspended particles, aggregates and material collected in sediment traps from coastal, shelf and open ocean domains. A description of a specific method for the analysis of marine sediment samples may be found in the recent article by *Verardo et al.* (1990). *Hurd* and *Spencer et al.* (1991) and the JGOFS Core Measurement Manuals (*UNESCO*, 1994) have described other recent methodical developments of POC and PON analysis. In addition, some alternative methods to measure POC and PON do exist. Among these, modern mass spectrometry adapted for oceanographic purposes is noteworthy, since it enables the combination of POC and PON detection and isotopic studies or tracer experiments.

17.6.1 Apparatus

Heraeus Elementanalysor CHN-O-Rapid.
Integrator: Hewlett–Packard Model 3394A or similar equipment or PC with suitable integration software.
Microbalance: For example, Cahn Model C-30 or similar.
Filtration unit: Suitable for 2.5 cm diameter glass fibre filters.

17.6.2 Reagents, analytical grade

Acetanilide (*reference material*).
Helium, 99.995 %.
Oxygen, 99.995 %.
Gas mix: 95 % N_2, 5 % H_2, for regeneration of Cu reductor (optional).
Gas regulators.
Copper (wire).
Copper oxide (wire).
Cerium dioxide, CeO_2.
Lead chromate, $PbCrO_4$.

Silver wool.
Phosphorus pentoxide, P_2O_5: Siccapent.

17.6.3   Principle of the analysis

The main components of the Heraeus CHN-O-Rapid analyser are an autosampler, a combustion column reactor, a reduction column, a gas chromatographic separation system, the detector unit and an output device for the analytical results. Helium is used as carrier gas. In the combustion reactor, oxygen gas and other oxidizing and catalyzing reagents support the complete high temperature combustion of organic carbon and nitrogen compounds to carbon dioxide, elemental nitrogen and N-oxides. Elemental copper in the reduction column reduces nitrogen oxides to N_2 and binds excess oxygen. Water and the combustion products CO_2 and N_2 are separated by gas chromatography, and N_2 and CO_2 are detected and quantified by thermal conductivity detectors (TCD). Other instruments may differ in some technical details, but the basic methodology remains the same.

17.6.4   Analytical procedure

The transport box with filters is thawed out in a dry environment (preferably in a desiccator). With a pair of clean tweezers the filters are folded into boats made of tin (Sn) foil and pelletized just before the analysis.

The autosampler of the Heraeus CHN-O-Rapid analyser has a capacity of 49 samples and/or standards. Sample pellets are first dropped through a ball valve into a chamber just above the combustion column which then is purged for about 1 min with helium to remove atmospheric nitrogen and carbon dioxide. After purging, the sample drops into the oxygen filled combustion column. Here it comes to rest on a plug of quartz wool which sits on top of a cerium dioxide filling. Downstream of the cerium dioxide the combustion tube is filled with bits of copper oxide (wire). The temperature of the combustion tube is maintained at 1050 °C but increases locally to up to 1800 °C during flash combustion of the tin foil encapsulating the sample. The combustion products, CO_2, N_2, various nitrogen oxides, halogens and sulphur compounds are swept with the helium carrier gas into the reduction column, which is maintained at 550 °C and filled with small bits of copper wire. Reaction with the hot copper reduces nitrogen oxides to N_2 and binds excess oxygen. Silver wool, loosely plugging the outlet of the combustion column, the in- and outlets of the reduction column and lead chromate at the top of the reduction column trap halogens and sulphur compounds which would interfere with analyte detection.

Thermal conductivity detectors are very sensitive to residual water vapour in the carrier gas stream. A gas chromatographic separation unit consisting of a short silver column and a longer copper column, both with silica gel as the stationary phase, separates the water vapour plus carbon dioxide from N_2. Residual traces of water vapour are trapped in a column filled with phosphorus pentoxide (Siccapent). The N_2 passes through all three columns undelayed. It is detected as the first peak with the TCD. After detection of the N_2 peak, heating of the silica gel to 85 °C releases the CO_2. Finally, the first silica gel column (silver) is heated to 250 °C to release the bulk of the water which bypasses the copper reduction column and the water absorbent (Siccapent) before it is vented to the outside atmosphere.

17.7   Calibration

The instrument is calibrated with high purity acetanilide (analytical-reagent grade reference material, US National Institute of Standards and Technology or equivalent). Acetanilide is used, because its elemental composition matches the elemental composition of particulate material obtained from seawater, *i.e.*, C:N = 8. The acetanilide is dried in a desiccator. Approximately 200 µg are weighed in a tin boat to the nearest 0.1 µg by means of a microelectrobalance. The tin boat is pelletized and analysed under the same conditions as a sample. At least ten standards are analysed.

17.8   Blanks

Extraneous contributions to the analytical signal may be caused by:
– dust in water samplers, flasks, *etc.*;
– contaminated filters;
– contamination of the tin boats;
– poor quality of the He carrier gas and the oxygen;
– contamination of flow controllers, gas lines, valves, *etc.*; and
– non-stoichiometric composition of the standard substance.

Great care must be taken to use clean sampling gear. Filters must be pre-ignited at 450 °C for 12 h (overnight), stored in clean metal or glass boxes and never be touched. Tin boats are best cleaned by ultrasonication in acetone followed by hexane and drying in a clean oven. Avoid contact with fingers. Gas lines may have to be flushed with acetone and hexane followed by thorough drying. Standard substances must be of certified high quality.

At least ten unused filters should be analysed to determine the procedural carbon and nitrogen blanks and the standard deviations from their mean values. The procedural blanks are represented by the ordinate intercepts, derived from the estimating equations for the relationship between peak areas and amounts of carbon and nitrogen in the standard substance. If a pure standard substance, high quality helium and oxygen, clean gas lines, clean tin boats and a clean instrument are used, the procedural blanks easily are kept below 5 µg of carbon and 0.5 µg of nitrogen.

References to Chapter 17

Alldredge, A.L., Silver, M.W. (1988). *Prog. Oceanogr.*, 20, 41.
Altabet, M.A. (1990). *Limnol. Oceanogr.*, 35, 902.
v. Bodungen, B., Wunsch, M., Fürderer, H. (1991). *Geophys. Monogr.* 63, 47.
Broecker, W.S., Peng, T.-H. (1982), *Tracers in the Sea*: Lamont-Doherty Geological Observatory, Columbia University, Palisades, New York, 1982, pp. 45–109.
Ehrhardt, M. (1969), *Kieler Meeresf.*, 25, 71.

Ehrhardt, M. (1983), in: *Methods of Seawater Analysis,* 2nd edn., Grasshoff, K., Ehrhardt, M., Kremling, K. (Eds.). Weinheim: Verlag Chemie, 1983, pp. 269–275.

Fernandez, E., Boyd, P., Nolligan, P.M., Harbour, D.S. (1993), *Mar. Ecol. Prog. Ser.*, 97, 271.

Fischer, G., Wefer, G. (1991), in: *Marine Particles: Analysis and Characterization*: Hurd, D.C., Spencer D.W. (Eds.). Geophys. Monogr. 63, pp. 391–397.

Gordon, D.C. (1969), *Deep-Sea Res.*, 16, 661.

Gordon, D.C. (1977), *Deep-Sea Res.*, 24, 257.

Gordon, D.D.J., Sutcliffe, W. H. J. (1974), *Limnol. Oceanogr.*, 19, 989.

Gordon, D.C., Wangersky, P.J., Sheldon, R.W. (1979), *Deep-Sea Res.*, 26, 1083.

Hedges, J.I., Stern, J.H. (1984), *Limnol. Oceanogr.*, 29, 657.

Hirota, J., Szyper, J.P. (1975), *Limnol. Oceanogr.*, 20, 869.

Holligan, P.M., Viollier, M., Harbour, D.S., Camus, P., Champagne-Philippe, M. (1983), *Nature, (London)*, 304, 339.

Honjo, S. (1996), in: *Particle Flux in the Ocean*: Ittekkot, V., Schäfer, P., Honjo, S., Depetris, P.J. (Eds.). Chichester: John Wiley & Sons, 1996, pp. 91–154.

Hurd, D.C., Spencer, D.W. (1991), *Marine Particles: Analysis and Characterization, Geophys. Monogr.* 63, 472 pp.

Karl, D.M., Harrison, W.G., Dore, J. (1991), in: *Marine Particles: Analysis and Characterization*: Hurd, D.C., Spencer D.W. (Eds.). *Geophys. Monogr.* 63, pp 33–42.

Kerambrun, P., Szekielda, K.-H. (1969), *Tethys*, 1, 581.

Knauer, G. (1991), in: *Marine Particles: Analysis and Characterization*: Hurd, D.C., Spencer D.W. (Eds.). *Geophys. Monogr.*, 63, pp 79–82.

Koike, I., Shigemitsu, T., Kasuki, T., Kazushiro, K. (1990), *Nature (London)*, 345, 242.

Kremling, K., Lentz, U., Zeitzschel, B., Schulz-Bull, D.E., Duinker, J.C. (1996), *Rev. Sci. Instrum.*, 67, 4360.

Lochte, K., Ducklow, H.W., Fasham, M.J.R., Stienen, C. (1993), *Deep-Sea Res. II*, 40, 91.

Longhurst, A.R., Harrison, W.G. (1989), *Prog. Oceanogr.*, 22, 47.

Michaels, A.F., Silver, M.W., Gowing, M.M., Knauer, G.A. (1990), *Deep-Sea Res.*, 37, 1285.

Redfield, A.C. (1934), *James Johnstone Memorial Volume*, Liverpool, pp. 176–192.

Riley, C.A. (1970), *Adv. Mar. Biol.*, 8, 1.

Sharp, J.H. (1974), *Limnol. Oceanogr.*, 19, 984.

Sellmer, C. (1995), MS Thesis, University of Kiel.

Smetacek, V., v. Bodungen, B., Knoppers, B., Peinert, R., Pollehne, F., Stegmann, P., Zeitzschel, B. (1984), *Rapp. P.-v. Réun. Cons. Int. Explor. Mer*, 183, 126.

Solorzano, L. (1977), *J. Exp. Mar. Biol .Ecol.*, 29, 81.

Trent, J.D., Shanks, A.L., Silver, M.W. (1978), *Limnol. Oceanogr.*, 23, 626.

UNESCO (1994), IOC/SCOR, *Unesco Manuals and Guides 29*, 170 pp.

Verardo, D.J., Froelich, P.N., McIntyre, A. (1990), *Deep-Sea Res.*, 37, 157.

Wangersky, P.J. (1974), *Limnol. Oceanogr.*, 19, 980.

Wefer, G. (1989), in: *Productivity of the Ocean: Present and Past*: Berger, W.H., Smetacek, V.S., Wefer, G. (Eds.). Chichester: John Wiley & Sons, 1989, pp. 139–153.

18 Preparation of lipophilic organic seawater concentrates

M. Ehrhardt and K.A. Burns

18.1 Introduction

Research since the late 1960s into the extent and nature of pollution of the marine environment by anthropogenic chemicals has led to the development of analytical techniques for qualitative and quantitative determinations of traces of organic contaminants in organisms, suspended particles, sediments and water.

Naturally occurring dissolved organic matter in seawater, usually referred to as DOM, is extremely complex in composition and spans a wide relative molecular mass range. No methods are yet available to separate all of its constituent compounds. The major difficulty with analyses of man made or man mobilised organic contaminants of seawater, however, is not usually the separation of complex mixtures or the chemical characterization of selected compounds. Powerful and efficient analytical methods are available to tackle these problems, *e.g.*, high-pressure liquid chromatography and capillary gas chromatography (one-or multi-dimensional) alone or in combination with various types of mass spectrometers. The greatest difficulty remains the collection of samples of sufficient volume for detailed analyses without the introduction of extraneous contaminants originating in the sampling and work-up procedures.

The most conspicuous organic contaminants of seawater are organochlorine compounds (*Broman et al., 1991*), fossil hydrocarbons and their oxidation products (*Ehrhardt and Burns*, 1993), linear alkylbenzenes, alkylphenols (*Marcomini et al.*, 1990; *Blackburn and Waldock*, 1995), and phthalate esters (*Waldock*, 1983). Concentration procedures for these lipophilic substances are described here. However, the analyst should be aware of the fact that contaminants are not the only lipophilic chemicals likely to be present in seawater. A large number of recently biosynthesised, as well as man made chemicals not listed above, will also be concentrated by these methods.

Lipophilic organic compounds in seawater usually occur at minute concentrations. The high sensitivity of modern analytical instruments would, in principle, permit analyses of the amounts present in small volume water samples. However, contamination of the sample by substances adhering to the surfaces of water samplers or laboratory glassware, by solvents, reagents, gases, *etc.*, becomes a severe problem at ultratrace levels of analytes usually present in seawater.

Numerous attempts have been made to overcome the limitations of small sample size and to accumulate lipophilic organic compounds from large volumes of water. Batch extractions with water-immiscible solvents have been described by *Gaul and Ziebarth* (1984) and *Theobald et al.* (1990). However, with increasing sample volume the volume of the liquid extractant must also increase because of its non-zero solubility in seawater. Thus, the attempt to increase the absolute amounts of trace contaminants by extracting larger sample volumes is offset by the disadvantage of increased solvent volumes. Solid adsorbents are better suited, because the only practical limitation to sample volume is their adsorptive capacity. For analyses of organic trace constituents of seawater the most popular adsorbent probably is Amberlite® XAD-2, a porous copolymer of styrene and di(vinyl) benzene (purified forms of XAD-2 are available as Supelpak-2® for air sampling and Supelpak-2B® for water sampling). Silica gel, chemically bonded with lipophilic alkyl groups, and polyurethane foam have also been used as solid phase extractants. *Gómez-Belinchón et al.* (1988) compared the efficiencies of liquid–liquid extraction *versus* adsorption on polyurethane foam and Amberlite XAD-2. With sample volumes of 300–400 L application of the three methods in parallel resulted in roughly the same recoveries for fatty acids and aliphatic, aromatic and chlorinated hydrocarbons. Qualitative differences were found, however, for high relative molecular mass aliphatic and aromatic hydrocarbons. Liquid–liquid extraction yielded the highest percentage followed by polyurethane foam and Amberlite XAD-2. *Vidal et al.* (1994) performed a similar study testing liquid–liquid extraction *versus* XAD-2 and n-octadecyl silanised silica gel with organochlorine pesticides spiked into river water. The three methods gave similar results at concentrations of 200 ng/L. Liquid–liquid extraction was more efficient at higher concentrations.

Ship-borne pumping systems (*Tokar et al.*, 1981; *de Lappe et al.*, 1983; *Schulz et al.*, 1988; *Baker and Eisenreich*, 1990; *Broman et al.*, 1991; *Iwata et al.*, 1993; *Ehrhardt and Petrick*, 1993; *Schulz-Bull et al.*, 1995) have been applied to collect sample water, but their depths range is limited. Also, ships may be sources of contaminants, and they are expensive to keep on station for extended periods of time. It is therefore simpler, safer and more convenient to pass water *in situ* through a bed of adsorbent. Instruments suitable for this purpose are available commercially such as the Seastar® *in situ* water sampler and its more modern version, the Infiltrex® system, both from Axys Environmental Systems Ltd., Sidney, B.C., Canada. The standard models have depth limitations of 300 and 200 m, respectively, but special units are available with greater maximum operating depths. The Seastar system has been used by, e.g., *Ehrhardt and Burns* (1990; 1993), the Infiltex II sampler *by Ehrhardt et al.* (1995). The Kiel *in situ* pump (KISP) of Ingenieurbüro Klaus Scholz, Fockbeck, Germany, may be operated at depths up to 6000 meters (*Petrick et al.*, 1996; see also Chapter 22). The method described below, optimised for the use of XAD-2 resin, may be employed with all suitable sampling systems, self-contained or otherwise.

18.2   Outline of the method

For analyses of dissolved organic constituents the water must be filtered before it enters the adsorption column, because beds of XAD-2 resin tend to retain small particles which usually cannot be removed by solvent extraction. Filtration through pre-ignited glass fibre filters separates 'particulate' material from 'dissolved'. Unusually high particle loadings in seawater will require frequent filter changes or increased filter areas. The filtered water is passed through a bed of XAD-2 resin at a flow rate depending upon the dimensions of the adsorption column. The loaded resin is countercurrent-extracted by refluxing with an appropriate solvent mixture in an apparatus specially designed for the purpose.

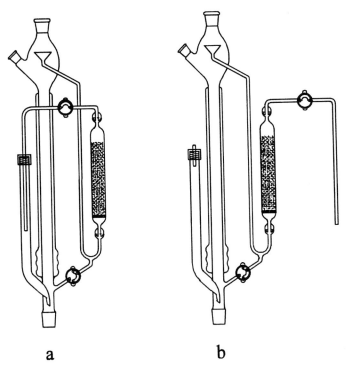

Fig. 18-1. (a) The glass continuous extractor is shown without boiling flask, reflux condenser and thermometer. To improve the extraction efficiency the vacuum-jacketed distillation column may be filled with glass helices. The lower flat stopcock is used do drain the column. (b) The extractor configured for fractionated distillation. The upper flat stopcock is used to adjust the reflux ratio.

Figure 18-1a shows an improved version of the extractor described by *Ehrhardt* (1987). The modification featuring a heat-insolated separation column instead of a simple steam duct may also be used for fractionated solvent distillation, if necessary through a bed of suitable adsorbent (activated charcoal, alumina, silica gel) for further product purification.

After extraction, lipophilic compounds are partitioned into n-hexane. More hydrophilic high relative molecular mass components of DOM remain in the more polar primary extract. If necessary, organic bases and acidic substances may be removed from the hexane phase by liquid extraction with aqueous acids and bases, respectively.

18.3 Reagents

All reagents and solvents must be distilled in glass, residue grade or equivalent. Cleaning procedures are described in *UNEP* (1995). To check solvents for contamination, 100 mL are rotary evaporated in a conical flask until a few mL remain. The volume is reduced further by blowing a slow stream of ultra-pure nitrogen through the flask. When the remaining vol-

ume is approximately 100 μL, 1 μL is analysed by gas chromatography under the same conditions as the samples.
Amberlite XAD-2 resin.
Acetonitrile.
Methanol: Alternative solvent.
Dichloromethane: Alternative solvent.
Water with low concentration of organic carbon: For example Milli-Q.
n-Hexane.
0.1 mol/L *hydrochloric acid.*
0.1 mol/L *NaOH solution.*
Anhydrous sodium sulphate.

18.4   Apparatus and glassware

Glass fibre filters: Usually 14 cm diameter (*e.g.*, Whatman GF/C, Schleicher & Schüll No. 6, Gelman AE), kiln-fired 4 h at 450 °C.
In situ extraction system.
Glass continuous extractor. Fig. 18-1.
250 mL *Round bottom flasks.*
100 mL *Conical flasks.*
250 mL *Heating mantle.*
Rotary evaporator: With solvent trap (Fig. 18-2) and 5 Å molecular sieve at the air inlet.
250 mL *Syringes.*
Pasteur pipettes.
250 mL *Separatory funnels.*
100 mL *Erlenmeyer flasks.*
50 mL Beakers.
Glass ampoules: Fig. 18-3.
Torch.

Fig. 18-2. Trap to prevent refluxing of solvent. To be inserted between flask and rotary evaporator.

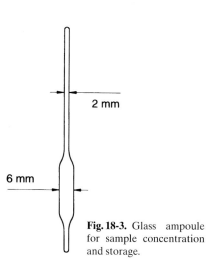

2 mm

6 mm

Fig. 18-3. Glass ampoule for sample concentration and storage.

18.5   Cleaning of the resin

XAD-2 resin must be cleaned thoroughly before being used for concentrating traces of organic compounds from seawater. Approximately 300 mL of the commercial product are put into a 2000 mL beaker. Milli-Q type water with a low concentration of dissolved organic carbon is used to wash the resin so that the nominal sized beads remain on the bottom of the beaker, whereas fines are removed with the overflowing water. Washing is continued until the supernatant water runs clear. This step of the cleaning procedure can usually be omitted, if the resin is a high purity product of uniform mesh size specifically labelled for organic trace analyses. Nominal size resin beads floating on the water surface have lost the water from their porous interior structure. To rehydrate the resin beads they are collected in a beaker and covered with methanol. Rehydration may take several hours. When water is added to replace the methanol, the resin beads must stay at the bottom of the beaker.

After the initial washing and rehydration of the resin the frit end of the glass column (shown in Fig. 18-1 as part of the glass continuous extractor) is connected to a water aspirator, and the resin slurry is drawn into the column. Gravity filling is also possible, but packing should be assisted by gentle aspiration. When the column is nearly full, the frit end should be capped and the column filled with Milli-Q water. After the resin has settled the top end is plugged with pre-extracted glass or quartz wool. Because the adsorption efficiency of a resin bed is a function of the water volume passed through it per unit time, its dimensions must be adjusted accordingly. *Harvey* (1972) found 4–5 bed volumes per minute to be the optimum flow rate for nearly 100 % recovery of DDT + PCBs (1,1,1-trichloro-2,2-bis(*p-chlorophenyl)ethane* + polychlorinated biphenyls) added to unfiltered seawater at concentrations of 5–10 mg/L. Our usual glass columns are 18 cm long with 2.5 cm inner diameter. They are filled with resin to a height of 15 cm above a coarse glass frit. The bed volume thus is $\approx$ 75 mL; 4.5 such bed volumes per minute are equivalent to 20 L/h. The maximum flow rate for the Infiltrex II samplers is 200 mL or 2.7 bed volumes per minute. This flow rate is adequate, but it takes longer to concentrate samples. In practice, a set of batteries will power the Infiltrex samplers for 30 hours. External battery packs and multiple filter units are available to increase sample volumes.

For cleaning the resin, the column is inserted into the glass continuous extractor (Fig. 18-1). This arrangement has several advantages over batch extraction. The extraction may be carried out in a protected, if necessary an inert gas atmosphere; a small solvent volume is sufficient for exhaustive extraction; the all glass system eliminates any possible contamination from plastic surfaces contacting solvents.

The resin is extracted with 200 mL of refluxing solvent for at least 12 h. A 1 : 1 mixture of methanol and dichloromethane is an excellent eluent. However, the high concentration of organically bound chlorine makes it a poor choice, if the resin is to be used for analyses of chlorinated hydrocarbons. In that case the acetonitrile-water azeotrope (16.3 % water) should be used despite lower recoveries of aliphatic hydrocarbons measured with surrogate standard additions. After cooling, extraction is resumed with fresh solvent for another 12 h period.

18.6   Resin blanks

After the heat is switched off, solvent is left in an extracted XAD-2 column for several hours (preferably overnight) so that any impurities remaining in the resin equilibrate with it. The boiling flask is then replaced with a clean flask, and the solvent is drained by opening the lower stopcock of the glass continuous extractor (Fig. 18-1). If aqueous acetonitrile has been used, the solvent is extracted three times with n-hexane in a 250 mL separatory funnel (10, 10, 30 mL). If the solvent is methanol plus dichloromethane, the dichloromethane is removed by rotary evaporation before extraction. Phase separation can be facilitated by adding 25 mL of low organic Milli-Q water to the methanol. The combined hexane extracts are dried with anhydrous sodium sulphate, reduced in volume and analysed as a sample. Common impurities of XAD-2 resin are naphthalene, *o-*, *m-* and *p*-diethylbenzene and *o-*, *m-* and *p*-ethylstyrene.

18.7   Storage of adsorption columns

Columns filled with clean XAD-2 resin and capped at both ends may be stored in a refrigerator. No blank deterioration was observed after storage periods of one year (*Petrick*, 1997). To guard against breakage if the temperature in the refrigerator drops below 4 °C, the columns may be filled with Milli-Q water saturated with pre-ignited sodium chloride.

18.8   Sample preparation

Water left in the column after recovery from sample collection is removed either with the vacuum of a water aspirator or, in cases demanding ultimate cleanliness, by pressing nitrogen through it. This precaution facilitates the initial passage of solvent through the column. Adsorbed material does not have to be eluted immediately after sample collection. Hydrocarbons adsorbed on XAD-2 resin have been found to be stable for periods up to 100 d even in the presence of oleophilic bacteria (*Green and Le Pape*, 1987).

The column is inserted into the extractor and eluted with approximately 200 mL of refluxing solvent (for solvent composition see Section 18.5) to which a suitable internal standard has been added. At the beginning of the extraction, the solvent passing through the column from the bottom up, removes a considerable volume of water from the interior of the resin beads (approximately 30 mL for a 75 mL resin bed). Depending upon the salinity of the sample water, dissolved sea salt may precipitate in the boiling flask. Four to six hours of continuous extraction are usually sufficient to remove all adsorbed materials. Completeness of extraction may be checked as described in Section 18.6.

After cooling to room temperature, the eluant is prepared and extracted as described in Section 18.6. If sea salt has precipitated, some ultra-clean water is added to dissolve it before extraction. If desired, organic acids and bases may now be partitioned into aqueous base and acid, respectively. The hexane phases are combined, washed with water free from organic carbon, dried over anhydrous sodium sulphate and concentrated to approximately 1 mL by rotary evaporation in a conical flask. A solvent trap (Fig. 18-2) should be inserted between the flask and the rotary evaporator to prevent refluxing of condensed solvent.

The concentrated sample solution is then transferred with a 250 mL syringe into a glass ampoule as shown in Fig. 18-3. Further concentration may be achieved by blowing a slow stream of ultra-clean nitrogen into the ampoule through a capillary held in the neck. During solvent evaporation the ampoule should be held well below room temperature to minimize the loss of volatile analytes. To check for completeness of desorption, the extracted resin is left in contact with the solvent which is then collected separately and prepared for analysis in like manner.

18.9 Sample storage

Glass ampoules (Fig. 18-3) as first described by *Ehrhardt and Derenbach* (1980) were found to be very useful. They are easily made from ordinary glass tubes of 3 mm inner diameter. These ampoules may be opened with a glass knife and re-sealed with a torch at least five times, because only a very short length of capillary needs be broken off to admit the needle of a microlitre syringe. In contrast to conventional vials they do not require rubber septa, which are frequently sources of contamination, they are light, which is helpful when determining the mass of the concentrate, and they are fairly inexpensive. Sealed under nitrogen and kept in a deep freezer, concentrates stored in them will last for at least one year. Conventional vials with Teflon lined septa may also be used; however, contact of the sample with the cap must be avoided, and long term storage must be monitored to ensure the samples do not evaporate. Prepared samples should be stored in a refrigerator or freezer depending on the time until analysis.

References to Chapter 18

Baker, J.E., Eisenreich, S.J. (1990), *Environ. Sci. Technol.*, 24, 342.

Blackburn, M.A., Waldock, M.J. (1995), *Water Res.*, 7, 1623.

Broman, D., Näf, C., Rolff, C., Zebühr, Y. (1991), *Environ. Sci. Technol.*, 25, 1850.

Ehrhardt, M., Derenbach, J. (1980), *Mar. Chem.*, 8, 339.

Ehrhardt, M. (1987), *ICES Techniques in Marine Environmental Sciences*, No. 4., 14 pp.

Ehrhardt, M., Burns, K.A. (1990), *J. Exp. Mar. Biol. Ecol.*, 138, 35.

Ehrhardt, M., Burns, K.A. (1993), *Mar. Pollut. Bull.*, 27, 187.

Ehrhardt, M., Petrick, G. (1993), *Mar. Chem.*, 42, 57.

Ehrhardt, M., Weber, R.R., Bícego, M.C. (1995), *Publção Esp. Inst. Oceanogr., S. Paulo*, 11, 81.

Gaul, H., Ziebarth, U. (1984), *Dt. Hydrogr. Z.*, 36, 191.

Gómez-Belinchón, J.I., Grimalt, J.O., Albaigés, J. (1988), *Environ. Sci. Technol.*, 22, 677.

Green, D.R., Le Pape, D. (1987), *Anal. Chem.*, 59, 699.

Harvey, G.R. (1972), *Special Report WHOI-72-86*, 19 pp.

Iwata, H., Tanabe, S., Sakai, N., Tatsukawa, R. (1993), *Environ. Sci. Technol.*, 27, 1080.

de Lappe, B.W., Risebrough, R.W., Walker II, W. (1983), *Can. J. Fish. Aquat. Sci.*, 40 (Suppl. 2), 322.

Marcomini, A., Pavoni, B., Sfriso, A., Orio, A.A. (1990), *Mar. Chem.*, 29, 307.

Petrick, G., Schulz-Bull, D.E., Martens, V., Scholz, K., Duinker, J.C. (1996), *Mar. Chem.*, 54, 97.

Petrick, G. (1997), I.f.M., Kiel, personal communication.

Schulz, D.E., Petrick, G., Duinker, J.C. (1988), *Mar. Pollut. Bull.*, 19, 526.

Schulz-Bull, D.E., Petrick, G., Kannan, N., Duinker, J.C. (1995), *Mar. Chem.*, 48, 245.

Theobald, N., Lange, W., Rave, A., Pohle, U., Koennecke, P. (1990), *Dt. Hydrogr. Z.*, 43, 311.

Tokar, J.M., Harvey, G.R., Chesal, L.A. (1981), *Deep-Sea Res.*, 11, 1395

Vidal, L.H., Trevelin, W.R., Landgraf, M.D., Rezende, M.O.O. (1994), *Int. J. Environ. Anal. Chem.*, 56, 23.

Waldock, M.J. (1983), *Chem. Ecol.*, 4, 261.

UNEP (1995), *Ref. Meth. Mar. Pollut. St.*, 65, 24 pp.

19 Adsorption chromatography of organic seawater concentrates

M. Ehrhardt and K.A. Burns

19.1 Introduction

Organic seawater concentrates obtained by liquid–liquid extraction with a water immiscible solvent or by liquid–solid absorption (see Chapters 18, 22) are usually rather complex in composition. Even the considerable resolving power of gas chromatographic capillary columns does not suffice to separate such extracts into single compounds. Especially when the compounds of interest belong to a structurally similar group comprising a small fraction of the total, it is advisable to separate raw extracts into compound groups before they are analysed at the single compound level. A physical property characterizing the members of a compound group is their polarity which, among other effects, controls the solubility in different solvents and interaction with polar particle surfaces. The degree of polarity may thus be used to separate compound groups from one another by adsorption chromatography.

19.2 Outline of method

Raw extracts in hexane solution are concentrated by rotary evaporation under reduced pressure below room temperature to <1 mL in a conical flask. The concentrate is transferred to a silica gel chromatography column. The mass ratio of extracted organic matter to silica gel should be 1 : 100 or less. Four fractions of increasing polarity are eluted with two silica gel bed volumes each of n-hexane, n-hexane plus 20 % dichloromethane, dichloromethane and methanol. The eluates are concentrated by rotary evaporation under reduced pressure below room temperature to <1 mL. Further concentration may be achieved by blowing a slow stream of purified nitrogen over the pre-concentrated sample solution inside a suitable ampoule.

19.3 Reagents

All reagents and solvents must be residue grade or equivalent. Cleaning procedures are described in *UNEP* (1995).

Silica gel, 70–210 mesh.
n-Hexane.
Dichloromethane.
Methanol.
Table salt.
Crushed ice cubes.

19.4   Apparatus and glassware

All glassware has to be cleaned meticulously (*UNEP*, 1995). Briefly, laboratory glassware is scrubbed with brushes in warm detergent and can be left to soak overnight. It is then rinsed with copious amounts of tapwater followed by deionised water and baked in a clean oven at 250 °C. Before use it is rinsed with hexane.
Separation column.
Soxhlet extractor.
Continuous extractor: Optional, Fig. 18-1a.
Extraction column: Optional.
10 mL *Glass ampoules*: Optional.
5 Å *Molecular sieve traps.*
Adapters.
50–100 mL *Conical flasks plus stoppers.*
Glass stoppered storage bottle for silica gel.
100–250 mL *Syringes.*
Pasteur pipettes.
Glass ampoules for sample storage.
Rotary evaporator.
Solvent trap for rotary evaporator, Fig. 18-2.
Water aspirator or similar vacuum pump.
Drying cabinet.
Tube oven: Optional
Tank of compressed nitrogen with regulator.

19.5   Cleaning of silica gel

Commercially available silica gel often is not clean enough for application in organic ultra-trace analyses. Silica gel may be cleaned by solvent extraction in a Soxhlet extractor followed by dehydration and activation in a drying cabinet. This method, however, is not without pitfalls as the cleaned product may pick up impurities from the air inside the drying cabinet. A safer and more convenient method involves the use of an extractor as described in Chapter 18.

19.5.1   Cleaning by Soxhlet extraction

A Soxhlet thimble, made of glass fibre which has been cleaned beforehand by heating to 450 °C for several hours, is filled with silica gel. The silica gel is extracted for several hours first with dichloromethane followed by extraction with n-hexane. The air inlet on top of the reflux condenser should be fitted with a 5 Å molecular sieve trap. After extraction the thimble with the silica gel in it is allowed to dry in a clean atmosphere. The silica gel is then activated in a clean drying cabinet at 250 °C for several hours and poured into a clean bottle with a ground-glass stopper before cooling. This method of storing the purified silica gel has serious disadvantages at ambient high relative air humidity, because each time the storage bottle is opened the silica gel will pick up some moisture and thus gradually lose its activity.

19.5.2   Cleaning by continuous extraction

Silica gel is filled into a clean and empty glass cartridge as used for liquid–solid adsorption when filled with XAD-2 resin (see Chapter 18). The cartridge is inserted into a continuous extractor (*Ehrhardt*, 1987; Fig. 18-1a), and the silica gel is extracted with dichloromethane followed by n-hexane, each time for several hours. After cooling and draining of the solvent the glass cartridge is separated from the extractor and inserted into a vertically mounted tube oven just wide and long enough to accommodate the cartridge. A tube oven for this purpose is easily made by winding a heating ribbon around a metal tube. At the downstream end the glass column is fitted with a straight distillation adapter the side tubulation of which is protected with a 5 Å molecular sieve trap. A small receiving flask is attached to the adapter. Using a suitable glass adapter purified nitrogen is passed at room temperature through the moist bed of silica gel to remove any remaining solvent. The tube oven heater is switched on only after the silica gel has dried to a free-flowing powder. Under a slow stream of nitrogen the silica gel is then activated at 250 °C for several hours. The heat is switched off, and without stopping the flow of nitrogen the cartridge is cooled until it can be touched and handled. The warm cartridge is removed, and 3–5 mL portions are poured into clean 10 mL glass ampoules which are then immediately flame-sealed.

19.6   Preparing a separation column

A glass separation column (Fig. 19-1) is capped at the downstream end and filled with n-hexane, except for the reservoir. A sorbent slurry is prepared either by pouring slighly more than the intended bed volume of silica gel into approximately twice its volume of n-hexane resting in a beaker; or a glass ampoule with clean and activated silica gel is opened and, using a Pasteur pipette, is covered with n-hexane immediately after so that exposure of the sorbent to ambient air is kept at a minimum. The slurry is transferred into the separation column with a Pasteur pipette, releasing the slurry underneath the surface of the solvent. The bulb of the Pasteur pipette is kept compressed until the silica gel particles have sunk away from its mouth. n-Hexane is then drawn back into the Pasteur pipette and added to

Fig. 19-1. Glass column for silica gel adsorption chromatography.

the slurry in the beaker or ampoule. The procedure is repeated until the level of silica gel has reached the upper end of the cylindrical part of the column. The reservoir is filled with n-hexane, the cap is removed and solvent is drained from the column under gravity flow. Gentle tapping will help to compact the bed of silica gel. Before the solvent meniscus has reached the silica gel the column is capped again and left standing to equilibrate for at least 1 h. The column is flushed with five bed volumes of n-hexane.

19.7   Sample separation

n-Hexane is drained from the column until the solvent level has reached the level of sorbent. The column is capped again and the concentrated sample solution is transferred with a 100–250 mL syringe. The plunger must be depressed gently so as not to inject the solution into the bed of silica gel. A receiving flask is placed underneath the column, and the cap is removed to allow the sample solution to sink into the sorbent bed. The first and least polar fraction is eluted with two silica gel bed volumes of n-hexane. Care must be taken not to disturb the sorbent bed when adding the solvent. If necessary, the flow rate is adjusted to approximately 1 mL/min with a head of nitrogen (around 10^4 Pa).

The second fraction is eluted with two bed volumes of a 20 % solution of dichloromethane in n-hexane followed by the third fraction eluting with two bed volumes of dichloromethane. Two bed volumes of methanol finally elute the most polar compounds from the column. All fractions are collected in 50–100 mL conical flasks.

The first fraction usually contains aliphatic hydrocarbons, polychlorinated biphenyls (PCB), and some monocyclic aromatic hydrocarbons; the bulk of aromatic hydrocarbons, 1,1,1-trichloro-2,2-bis(*p*-chlorophenyl)ethane (DDT) and 1,1-dichloro-2,2-bis(*p*-phenyl)-ethene (DDE) are found in the second fraction. Carbonyl compounds such as phenylalkanones, anthraquinone and fluorenone elute with dichloromethane. The total mass of the most polar fraction eluting with methanol usually surpasses the mass of all other fractions combined by about a factor of ten. It comprises fatty acids, alcohols, phenols, γ-lactones and often a large variety of unknown compounds.

After addition of suitable internal standards the fractions are concentrated by rotary evaporation to >1 mL. The rotary evaporator should be fitted with a solvent trap as shown in Fig. 18-2 to prevent refluxing solvent to flow back into the attached flask. The rotary evaporator air inlet should either be connected to a tank of compressed nitrogen or be fitted with a 5 Å molecular sieve trap. The concentrated samples are transferred with a 100–250 mL syringe into glass ampoules as shown in Fig. 18-3 which are flushed with nitrogen and flame-sealed while being cooled to well below room temperature with a mixture of crushed ice cubes and table salt.

References to Chapter 19

Ehrhardt, M. (1987), *ICES Techn. Mar. Environ. Sci.*, No. 4, 14 pp.
UNEP (1995), *Ref. Meth. Mar. Pollut. St.*, 65, 24 pp.

20   Clean-up of organic seawater concentrates

D.E. Schulz-Bull and J.C. Duinker

20.1   Introduction

Individual concentrations of many natural and anthropogenic organic compounds in seawater are in the low femtogram (10^{-15} g) to nanogram (10^{-12} g) per litre range. The concentrations are minute, yet the number of components, unknown but certainly very large, renders even the lipophilic material dissolved in seawater an exceedingly complex mixture. The determination of organochlorines such as polychlorinated biphenlys (PCBs), polychlorinated dibenzodioxines/furanes (PCDD/Fs) or chlorinated pesticides, all in themselves complex mixtures with very low concentration levels in seawater, therefore, requires efficient sample clean-up.

Clean-up methods in common use include liquid–solid adsorption- and gel-permeation chromatography as well as chemical treatment (see Chapter 19). Clean-up procedures for specific groups of substances, *e.g.*, chlorobiphenyls (CBs), pesticides, polycyclic aromatic hydrocarbons (PAHs), are often improved modifications of adsorption chromatography under hydrostatic pressure. Adsorbents such as silica gel, Florisil and alumina are used in combination with various eluents (*Lang*, 1992). However, poor reproducibility and low separation efficiency are serious drawbacks to these techniques relying on chromatography with open columns. To overcome these problems *Petrick et al.* (1988) applied high-performance liquid chromatography (HPLC) to clean-up seawater extracts. This was an important contribution to increasing the sensitivity of determinations of organochlorine compounds in seawater samples by capillary gas chromatography with electron capture detection (GC-ECD) (*IOC*, 1993; *Schulz-Bull et al.*, 1995).

20.1.1   Particular applications

The determination of PCCD/Fs usually involves extensive clean-up to remove co-extracted compounds, typically present in greater amounts. The PCDD/Fs and the most toxic non-ortho chlorine substituted CBs (Nos. 77, 126 and 169) are planar molecules. They can be separated from bulk PCBs and other interfering compounds by chromatography on activated charcoal (*Jensen and Sundström*, 1974; *Huckins et al.*, 1980; *Smith*, 1981; *Kannan et al.*, 1991). The advantages of HPLC techniques on activated carbon (PX21) dispersed with n-octadecane (C_{18}) was shown by *Kannan et al.* (1993); *Feltz et al.* (1995) described an automated HPLC fractionation of PCDDs and related compounds. Modern clean-up procedures comprise HPLC on 2-(1-pyrenyl)ethyl dimethyl silylated silica (PYE) columns (*Haglund et al.*, 1990; *Kannan et al.*, 1998), a two-dimensional HPLC system for the separation of PCDD/Fs, PCB and PAHs (*Zebühr et al.*, 1993) and the automated HPLC method by *Bandh et al.* (1996) employing a nitrophenylpropyl silica column coupled with a PYE column for separations of PCBs, PCDD/Fs and PAHs according to aromaticity and planarity.

PAHs, PCBs, PCDD/Fs and pesticides are related classes of organic contaminants. They are therefore isolated together. Unfortunately, these compound groups interfere among themselves in gas chromatographic analyses with electron capture or mass spectrometric detection. As their chemical properties are fairly similar, no technique has yet been found for clean separation by adsorption chromatography on a single column. The HPLC method described here may be modified to solve specific problems, *e.g.*, the separation of methylesters from long-chain unsaturated methylketones or the fractionation of unsaturated wax esters. The clean-up of seawater samples for the determination of PAHs by HPLC has been outlined by *Witt* (1995) and by *Schulz-Bull et al.* (1998).

Another current problem in environmental studies is the chromatographic separation of enantiomers of marine pollutants (*Hühnerfuss and Kallenborn*, 1992). HPLC with chiral stationary phases is a new tool to investigate the fate of chiral organic compounds and their degradation products in the environment (*Ludwig et al.*, 1992).

The following is a detailed description of a clean-up procedure for seawater extracts (see also Chapters 18 and 22) into fractions of specific organochlorines, prior to gas chromatographic analysis with electron capture detection. Special attention is given to chlorobiphenyls (CBs).

20.1.2   General remarks

Often the concentrations of organics in seawater extracts are so low that sample extracts can be injected directly on analytical grade HPLC columns. If higher amounts of interfering substances are present such as pigments, alcohols or acids, a pre-clean-up by low-pressure column chromatography on silica gel or alumina is useful to protect the HPLC column (see Chapter 19). Extracts of marine organisms must be freed of the biogenic lipids. Several chemical treatments (KOH, sulphuric acid) and chromatography on alumina or gel permeation chromatography are commonly applied for this purpose (*Lang*, 1992). When sediments or related matrices are analysed, it is often necessary to remove elemental sulphur. This is accomplished by reaction with activated heavy metals such as Cu/Hg (*Smedes and de Boer*, 1997).

20.2   Reagents and equipment

Acetone, methanol, dichloromethane and n-pentane: All solvents must be analytical reagent grade and redistilled.
Pulse-free HPLC pump: Constametric III or similar.
Rheodyne injection valve: With 250 μL loop capacity.
Stainless-steel column: 200×4 mm i.d., Nucleosil 100-5, Macherey & Nagel, Germany.
Guard column: 30×4 mm i.d., Nucleosil 50-5, Macherey & Nagel, Germany.
Three-way valve: Latek, Germany.
Syringes, 50–250 μL.

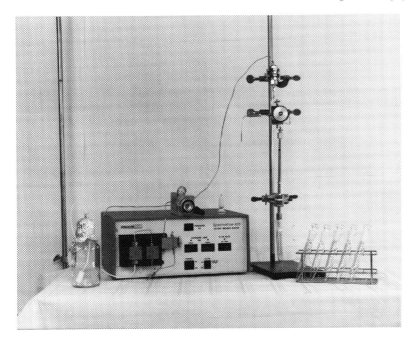

Fig. 20-1. High-performance liquid chromatography system with sample collection.

Flasks plus stoppers, 250 mL.
Stoppered glass tubes, 10 mL: Graduated with 0.1 mL markings.
Glass ampoules: 1.0 mL for sample storage.
Pasteur pipettes.
Rotary evaporator.
Splash guard with male standard taper 14 and female standard taper 29 ground joints (see Fig. 18-2).
Standard reference compounds: Promochem, Germany.
Ultrapure nitrogen.

Many HPLC systems are available commercially (*Snyder and Kirkland*, 1979). The sytem described here is depicted in Fig. 20-1.

Introduction of a guard column between pump and separation column (*Utschakowski*, 1998) is a useful modification of the system; it becomes a necessity, if extracts are more concentrated. The guard column protects the main column from polar lipids and small particles present in extracts. The guard column can be rinsed with acetone or methanol; the waste is drained through the valve in front of the main column.

20.2.1 Cleaning procedures

Solvents and laboratory ware must be cleaned meticulously to avoid sample contamination and to achieve low procedural blanks. Glassware should be washed with detergents, rinsed several times with clean water and baked at 250 °C overnight. All solvents are distilled in an all-glass distillation apparatus flushed with a gentle stream of nitrogen to exclude contaminant vapours potentially present in the laboratory atmosphere. The quality of distillates is checked by concentrating 100 mL to 50 μL of which 2 μL are injected into a GC-ECD (see Section 22.3). Batches of 250 mL of clean solvent can be stored (under nitrogen) in a refrigerator for no longer than one week. A new HPLC column should be rinsed with 250 mL each of acetone, dichloromethane, dichloromethane–pentane (1:1) and finally with 500 mL of n-pentane. The pure solvents are stored and the mixed solvents are both prepared and stored in 250 mL bottles.

20.2.2   Calibration of the HPLC system

Standard mixtures are run on the HPLC system for evaluation of the separation volume, reproducibility and accuracy of the system. The composition of the standard mixture depends on the compound class under study. The amounts injected on the HPLC column are from 100 pg (organochlorines) to 1000 ng (hydrocarbons) each in 200 μL of solvent (hexane). The solvent volumes required to elute selected groups of organic contaminants are summarized in Table 20-1. The solvent flow rate is 0.5 mL/min. The eluates are collected in fractions of 0.5 mL. Each fraction is analysed by GC-ECD or GC-FID (flame ionization detection). Recoveries are from 90 to 100 %.

To calibrate the system for the separation of chlorobiphenyls from interfering substances the standard mixture should at least contain:
One alkane (n-$C_{20}H_{42}$,);
hexachlorobenzene (HCB);
one trichlorobiphenyl (28);
one hexachlorobiphenyl (180);
hexachlorocyclohexane (HCH);
1,1–bis–(4-chlorphenyl)–2,2,2–trichlorethan (p.p'-DDT);
1,1–bis–(4-chlorphenyl)–2,2–dichlorethan (p.p'-DDE);
methylnaphthalene;
anthracene; and
phenanthrene.

Calibrations with the standard mixture should be repeated after 20–30 sample runs or at least once a week.

20.2.3   Procedural blanks

A procedural blank of the HPLC system involves elution of the HPLC column with the solvent(s) to be used for subsequent sample separations, concentration of the solvent fractions to be collected and injection into the GC. A procedural blank of the HPLC system in combination with sample concentration by, e.g., liquid–solid adsorption on XAD-2 resin (see Chapter 18), involves extraction of a clean resin column, concentration of the extract, separation with the HPLC system using the proper combination of solvents, concentration of the eluates to about 50 μL (the volume is checked with a 100 μL syringe) and gas chromatographic analysis of 2 μL injections with ECD (organochlorines) or FID-MS (PAH, alkanes) detection.

Table 20-1. Volume of solvents required to elute the constitutents of the synthetic mixtures (from *Petrick et al.,* 1988). Elution solvents: n-pentane (0–11.0 mL); 20 % dichloromethane in n-pentane (11.0–15.0 mL); 100 % dichloromethane (from 15 mL onwards). Chlorobiphenyls numbers according to *Schulz et al.* (1989)

Compound	Elution volume (mL)
Alkanes C_{10}–C_{28}; pristane, phytane	0.5–2.0
Alkenes $C_{16:1}$–$C_{20:1}$	0.5–2.0
Hexachlorobenzene	2.5–3.0
PCB: Clophen A30, A60	2.5–4.5
Chlorobiphenyls:	
Congeners No. 101, 153, 180	2.5–3.5
Congeners No. 28, 52, 138	3.5–4.5
p.p'-DDE	4.0–4.5
1-Methylnaphthalene	4.0–4.5
Dimethylnaphthalenes	4.5–5.0
Acenaphthylene	4.5–5.0
Anthracene	4.5–5.0
Phenanthrene	5.0–6.0
Fluorene	5.0–6.0
Pyrene	5.0–6.0
9-Methylanthracene	6.0–7.0
Fluoranthene	6.0–7.0
Toxaphene (polychlorinated camphenes)	6.0–12.5
Chrysene	7.0–8.0
α-Hexachlorocyclohexane	11.0–12.0
p.p'-DDT	11.0–12.0
β-Hexachlorocyclohexane	11.0–12.0
γ-Hexachlorocyclohexane	11.0–12.0
Dieldrin	15.0–18.0

20.3   Compound group separation of seawater samples

The preparation of seawater extracts is described in Chapters 18 and 22. If large amounts of interfering compounds such as biogenic lipids are present in the extracts a pre-clean-up by silica gel adsorption chromatography is recommended (see Chapter 19).

Before an extract is analysed, 10 mL of n-pentane are used to flush the system and to adjust the solvent flow-rate to 0.5 mL/min. Extracts, concentrated to 200 μL, are injected with the Rheodyne injector. The fractions, determined from the calibration procedure (Table 20-2), are collected in 10 mL tapered, thick-walled test tubes with ground joints that

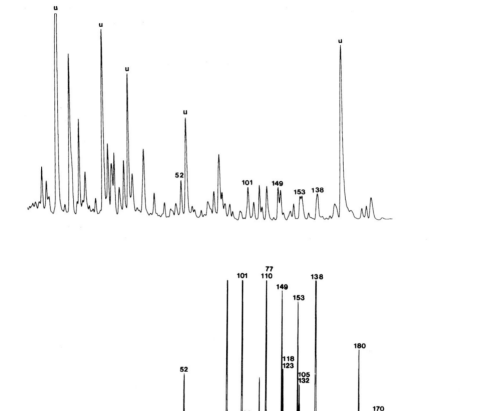

Fig. 20-2. ECD chromatograms of seawater extracts. After silica gel chromatography, top (u = unknown non-CB compound); after clean-up by the HPLC method (second fraction containing the CBs), bottom.

can be connected directly to a splash guard serving as a reducing union to a rotary evaporator. The fractions are concentrated to about $200 \, \mu L$ for injection into the GC ($1–2 \, \mu L$) or further concentration after transfer into glass ampoules as shown in Chapter 18, Fig. 18-3. The second fraction will include the chlorobiphenyls and the DDE quantitatively (chromatogram is shown in Fig. 20-2).

Table 20-2. Fractions, optimized for the separation of CBs, eluting from the HPLC column using synthetic mixtures and samples (from *Petrick et al.,* 1988). Elution solvents and synthetic mixtures see Table 20-1

Fraction No.	Volume (mL)	Synthetic mixtures	Samples
1	0.5–2.0	alkanes, alkenes	hydrocarbons
2	2.0–4.5	PCBs, HCB, DDE, 1-methylnaphthalene	PCBs, DDE, alkylbenzenes
3	4.5–11.0	PAHs, Toxaphene	PAHs, Toxaphene
4	11.0–15.0	pesticides, Toxaphene	pesticides, Toxaphene
5	15.0–25.0		alcohols, steroles

References to Chapter 20

Bandh, C., Ishaq, R., Broman, D., Näf, C., Rönquist-Nii, Y., Zebühr, Y. (1996), *Environ. Sci. Technol.,* 30, 214.

Feltz, K.P., Tillitt, D.E., Gale, R.W., Peterman, P.H. (1995), *Environ. Sci. Technol.,* 29, 709.

Haglund, P., Asplund, L., Järnberg, U., Jansson, B. (1990), *J. Chromatogr.,* 507, 389.

Huckins, J.N., Stalling, D.L., Petty, J.D. (1980), *J. Assoc. Off. Anal. Chem.,* 63, 750.

Hühnerfuss, H., Kallenborn, R. (1992), *J. Chromatogr.,* 580, 191.

IOC (1993), Chlorinated biphenyls in open ocean waters: sampling, extraction, clean-up and instrumental determination, IOC Manuals and Guides No. 27, Paris: UNESCO.

Jensen, S., Sundström, G. (1974), *Ambio,* 3, 70.

Kannan, N., Petrick, G., Schulz-Bull, D.E., Duinker, J.C., Boon, J., van Arnhem, E., Jansen, S. (1991), *Chemosphere,* 23, 1055.

Kannan, N., Petrick, G., Schulz-Bull, D.E., Duinker, J.C. (1993), *J. Chromatogr.,* 642, 425.

Kannan, N., Petrick, G., Bruhn, R., Schulz-Bull, D.E. (1998), *Chemosphere,* 37, 2387.

Lang, V. (1992), *J. Chromatogr.,*595, 1.

Ludwig, P., Gunkel, W., Hühnerfuss, H. (1992), *Chemosphere,* 24, 1423.

Petrick, G., Schulz, D.E., Duinker, J.C. (1988), *J. Chromatogr.,* 435, 241.

Schulz, D.E., Petrick, G., Duinker, J.C. (1989), *Environ. Sci. Technol.,* 23, 852.

Schulz-Bull, D.E., Petrick, G., Kannan, N., Duinker, J.C. (1995), *Mar. Chem.,* 48, 245.

Schulz-Bull, D.E., Petrick, G., Bruhn, R., Duinker, J.C. (1998), *Mar. Chem.,* 61, 101.

Smedes, F., de Boer, J. (1997), *Trends Anal. Chem.,* 16, 503.

Smith, L.M. (1981), *Anal. Chem.,* 53, 2152.

Snyder, L.R., Kirkland, J.J. (1979), *Introduction to Modern Liquid Chromatography,* 2nd edn. Wiley: New York.

Utschakowski, S. (1998), Dissertation, Institut für Meereskunde an der Universität Kiel.

Witt, G. (1995), *Mar. Pollut. Bull.,* 31, 237.

Zebühr, Y., Näf, C., Bandh, C., Broman, D., Ishaq, R., Pettersen, H. (1993), *Chemosphere,* 27, 1211.

21 Determination of petroleum residues dissolved and/or finely dispersed in surface seawater

M. Ehrhardt and K.A. Burns

21.1 Introduction

Oil pollution and its physical, chemical and biological effects appear to be inevitable consequences of the production, transport and use of mineral oil and its products. A vast number of publications, too numerous to be cited here, demonstrate that petroleum-derived hydrocarbons have been distributed widely in the marine environment.

Crude oils and oil products are extremely complex mixtures predominantly of hydrocarbons comprising structural elements such as straight and branched carbon chains, saturated and aromatic rings, in all possible combinations. Compounds containing S, N, O, V and Ni are minor constituents of crude oils. Olefinic hydrocarbons, if present at all, are trace constituents of crude oils but may reach detectable concentrations in refined products.

Hydrocarbons are present in seawater not only as a result of human activities, but also as a consequence of biological processes. In contrast to mineral oils and their products, biogenic hydrocarbon mixtures are much less complex in composition. Olefinic and aliphatic hydrocarbons are principal components (*Clark and Blumer*, 1967; *Youngblood and Blumer*, 1973). A small number of monocyclic aromatic hydrocarbons have been shown to be biosynthesised by marine organisms, *e.g.*, the substituted benzene (laurene) by *Laurencia sp.* (*Irie et al.*, 1965), some carotenes with benzoid terminal groups by the sponge *Reniera japonica* (*Yamaguchi*, 1957, 1958), as well as olefins and alkanones substituted with 2–3 benzene rings by a starfish (*Yayli*, 1988).

However rare aromatic hydrocarbons are as products of biosynthesis, among fossil hydrocarbons they are abundant. Because aromatic hydrocarbons are more polar than aliphatic hydrocarbons, they usually dominate the water soluble fraction of crude oils and their products.

Incomplete combustion of oil and other carbonaceous materials, *e.g.*, in internal combustion engines, industrial processes (*Bjørseth and Eklund*, 1979), domestic heating, forest fires and intentional biomass burning (*Greenberg et al.*, 1984) is another source of polycyclic aromatic hydrocarbons (PAH), primarily to the atmosphere. Dry deposition and atmospheric precipitation are the principal vectors of their transport into surface seawater. *Gustafsson et al.* (1997) discussed the export of deposited PAH from surface waters by association with sinking particles.

Fossil aromatic hydrocarbons and those generated by pyrolysis are distinguished by a particular structural detail: most fossil aromatic hydrocarbons carry alkyl substituents which are usually missing in the structures of pyrogenic aromatic hydrocarbons, due to the much higher temperature of formation (*Sportsøl et al.*, 1983; *Peters and Moldowan*, 1993). The predominance of unsubstituted PAH is not necessarily an indicator of pyrogenic origin. *Ehr-*

hardt et al. (1992) showed that alkyl substituted PAH are photooxidized faster than the unsubstituted parent compounds so that sunlight illumination may gradually change the compositional signature of the water soluble fraction of an oil until it resembles a product of pyrolysis.

Detrimental effects on marine life have prompted the development of a rapid, sensitive and inexpensive method to assess concentrations and to follow the distribution of petroleum oil dissolved and/or finely dispersed in seawater, particularly in cases of accidental discharge. A method to measure concentrations of contaminating hydrocarbons must be insensitive to the presence of recently biosynthesised hydrocarbons, or it must provide a means of discriminating between recent biogenic and other sources. The specificity of the fluorimetric method described in this section is based on the much higher abundance of aromatic structures in fossil and/or pyrogenic hydrocarbon mixtures than in those of recent biosynthetic origin.

UV fluorimetry (UVF) is a sensitive method for the detection and quantification of aromatic hydrocarbons; and it is not affected by aliphatic hydrocarbons. Thus it has proven to be a useful tool for measuring fossil hydrocarbon concentrations in seawater. UV fluorimetry does not distinguish between alkyl substituted and unsubstituted aromatic hydrocarbons and therefore is not suitable for discriminating between fossil and pyrogenic sources. Hydrocarbon degradation products with aromatic structure often present in oil contaminated waters will also fluoresce.

UV fluorimetry has found wide-spread application for investigating and monitoring the degree of aquatic oil contamination because of its simplicity, sensitivity and ease of application. For example, *Wattayakorn et al.* (1998) used the UVF method to determine the distribution of petroleum hydrocarbons in the Gulf of Thailand and to provide sufficient data points for use in a three dimensional hydrographic circulation model.

21.2   Principle of the method

Aromatic hydrocarbons, when irradiated with UV light of a suitable wavelength, are excited to an elevated level of electronic energy from which they return to the electronic ground state by emitting electromagnetic radiation. When spectral multiplicities remain unchanged during transitions, the radiation is known as fluorescence. Its frequency distribution reflects the energy differences between the electronic ground state and excited levels that, when recorded with a suitable instrument, result in a fluorescence spectrum characteristic of the excited molecule. Appropriate textbooks should be consulted for a detailed account of the physical principles (*e.g., Calvert and Pitts,* 1967; *Carey and Sundberg,* 1990; *Wayne and Wayne,* 1996). The fluorescence spectra of petroleum residues are rather featureless humps, because the spectra of many compounds overlap. For each individual compound the intensity of its fluorescence is a linear function of its concentration unless concentrations are so high that self-absorption occurs. Because fluorescence quantum yields are not the same for different compounds, measurements are strictly comparable only for constant sample composition. Thus for quantification, a standard must be used that shows similar spectral patterns to the samples.

Aromatic hydrocarbons together with other lipophilic compounds are extracted from a known volume of seawater with n-hexane or cyclohexane in which Gelbstoff and humic substances resulting from natural decay of recent biological material are insoluble. This relative insolubility is important, because many of the latter substances can also be excited to fluoresce and would thus interfere with the determination of oil concentrations. After extraction, the sample solution is dried, brought to a known volume and the intensity of its fluorescence is compared with that of a suitable standard solution.

For oil spill monitoring, it is useful to calibrate the instrument against solutions of the spilled oil. However, even when the whole oil is available for use as a standard, the water samples may contain only the water soluble components of the oil. The water soluble components are predominantly the benzene and naphthalene fractions, which have spectral maxima at lower wavelength than the higher relative molecular mass aromatics that are the most highly fluorescent fractions of the original crude oil. In this case it would be necessary to use a water soluble fraction of the crude to calibrate the fluorimeter at lower wavelengths than for the whole oil.

21.3 Reagents

All reagents and solvents must be residue grade or equivalent distilled-in-glass quality. Cleaning procedures are described in *UNEP* (1995).
n-Hexane or cyclohexane.
Diethyl ether.
Anhydrous sodium sulphate.
Ultrapure nitrogen.
5 Å Molecular sieve.

21.4 Apparatus

Weighted metal frame (UNESCO, 1984): To accommodate a reagent bottle (4 L or similar) with nylon retrieving line and float (Fig. 21-1).
Spectrofluorimeter: Preferably capable of synchronous scanning, with wavelength reproducibility of excitation and emission monochromators ± 2 nm or better; typical monochromator gratings of 600 lines/nm blazed at 300 nm or better, spectral resolution with bandpass of 2.5 nm or less, maximum bandpass of 10 nm and a response of S-20 or S-5 on the photomultiplier tube.
Rotary evaporator with reflux trap (Fig. 18-2).
Soxhlet extractor.
4 L *Amber reagent bottles*: With Teflon-lined screw caps.
100 mL *Erlenmeyer flasks.*
50 mL *Beakers.*
Pasteur pipettes.
100 mL *Conical flasks.*
10 mL *Volumetric flasks.*

Teflon screw cap: Fitting the sampling bottle with a three way flat, ground glass, or Teflon-plug stopcock and long-stem funnel (Fig. 21-2). This apparatus can be manufactured easily by any competent workshop (optional).

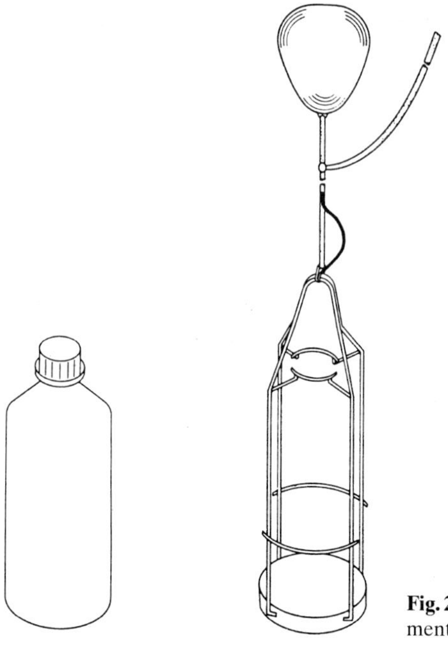

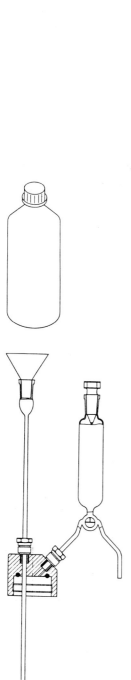

Fig. 21-1. Sampler recommended by the Intergovernmental Oceanographic Commission (IOC).

Fig. 21-2. Diagram of Teflon screw cap with long-stem funnel and separator funnel to be used for phase separation.

21.5 Cleaning of glassware

It is essential to clean the sampling bottles and all glassware used for UVF analyses in the most meticulous manner (*UNEP*, 1995). Briefly, laboratory glassware is soaked in detergent solution at least overnight and then scrubbed with brushes. It is rinsed with copious amounts of tapwater followed by deionized water and baked in a clean oven at 200 °C.

21.6 Sampling and sample storage

Not infrequently, even research vessels are sources of petroleum which rapidly forms a plume around the ship when on station. Therefore, samples for the determination of petroleum-derived hydrocarbons in seawater should be taken immediately after the ship has come to a stop. If this is not possible, one might consider rowing a rubber dinghy some distance upwind from the ship to take the samples (do not use an outboard engine).

Immediately before the sample is taken, the sampling bottle is fastened in the weighted metal frame and the screw cap is removed. The assembly is thrown into the water in such a way that the bottom of the sampler hits the water surface first. By doing so, the surface film is destroyed before the bottle dips into the water, thus ensuring that only bulk water is sampled. The sampler rapidly sinks to the depth determined by the length of line between the sampler and the float (usually 1 m). When no more bubbles are seen to rise, the sampler is retrieved. A measured volume (*e.g.*, 100 mL) of water is spilt to facilitate vigorous agitation of water and extractant in the bottle, and 25 mL of spectroscopic grade n-hexane or cyclohexane are added. Samples may be extracted either by vigorous manual shaking (5 min) or with a mechanical wrist-action shaker (15 min).

After extraction and phase separation the hexane must be recovered. The use of a large separatory funnel for this purpose is discouraged, because it is difficult to keep clean. Instead, pre-extracted water is added carefully until the hexane collects in the narrow neck of the sample bottle. Portion by portion it is drawn carefully into a Pasteur pipette and transferred into a 100 mL Erlenmeyer flask. It is useful to pre-determine the level of the interface between the water and the solvent in the neck of the bottle and to use a pinch clamp on the pipette to position it just above this interface.

Much more convenient is the following method: the normal screw cap is replaced with the assembly shown in Fig. 21-2. Extracted seawater is added to the bottle through the long-stem funnel which ends near the bottom. With proper positioning of the three-way stopcock the hexane layer is thus forced into the separatory funnel, usually together with some water and sometimes as a froth. When the phases refuse to separate, a few mL of diethyl ether are added to the solvent in the separatory funnel which is then removed from the screw cap and shaken. The aqueous phase is drained and the organic phase collected in a 100 mL Erlenmeyer flask.

Experience has shown that one extraction per sample removes > 90 % of dissolved hydrocarbons from moderately contaminated water. A second extraction thus is not recommended, because the subsequent concentration of a larger solvent volume is likely to result in evaporative losses of analytes.

The extract is dried with anhydrous sodium sulphate, while the flask is capped with a 50 mL beaker. The dried extract is transferred into a 100 mL conical flask and reduced in volume to a few mL with a rotary evaporator (reduced pressure, water bath at room temperature). It is highly recommended to insert a solvent trap as depicted in Fig. 18-2 between the flask and the rotary evaporator. Such traps are effective safeguards against possible contamination by solvent refluxing from the rotary evaporator.

After partial evaporation the pressure inside the rotary evaporator is equilibrated with the atmosphere through a 5 Å molecular sieve trap. The solution is transferred into a 10 mL volumetric flask with a Pasteur pipette and made up to volume. This solution is ready for measurement in the UV spectrofluorimeter. Should volumetric flasks not be available, the volumes of individual extracts may be determined by weighing on a top-loading balance. The densities are 0.6603 g/cm^3 (n-hexane) and 0.77855 g/cm^3 (cyclohexane), both at 20 °C.

When samples are taken from a biologically highly productive area or in an estuary, it might be necessary to remove interfering non-hydrocarbon material by a short column chromatographic clean-up. At very low levels of petroleum pollution it has been found, however, that the clean-up procedure is more prone to contaminate the sample than to remove unwanted material.

21.7 Column chromatographic clean-up

Basically the same procedure is applied as described in Chapter 19. Three bed volumes of 20 % dichloromethane in hexane are used to elute the sample which is reduced in volume by rotary evaporation. The concentrated solution is transferred into a volumetric flask as described above and made up to volume. This solution is ready for measurement.

21.8 Blanks

To ensure that samples were not contaminated during work-up, 25 mL n-hexane or cyclohexane rinses of the sampling bottles should be taken through the entire analytical procedure.

21.9 Spectrofluorimetric analysis

Before a sample is analysed, a solvent baseline is recorded by exciting spectrograde n-hexane or cyclohexane at 310 nm, while its fluorescence spectrum is recorded between 320 and 500 nm. Recording a fluorescence spectrum should always start approximately 10 nm upscale from the excitation wavelength to prevent the intense Rayleigh-scattered light from impinging on the photomultiplier.

Exciting the sample at 310 nm usually results in a rather featureless spectrum consisting of a broad elevation with a fluorescence maximum near 360 nm. Therefore, the intensity of the fluorescence emission at 360 nm is often used to calculate the sample concentration. Under certain conditions, *e.g.*, when a major portion of the pollutant is a diesel fuel rich in monocyclic and dicyclic aromatics but poor in more highly condensed PAH, higher sensitivities may be obtained by selecting lower excitation and emission wavelengths (*e.g.*, 280 ex/ 330 em).

21.10 Calibration and quantification

When the source of the fluorescing material extracted from the water is known and this material can be obtained in substance, *e.g.*, in the case of an accidental discharge of oil, the oil should be used as the reference. When the source is unknown and the spectrum has the usual shape of a broad elevation with a maximum near 360 nm, experience accumulated over several years indicates that a light Arabian crude oil is a useful reference material. For calibrating analyses of moderately oil contaminated waters, a solution is prepared of approximately 20 mg of this oil per 10 mL spectroscopic grade n-hexane.

Exactly 2 mL of the clean solvent is pipetted into the quartz cell of the spectrofluorimeter. Its fluorescence intensity (FI) at the selected wavelengths is recorded. For calibrating measurements in moderately contaminated water the standard solution is added in 10 μL increments, each addition representing 1.0 μg of oil per mL of hexane in the fluorimeter cell. After each 10 μL addition to the solution in the cell, its contents are mixed by cautious swirling, care being taken not to spill any solvent and FI recorded. The detector outputs at the selected wavelengths are plotted against corresponding oil concentrations, and the slope S [μg $\cdot$ mL^{-1} $\cdot$ FI^{-1}] of the regression is calculated. The detector response is a linear function of concentrations between 0 and approximately 10 μg/mL oil in hexane (Fig. 21-3).

Sample oil content is determined from the FI readings (mV detector output) at a dilution within the linearly calibrated range. The sample $FI_{(S)}$ is corrected for the reading of the blank hexane $FI_{(h)}$, then multiplied by the dilution used to take the reading and by the slope (S) of the regression line. The concentration is then calculated by taking into acount the toatl volume V_h [mL] of the sample extract and the volume V_S[L] of the extracted water.

$$(FI_{(S)} - FI_{(h)}) \times \text{dilution} \times S \times V_h/V_S = \mu\text{g oil per litre of seawater.}$$

For analysis of dilute samples, 1 mL of the sample is transferred into the quartz cell. Its FI is read. Then 1 mL of hexane is added and the solution is re-read. The second reading should be approximately ½ of the first. If the solution is too concentrated it will have to be diluted until it reads linearly with dilution. For more concentrated solutions it is useful to add 1 mL of hexane to the cell and then read 10 μL additions of the sample extract.

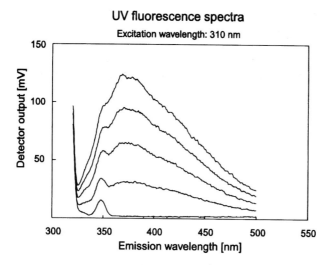

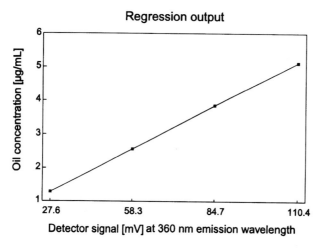

Fig. 21-3. Example of a calibration graph made from Kuwait export blend crude oil. The peak centered around 348 nm is caused by Raman scattering of the solvent.

21.11   Sources of error

The major source of error in quantification is the choice of a suitable calibration standard. Pre-scanning the samples to choose a suitable standard and optimum wavelengths for the FI measurement will reduce this source of error. It is equally important to ensure that the solution is dilute enough to prevent inner filter effects. The dilution procedure described here minimizes this source of error.

All substances other than oil extractable from seawater into a non-polar solvent and emitting fluorescent light at 360 nm when excited at 310 nm are possible sources of positive errors. These include components of creosote, which is used as a wood preservative.

Sources of negative error are substances which quench the fluorescence of PAH. High concentrations of phthalate esters, a group of ubiquitous environmental contaminants, have this effect. However, since a concentration of 10 ng of phthalate ester per mg of oil quenches the fluorescence by approximately 1 %, only very high concentrations of phthalates would impair determinations of seawater-accommodated oil residues.

21.12 Sensitivity and range of the method

With a sample volume of 4 L, a high quality grating spectrofluorimeter will produce interpretable signals from total oil concentrations of approximately 50 ng/L. At these very low concentrations extreme precautions to avoid sample contamination are of paramount importance. They determine the successful application of the method to a higher degree than the instrument sensitivity.

The upper concentration limit as defined by the deviation of the detector response from linearity caused by auto-quenching is approximately 8 mg oil/mL hexane. Thus, at very high concentrations of oil residues in the water that might be encountered under special circumstances in the immediate vicinity of an oil spill, samples may have to be dissolved in larger volumes of n-hexane for measurement or the extract diluted.

21.13 Spectral scanning for source identification

Oils from different sources have different molecular compositions which can often be distinguished on the basis of their fluorescence spectra. *ASTM* (1993) recommends the following procedure for determining emission spectra. An oil solution of approximately 10 mg/mL in hexane is placed in the quartz cell. Fingerprints are wiped off and the cell is placed in the fluorimeter. Bandpasses are set at 10 nm or less for excitation and at 2.5 nm or less for emission monochromators. The excitation wavelength is set to 254 nm. The emission wavelength is set to achieve maximum FI and the instrument or the solution adjusted to achieve 90 % of full scale deflection on the recorder or computer screen. If too concentrated, the solution should be diluted. If too dilute, the sample could be further concentrated or the bandpass widths could be increased. When optimum conditions are set, the emission is scanned between 280 and 600 nm. Then at the same instrument settings, a spectrum is recorded of the clean solvent. Usually a single scan at 254 nm is sufficient for most oils, but lighter oils or samples with heavier PAH will require shorter or longer excitation wavelengths. The emission spectrum of an unknown oil can then be compared with reference oils for general source determination.

Synchronous excitation/emission scanning is used to obtain better signal resolution of the aromatic hydrocarbons. In this technique excitation and emission are scanned simultaneously with a fixed wavelength offset. As a consequence of the Franck–Condon principle and because energy is lost by thermal relaxation of the excited states, the fluorescence light has a longer wavelength than the exciting radiation. *UNEP/IOC/IAEA* (1992) described a procedure which uses 25 nm differences and is useful for differentiating broad classes of oils. *ATSM* (1993) described a procedure which uses 6 nm differences and produces higher resolution. Bandpass widths and wavelength differences can be adjusted to obtain sufficient resolution to differentiate many types of oil. Figure 21-4 (*NAS*, 1985) illustrates the discrimination of synchronous excitation spectrofluorimetry. It shows the spectra of a whole oil (top) and its water soluble fraction (bottom).

Synchronous excitation spectra

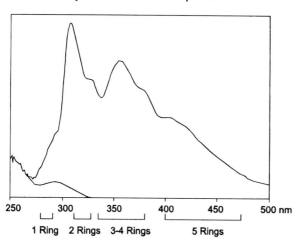

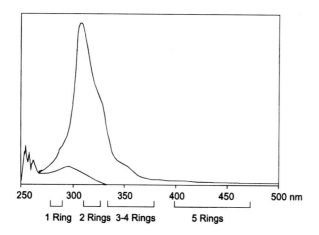

Fig. 21-4. Representative synchronous excitation fluorescence spectra of a whole oil (top) and its seawater soluble fraction (bottom) collected near the IXTOC1 blow-out in the Gulf of Mexico (From: *NAS*, 1985).

UVF spectrometry, especially in synchronous excitation mode, is a rapid and relatively inexpensive means of differentiating many oils. However, the technique is only one in the arsenal of analytical methods available for this task. Other methods are described in *UNEP/IOC/IAEA* (1992), and a detailed discussion on molecular markers has been given by *Peters and Moldowan* (1993).

References to Chapter 21

ASTM (1993). *Standard Test Method for Quantification of Complex Polycyclic Aromatic Hydrocarbon Mixtures or Petroleum Oils in Water.* Designation D5412-93. Philadelphia: American Society of Testing Materials, 1916 Race St. Philadelphia, PA. 19103, USA, 1993.

Bjørseth, A., Eklund, G. (1979), *HCR & CC*, 2, 22.

Calvert, J.G., Pitts, N.J. (1967), *Photochemistry.* New York: John Wiley & Sons, Inc..

Carey, F.A., Sundberg, R.J. (1990), in: *Advanced Organic Chemistry, A*, 3rd. ed., New York: Plenum Press.

Clark, R.C., Blumer, M. (1967), *Limnol. Oceanogr.*, 12, 79.

Ehrhardt, M., Burns, K.A., Bícego, M.C. (1992), *Mar. Chem.*, 37, 53.

Greenberg, J.P., Zimmerman, P.R., Heidt, L., Pollock, W. (1984), *J. Geophys. Res.*, 89, 1350.

Gustafsson, Ö., Gschwend, P.M., Buesseler, K.O. (1997), *Mar. Chem.*, 57, 11.

Irie, T., Yasunari, Y., Suzuki, T., Imai, N., Kurosawa, E., Masamune, T. (1965), *Tetrahedron Lett.*, 3619.

NAS (1985), *Oil in the Sea: Inputs, Fates and Effects*: Farrington, J.W. (Ed.). Washington, D.C.: National Academy of Sciences.

Peters, K.E., Moldawan, J.M. (1993), *The Biomarker Guide: Interpreting Molecular Fossils in Petroleum and Ancient Sediments.* Englewood Cliffs, N.J.: Prentice Hall.

Sporstøl, S., Gjøs, N., Lichtenthaler, R.G., Gustavsen, K.O., Urdall, K., Oreld, F. Skel, J. (1983), *Environ. Sci. Technol.*, 17, 282.

UNEP (1995), *Reference Methods for Marine Pollution Studies* No. 65.

UNESCO (1984), *Manual for Monitoring Oil and Dissolved/Dispersed Petroleum Hydrocarbons in Marine Waters and on Beaches.*

UNEP/IOC/IAEA, *Determination of Petroleum Hydrocarbons in Sediments*: Reference Methods for Marine Pollution Studies No. 20. UNEP, 1992.

Wattayakorn, G., King, B., Wolanski, E. Suthanaruk, P. (1998), *J. Coastal Res.* 14, 472.

Wayne, C.E., Wayne, R.P. (1996) *Photochemistry.* Oxford: Oxford University Press.

Yamaguchi, M. (1957), *Bull. Chem. Soc. Jpn.*, 30, 979.

Yamaguchi, M. (1958), *Bull. Chem. Soc. Jpn.*, 31, 51.

Yayli, N. (1994), *Indian J. Chem. B. Org.-Med.*, 33, 556.

Youngblood, Y.Y. Blumer, M. (1973), *Mar. Biol.*, 21, 163.

22 Determination of selected organochlorine compounds in seawater

J. C. Duinker and D. E. Schulz-Bull

22.1 Introduction

Several man-made organochlorine compounds have become ubiquitous environmental contaminants. Considerable concern about their presence even at extremely low levels in seawater stems from their persistence, solubility in lipid tissue (and thus high bio-concentration and bio-magnification factors) and direct or long-term harmful effects (*Safe and Hutzinger*, 1987; *Kimbrough and Jensen*, 1989). Because of their very persistence, distribution patterns in seawater and in suspended particles may also be used to study oceanic processes.

The world ocean, especially the North Atlantic, is supposed to be the largest sink for persistent anthropogenic chemicals such as polychlorinated biphenyls (PCBs). Concentrations and distribution patterns in seawater for such compounds therefore are urgently needed to understand the environmental response of these anthropogenic contaminants. The information available at present, however, is very limited (*Iwata et al.*, 1993; *Schulz-Bull et al.*, 1998). This chapter represents a detailed description of problems associated with the determination of organochlorine compounds in seawater and ways to overcome them, as so far they have not been treated concisely and comprehensively in earlier publications (*Dawson*, 1976; *Duinker and Hillebrand*, 1983a). Significant progress has been made, both with respect to instrumental analytical methods and sampling at sea. It is now feasible to determine reliably the extremely low concentrations of, *e.g.*, polychlorinated biphenyls, chlorobenzene, hexachlorocyclohexanes and the DDT-family in open-ocean waters (*IOC*, 1993). The methods presented here are also applicable to natural waters (rivers, lakes) with higher concentrations of the compounds of interest.

To understand the mechanisms of distribution, transport, degradation and possible detrimental effects of organic compounds in the marine environment, access to accurate analytical data on well defined compounds is essential. In this chapter attention will thus be focused on the application of high-resolution capillary gas chromatography.

22.1.1 Organochlorines in seawater

The distribution of trace organic compounds such as chlorobiphenyls (CBs) and polycyclic aromatic hydrocarbons (PAHs) in seawater and marine particulate matter is determined by the complex influence of physical, biological and chemical processes. The interpretation of horizontal and vertical distribution patterns of these substances requires information on the corresponding characteristics of the water bodies, *e.g.*, temperature, salinity, depth and the concentrations of nutrients and dissolved oxygen.

The chlorobiphenyls are chemically related compounds with a wide range of physico-chemical properties such as vapour pressure, water and lipid solubility, and particle/solution distribution coefficients. They can thus be used as model compounds to forecast the behaviour and distribution of other, less well studied organics. The CB distribution patterns can be used as a basis for theoretical models to evaluate the reliability of experimentally determined distribution patterns of a variety of compounds.

Experience gained during the last decade in the determination of CBs in off-shore surface and deep waters has shown that concentrations are extremely low, much lower than reported earlier (see Table 22-1). Concentrations reported for the Mediterranean Sea (*Tolosa et al.*, 1997; *Schulz-Bull et al.*, 1997), the North Sea (*Schulz-Bull et al.*, 1991) and the Baltic Sea (*Schulz-Bull et al.*, 1995) were well above those found in surface waters of the open ocean (*Iwata et al.*, 1993; *Schulz-Bull et al.*,1998). In deep-ocean water (*Schulz et al.*, 1988; *Petrick et al.*, 1996; *Schulz-Bull et al.*, 1998) much lower concentrations were found than in surface waters. In North Atlantic Deep Water, values of individual CBs were found to be <0.01 pg/L, yet concentrations in solution were higher than those in suspended material on an equal volume basis. It turns out that the distribution of CBs between solution and suspension is determined primarily by molecular properties (characterized by octanol/water distribution coefficients). However, biological processes disturb the establishment of equilibria. This phenomenon has been observed in river water, in estuarine and coastal waters and during biologically active periods in the surface layer of the open ocean.

The determination of the very low concentrations of organic compounds in ocean water is fraught with problems. Because of the hazard of contamination, the sampling procedure probably is the most critical step in maintaining sample integrity; this has not changed despite significant methodological improvements. Application of pumping systems which sample surface water in front of the moving vessel followed by filtration and extraction on-board ship has resulted in oceanographically consistent data sets on CBs in surface waters of various seas. *In situ* filtration/extraction systems were extremely helpful in establishing some vertical profiles in the open ocean.

To obtain reliable data on concentrations of organochlorine compounds, in particular of individual CBs, extreme precautions are necessary to avoid sample contamination from equipment and the atmosphere. This is a serious hazard in all steps of the analytical procedure: pre-cruise preparations, activities on-board ship, sample storage and final instrumental analysis in the laboratory. Emphasis will be placed on problems associated with the determination of chlorobiphenyls, but the methods are also applicable for determining several other relatively apolar organic compounds in seawater. For determining more polar components, procedures for sampling, high-performance liquid chromatography (HPLC) and gas chromatography (GC) have to be modified.

Because concentrations of chlorinated organics in seawater are extremely low (typically in the pg/L or fg/L range), their determination involves concentration steps over several orders of magnitude so as to increase the amounts of substance available for determination to a level compatible with the sensitivity of the analytical instrument. At the same time they must be separated from other, interfering, seawater constituents present at much higher concentrations.

Not all peaks in ordinary gas chromatograms of seawater extracts trace single compounds, many are composite peaks of co-eluting substances. Emphasis will be placed here on a method to separate compounds thus characterized, *e.g.*, chlorinated biphenyls, chlorobenzenes, members of the DDT family and hexachlorocyclohexane isomers. Its resolving

Table 22-1. Concentrations of chlorobiphenyls (pg Σ CB per litre seawater) reported for various seas and open ocean

Sample	Dissolved	Suspended	Reference
I Baltic Sea			
Nov. 1988	26.0–78.0	no data	*Schulz-Bull et al.,* 1995
Nov. 1989	14.0–237.0	4.0–26.0	
April 1991	3.0–146.0	2.0–2859.0	
II North Sea			
Febr. 1988	13.7–174.7	–	*Schulz-Bull et al.,* 1991
Aug. 1988	28.1–415.0	5.4–155.3	
1975	100–5000	–	*Dawson and Riley,* 1977
III Laptev-Sea			
Aug. 1993	0.79–1.08	1.7–63.0	unpublished
IV Mediterranian Sea			
April 1982	600–13300	100.0–6400	*Burns and Villeneuve,*1987
Nov. 1988	1.7–43.9	190.0–615.0	*Schulz-Bull et al.,* 1997
V North Atlantic			
Mai 1986 [a]	1.56–21.0	no data	*Schulz et al.,* 1988
Nov. 1987 [b]	< 1–12.79	no data	*Schulz,* 1990
Mai 1992 [c]	0.95–6.02	0.05–3.76	unpublished
Aug. 1993 [d]	0.03–3.22	0.29–11.24	*Schulz-Bull et al.,* 1998
not specified [e]	21–29	no data	*Iwata et al.,* 1993

[a–c] 47°N, 20°W; [d] around Island; [e] *ca.* 30–35°N.

power does not suffice, however, to separate Toxaphene, Chlordane and chlorinated paraffins. Of these complex mixtures of chlorinated substances only a few components have been identified (*Muir and de Boer,* 1993; *Kimbrough and Jensen,* 1989). Although analysis is possible in terms of technical formulations, these organochlorines, therefore, will not be discussed in this chapter. However, Toxaphene and other chlorinated compounds are now produced in large amounts to replace PCBs and DDT; more detailed information on organochlorines, therefore, is urgently needed.

22.2   Sampling

22.2.1   General remarks

Until recently, sampling was the most problematic part of the complex procedure necessary for determining trace organic contaminants in seawater, mainly due to sample contamination and difficulties of separating solutes and particles (see also Chapters 1 and 2). In fact, the sampling procedures caused the largest problems and uncertainties in the determination of PCBs in open-ocean waters (*IOC*, 1993).

There are many more pitfalls likely to endanger sample integrity. Conditions and activities on-board ship represent the largest potential source of contamination for collected seawater samples, in particular the engine room, the storage rooms for fuel and oils, and painting and waste incineration. During extended station periods exhaust particles accumulate on deck, from where they are distributed over the entire ship. We have carried out stringent tests of samples and sampling procedures during cruises with several research vessels and came to the conclusion that clean analyses of open-ocean samples on-board ship are essentially impossible. Unless clean laboratories (*e.g.*, clean-room containers) are available, filters charged with suspended particles and resin columns used for concentrating dissolved constituents should therefore be returned to the home laboratory for further treatment.

22.2.2   Sampling and filtration of seawater

Like other chemicals, organochlorines occur in seawater in a continuum of dissolved, colloidal and discrete particulate forms (*Gustafsson and Gschwend*, 1997). Operationally, 'dissolved' and 'particulate suspended' forms may be distinguished on the basis of separation techniques such as filtration or centrifugation. The separation may depend on size and density of the particles; it may also depend on the composition of the suspension. For instance, particles smaller than the nominal filter pore size are retained on a clogged filter.

As organochlorine compounds usually occur at extremely low concentrations in seawater solution, they must be extracted into an organic medium and be concentrated over many orders of magnitude prior to gas chromatographic analysis with electron capture detection (GC-ECD). The required water volume depends on:
1)   the concentration levels in the sample;
2)   the detection limits of the analytical methods; and
3)   the contribution of contamination during sampling and clean-up procedures.

Even if contamination could be avoided entirely, sample volumes must approximate 1000 L to enable analysis, at an instrumental detection limit of 0.1 pg (ECD), of components at the 0.01 pg/L concentration range typical for open ocean waters. Sample contamination aggrevates the problem. This limitation seems to have been overlooked in several earlier studies (see review by *Harding*, 1986).

Further difficulties originate in the presence of particles (*Duinker*, 1986; *Brownawell and Farrington*, 1986). The transport mechanisms of water and particles are different. The environmental fate of a chemical, therefore, depends on its distribution between solution and suspension, which is determined not only by its liquid–solid distribution coefficient but also by the characteristics of particles and their concentrations.

22.2.2.1 Surface water sampling

Pumping systems are capable of delivering large volumes of surface seawater for subsequent filtration/extraction on deck. A system for sampling surface water from the bow while the ship is in transit (see Chapter 1, Fig. 1-2) has been applied successfully by *Schüßler and Kremling* (1993) and *Schulz-Bull et al.* (1995).

22.2.2.2 Deep-water sampling

Batch samples

A spherical glass vessel has been suggested as a low contamination device to sample seawater and extract it with a water immiscible organic solvent (*Gaul and Ziebarth*, 1983; *Theobald et al.*, 1990). Its volume, however, is limited to 20–100 L, and no separation of suspended and dissolved phases is feasible.

Up to 400 L samples can be collected at arbitrary depths with modified stainless-steel Bodman/Hydrobios samplers (*Schulz et al.*, 1988). These large volumes may be filtered on-board ship. However, there are some serious problems with the systems. The extent of contamination during sampling as well as during filtration on-board cannot be estimated, nor can it be controlled. Also, estimations of partitioning between dissolved and particulate forms may be far from correct, as particles may settle on the bottom of the sample container during the long period required for filtration of such large water volumes and may thus escape identification and quantification.

In situ sampling systems

In situ sampling/filtration/extraction systems offer the most favourable conditions for accurate determinations of organic trace compounds in seawater.

The Seastar *in situ* sampler, designed specifically for organic compounds, could be used to depths of 400 m and for sample volumes of up to about 200 L (*Green et al.*, 1986; *Ehrhardt and Burns*, 1990). It has been replaced, in the meantime, by the Infiltrex system with options for increased sampling capacity and greater maximum depth of operation (see Chapter 18). Other systems have been designed for the collection of large amounts of particles for the determination of inorganic trace elements and radioisotopes (*Simpson et al.*, 1987); they have not been designed nor tested for sampling trace organics. As this requires special precautions to eliminate contamination (*e.g.*, in the choice of construction materials), they probably do not meet the required specifications.

The Kiel *in situ* pump system (KISP) for filtration and extraction of trace organics at the depth of sampling (*Petrick et al.*, 1996) is suitable for volumes of up to 2000 L or more (depending on particle concentration) and depths in excess of 6000 m. Sufficient amounts of PAH and CB may thus be collected in open-ocean waters to allow their analytical determination at concentration levels around or below 0.01 pg/L. The sampler is depicted in Fig. 22-1 (technical details can be found in Chapter 2).

The operation of the unit is software controlled, and essential data are recorded for later evaluation. The sample water only comes in contact with Teflon, stainless steel and polyethylene. Before the water passes through the adsorption column it is filtered to collect suspended particles for separate analysis. Another reason is that no particles should collect on

Fig. 22-1. Kiel *in situ* pump. See also Fig. 2-2.

the resin bed; such particles will cause clogging, greatly reducing the flow rate through the column. Furthermore, particles collected on the resin bed will later be solvent extracted together with it and their organic contents erroneously included in the dissolved phase. 'Dissolved' material (defined operationally by the pore size of the filter material) is concentrated with macroreticular resins such as Amberlite XAD-2 or Serdolit. A limiting factor in obtaining sufficient material on the resin is filter clogging. The obstacle is most serious in surface waters and in coastal regions, but has not been observed in the deep ocean.

After retrieval of the sampler, the filter and resin cartridges are removed. Filters are stored at $-20\,°C$, resin cartridges at $+4\,°C$. Data are read from the system's on-board computer, and the batteries are recharged. Several units can be attached to the hydrographic wire to obtain vertical profiles of concentrations. Full details of the procedure have been published by *Petrick et al.* (1996). As blank determinations are part of the standard procedures, contamination can be checked and eliminated, if necessary. Procedural blanks may be kept as low as 0.001 pg/L in a 1000 L sample. Solutes may thus be determined reliably at the 0.005 pg/L concentration level with a 5:1 signal to noise ratio.

22.3   Laboratory procedures

Great care has to be exercised in the preparation of materials for sample extraction and concentration/clean-up of sample extracts so as to detect any source of contamination and to reduce it to an acceptably low level. All work should be carried out in clean-room laboratories, in clean benches or under pure gas protection.

22.3.1   Cleaning and storage of glassware, filters, chemicals and solvents

All glassware should be washed with detergents overnight, rinsed several times with clean water and heated to 250 °C for 12 h. No organic solvents should be used to rinse the glassware prior to use.

Glass fibre filters retaining particles > 1.2 μm (Whatman GF/C) or > 0.7 μm diameter (Whatman GF/F) are used for filtration. Solvent extraction is impracticable for filter cleaning as large amounts of solvent are required, and the filters become brittle when wetted with water at a later stage. An effective cleaning method is calcination at 350–370 °C. The filters should be well separated from each other in the furnace to allow efficient evaporation of contaminants. The cleaned filters can be stored in clean Petri dishes or wrapped in aluminium foil.

Sodium sulphate is cleaned conveniently by baking for 10 h in an oven at 350 °C. It is stored in glass-stoppered bottles.

Silica gel and aluminium oxide have to be treated chemically (see also Chapter 19). They are extracted with dichloromethane in a Soxhlet apparatus, then with n-hexane, each time for about 24 h. After extraction the wet grains are dried in a rotary evaporator. The flask should rotate at minimum speed to avoid mechanical breaking off the particles into smaller units. Rotary evaporation is stopped as soon as the silica gel or aluminium oxide start to 'rain' down the side of the flask. The oxides are transferred into and dried in a drying pistol at 0.01 hPa. The temperature is raised from 30 °C in steps of 30 °C to 120 °C, held at each plateau for 1 h and then returned to room temperature under vacuum. To avoid contamination during storage the clean adsorbents are flame-sealed in glass ampoules immediately after cooling in portions just enough to fill the separation column. Care must be taken to restrict exposure to the atmosphere to a minimum prior to and during flame-sealing, because active silica gel or aluminium oxide attract water and contaminant vapours. If required, 10 % by mass of water is added to the fully active silica gel or aluminium oxide for controlled partial deactivation; they are then less liable to undergo activity changes during storage ensuring good reproducibility.

Commercially available solvents such as acetone, acetonitrile, dichloromethane, hexane and pentane are invariable contaminated with ECD active substances. Concentrations vary with the quality ordered, the batch number and the supplier. Contamination should be checked by concentrating a 100 mL portion of the solvent to 50 μL in a rotary evaporator and injection of 2 μL of the concentrate into a GC with electron capture detection. No peak in the ECD chromatogram (after 10 min) should be larger than that of 0.1 pg of chlorobiphenyl. Otherwise, the solvent must be distilled.

The following procedure has been found to be very efficient; it also is cost-effective, because technical grade solvents may be used as feed stock. An all-glass distillation apparatus is equipped with 130–150 cm vacuum-jacketed separation columns. The column fillings must be glass to allow cleaning with nitric acid; glass helices are particularly useful. The still is then freed of acid by distilling 500 mL of water through it twice. To exclude room air from contact with the organic solvent it is essential that during distillation nitrogen gas (15 mL/min) flows from the distillation flask to the condenser through which it escapes to the outside. Static methods of excluding room air (*e.g.*, activated charcoal or molecular sieve filters) are not recommended, because their effectiveness decreases with increasing saturation of the sorbent. The solvent is distilled at a reflux ratio of 1 : 20, and the condensate is collected in a receiving flask of 1 L capacity. From here it is transferred into glass containers holding no more solvent than is necessary for two analyses.

The reason is that a bottle with enough solvent for 10–15 analyses will be opened and closed many times. Even in stoppered bottles the contents are contaminated from the surrounding atmosphere. The following example will illustrate this: Dichloromethane was distilled in a nitrogen atmosphere as indicated above. A 100 mL volume was concentrated to 100 μL and 2 μL were injected on-column. The ECD response was equivalent to 0.1 pg per CB congener. The solvent was left in a glass-stoppered flask in the laboratory for 4 h. The same procedure then resulted in a total amount of 10 ng total CBs in 100 mL of solvent, rendering the solvent unsuitable for further use.

Screw-capped bottles also show leakage because of 'breathing' under variable temperature conditions which results in amazingly rapid contamination of solvents. This effect can be minimized by storage at a constant low temperature ($-20\,°C$). Contamination from the atmosphere is one of the more serious threats to solvent integrity. We have encountered this problem invariably in the laboratory, in the outside atmosphere and particularly on research vessels.

The only way to store ultrapure organic solvents for unlimited periods of time, including transportion, is in flame-sealed glass ampoules. The solvent is transferred into the ampoule directly from the receiving flask. The filled ampoule must be cooled with liquid air to solidify the solvent. The filled ampoule is flushed with nitrogen gas to remove oxygen and sealed while still in the Dewar flask to avoid volume increases of the vapour phase, which would impede proper sealing. The ampoules can be stored safely in appropriate boxes for unlimited periods of time. The ampoules must be opened carefully, because at room temperature pressure will build up inside. Use of safety goggles and wrapping of the ampoule in a towel is strongly recommended.

22.3.2   Analyte enrichment

Basically, two methods are available for concentrating organic compounds from seawater: solvent extraction and sorption onto a solid adsorbent.

The advantages and disadvantages of solvent extraction have been discussed in the second edition of this book. It becomes inconvenient when large volumes have to be extracted, as is necessary for determinations of organochlorine compounds in seawater; other more suitable methods being available it will not be discussed here.

Various materials have been used to adsorb organochlorine compounds from natural waters. They include: activated charcoal; urethane foam plugs; polyurethane foam coated with adsorbents; a porous polymer (Tenax GC); a mixture of activated charcoal powder, MgO powder and refined diatomaceous earth; Carbowax 4000 and n-undecane on Chromosorb DMCS; Serdolit and Amberlite XAD resins (references were given in the second edition). The last material has been used most successfully (*IOC*, 1993). The use of XAD resin has been described in some detail by *Dawson* (1976). Problems initially encountered with cleaning of this adsorbent have been solved in the meantime. Extremely low blank values can now be obtained for XAD-2 columns. Their use allows reliable determinations of CBs present in seawater at levels as low as 0.005–0.1 pg/L.

The advantages of using XAD-2 resin, as we see them today, are the large volumes that can be processed per unit time (90 L/h, *i.e.*, 5 bed volumes/min), and the use of several extractors in series, if necessary. The extraction efficiency of a single column is 70–90 %, if the capacity of the column is not exceeded (*Schulz*, 1990). This is in good agreement with the results of another study (*Gómez-Belinchón et al.*, 1988).

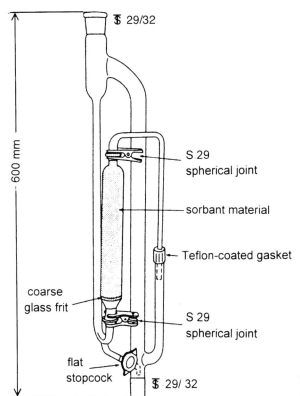

S 29/32

S 29
spherical joint

sorbant material

Teflon-coated gasket

coarse
glass frit

S 29
spherical joint

flat
stopcock

S 29/32

600 mm

Fig. 22-2. Extractor for XAD-2 columns according to *Ehrhardt* (1987).

Procedures for cleaning the resin, blank determination and analyte desorption are described in Chapter 18. A simpler version of the extractor shown in Fig. 18-1a may also be used; it was described by *Ehrhardt* (1987) and is shown in Fig. 22-2.

As an additional precaution for the very delicate CB determinations a nitrogen gas flow of 10 mL/min through the condenser should be maintained during extraction. The flask is then cooled to approximately 0 °C, preferably in a freezer, before the solvent is removed by rotary evaporation at 0.1 hPa. The boiling chips stay in the flask, because removal is a potential source of contamination. At the low temperature of the solvent they will not induce boiling. A layer of ice growing on the outside of the flask retards the evaporation of the solvent (boiling point 76 °C at atmospheric pressure). It may be removed by immersion in a water-bath at 30 °C after 15 min. It is important to allow for this delay as aqueous acetonitrile tends to 'bump' or boil explosively.

Note: For analyses of compounds more polar than CBs it is important to remove acetonitrile quantitatively. Complete removal is indicated by water droplets appearing on the inner wall of the solvent trap. Residual acetonitrile would seriously impede the extraction with n-hexane of more polar organic seawater constituents from the remaining water phase.

22.3.3   Volume reduction of organic solvents and solutions

The volume of extracts dissolved in, *e.g.*, hexane is reduced conveniently with a rotary evaporator. We have found insertion of a solvent trap (Chapter 18, Fig. 18-2) between the rotating flask and the steam duct extremely beneficial to control and eliminate the possible loss of solutes during fast or 'explosive' distillation of the solvent. The initial temperature of the solvent should be lower than 10 °C, to avoid rapid initial boiling, particularly when using pentane and dichloromethane as solvents, to avoid losses of low boiling solute fractions. The hazard diminishes as volume reduction progresses, because the heat of evaporation lowers the temperature of the solvent. With these precautions, volume reduction from 100 mL to 0.5 mL of a C_{10}–C_{24} n-alkane mixture in hexane is possible without discrimination against low boilers. The procedure takes about 30 min at 0.1 hPa. A blanket of N_2 gas is required to avoid contamination of the solvent from the atmosphere. Using activated charcoal or a molecular sieve for cleaning ambient laboratory air before it enters the rotary evaporator in many cases is inappropriate, as the contact time between the air and the adsorbent often is too short for effective removal of contaminants. Moreover, it is difficult to observe the condition of the adsorbent which changes with time at a rate depending on the quality of the laboratory air. A rotary evaporator vented with N_2 gas can easily remain clean for periods of 1–2 years as was shown by running blanks; this option, therefore, is preferred.

Volume reductions from 0.5 ml to 20 μL may be accomplished in narrow glass ampoules as described in Chapter 18. The removal of 500 μL at 0 °C can take around 15 min. Too rapid removal results in discrimination against low boilers. This effect can be detected by concentrating a C_{10}–C_{24} n-alkane mixture in hexane. Cooling is applied during solvent removal. The ampoules can then be sealed immediately after the desired final volume has been reached. As soon as the tip of the ampoule takes the form of a sphere, the sealing procedure is completed successfully, and the contents of the ampoule can be stored safely. The ampoules are made from 4 mm inner diameter glass tubes (see Chapter 18, Fig. 18-3) or normal Pasteur pipettes. The use of these ampoules seems the only way to store samples or blanks without any contamination or loss over prolonged periods of time.

22.3.4   Extraction of filters

The filter and material collected on it are extracted for 6 h with 50 mL of acetonitrile. The procedure to which the extracts are then subjected for cleaning and separation of fractions of different polarities are the same as those described for water extracts.

22.3.5   Compound-class separation of extracts by HPLC

Most environmental sample extracts in an organic solvent require clean-up before they can be analysed by GC-ECD reliably and without causing damage to the separation column. Failure to remove interfering compounds, usually present at much higher concentrations than the organochlorine compounds of interest, will lead to interferences with GC-ECD procedures.

Normal treatment of the sample extracts involves clean-up over an Al_2O_3 micro-column, followed by class separation with a silica gel micro-column. In our experience Al_2O_3 clean-up (4×0.5 cm i.d. column, 2 g of Al_2O_3, deactivated with 10 % water), followed by a simple HPLC method, is very effective for eliminating interfering compounds. Increasingly polar fractions of organic compounds are obtained (aliphatics, PAHs, pesticides, Toxaphene, CBs). These can be determined accurately by GC methods, without contamination or deterioration of injector, columns and detector. The recommended procedure is described in detail in Chapter 20.

22.4 Gas chromatographic separation and detection

22.4.1 Capillary columns and operating conditions

Ideally, each compound should elute as a single peak. This is difficult to achieve when dealing with complex environmental samples. The use of high-efficiency fused silica capillary columns therefore is obligatory. Good quality columns are available commercially as narrow or wide-bore (0.2–0.5 mm i.d.) fused silica, in various lengths (up to 100 m) and with a wide choice of coating and film thickness (0.05–0.25 µm). Capillary columns have a high specific gas permeability and a very small amount of liquid phase. Pressure-regulated flow through the column is only a few mL per minute. Efficiency is high: a typical number for total effective plates is 400 000 for a 50 m capillary column. Capacity is about 100 pg per component.

Several column coatings have been used for determinations of organochlorines in environmental samples (*Lang*, 1992; *Galceran et al.*, 1993). High-resolution chromatograms of environmental samples usually show many peaks. Each analyst selects his or her columns and experimental conditions as a compromise between resolution and analysis time usually determined by trial and error. Optimum conditions for separation of one pair of peaks may be different from those for another pair. It may therefore be impossible to optimize conditions for all components of interest with just one column. Depending on the problem, the coating is selected from various possibilities such as hydrocarbon Apiezon-L and methyl silicone or methyl phenyl silicone columns (SE-30, SE-52, SE-54, CP-SIL-5, SIL-7) and others.

Column life and efficiency are maintained as long as the liquid phase remains as a thin, evenly distributed film. The column performance usually decreases, when the liquid phase is repelled by the surface. Thus, displacement of the liquid phase at the inlet end of the column can occur after a large number of splitless and in particular on-column injections (removal of a few coils may bring back the original efficiency without significantly modifying retention behaviour). Deterioration of the entire column is accelerated by continued exposure to a high temperature; this is particularly the case with reduced or zero carrier gas flow. Column quality is also determined by the nature of injected samples (it is particularly sensitive to materials that are more strongly adsorbed than the liquid phase) and to carrier gas impurities (water and oxygen). Columns coated with methyl silicone (SE-30), 5 % phenyl (SE-52) and 1 % vinyl 5 % phenyl (SE-54) methyl silicone gums can tolerate short-term exposure at 320 °C; they are also to some extent resistant to water and oxygen. The use of an SE-54 column is highly recommended as the retention properties of all CB congeners have been determined for this column type (*Mullin et al.*, 1984).

Some additional remarks on operational conditions may be useful to maintain column integrity when determining extremely low concentrations of, *e.g.*, CBs in seawater at concentrations as low as 0.1–2 pg/L.

New capillary columns have to be conditioned to remove residual traces of solvent and lower relative molecular mass fractions of the liquid phase. Carrier gas should flow at room temperature for some time to remove oxygen; the column is then exposed to moderate temperatures (80–100 °C) for some hours before the temperature is increased to a value that must be a compromise between minimum time required to achieve a stable baseline and maximum column life time. The normal temperature is the maximum temperature required for the analysis. To avoid destruction of the column at higher temperatures, a sufficient flow of carrier gas through the column should be maintained. During conditioning, the column should be left disconnected from the ECD so as to minimize detector contamination.

Older columns may have to be subjected to higher temperatures periodically to remove carrier gas impurities that have accumulated. The ECD may remain connected to the column provided that it is kept at an elevated temperature (300–320 °C).

The carrier gas must be of high purity (H_2, at least 99.999 %). Impurities can saturate molecular sieve traps, gas lines and other materials and, if not removed, result in bleeding at higher temperatures which causes baseline instability and shortens the life of the column.

22.4.2 Single column and multi-dimensional gas chromatography

Most laboratories use high-resolution capillary columns for analyses of complex environmental samples. General aspects of GC equipment requirements have been discussed in the second edition of this book.

The composition of chlorobiphenyls in commercial mixtures and environmental samples is so complex that no single GC column is available which separates all constituents of CB mixtures into single, well-separated peaks. Until recently, it was thus impossible to analyse CB mixtures unambiguously in terms of individual congeners. *Mullin et al.* (1984) have published the data required to approach this goal by reporting the retention properties of all theoretically possible 209 CB congeners on one particular column, *i.e.*, SE-54. This allows the identification of CBs that are well separated on an SE-54 column from other CBs present in the mixture. The data of *Mullin et al.* (1984) do not allow the determination of single CBs eluting from an SE-54 column in composite peaks. This problem has been solved by application of a multi-dimensional GC technique (MDGC-ECD).

A short description of the technique is given here. Detailed information is available in the literature (*Duinker et al.*, 1988). In the MDGC-ECD mode, two capillary columns of different polarities are arranged in series so that the second column receives only small pre-selected fractions eluting from the first column. The columns are located in two independent ovens with two separate [63]Ni electron capture detectors (the main and the monitor detector). A valveless pneumatic control system regulates the sample flow. It is led either to the monitor detector or to the second column and the main detector, depending on pressure settings. An example of the power of MDGC-ECD is given in Fig. 22-3. With properly selected columns, all 209 CB congeners can be baseline separated allowing their accurate identification and quantification, even at trace concentration levels (*Schulz et al.*, 1989).

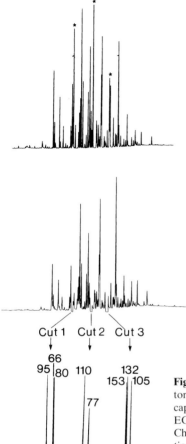

Fig. 22-3. Chromatogram of Aroclor 1254 recorded by the monitor-ECD in the MDGC mode without heart cuts (top), SE-54 capillary column. Chromatogram re-recorded by the monitor-ECD in the MDGC mode, reflecting the cut domains (middle). Chromatogram recorded by the main-ECD reflecting the separation of all possible congeners in the three cuts (lower), with the aid of a synthetic mixture (*Schulz*, 1990).

22.4.3 Gases and gas supplies

The quality of external gas supplies requires serious attention when determining chlorobiphenyls and pesticides at high instrumental sensitivities. Treatment of metal tubing for gas transfer by a simple rinsing procedure with solvents may be insufficient because some contaminants are not easily removed quantitatively. Metal tubing has to be heated (*e.g.*, with a butane burner) under a moderate stream of N_2 gas (30 mL/min) starting at the gas supply side. A smell test at the end of the tubing is usually sufficient to check if the treatment was effective. Stainless steel is the preferred material: it has better mechanical stability than copper and diffusion of gases (especially of H_2) through the wall is considerably less.

22.4.3.1   Selection of carrier gas

The selection of carrier gas (N_2, He, H_2) is usually a compromise between several aspects, *e.g.*, resolution and analysis time. Nitrogen results in higher column efficiency than both helium and hydrogen, but its average linear velocity (u') at elevated temperatures is considerably lower than that of He and H_2. Also, the change in efficiency with u' is smaller for the latter which thus are the preferred carrier gases.

Nitrogen and helium are not dangerous because they are inert; the increase of their viscosity with temperature, however, is a drawback. Unless a constant flow regulator is used, the carrier gas velocity in the column decreases with increasing temperature, when the pressure is kept constant (as is the norm). The flow velocity of He decreases from the optimum of about 2.0 mL/min at 60 °C to 0.5 mL/min at about 250 °C (head pressure 0.4×10^5 Pa (0.4 bar), column length 30 m, 0.32 mm i.d.). Thus, the optimum carrier gas velocity is not maintained during a temperature programmed run, resulting in reduced column efficiency (increase of peak width, increased retention times, poor separations). The same applies to N_2. These factors become apparent and critical at concentration levels approaching detection limits, when signal peaks start to disappear in the baseline noise. The problem can be solved partly with an automatic flow controller. However, such problems do not occur when H_2 is used as carrier gas. Its viscosity is practically constant in the temperature range used during a GC run. Moreover, its high diffusion velocity allows rapid re-establishment of equilibria between vapour phase and stationary phase. Peaks are thus narrower and taller, and retention times are shorter. Unfortunately, its use may cause reduction reactions of chlorinated compounds catalysed by hot stainless-steel surfaces.

Injection in the splitless mode may enhance this effect. The sudden evaporation of the injected volume of liquid hexane ($2 \mu L$) results in about 0.5 mL of vapour. Ideally, this volume will just fill the glass insert of the injector. However, part of it may escape and come into contact with hot metal surfaces, potentially inducing reactions; *p,p'*-DDT, for instance, may disappear to a large extent. This does not happen with on-column injection; this mode of injection therefore is recommended when H_2 is used as carrier gas.

22.4.3.2   Carrier gas impurities

Impurities in the carrier gas may cause irregularities in the baseline, ghost peaks and generally poor chromatograms. If the septum and the glass liner in the injector have been excluded as possible sources, such impurities may be detected by temperature-programmed blank runs after the column has been kept at a low temperature for some time (*e.g.*, overnight). Septum bleed may cause similar problems; this source can be eliminated by maintaining a septum purge gas flow. Septa should be changed at regular intervals according to instructions given by the manufacturer.

Even ultrapure gases for GC applications sometimes contain oxygen in the ppm concentration range. This leads to slow but certain deterioration of capillary columns at higher temperatures causing shifts in retention times, reduced column efficiency and baseline drift. This is a particular hazard for columns with thin layers of stationary phase (0.15–0.25 μm), as are used for CB determinations.

Polyethylene and Teflon are not allowed as gas supply tubing materials, because oxygen may enter the gas by diffusion or through small leaks. It is essential to insert a combination of an activated charcoal filter and oxygen scrubber (Gas Clean filter) into the gas supply line just before the inlet of the GC. Such filters eliminate traces of oxygen and water. They also prevent contamination, should the gas supply tank be emptied completely by accident. A change in colour indicates the need for replacement of the filter/scrubber. The gas cleaning filters should not be made of glass as this material breaks easily. If this happens, the laboratory may then, within a short time, fill with hydrogen because of the high velocity of the outflowing gas. This may result in an explosion in rooms not properly ventilated.

22.4.3.3 Leaks

Another potential source of trouble is gas leaks in the GC. Diluted detergent solutions may be used to detect them, but tightening the fittings will squeeze the liquid from between the front ferrule and the conus into the gas supply system. This causes sharp peaks as well as baseline drift and variations in detector sensitivity and thus is not encouraged.

A better method to search for leaks is dripping about 0.5 ml of a volatile chlorinated solvent, *e.g.*, dichloromethane slowly from a syringe onto the suspected leak site, starting from near the detector (ECD). A clear peak signals the presence of a leak. The time lag between solvent application and signal detection increases with increasing distance between the leak and the ECD (*e.g.*, from seconds to minutes).

The connections of the capillary to the detector, injector and the insert are additional and frequent trouble spots. Injection of 2 mL of dichloromethane into the closed oven is a useful overall leak test. A rapid response of the detector (10–20 s) indicates a leak near the detector, a signal delayed by the residence time in the column (*e.g.*, 60 s for a 30 m column length) indicates a leak near the injector. The more specific signal test is then applied as discussed above. Ferrules for the capillary are frequent sources of problems with some equipment. Graphite or vespel ferrules must have a 0.5 mm bore for a 0.32 mm i.d. column.

A GC system that has been cured and tested in this way can be used without problems over long time periods (apart from electronic errors). We have used columns even with thin layers of stationary phase in such pure and leak-free systems for more than 2000 temperature programmed runs (140–250 °C) without appreciable changes in separation efficiency and sensitivity. In this ideal case the quality of the injected samples is the limiting factor. Samples can be prepared so that the quality of the final solutions corresponds to that of standards. In such cases, capillary columns or detectors are not destroyed or contaminated, but of course they are subject to natural ageing. Clean-up involving the use of aluminium oxide and silica gel chromatography as well as HPLC assist in obtaining such sample solutions (see Chapter 20).

22.4.4 Sample injection

For trace analysis, sample introduction without splitting part of the sample to a vent is the preferred mode for obtaining maximum sensitivity. This can be achieved with splitless or on-column injection techniques (*Schomburg*, 1987).

In the splitless mode, full efficiency of the column is realized by concentrating the sample components in a narrow band near the front end of the column prior to analysis, either by utilizing the solvent effect or by condensation of the solutes. The latter mechanism operates effectively for compounds boiling about 150 °C above the column temperature. Compounds with lower boiling points are concentrated by the solvent effect which requires a high solvent concentration at the column inlet. The solvent effect, based on stronger retention of the front than the rear of the sample plug, when encountering a liquid phase mixed with retained solvent at the inlet end of the column, is most efficient at a column temperature of 10–30 °C below the boiling point of the solvent. The temperature of the injector should cause rapid evaporation of solvent and solutes but it should be low enough to minimize septum bleed and avoid destruction of sensitive components.

The splitless injection method is often used, because it is convenient. However, there are a number of risks that are not obvious. The split and the splitless modes of operation require rather different volumes of the insert liner. In the splitless mode, the vapour volume of 2 mL of solvent has to be accommodated quantitatively to avoid loss of analytes. An insert with too small a volume may force part of the relatively large volume of vapour formed into the injector block or the gas tubing (*Grob*, 1981). This may result in memory peaks appearing at odd places in the chromatogram. High-boiling contaminants remaining after insufficient sample clean-up are another source of problems. They tend to accumulate in the glass insert where polymerization or other reactions cause the formation of undesired coatings, affecting the evaporation of sample components. Low-boiling analytes may be late in reaching the separation capillary with consequent changes in retention characteristics. Continuous bleeding from the insert seriously affects the quality of the column resulting in poor stability of retention times over longer time periods. These problems may be eliminated by replacing the insert liner which is usually more effective and cheaper than cleaning the old insert.

Finally, the position of the column in the insert is critical for column performance and for the amount of sample that can be transferred to the column in, *e.g.*, 30 s. It must be determined experimentally in a series of repeated injections of the same solution with varying column positions until maximum detector response is attained.

In the on-column technique the sample is introduced directly into the column. The sample must be injected into a cold injector (or any other specific device, *e.g.*, a programmable temperature evaporator), so that the evaporated solvent is retained in the column quantitatively. The injection must be slow enough (say in 10 s) to avoid a (short-term) large solvent vapour : carrier gas ratio. The needle of the syringe must reach far enough into the column to avoid escape of sample components into the injector block. As the presence of the needle causes a volume reduction to <0.5 mL, the pressure of the carrier gas must be high enough to enable the gas to enter the column.

Experience obtained over several years has shown that during slow and careful on-column injection the solvent and sample components are distributed over a distance of ≈40 mm in the column before they evaporate completely. To utilize the full efficiency of the column, the sample components are transferred to a narrow band at a cold spot of the column after flash evaporation.

22.4.5   Detectors

22.4.5.1   The Electron Capture Detector (ECD)

The electron capture detector is an essential component in the determination of trace amounts of organochlorines for which its sensitivity is roughly five orders of magnitude higher than for hydrocarbons. For instance, the detection limit for lindane may be as low as

0.02 pg/s using capillary columns. High-energy electrons, emitted from a source within the detector (*e.g.*, a ^{63}Ni foil, half-life 92 years), repeatedly collide with carrier gas molecules, producing secondary electrons of thermal energy. These electrons can be captured by sample molecules. The resulting reduction in cell current is the output signal of an ECD. However, the response function of current *versus* concentration of electron capturing molecules is non-linear. The useful linear range of an ECD is greatly improved if it is operated in the constant-current pulsed mode. Short voltage pulses are applied to the cell electrodes to collect the electron population in the ECD cell. The current generated by the detector cell is automatically regulated by the frequency of the polarizing pulses to maintain a certain standing current. An increase in concentration of electron-capturing molecules in the cell causes a change in the polarizing pulse frequency necessary to restore the balance between the detector cell current and the standing current. The response over a voltage/frequency converter is linear with concentration over a wide range. The dynamic range (covering 4–5 decades of concentration) depends on various parameters such as detector temperature, pulse width and standard current level. The optimum carrier gas flow for an ECD (about 30 mL/min) is much higher than the flow through the column. Thus, an additional make-up gas is needed (normally nitrogen) to purge the detector. Operational conditions should be optimised for all these parameters. High-boiling organic compounds eluting from the column may contaminate the detector causing reduction of sensitivity. The effect is less serious at higher detector temperature. Periodic heating to 350 °C overnight is helpful in maintaining good detector performance. Operating the ^{63}Ni ECD at 320 °C results in relatively limited contamination.

Response factors

Response factors of individual chlorobiphenyls on ECDs depend on the number and the positions of the chlorine atoms in the molecule, and also on the analytical instrument (*Frame*, 1997). The ECD response is affected in particular by characteristics of the detection and injection systems, such as temperature, state of contamination and geometry. Considerable differences exist between detector responses of on-column and split/splitless injection, in particular the dependence on the amounts injected (*i.e.*, the degree of linearity). It is not surprising, therefore, that apparently conflicting data are cited in the literature (*Mullin et al.*, 1984) and also that significant differences have been found for response factors even between carefully prepared standards.

The use of published data on response factors (*e.g.*, such as the detailed list in *Mullin et al.* (1984)) for estimating the responses of other CBs (*e.g.*, those not available as reference materials) has limited value. The only reliable way to quantitate CBs in samples uses reference materials containing known amounts of the congeners of interest.

22.4.5.2 Mass spectrometry

The use of a mass spectrometer as detector for organochlorines is becoming more popular. The most common ionization method is by electron impact. Negative chemical ionization mass spectrometry is extremely sensitive for higher chlorinated compounds. Single ion monitoring mass spectrometry (MS-SIM) may not be quite as sensitive as an ECD, but the selectivity for CBs is better, often enabling distinction between co-eluting congeners with different numbers of chlorine atoms. The correlation of MS-SIM relative response factors (RRFs) is inverse to that of the ECD, the ECD RRFs increase with the extent of chlorination. Excellent results can be obtained with high-resolution quadrupole instruments (*Frame,* 1997). Quantification may be based on ^{13}C labelled internal standards.

22.4.6   Identification

The most widely used attribute to characterize a chromatographic peak is its retention time or its relative retention time, *i.e.*, the adjusted retention time relative to the adjusted retention time of a selected reference compound. The much higher resolution of capillary columns than that of packed columns results in a considerably higher probability of separating closely eluting components. The labelling system proposed by *Ballschmiter and Zell* (1980) is extremely useful for characterizing individual components. However, for some peaks the numbering is inconsistent (nos. 199, 200, 201 in the last paper). The revised list of all environmentally relevant CBs (*i.e.*, containing those that have been detected at concentration levels >0.05 % (m/m) in commercial mixtures (Clophen and Aroclors) was published by *Schulz et al.*, (1989).

Some techniques are available to obtain additional information on the identity of a certain peak. Mass spectrometric methods, in particular mass fragmentography, are extremely useful to identify and distinguish components with even retention properties (*Duinker and Hillebrand*, 1983b). MS techniques are specific and their sensitivity (femtomole range) is comparable to that of GC-ECD techniqes. Many marine laboratories are equipped with GC-MS systems, mostly used for analysing biological samples. It is expected that the number of applications for samples with much lower concentrations of organochlorine compounds, in particular in seawater, will increase significantly in the near future. An additional identification technique is offered by multi-dimensional GC-ECD (see Section 22.4.2). Only a few laboratories are presently equipped with multi-dimensional gas chromatographs, but many CB congeners can be determined with a single SE-54 column provided that they elute as single peaks, well separated from adjacent congeners. These congeners are listed in Table 22-2.

Table 22-2. List of CB congeners that elute as single peaks from an SE-54 column. They are identified by their IUPAC numbers (*Schulz et al.*, 1989) and chlorine numbers (*n*Cl)

*n*Cl	IUPAC number
1	1,2,3
2	6
3	19, 25, 26, 29, 34, 35
4	40, 44, 45, 46, 49, 52, 63, 69, 70, 74
5	83, 84, 85, 88, 91, 92, 96, 97, 99, 107, 119
6	128, 130, 134, 135, 136, 146, 167, 169
7	172, 174, 175, 177, 180, 183, 185, 187, 189, 191, 193
8	194, 197, 198, 199, 200, 205
9	206, 207
10	209

22.4.7   Standards and quantification

22.4.7.1   Standards

A stock solution (about 10 mL) of each compound of interest is prepared at a concentration of 1–2 mg per 10 mL by gravimetry, in, *e.g.*, isooctane. Less concentrated stock solutions (50–100 mL) are prepared by dilution (vol/vol) of an aliquot of the concentrated stock solution. Working standards are prepared by mixing subsamples of the dilute stock solutions in portions roughly inversely proportional to the response factors of the various compounds. Peak heights (or peak areas) in the final mixture are then approximately the same. The final volume of this mixture is adjusted with isooctane to yield concentrations that are within the dynamic range of the detector, *e.g.*, 1–100 pg/µL for ECD. Calculation of the absolute amounts injected with 1 µL sample volumes should take into account the density of the solvent, when the original stock solution has been made up on a mass/mass rather than on a mass/volume basis. All standard solutions are sealed in glass ampoules and stored at −20 °C. Masses are recorded on the bottles and checked before subsamples are removed later on. The quality of all concentrated stock solutions of the individual compounds should be checked by GC with flame ionization detection (FID) for major non-electrocaptive compounds and of all diluted stock solutions by GC-ECD.

22.4.7.2   Calibration

Several injections should be made of standard solutions of different concentrations. The responses are plotted against injected volumes to determine the linear range of detector response and the response factor for each component (peak area or peak height). A more sensitive method is to plot response/injected mass against injected mass (*Wells et al.*, 1988). Peaks in sample chromatograms for which the response is not too different from that of the standard, within the linear range, may be quantified with the same standard. The relative areas or heights of different peaks in any sample chromatogram can differ widely so it may be necessary to inject the sample extract several times in succession after the appropriate concentration or dilution.

Many compounds are resolved completely from each other using capillary columns. These compounds can be quantified accurately when standards of sufficient purity are available. The calibration mixture must be analysed under the same instrumental conditions which are used for the sample. Differences in sensitivity of the detector for different components in the sample are accounted for by the response factors during automatic calculation. A peak must be identified correctly before it is quantified. Identification is based on the retention times of designated peaks in the calibration table (the reference peaks) and those of the peak of interest.

The external standard method uses absolute response factors, the internal standard method is calibrated in terms of response ratios. In both calibration methods, each peak is calculated independently. In external standard methods, the amount of sample injected must be highly reproducible. The method is well suited for automatic mechanical methods of injection. Optimum system performance must be maintained by frequent checks and regular recalibration. The internal standard method is independent of sample size and compensates for any slight instrumental drift. When used properly, it is the most accurate method of quantification. However, the internal standard must be added to each sample in a highly reproducible way, and it should include components with both low and high volatility.

An additional problem with seawater is that any added CB spike may not be in the same form as was originally present in the sample. The role of colloids is uncertain, and partitioning problems may result in differences between liquid–liquid and solid adsorbent extractions of seawater. Until more information is available, we recommend the internal standard method with the whole range of CBs encountered in the samples to check the procedure for recoveries. Such tests should be repeated on a regular basis, for example, once per month. For routine analyses the external standard method suffices. With this issue we are facing a dilemma: without an effective internal standard we cannot know the exact phase distribution of the analyte; on the other hand, without knowledge of the phase distribution we cannot be certain if we add the correct internal standard.

22.5   Interpretation and presentation of results

There are a few approaches to facilitate the interpretation of data on the complex and variable mixtures of chlorobiphenlys in marine environmental samples. The composition of mixtures can be represented as mole percent contributions of individual CBs to their sum (*Duinker et al.*, 1980) or as molar ratios, *e.g.*, CBx/CB153 (*Boon et al.*, 1992). This allows visual, qualitative and quantitative comparisons between samples with widely different overall compositions. Quantitative and less arbitrary comparison is possible with statistical methods such as principal component analysis (*Jackson*, 1991). Finally, distribution patterns between the CB mixtures in solution and in suspended particles can be studied by plotting concentration ratios in these compartments (*i.e.*, distribution coefficients) against known molecular properties, *e.g.*, octanol–water distribution coefficients (*Schulz-Bull et al.*, 1998).

References to Chapter 22

Ballschmiter, K., Zell, M. (1980), *Fresenius Z. Anal. Chem.*, 302, 20.
Boon, J.P., Arnhem, E.V., Jansen, S., Kannan, N., Petrick, G., Schulz-Bull, D.E., Duinker, J.C., Reijnders, P.J.H., Goksøyr, A. (1992), in: *Persistent Pollutants in Marine Ecosystems*: Walker, C. H., Livingstone, D.R. (Eds.). Oxford: Pergamon Press, 1992; pp. 119–159.
Brownawell, B.J., Farrington, J.W. (1986), *Geochim. Cosmochim. Acta*, 50, 157.
Burns, K.A., Villeneuve, J.P. (1987), *Mar. Chem.*, 20, 337.
Dawson, R. (1976), in: *Methods of Seawater Analysis:* Grasshoff, K. (Ed.). Weinheim: VCH, 1976; pp. 234–255.
Dawson, R., Riley, J.P. (1977), *Estuar. Coast. Mar. Sci.*, 4, 55.
Duinker, J.C., Hillebrand, M.T.J., Palmork, K.H., Wilhelmsen, S. (1980), *Bull. Environ. Contamin. Toxicol.*, 25, 956.
Duinker, J.C., Hillebrand, M.T.J. (1983a), in: *Methods of Seawater Analysis*, 2nd edn: Grasshoff, K., Ehrhardt, M., Kremling, K. (Eds.). Weinheim: VCH, 1983; pp. 290–309.

Duinker, J.C., Hillebrand, M.T.J. (1983b), *Environ. Sci. Technol.,* 17, 449.

Duinker, J.C. (1986), *Netherl. J. Sea Res.,* 20, 229.

Duinker, J.C., Schulz, D.E., Petrick, G. (1988), *Anal. Chem.,* 60, 478.

Ehrhardt, M. (1987), *Techniques in Marine Environmental Sciences,* No. 4, Copenhagen: ICES.

Ehrhardt, M., Burns, K.A. (1990), *J. Exp. Mar. Biol. Ecol.,* 138, 35.

Frame, G., (1997), *Anal. Chem.,* 69, 468A.

Galceran, M.T., Santos, F.J., Barceló, D., Sanchez, J. (1993), *J. Chromatogr., A,* 655, 275.

Gaul, H., Ziebarth, U. (1983), *Dtsch. Hydrogr. Z.,* 36, 191.

Gómez-Belinchón, J.I., Grimalt, J.O., Albaigés, J. (1988), *Environ. Sci. Technol.,* 22, 677.

Green, D.R., Stull, J.K., Heesen, T.C. (1986), *Mar. Pollut. Bull.,* 17, 324.

Grob, K., Jr. (1981), *J. Chromatogr.,* 213, 3.

Gustafsson, Ö., Gschwend, P.M. (1997), *Limnol. Oceanogr.,* 42, 519.

Harding, G.C. (1986), *Mar. Ecol. Progr. Ser.,* 33, 167.

IOC (1993), *Chlorinated Biphenyls in open Ocean Waters.* IOC Manuals and Guides No. 27, Paris: UNESCO.

Iwata, H., Tanabe, S., Sakai, N., Tatsukawa, R. (1993), *Environ. Sci. Technol.,* 27, 1080.

Jackson, J.E. (1991), *A User's Guide to Principal Components.* New York: Wiley.

Kimbrough, R.D., Jensen, A.A. (Eds.) (1989), *Halogenated Biphenyls, Terphenyls, Naphthalenes, Dibenzodioxins and Related Products,* 2nd ed. Amsterdam: Elsevier.

Lang, V. (1992), *J. Chromatogr.,* 595, 1.

Muir, D.C.G., de Boer, J. (1993), *Chemosphere,* 27, 1827.

Mullin, M.D., Pochni, C.M., McCrindle, S., Romkes, M., Safe, S.H., Safe, L.M. (1984), *Environ. Sci. Technol.,* 18, 468.

Petrick, G., Schulz-Bull, D.E., Martens, V., Scholz, K., Duinker, J.C. (1996), *Mar. Chem.,* 54, 97.

Safe, S., Hutzinger, O. (Eds.) (1987), *Environmental Toxin Series 1.* Berlin: Springer.

Schomburg, G. (1987), *Gas chromatography.* Weinheim: VCH.

Schüßler, U., Kremling, K. (1993), *Deep-Sea Res.,* 40, 257.

Schulz, D.E., Petrick, G., Duinker, J.C. (1988), *Mar. Pollut. Bull.,* 19, 526.

Schulz, D.E., Petrick, G., Duinker, J.C. (1989), *Environ. Sci. Technol.,* 23, 852.

Schulz, D.E. (1990), Dissertation, Institut für Meereskunde an der Universität Kiel.

Schulz-Bull, D.E., Petrick, G., Duinker, J.C. (1991), *Mar. Chem.,* 36, 365.

Schulz-Bull, D.E., Petrick, G., Kannan, N., Duinker, J.C. (1995), *Mar. Chem.,* 48, 245.

Schulz-Bull, D.E., Petrick, G., Johannsen, H., Duinker, J.C. (1997), *Croat. Chem. Acta,* 70, 309.

Schulz-Bull, D.E., Petrick, G., Bruhn, R., Duinker, J.C. (1998), *Mar. Chem.,* 61, 101.

Simpson, W.R., Gwilliam, T.J.P., Lawford, V.A., Fasham, M.J.R., Lewis, A.R. (1987), *Deep-Sea Res.,* 34, 1477.

Theobald, N., Lange, W., Rave, A., Pohle, U., Koennecke, P. (1990), *Dtsch. Hydrogr. Z.,* 43, 311.

Tolosa, I., Readman, J.W., Fowler, S.W., Villeneuve, J.P., Dachs, J., Bayona, J.M., Albaigés, J. (1997), *Deep-Sea Res.,* 44, 907.

Wells, D.E., de Boer, J., Tuinstra, L.G.M.T., Reutergardh, L., Griepink, B. (1988), *Fresenius Z. Anal. Chem.,* 332, 591.

23 Determination of volatile halocarbons in seawater

E. Fogelqvist

23.1 Introduction

Seawater contains considerable amounts of dissolved halogenated organic substances. The following chapter deals with the low relative molecular mass fraction, *i.e.*, substances with one to four carbon atoms. This group of compounds is often called halocarbons or volatile halogenated organic compounds (VHOC).

Until a few decades ago, it was generally believed that halocarbons are exclusively man-made substances appearing in the marine environment as contaminants originating in industrial activities. The pioneering works of *Lovelock* (1975), *Burreson and Moore* (1975) and *Fenical* (1975) at the beginning of the 1970s, followed by *Dyrssen and Fogelqvist* (1981), *Gschwend et al.* (1985), *Singh et al.*, (1983), *Manley and Dastour* (1987), and others, about a decade later, showed, however, that marine plants produce significant amounts of brominated and iodinated organic compounds. Many reports have followed (for review see *Gribble*, 1994), and now a large number of biogenic halocarbons, including chlorinated substances, have been identified (*Nightingale* et al., 1995).

Most of the halocarbons produced by industry are intermediates for the manufacture of various products. Halocarbons are also used as solvents for dry cleaning and degreasing in the mechanical industry. Much of this production ends up in waste waters. Others, such as chlorofluorocarbons (CFCs) have been used as aerosol propellants which were discharged to the atmosphere on a short time scale. Some halocarbons are slowly leaking to the environment for other reasons, as for example CFC-12 used as heat exchanger in refrigerators. CFC-12 concentrations in the atmosphere therefore are still increasing, while concentrations of many other CFCs decrease as a result of restrictions on production and use.

Halogenated organic substances are a potential risk to the stratospheric ozone, provided their residence times in the atmosphere are long enough for them to reach the stratosphere. The impact on the ozone chemistry increases with atomic number, *i.e.*, bromine is more aggressive than chlorine. The atmospheric residence times of the most stable compounds are of the order of a hundred years, while others break down within a few days. Residence times are longer in seawater, except in anoxic waters (*Bullister and Lee*, 1995; *Tanhua et al.*, 1996).

Table 23-1 lists the anthropogenic and biogenic halocarbons which have been found dissolved in seawater. The large number of compounds, all with similar physico-chemical properties and concentration ranges in the fmol/L to μmol/L range, indicate that an analytical method capable of measuring all of them simultaneously must combine the highest possible separation power with extreme sensitivity.

Table 23-1. Volatile halocarbons occurring in seawater, chemical formulae and boiling points (B.P.).

Halocarbon	Formula	B.P.[a] (°C)
Chloromethane, methyl chloride	CH_3Cl	−24
Bromomethane, methyl bromide	CH_3Br	4
Iodomethane, methyl iodide	CH_3I	42
Dichloromethane	CH_2Cl_2	40
Dibromomethane	CH_2Br_2	97
Diiodomethane	CH_2I_2	182
Chloroiodomethane	CH_2ClI	109
Trichloromethane, chloroform	$CHCl_3$	62
Bromochloromethane	CH_2BrCl	68
Bromodichloromethane	$CHBrCl_2$	90
Dibromochloromethane	$CHBr_2Cl$	120
Tribromomethane, bromoform	$CHBr_3$	150
Dichlorodifluoromethane, CFC-12	CCl_2F_2	−30
Trichlorofluoromethane, CFC-11	CCl_3F	24
Tetrachloromethane, carbon tetrachloride	CCl_4	77
Iodoethane	$CH_3–CH_2I$	72
1,1,1-Trichloroethane, methyl chloroform	$CH_3–CCl_3$	74
1,1,2-Trichloro-1,2,2-trifluoroethane, CFC-113	$CCl_2F–CClF_2$	48
Trichloroethene, trichloroethylene	$CHCl = CCl_2$	87
Tetrachloroethene, perchloroethylene	$CCl_2 = CCl_2$	121
1-Iodopropane	$CH_2I–CH_2–CH_3$	102
2-Iodopropane	$CH_3–CHI–CH_3$	89
1-Iodobutane	$CH_2I–CH_2–CH_2–CH_3$	131
2-Iodobutane	$CH_3–CHI–CH_2–CH_3$	118

[a] From *Handbook of Chemistry and Physics, 73rd edn., CRC Press*, Boca Raton, 1992–1993.

Three of the CFCs and carbon tetrachloride are extensively used as transient tracers for oceanographic studies of large scale water mixing and movement. The tracer data are used together with other oceanographic parameters for oceanographic models, and thus demand measurements of the highest possible precision and accuracy. Techniques for such high-performance determinations of halocarbons will be described in this chapter. They all derive from the work of *Bullister and Weiss* (1988) and are based on purge and trap (P&T) sample concentration, gas chromatographic (GC) separation of the halocarbons and electron capture detection (ECD).

Simpler and less expensive methods, which might be adequate for purposes other than oceanographic modelling, utilise a pentane–water extraction technique either batchwise (*Eklund et al.*, 1978; *Fogelqvist and Larsson*, 1983; *Abrahamsson and Klick*, 1990) or by a segmented flow technique which allows for liquid extraction of the seawater samples on-line with subsequent gas chromatographic analysis (*Fogelqvist et al.*, 1986). However, the most volatile halocarbons, *e.g.*, the methyl halides and some of the CFCs, cannot be measured following pentane extraction, because the solvent peak interferes with the first analyte peaks in the chromatograms. Even though liquid extraction is suitable for measurements of selected halocarbons in chlorinated tap water and in contaminated sewage waters, and is recommended by environmental protection agencies in several countries, the method has generally been abandoned for measurements in seawater. This chapter, therefore, will only describe the P&T work-up technique.

23.2   Sampling of seawater

The demands on sampling methodology are set by the high volatility of the halocarbons and the risk of contamination of water samples from ambient air and sampling equipment. Usually, the samplers are of the Niskin type with Luer taps. Niskin samplers are made of poly(vinyl chloride) (PVC) which is a polymer of the halogenated hydrocarbon monomer vinyl chloride. Especially when they are new, they must be cleaned thoroughly with a strong detergent and rinsed with copious amounts of seawater. The O-rings in both ends of the samplers are other sources of contamination. Viton O-rings, heated in an oven at 50–60 °C for a few hours, have proven to be free of contamination. The strings inside Niskin type samplers should be made of stainless-steel, with or without a Teflon cover, and rinsed the same way as the samplers.

Samplers should be as large as possible, because the increasing ratio with size of volume over potentially contaminated area reduces the risk of sample contamination. The volume should be at least 5 L, preferably 10 L, even if less than 1 L is used for the halocarbon measurements.

To ensure that the samplers are clean, halocarbon free water should, if possible, be analysed on a regular basis. Such water is not easy to obtain. Distilled water as well as water deionized by other means is usually heavily contaminated with one or another halocarbon, and therefore is unsuitable. Seawater sparged with purified nitrogen is virtually non-contaminated, but enough water to fill a set of up to 24 samplers is hard to purify this way. Another option is to monitor seawater of the lowest possible concentration regularly, and make sure that the scatter of results lies within a reasonable range as compared with the analytical uncertainty.

Volatiles with low water solubility equilibrate with the head-space air in the sampler. It is important, therefore, that halocarbons are sampled immediately after retrieval of the sampler and before any other water is withdrawn.

The water is drawn from the samplers with 100 mL ground-glass syringes with Luer-lock fittings (Philip Harris Int., UK) and equipped with three-way stainless-steel stopcocks (made to order by Millipore). Metal stopcocks made of brass, even if plated, are inapt, because they are corroded by seawater. Stopcocks made of plastic materials (Millipore) are inexpensive and easy to handle but susceptible to adsorbing organic substances, and should therefore be avoided, unless proven fit for the purpose of the measurements. Whichever type of stopcock is used, they must not be left to dry containing seawater, because precipitated salt crystals will damage them when turned.

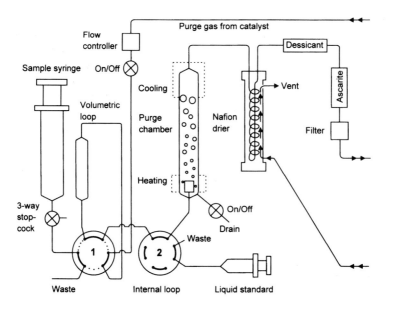

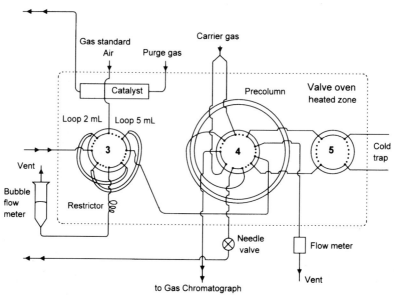

Fig. 23-1. The manifold scheme of the sample work-up system, excluding the cold trap (see Fig. 23-2), with all valve positions as in the first phase of the analysis, *i.e.*, the P&T phase. The figure illustrates the injection of a water sample through valve 1 with the volumetric loop, the purge chamber and drying of the purge gas. Internal and liquid standards are injected through valve 2 (optional). Valve 3 is used to inject gaseous standard mixtures from a gas tank (not shown) or air samples. Valve 4 directs the purge gas flow either to the trap or to vent. It also supports the pre-column through which carrier gas is either fore-flushed (clockwise) to the gas chromatograph or back-flushed (counter-clockwise) to vent. While in the fore-flush mode, the carrier gas passes through valve 5 and the trap. Valve 5 closes the trap while other valves are turned, gas flows are changed, *etc.* Valves 3, 4 and 5 are kept at a constant temperature of 50–70 °C.

For subsampling from a water bottle, the syringes are first rinsed three times with the sample water and then filled carefully so as not to leave any air bubbles inside. The syringes are stored immersed in seawater to prevent any possible exchange between the water inside and ambient air. The water in a bucket, taken from the surplus water in the samplers, is replaced regularly. Also, the syringes should be kept cold before analysis, preferably with crushed ice, or in a cold store. The reason is that seawater of low temperature inevitably creates small gas bubbles inside the syringes when the temperature increases.

If the above routines are followed, the samples can be stored for several hours before analysis, but not long enough to bring them back to a shore based laboratory. For transportation, *Busenberg and Plummer* (1992) developed a technique using glass ampoules filled with halocarbon-free nitrogen. After filling the ampoules, they are flame-sealed. This technique has been tested successfully for storing CFC samples for up to 1 year and is used by W. Smethie's group for field work (*Mensch et al.*, 1998).

23.3   Purge and trap (P&T) work-up

The following sections refer to the P&T system illustrated in Figs. 23-1 and 23-2.

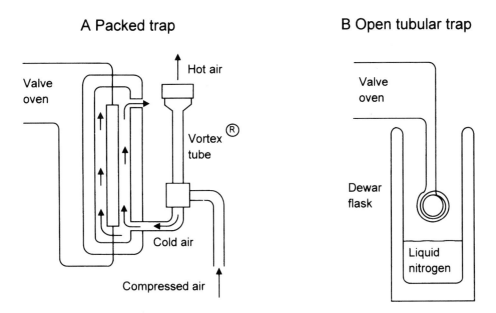

Fig. 23-2. Two examples of traps. A shows a trap packed with Porapak N cooled to $-15\,°C$ with air from a Vortex tube and heated during desorption by an electrical current through the trap. B shows an open tubular trap cooled in the vapours of liquid nitrogen. The trap is heated for desorption of the halocarbons either by immersing it into boiling hot water or by an electrical current through the tube. See text for details.

23.3.1   Materials

All surfaces to which the halocarbons are exposed must be as non-adsorptive as possible and must not catalyse their chemical breakdown. Thoroughly cleaned and deactivated stainless-steel and glass are the optimal choices. Most stainless-steel tubing, when leaving the manufacturer, is covered with a thin oil film which has to be removed by rinsing with a suite of solvents of increasing polarity, *e.g.*, hexane, acetone and last methanol. Solvent residues are dried out with a flow of nitrogen gas while the tubing is heated to 400–600 °C. Glassware is cleaned the same way, but should not be heated to more than 200 °C. Ferrules softer than stainless-steel, *e.g.*, in connections between glass and stainless-steel tubing, should be made of Teflon.

Stainless-steel tubing and fitting valve ports of 1/8 in outer diameter (o.d.) are used in all parts where water flows, *i.e.*, from the sample syringe through valves 1 and 2 (Fig. 23-1) and to the purge chamber. Downstream of the purge chamber the tubing is 1/16 in (1.6 mm) o.d. or even narrower. Stainless-steel tubing, with 0.53 mm inner diameter (i.d.), 0.7 mm o.d. (Frontier Laboratories, Fukushima, Japan), has the same dimension as the commonly used megabore GC columns.

The rotary valves are two-position stainless-steel valves (Valco Instruments, Inc.) with four to ten ports. The first two valves, 1 and 2 in Fig. 23-1, which are exposed to seawater, are made of Hastelloy C-22, a seawater resistant nickel–chromium–molybdenum alloy. The three valves in the heated zone (3, 4 and 5), if installed inside the GC oven, are equipped with high-temperature rotors, specified by the manufacturer (Valco) to operate at temperatures up to 340 °C. All valves can be actuated electrically for various degrees of automation.

23.3.2   Volumetric loops

The exact amount of water to be analysed is determined by a loop connected to valve 1, a glass tube approximately 10 cm long (30 mm o.d.) narrowed at the ends to 6 mm ($\approx 1/4$ in). The volume including connections should be in the range of 30–40 mL. Liquid standards are injected through a low volume (1 or 2 μL) internal loop in valve 2. Gaseous injections of standard mixtures or air are made by various combinations of subsequent injections through valve 3, with its two loops of 2 and 5 mL volume, respectively (see Section 23.3.5).

The exact volumes of the loops for water and gaseous injections are determined following a procedure described by *Wilke et al.* (1993). One loop at a time is filled with distilled and degassed water, and the mass difference between the filled and empty loop (while connected to the valve body) is measured to an accuracy and precision of ±0.05 %. The internal loops in valve 2 are calibrated against gaseous injections into valve 3 of easily quantifiable halocarbons available as both liquid and gaseous standards, *e.g.*, carbon tetrachloride and CFC-11. The volume of the loop can hardly be determined with an accuracy better than ±5 %. On the other hand, these loops are used for applications where this level of data quality suffices.

23.3.3   Purging of halocarbons from the water

In this section, the details of the P&T system will be described following the pathway of the purge gas, starting from the inlet in the upper right part of Fig. 23-1. Nitrogen gas of highest possible purity (classified as N_2 5.5, 99.999 % nitrogen or more) is supplied from a gas tank equipped with a metal bellows pressure regulator, which is kept at an outlet pressure of ≈ 5 bar (5×10^5 Pa). To aid in the catalytic removal of halocarbon impurities (see below), the purge gas is doped with 0.5–1 % ultrapure hydrogen (purchased as a special gas mixture from the gas manufacturer).

When a sample volume of 30 mL is purged for 8 min at a flow rate of 80 mL/min, and 640 mL purge gas thus passes through the sample, concentrations of volatile organic impurities in the purge gas must be close to zero. To keep the purge gas halocarbon-free, it is first passed through a stainless-steel trap containing ≈ 50 mL of Molecular Sieve 13X (Alltech Associates, Inc., 40–60 mesh). The trap needs regular conditioning at 300 °C with nitrogen flowing through it.

In order to remove impurities not trapped in the molecular sieve, the purge gas is then passed through a tube (≈ 10 cm long, o.d. 1/4 in) containing a hydrogen-palladium catalyst on alumina (Aldrich, Germany). The reactant is hydrogen, and the palladium catalyst is kept saturated with it by the hydrogen admixture of the purge gas. This eliminates all remaining halocarbons. The tube containing the catalyst is held at an elevated temperature, preferably 250 °C, in a separate heating block, but is also reasonably effective at 50–70 °C, if mounted in the GC or the valve oven.

The purge gas flow rate is kept constant at ≈ 80 mL/min by a metal bellows flow controller (Porter Instruments, VCD-1000, see Fig. 23-1). A toggle valve downstream of the flow controller is used to stop the gas flow for maintenance of the system or for determination of extraction efficiencies, *etc.* The purity of the purge gas is checked by running the analytical programme without injection of seawater.

If the sample water contains particles, *e.g.*, planktonic algae in surface waters or other solids, there is a risk of scratching valve rotors or blocking the narrow passages. Such particles may be removed with glass fibre filters (Whatman GF/F) of 2.4 cm diameter, provided they are cleaned properly. The filters are inserted between the syringe and valve 1.

Most of the 100 mL of seawater in the sample syringe (about 2/3) is used to flush the sample volumetric loop, before valve 1 is turned. The purge gas backflushes the water out of the loop and *via* valve 2 to the purge chamber.

Valve 2 is an internal loop valve used for the injection of liquid standards; it may be disconnected when the system is used for gaseous halocarbons only. Valve 2 can also serve as a device for introducing internal standards into the sample stream which is especially useful for quantifying late peaks in the chromatogram. A suitable internal standard is bromotrichloromethane ($CBrCl_3$) dissolved in methanol.

The purge chamber is a glass tower with a glass frit (micro-filter candle, porosity 2, Duran) at the bottom, through which first the seawater enters and then the purge gas. Gentle heating of the lower part of the chamber increases the extraction efficiency for less volatile halocarbons. Minimising the head-space volume in the tower also increases the extraction efficiency. However, bursting bubbles create a spray of water droplets adding to the problem of drying the purge gas before the next step of the analysis. The head-space, therefore, must not be too small.

23.3.4 Drying the purge gas

Drying the gas before it enters the heated zone is essential for three reasons. First, chromatographic columns are liable to deterioration, if the carrier gas contains traces of water vapour. Second, all droplets contain seasalt which, when the water evaporates, forms salt particles causing damage to the valves downstream in the system. Third, salt particles may block narrow passages, such as an open tubular trap of small dimensions or a microtrap.

There are several ways to remove water from the gas downstream of the purge chamber. When its upper part is cooled, most of the water condenses on the surface and flows back into the bulk liquid. To this end, either a metal block is mounted around the top, kept cold by water from a cooling bath, or it is more efficient to lead the cooling water through a few coils of glass tube inside the upper part of the chamber.

After the purge chamber, the gas stream passes through ≈ 1 m of a Nafion tube (1/16 in o.d., Perma Pure Inc., USA). Nafion is a semipermeable membrane through which water vapour and polar compounds pass freely, but none of the halocarbons. The Nafion is coiled in a Perspex (Plexiglass) tube (≈ 20 cm long, o.d. 30 mm) through which a counterflow of dry gas is led, *e.g.*, the effluent from valve 4 (see Fig. 23-1).

A complement to condensation in the purge chamber and the Nafion drier is a glass tube (1/4 in or 6 mm o.d.) containing anhydrous magnesium perchlorate as desiccant [$Mg(ClO_4)_2$, Merck, Germany]; the glass tube is plugged at both ends with silanised glass wool. The desiccant absorbs the remaining water vapour, it also indicates visibly at an early stage if the gas is too humid. If water is not effectively removed by condensation in the purge chamber and by a Nafion drier, the desiccant has to be replaced regularly after a few samples, otherwise once a week, after several hundred samples. Condensation in the purge chamber is more effective than increasing the amount of desiccant.

Hydrogen sulphide, present in anoxic waters, co-elutes with the halocarbons in the first part of the chromatogram; it has a moderate ECD response, but high concentrations make any measurements of methyl halides and the earliest CFCs impossible. The Ascarite tube shown in Fig. 23-1 (dimensions about the same as the desiccant tube) is an optional device for removing hydrogen sulphide; it is used only for analyses of anoxic/suboxic waters. Ascarite is sodium hydroxide coated silica, which binds acidic gases such as hydrogen sulphide and carbon dioxide.

As mentioned above, any fine particles such as seasalt crystals, magnesium perchlorate or Ascarite would damage rotary valves by scratching the rotor surfaces. The gas line filter (Supelco, pore size 0.5 µm) shown in Fig. 23-1 after the Ascarite tube guards against transfer of particles to the set of three valves in the heated zone. The filter is removed easily for cleaning by rinsing and heating in an oven for a few hours. HPLC frits, available for all actual tube diameters, cause a noticeable pressure drop, and must be used sparingly.

23.3.5   The valve oven

After the line filter, the gas passes three valves, which are all held at an elevated and constant temperature of 50–70 °C, partly for controlling the amount of gas injected and partly to avoid condensation/adsorption of unwanted compounds in the lines and the valves. If the chromatographic programme is isothermal, and if the GC oven is large enough, the valves can be mounted in it. If not, an insulated valve oven (Heated valve enclosure, Valco) has the advantage that the temperature, controlled by one of the GC auxiliary ports, can be set independent of the oven temperature. The distance between the valve and the GC oven should be as short as possible. If needed, an electrically thermostated transfer line (*e.g.*, Varian Associates Inc.) keeps the tube at a constant elevated temperature. To avoid trapping of less volatile compounds and carry-over effects, the gas flow should never be exposed to a temperature drop on its way from the purge chamber to the GC column.

Valve 3, the first one of three valves in the valve oven, is used for gaseous injections. The two loops of 2 and 5 mL capacity allow for injection volumes of 2 mL, 5 mL, 7 mL (2+5), 9 mL (2+5+2), 12 mL (5+2+5), 14 mL (2+5+2+5), 16 mL (2+5+2+5+2), 19 mL (5+2+5+2+5), *etc.*, through valve 4 into the cold trap where the halocarbons accumulate. After a loop has been flushed with several loop volumes of injected gas, indicated by a bubble flowmeter at the outlet, it is left for pressure equilibration with ambient air for about 10 s before the valve is turned and the gas injected. Between the valve and the bubble meter is a coil of 1/16 in stainless-steel tubing, about 0.5 m long, which prevents penetration of ambient air into the loop during pressure equilibration. If gaseous standards are injected into the P&T system, not only must the volume be known of the sample loop to calculate the masses of the halocarbons, but also the temperature and the pressure. Keeping valve 3 in a heated zone of known temperature leaves only the air pressure to be recorded at the time of injection (see Section 23.6.1).

Valve 4 controls the directions of the purge and carrier gases through the trap and to the GC. In one position of the valve, the purge gas passes the trap. When turned, the carrier gas back-flushes the halocarbons from the trap through a pre-column to the GC. The gas chromatographic separation begins in the short pre-column (≈ 7 m long fused silica, DB 624, J&W) which is of the same kind as the 75 m main column in the gas chromatograph. The pre-column is coiled a few wide turns around the valve body. The analytes continue to the main column until the valve is turned, after which the later eluting compounds are vented out by a backflush flow of carrier gas. There is always a constant flow of carrier gas through the pre-column, whichever way the valve is turned, and a needle valve or a flow controller at the vent maintains this flow. (If the carrier gas is either helium or hydrogen, a needle valve might not be reliable enough.)

Preferably, the GC column is connected directly to valve 4, even if the valve is mounted outside the GC oven in a separate compartment, otherwise a silanised stainless-steel tubing, as short as possible, of the same dimension as the GC column should be used as transfer line.

Valve 5, the trap valve, opens and closes the trap. If the system is automated and computer controlled, and if the trap is electrically and thereby quickly heated in a reproducible way, this valve is unnecessary. If installed, it should be closed at the end of the purging phase and opened a few seconds after the carrier gas flow has flushed the lines to it, *i.e.*, a few seconds after turning valve 4 to the GC injection mode.

23.3.6 Trapping and desorption

The trap serves not only as pre-concentration device, but also for band sharpening which is a prerequisite for optimal separation in the chromatographic column. When the gas from the purge chamber passes through the trap, the halocarbons are either absorbed by a resin or adsorbed on to the walls of a tube. The halocarbons are trapped at low temperatures and desorbed by heating the trap.

Figure 23-2 shows two examples of cold traps, one packed and one an open tubular trap. In both cases the gas flows through the trap in one direction during the purge phase and in the opposite direction when the trap is heated and the halocarbons desorbed, *i.e.*, the halocarbons are back-flushed out of the trap. The back-flushing ensures that band sharpening is maintained, and that no accumulation of heavier compounds in the trap will cause its deterioration.

23.3.6.1   Packed trap

Porapak N (80–100 mesh, Alltech Associates, Inc.) is a commonly used absorption material for measurements of halocarbons in seawater. Although it traps even the most volatile halocarbons at temperatures above zero, a lower temperature, -10 to $-15\,°C$, is recommended. Lower temperatures increase the retention time of the halocarbons in the resin just as in a packed gas chromatographic column. This leads to enhanced band sharpening and thus to better subsequent gas chromatographic separation. The trap is a 1/8 in stainless-steel tube 10–15 cm long, filled with the resin and plugged at both ends with deactivated (silanised) glass wool.

The trap may be cooled either by recirculating cold water from a water bath through a metal block surrounding the trap (*Pankow*, 1991) or with a Vortex tube. The Vortex tube (Vortec Corp.) uses compressed air to produce two distinct vortices travelling at different speeds which causes a simple heat exchange. Minimum inlet pressure of the compressed air is 5.5 bar (5.5×10^5 Pa), and minimum flow rate 200 L/min. The trap is mounted inside a 3/8 in glass tube through which flows the cold air from the Vortex tube (*Happell et al.*, 1996).

The cables from a variable transformer are clamped to the ends of the trap to pass a low-voltage, high amperage AC current ($\approx 25\,A$ at $\approx 3\,V$) through it, which heats the trap to $120\,°C$ and thus desorbs the halocarbons. A thermocouple is attached to the trap, and a temperature controller (Omega CN370) turns the electrical power on and off by way of a solid-state relay. The trap is electrically insulated by deactivated fused silica tubing inserted between the trap and the valve.

23.3.6.2   Open tubular trap

The open tubular trap is a $\approx 0.5\,m$, 0.75 mm i.d. stainless-steel tube, coiled a few turns, to give a total distance from the valve of 15–20 cm (Fig. 23-2). Properly cooled, the open trap offers very sharp bands of condensed halocarbons, and thereby high chromatographic resolution, high narrow peaks and improved sensitivity. Effective condensation of halocarbons on the internal surface of the tube demands temperatures below $-140\,°C$; the cooling medium therefore has to be liquid nitrogen which is kept in a small Dewar flask. If the temperature is too low, below $-170\,°C$, the oxygen, which is purged out of the seawater together with other volatiles, may condense and plug the narrow tube. The trap, therefore, must be positioned in the vapours just above the surface of the liquid nitrogen. Desorption can be accomplished either by immersing the trap into boiling hot water or electrically (see Section 23.3.6.1), provided the coils of the trap are electrically insulated.

23.3.6.3   Microtraps

During the desorption phase, when the GC carrier gas passes through the trap, the optimal flow rate is determined by the GC column used. The P&T system described in this chapter was built for a megabore capillary column with an optimal gas flow of ≈ 8 mL/min. However, as mass spectrometers become increasingly robust and suitable for field work, packed traps have been developed for the lower carrier gas flow rates (1–2 mL/min) which these instruments demand (*O'Doherty et al.*, 1993; *Mitra and Yun*, 1993). Because the GC columns are narrower, 0.32 mm i.d., the trap must be miniaturized accordingly, and thus holds much less packing material. To avoid the risk of breakthrough, the temperature and the gas flow rate through the trap during the purge phase will have to be kept lower, *i.e.*, low enough to guarantee 100 % retention of the methyl halides and CFC-12 which are the most volatile halocarbons.

23.3.7   Automation

Automation not only makes the system more user-friendly, it generally improves the data quality. The higher the level of automation, the higher is the reproducibility of timed events, and thereby the measurements. Gas chromatographs are usually equipped with a software package for controlling the GC, integration and evaluation of peaks *etc.* Also, the software can usually control a few external events such as the turning of valves in the P&T system, the temperature in certain external compartments such as a trap, or a heated zone. If the GC software package does not meet all the demands of the P&T system, a PC operated relay card with, *e.g.*, 16 relay outputs can be installed in the computer and programmed to control the external events. The entire analytical procedure, P&T work-up and GC analysis, can thus be automated to various degrees. For example, a P&T system with a trap, which is cooled by a Vortex tube and electrically-heated can be fully automated. A single command by the operator, after the sample is loaded, starts the analytical sequence. An improvement is a construction with several water samples loaded in a multi-port valve which allows a series of seven samples to be analysed unattended by the operator (*A. Putzka*, personal communication).

23.4   Gas chromatography

23.4.1   Separation column

Modern gas chromatographs are equipped with capillary columns for better separation of complex mixtures of organic compounds than can be achieved with packed columns. The halocarbons in the examples shown in Fig. 23-3 and Fig. 23-4 were separated with a fused silica column (DB 624, J&W, 75 m long, 0.53 mm i.d., film thickness 3 μm). The column was developed for halocarbon separation and has been used extensively for this purpose by many research groups for several years. The i.d. is a compromise between a packed and a capillary column. It allows for a carrier gas flow rate of 8 mL/min as compared with 2 mL/min for narrow bore columns, and it offers the high resolution of open tubular columns. An example of a column with smaller dimensions, *e.g.*, for the use of microtraps in combination with mass spectrometry, is a Restek 502.2 column with 0.32 mm i.d. and a film thickness of 1.8 μm (Restek Corp., USA).

The best carrier gases for capillary GC are hydrogen or helium, although nitrogen can be used with megabore columns. For highest possible sensitivity and sustained high detec-

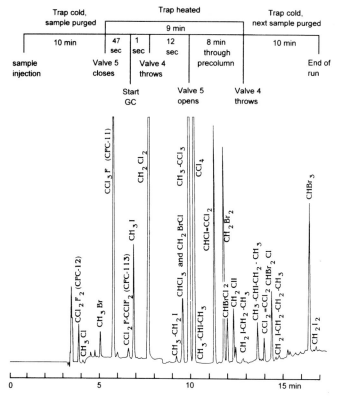

Fig. 23-3. A chromatogram of the entire range of halocarbons from an analysis of southern ocean surface seawater according to the 'Halocarbon mode' described in detail in the text. The upper part of the figure illustrates the timing of events (time axis not to scale). An open tubular trap as shown in Fig. 23-2 heated in boiling water was used.

tor performance, the carrier gas is thoroughly purified with a molecular sieve trap (Molecular Sieve 13X, the same as is used for purge gas purification, see Section 23.3.3) and an oxygen trap.

23.4.2   Two modes of halocarbon analysis

The two chromatograms shown in Figs. 23-3 and 23-4 demonstrate two modes of halocarbon analysis. They differ only in the timing of P&T events and the temperature programme of the GC oven, parameters that are all software controlled.

The first mode (Fig. 23-3), here called the 'Halocarbon mode', measures the whole range of halocarbons. A programmed temperature increase in the GC oven ensures the elution of all compounds as narrow peaks within a reasonable time span. The chromatographic run starts with the oven temperature of 70 °C held constant for 2 min, then increased at 5 °C/min to 140 °C followed by a 10 °C/min temperature rise to 180 °C for conditioning. The total run

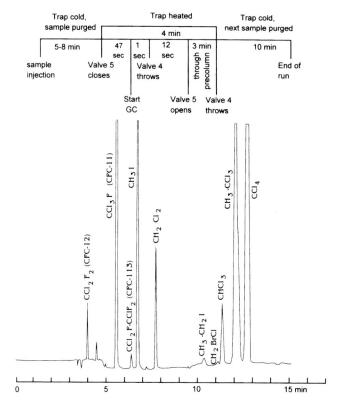

Fig. 23-4. A chromatogram from a surface seawater sample analysed according to the 'Tracer mode' uti-
lising the pre-column to get a heart-cut chromatogram (see text for details). The time axis (not to scale)
is illustrated in the upper part of the figure. An open tubular trap as in Fig. 23-2 B heated in boiling
water was used.

time is 29 min, but a new sample can be injected after 20 min (allowing 1 min for cooling of
the trap) yielding a sample throughput of 3 per hour.

The second chromatogram (Fig. 23-4), run in the 'Tracer mode', is a heart-cut of the chromatogram in Fig. 23-3 returning data on a selection of halocarbons only, in this case the
CFC tracers. In this mode, the GC oven temperature is kept constant at 70 °C, which also
enhances the stability of the detector signal. The total run time is 19–22 min, injection of a
new sample is possible after 10–13 min (1 min for cooling of the trap), but overlapping of
samples is determined by the chromatogram, which takes 13 min. The purge time may thus
be varied between 5 and 8 min.

By heart-cut chromatography, the chromatographic conditions can be optimised for the
target compounds, the time needed for one analysis is shorter leading to higher sample
throughput, and the main column and the detector is protected against late eluting compounds, *i.e.*, there is less need for conditioning.

The two chromatograms in Figs. 23-3 and 23-4 reveal that there is a base-line shift
after the CFC-12 peak. The base-line returns to the original level after the CH_2Cl_2 peak.
This shift is due to a pressure change at the turn of valve 4, and shows that the fore-

and back-flush rates of carrier gas through the pre-column were not exactly the same. Changing the timing of valve actuation moves the shift in either direction along the time axis.

23.5 Detection

23.5.1 Electron capture detection

The electron capture detector (ECD) has two major advantages over other GC detectors, which makes it ideal for the measurements of halogenated organic substances. It is highly selective and orders of magnitude more sensitive than other GC detectors for molecules containing electronegative atoms such as fluorine, chlorine, bromine, iodine and to some extent oxygen and sulphur. Only a mass spectrometer in single-ion monitoring (SIM) mode may match its sensitivity. However, the detector response varies significantly not only with the number and type of halogen atoms, but also with the molecular structure, from the very efficient absorption of electrons by carbon tetrachloride to the monohalides, for which the mass specific detector signal can be orders of magnitude lower. A drawback of the ECD is its non-linear response except over a limited range of analyte concentrations, and this has to be taken into account for proper standardization (see Section 23.6.2.2 and Fig. 23-5).

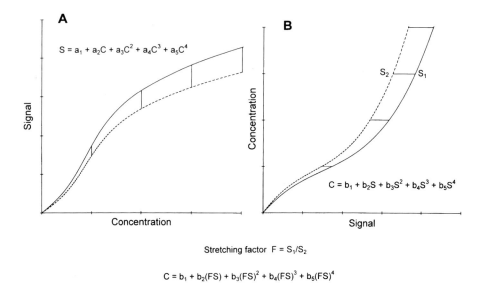

Fig. 23-5. Non-linear standard curves of a hypothetical halocarbon. The solid line is the original standard curve, the broken line illustrates the same curve measured later with lower detector response. From the lines between the curves, the correction factor $F = S_1/S_2$ is used to transform the new curve of lower response to the original. A shows the traditional way of plotting the detector signal against concentration and the equation for the curve, where S is the signal, C is the concentration. B shows the same curves with reversed axes and the stretching technique, which allows for a straightforward calculation of C once F is determined. The constants a and b are determined by curve fitting.

The size of an ECD cell, determined by the flight distance from the nickel foil to the collector, is not optimal for the low gas flow rates of capillary GC. A make-up gas is therefore added to the carrier gas to counteract band broadening within the cell which would ruin the good separation of the column. The make-up gas is nitrogen purified the same way as the carrier gas (see Section 23.4). After optimising the flow rates of carrier and make-up gases, these must be kept constant with reliable pressure regulators. Otherwise the reproducibility of measurements decreases. The pressure in the feed line should be $\approx 2\,\mathrm{bar}$ $(2 \times 10^5\,\mathrm{Pa})$ higher than downstream of the pressure regulator.

The detector temperature may be set to anywhere between 250 and 350 °C; the higher the temperature, the higher is the signal intensity. However, high detector temperatures make the radioactive foil vulnerable to reaction with traces of oxygen in the carrier and the make-up gases. It is therefore essential for a long lifetime of the detector that both gases are purified with a Molecular Sieve 13X trap followed by an oxygen trap.

23.5.2   Mass spectrometry

In recent years mass spectrometers have become increasingly robust, compact and less expensive, at least on a gathered information/price basis. Bench-top instruments are available with either quadrupole or ion trap mass filters. Applying single-ion monitoring (SIM) of a few characteristic fragments of the halocarbon molecules, the sensitivity is equal to or better than that of the ECD.

Depending on the efficiency of the high vacuum pump, one major modification of the P&T method may be necessary to keep the pressure in the ion source within acceptable limits. The optimal gas flow into the mass spectrometer is usually 1–2 mL/min, which means that a miniaturized P&T system with a microtrap and a narrow-bore GC column is needed. Techniques for measuring halocarbons in the atmosphere by mass spectrometry have been used for several years (*e.g.*, *O'Doherty et al.*, 1993), whereas systems for seawater measurements were developed more recently. Either a constant flow of seawater equilibrates with a gas which is subsampled for analysis (*J. Butler*, personal communication), or a modified P&T system is used which allows for a larger range of halocarbons to be determined (*Ekdahl and Abrahamsson*, 1997).

23.6   Calibration

The intensity of a peak is usually determined by integrating the area underneath. However, measuring the peak height may yield better results in cases when the base of the peak is influenced by nearby eluting peaks. Examples are the CH_2ClI peak in Fig. 23-3 and its neighbour.

23.6.1 Standard mixtures

23.6.1.1 Gaseous standards

High precision gaseous standards contain the most volatile halocarbons, *i.e.*, CFCs-11, -12, -113, carbon tetrachloride and methyl chloroform (1,1,1-trichloroethane). Preparing an accurate mixture of these gases is a demanding task. Extreme control of the masses of the various components of the mixture is imperative, which means that the pressure and the temperature at which they are mixed must be measured exactly (*Happell and Wallace*, 1997). Very few laboratories have the facilities to do this, and even fewer can offer high precision standards at the concentration ratios of seawater for purchase (Brookhaven National Laboratory, Upton, NY, USA). Other laboratories prepare less accurate secondary standards, which are then calibrated against a high precision standard.

A secondary standard is prepared by mixing the substances in a dilution chamber with a volume of more than $10 \, m^3$. The larger the better, since the final concentrations of individual components have to be very low. Most of the halocarbons are liquids at room temperature, and are mixed at about the same relative ratios as found in seawater. The gaseous methyl chloride, methyl bromide and CFC-12 are released into the dilution chamber from glass ampoules. The toxicity of methyl halides deserves special attention. An evacuated stainless-steel cylinder, electropolished inside, with a volume of 1–5 L, is opened in the chamber. The gas mixture let into the cylinder is then pressurised with nitrogen gas to reach the required concentration level (mixing ratio).

A new secondary standard is not stable the first weeks after preparation. In particular carbon tetrachloride adsorbs to the cylinder walls and decreases in concentration. Water vapour in the standard mixture seems to prevent the decrease of carbon tetrachloride concentrations and should therefore be added to the mixture. After calibration, the secondary standard is left to stabilize for at least one month, after which time it is re-calibrated.

23.6.1.2 Liquid standards

Less volatile halocarbons are standardised against liquid standards injected through the small internal loop of valve 2 (Fig. 23-1). Stock solutions in either methanol or acetone are prepared gravimetrically by the mass difference of an HPLC glass syringe filled with the halocarbon before and immediately after emptying the syringe directly into the solvent in a volumetric flask.

23.6.2 Calibration curves

23.6.2.1 Linear calibration curves

The simplest calibration curve is a straight line through the origin. Unfortunately, this is not very common due to the non-linear response of the ECD. Wider linear ranges are obtained by mass spectrometric detection. Linear or not, a calibration curve not passing through the origin indicates a blank signal of the P&T system, usually leakage in the plumbing or in the valves. Even a small leak may cause considerable blank signals.

Whichever detector is used, linearity of its response has to be checked by a series of seven or more standard measurements, evenly spaced over the expected range of concentrations. Once linearity is proven, only two or three points are measured for subsequent calibrations.

23.6.2.2   Non-linear calibration curves

The response of the ECD is not linear over a large concentration range, and the curvature varies with the compound. Therefore, calibration curves have to be determined, one for each halocarbon, again each one with at least seven points. The calibration values can be fitted to polynomial equations (*e.g.*, Microsoft Excel Solver). The curve fitting should be done stepwise, first assuming a second order equation and then refining to higher orders if needed, until the deviations of standard points from the curve are within acceptable limits.

Chromatographic integration software usually has facilities to evaluate peaks, even if the calibration curve is quadratic or cubic. If not, it is recommended that the axes are switched and the concentration is plotted against detector signal as in Fig. 23-5 B.

The shape of the calibration curve for a specific compound changes less with time than the detector response. It is thus possible to use the same curve for calibration over a reasonably long period, as long as it is adjusted for variations of detector response. The method is called stretching. This is especially useful when time does not permit a seven to nine point standard curve to be determined often enough. A few standards are measured to obtain the correction factor, and the concentration is calculated as in Fig. 23-5.

23.6.3   Determination of extraction efficiency

For correct measurements, the extraction efficiency for each analyte must be known and compensated for. To determine the extraction efficiency, a natural seawater sample at a high concentration level is purged, and the water is left in the purge chamber after the flow of purge gas has been stopped. The first measurement is made using the method to be evaluated. Then, the purge gas is turned on again, the procedure is repeated and a second measurement is made.

The original concentration C, the extraction efficiency A ($0 < A \leq 1$), the concentration C_1 recorded after the first, and C_2 after the second measurement follow the equations

$$A = 1 - \frac{C_2}{C_1} \quad \text{and} \quad C = \frac{C_1}{A}$$

23.7   Quality assessment

23.7.1   Sensitivity

Generally, the limits of detection (LD, see Chapter 1) of halocarbons are in the fmol/L to pmol/L range. Carbon tetrachloride has the highest detectability (the strongest mass specific response in the ECD) and an LD of about 10 fmol/L; the other tracers are of the same order of magnitude; all but a few of those listed in Table 23.1 have LDs of >1 pmol/L. The late eluting and less volatile bromoform and diiodomethane and the monohalogenated propanes and butanes have the highest LDs.

23.7.2   Reproducibility

The reproducibility (R) or precision of an analytical method is a multi-level concept, expressed as the relative standard deviation (RSD) of several measurements of one sample (see Chapter 1).

The reproducibility of the analytical procedure, involving all steps from sampling of the water to the final data, can be obtained in one single experiment or on the basis of periodically repeated experiments. A common method is to fill all 12 or 24 samplers of a rosette with water from the same water mass, *i.e.*, at the same depth in a homogeneous layer of medium concentrations. Normally, the RSDs for the transient tracers measured this way are 1 % or lower for all but CFC-113, for which the standard deviation (SD) might be of the same magnitude, but the low concentrations result in RSDs of up to 2–3 %. At very low concentrations, close to the LD, a relative measure of reproducibility is meaningless, and therefore R should be given as the highest value of LD and RSD, *e.g.*, '$R = 10$ fmol/L or 1.5 %, whichever is larger.'

23.7.3   Accuracy

The accuracy of measurements is the quality criterion most difficult to determine (see Chapter 1). However meticulously prepared the standards might be, the volumetric loops prepared and calibrated, and however carefully other easily identified sources of systematic errors are taken into account, there are no direct measurements or calculations of the analytical accuracy. Intercomparison with other laboratories, exchange of standard mixtures, *etc.*, give some indication of the correct concentration range. In international work, the demands are about twice the reproducibility limits, *i.e.*, within 2–5 %. Such demands necessitate the exchange of highly reliable standard mixtures which are difficult to prepare and expensive to purchase. For halocarbon studies others than tracer work and data collection for international data inventories the demands are less stringent.

23.8 Acknowledgements

Cordial thanks are given to Toste Tanhua for his dedication to the development of techniques for the halocarbon measurements.

Figures 23-1 and 23-2 were fashioned and drawn by Bengt Molander, for which I owe him sincere and friendly thanks.

References to Chapter 23

Abrahamsson, K., Klick, S. (1990), *J. Chromatogr.*, 513, 39.

Bullister, J.L., Weiss, R.F. (1988), *Deep-Sea Res.*, 35, 839.

Bullister, J.L., Lee, B.-S. (1995), *Geophys. Res. Lett.*, 22, 1893.

Burreson, B.J., Moore, R.E. (1975), *Tetrahedron Lett.*, 7, 473.

Busenberg, E., Plummer, L.N. (1992), *Water Resources Res.*, 28, 2257.

Dyrssen, D., Fogelqvist, E. (1981), *Oceanol. Acta*, 4, 313.

Ekdahl, A., Abrahamsson, K. (1997), *Anal. Chim. Acta*, 357, 197.

Eklund, G., Josefsson, B., Roos, C. (1978), *J. HRCC&CC*, 1, 34.

Fenical, W. (1975), *J. Phycol.*, 11, 245.

Fogelqvist, E., Larsson, M. (1983), *J. Chromatogr.*, 279, 297.

Fogelqvist, E., Krysell, M., Danielsson, L.-G. (1986), *Anal. Chem.*, 58, 1516.

Gribble, G.W. (1994), *Environ. Sci. Technol.*, 28, 310A.

Gschwend, M.P., MacFarlane, J.K., Newman, K.A. (1985), *Science*, 227, 1033.

Happell, J.D., Wallace, D.W.R., Wills, K.D., Wilke, R.J., Neill, C.C. (1996), *Report BNL-63227*. Brookhaven National Laboratory, Upton, Long Island, NY, USA.

Happell, J.D., Wallace, D.W.R. (1997), *Deep-Sea Res.*, 44, 1725.

Lovelock, J.E. (1975), *Nature* (London), 256, 193.

Manley, S.L., Dastour, M.N. (1987), *Mar. Biol.*, 98, 477.

Mensch, M., Smethie, Jr., W.M., Schlosser, P., Weppernig, R., Bayer, R., (1998) in: *Transient Tracer Observations During the Drift and Recovery of Ice Station Weddell*: Jacobs, S., Weiss, R.F. (Eds.). Antarctic Research Series, Antarctic Coastal Oceanology, 75, pp. 241–256.

Mitra, S., Yun, C. (1993), *J. Chromatogr.*, 648, 415.

Nightingale, P.D., Malin, G., Liss, P.S. (1995), *Limnol. Oceanogr.*, 40, 680.

O'Doherty, S.J., Simmonds, P.G., Nickless, G. (1993), *J. Chromatogr.*, 657, 123.

Pankow, J.F. (1991), *Environ. Sci. Technol.*, 25, 123.

Singh, H.B., Salas, L.J., Stiles, R.E. (1983), *J. Geophys. Res.*, 88, 3684.

Tanhua, T., Fogelqvist, E., Bastürk, Ö. (1996), *Mar. Chem.*, 54, 159.

Wilke, R.J., Wallace, D.W.R., Johnson, K.M. (1993), *Anal. Chem.*, 65, 2403.

24 Determination of dimethyl sulphide in seawater

G. Uher

24.1 Distribution of dimethyl sulphide in seawater

In the early 1970s, the pioneering work of *Lovelock et al.* (1972) revealed that dimethyl sulphide (DMS) is the predominant reduced sulphur compound in marine surface waters. Subsequent work confirmed that the global average DMS concentration (3 nmol/L) in sea surface water exceeds the levels of other volatile sulphur compounds by roughly three orders of magnitude. Today it is widely accepted that the sea-to-air flux of DMS from the ocean surface (*ca.* 1.2 Tmol sulphur a^{-1}, *Andreae*, 1990) is the principal source of natural sulphur to the atmosphere, accounting for about a quarter of the global sulphur emissions. After the dominant role of DMS in the natural sulphur cycle had been firmly established, the hypothesis that oceanic DMS is part of a climate feedback loop in which DMS-derived cloud condensation nuclei affect the Earth's radiation balance by changing the reflectivity of marine clouds (*Charlson et al.*, 1987), has again drawn attention to the marine biogeochemistry of DMS.

Ocean DMS is primarily formed by the enzymatic cleavage of dimethylsulphonium propionate (DMSP), which is several orders of magnitude faster than the abiotic decomposition of DMSP (*Dacey and Blough*, 1987).

$$(CH_3)_2S^+CH_2CH_2COO^- \rightarrow (CH_3)_2S + CH_2=CHCOOH \qquad (24\text{-}1)$$
$$\text{DMSP} \qquad\qquad\qquad \text{DMS} \qquad \text{Acrylic Acid}$$

DMSP is produced by specific marine phytoplankton species (*Keller et al.*, 1989), probably as an osmoregulatory substance and cryoprotectant (*Kirst et al.*, 1991; *Vairavamurthy et al.*, 1985). Since DMSP production is confined to the upper ocean layer, the vertical distribution of DMS is generally characterized by a maximum near the surface, and a sharp decrease in concentration at the base of the photic zone. In sea surface water, DMS concentrations typically span the range of 0.5–20 nmol/L, dependent on the spatial and seasonal distribution pattern of DMSP producing algae (*Turner et al.*, 1988). However, DMS levels may even exceed 100 nmol/L in the course of blooms of prymnesiophytes, in particular *Phaeocystis* species (*Liss et al.*, 1994). Despite the pronounced patchiness associated with episodic bloom events, the average DMS concentrations of the major biogeographical regions of different primary productivity fall in the rather narrow range of 2.1–5.4 nmol/L, and display similar values for open ocean and coastal regions (*Andreae,* 1990). These findings have been attributed to algal speciation, as the phytoplankton biomass of productive ocean areas is often dominated by minor DMS producers (*e.g.*, diatoms), whereas the taxonomic groups which include the most significant DMS producers (*e.g.*, prymnesiophytes), are more abundant in the oligotrophic ocean (*Andreae*, 1990; *Iverson et al.*, 1989).

Besides the effects of algal speciation, DMS production is also affected by a variety of biological processes, including zooplankton grazing (*Dacey and Wakeham*, 1986), and cell lysis (*Nguyen et al.*, 1988). DMS removal from the mixed layer is probably dominated by bacterial consumption (*Kiene*, 1992). However, air–sea gas exchange and photodegradation may also be important, dependent on the prevailing meteorological conditions (*Kieber et al.*, 1996). The variety of processes involved in ocean DMS cycling form an extremely dynamic system, resulting in highly variable DMS concentrations. Consequently, investigations of the marine sulphur cycle require analytical methods which are able to resolve the pronounced small-scale variability of nanomolar DMS levels in seawater.

24.2   Principle of the method

Since the introduction of the flame photometric detector (FPD) (*Brody and Chaney*, 1966) and its first application to marine DMS (*Lovelock et al.*, 1972), gas chromatography with flame photometric detection (GC-FPD) has become the standard technique for the determination of dissolved DMS. Among other sulphur-selective detectors, thus far only the sulphur chemiluminescence detector (SCD) (*Benner and Stedman*, 1989; *Shearer*, 1992) has also been used for the determination of dissolved DMS (*e.g.*, *Ledyard and Dacey*, 1994). However, detailed descriptions of this emerging analytical technique are still lacking. In contrast, the popularity of the FPD resulted in the publication of a variety of methods for the determination of oceanic DMS (*e.g.*, *Andreae and Barnard*, 1983; *Leck and Bågander*, 1988; *Turner and Liss*, 1985), two of which have been compared during an inter-laboratory calibration (*Turner et al.*, 1990). In the following, some principles of DMS determination by GC-FPD will be discussed, before the analytical procedure is described.

24.2.1   The flame photometric detector in the sulphur mode

The flame photometric detector is the principal component in the determination of sulphur compounds for which it offers a selectivity of about five orders of magnitude with respect to hydrocarbons. The selective sulphur detection is based on the formation of electronically excited S_2^* molecules in a hydrogen-rich flame. These short-lived species revert to their ground state and emit characteristic molecular band spectra with peak wavelengths at 384 and 394 nm. This chemiluminescent radiation passes an optical filter and is monitored by a UV-sensitive photomultiplier.

An outstanding feature of the FPD in the sulphur mode is its non-linear response function. In general, the relationship between peak area, A, and analyte mass, m, is described by an exponential form:

$$A = k\,m^n$$
$$\log A = \log k + n \log m$$

(24-2)

where k is a proportionality constant and the exponent n (commonly referred to as either the n-value or the linearity factor) represents the slope in the logarithmic form of Eq. (24-2). One might assume that the FPD response is exactly quadratic, because the formation of the chemiluminescent species S_2^* in the burner flame is a second-order reaction with respect to the sulphur concentration entering the detector. However, reported linearity factors range from about 1.5 to slightly greater than 2. These deviations from the theoretical n-value of 2 have been attributed to the intricate effects of sulphur flame chemistry and quenching processes on the chemiluminescent emission intensity. In particular, both n-value and detector response are affected by detector gas flow rates and the O_2/H_2 ratio within the flame. Optimum O_2/H_2 ratios usually vary between 0.1 and 0.4, and are strongly dependent on detector design and carrier gas flow. Thus, the gas flow rates should be optimised individually for any specific GC-FPD system (*e.g., Cardwell and Marriott*, 1982).

The dynamic range of the exponential sulphur response function is limited to about 2–3 orders of magnitude. The response reduction at the upper limit is probably caused by collisional quenching of the chemiluminescence emission at high analyte levels. In addition, sulphur response quenching by co-eluting hydrocarbons has been reported for conventional single-flame FPD, a problem which is minimized by using the dual-flame design (*Patterson*, 1978); however, it is unlikely to interfere with DMS determination at the low hydrocarbon levels present in seawater. Deviations from the response function at the lower limit of the dynamic range are mostly related to the presence of low-level sulphur contamination from gas cylinders, O-rings or diaphragms. At analyte levels in the order of background sulphur, the dependence of the detector response on analyte-derived sulphur decreases with decreasing analyte concentration, *i.e.*, the n-value decreases towards the detection limit. Consequently, the n-value may vary with operating conditions, system impurities, matrix effects and concentration range covered. Thus, n should always be determined under the relevant conditions and over the concentration range of interest. Given the undefined n-value, the presence of an analytical blank might be irritating in trace level determinations of DMS, because the blank can only be calculated from Eq. (24-2) if both n-value and proportionality constant k are known. An iterative method for the blank estimation is given in Section 24.6.

The above mentioned practical aspects of FPD detection as well as the phenomenological FPD response models have been reviewed excellently by *Farwell and Barinaga* (1986).

24.2.2 Purge and trap preconcentration

The detection limits of most GC-FPD systems fall within the range of about 0.5–2 pmol DMS. Consequently, the detection of nanomolar DMS levels in seawater requires sample volumes of the order of 10 mL or higher. Therefore, a purge and trap procedure is the common feature of all GC-FPD methods, *i.e.*, dissolved DMS is extracted from a discrete water sample by gas stripping, and concentrated onto a cryofocusing trap prior to injection into the gas chromatographic system. With few exceptions (*e.g., Leck and Bågander*, 1988), most workers remove the water vapour from the purge gas stream in order to avoid clogging of the cryotrap and interferences of excess water with trapped sulphur gases. Extensive tests by *Andreae and Barnard* (1983) indicated that both K_2CO_3 drying tubes and a cold trap (operated at $-35\,°C$) are equally reliable and effective. Others found the Nafion dryer a convenient solution (*Turner and Liss*, 1985), the efficiency of which has been shown to be comparable to that of desiccants and cold traps (*Leckrone and Hayes*, 1997).

24.3   Analytical system

The reported performance characteristics of the earlier methods for dissolved DMS indicate that similar analytical precision can be obtained with different purge and trap systems, and that major analytical problems are associated with sample handling and storage (see Section 24.8). Unlike the inter-comparison of aircraft instrumentation for atmospheric DMS (*Gregory et al.*, 1993), no such comprehensive inter-calibration of analytical procedures for dissolved DMS has been carried out. Therefore, the choice of a particular method may reflect the personal experience of the author. Here, a semi-automated method modified from *Andreae and Barnard* (1983) is described. It provides adequate precision (better than

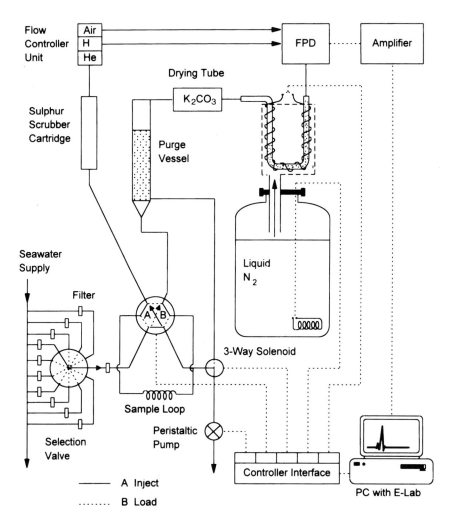

Fig. 24-1. Schematic diagram of the automated analysing system for DMS in seawater. Electrical connections are shown as bold, dotted lines. Gas flows and sample flow are shown as solid lines.

5 % relative standard deviation) over the DMS concentration range in sea surface water, and has proved reliable in seagoing work (*Uher et al.*, 1996). The method allows unattended operation over limited time periods and rapid sample analysis (20 min sampling intervals). Thus, it is highly suitable for the study of the small-scale variability of sea surface DMS.

A schematic diagramme of the automated system is shown in Fig. 24-1. Major parts of the analytical apparatus are custom-built by the workshops of the Max Planck Institute for Chemistry, Mainz, Germany. However, the main features of the method are easily adapted with commercially available equipment.

24.3.1 Purge and trap unit

The arrangement of the individual components of the purge and trap unit is depicted in Fig. 24-2.

The purge vessel is made of Duran (Pyrex) glass [20 mm outer diameter (o.d.), 20 cm length] and accommodates a glass frit at the bottom (Schott, pore size G1); the top 5 cm are widened to 40 mm o.d. to prevent bubbles entering the helium outlet. The above dimensions (*ca.* 50 mL internal volume) are chosen to minimize purge gas flushing time and are suitable

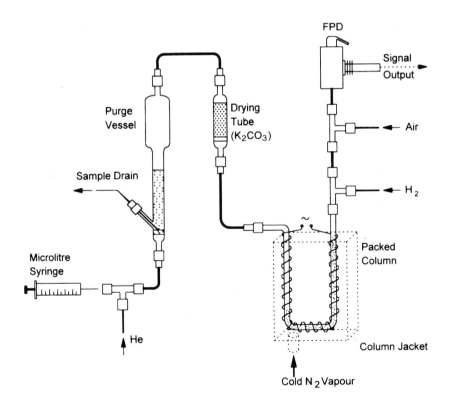

Fig. 24-2. Diagram of the purge and trap unit. The dashed lines indicate the Teflon-coated aluminium jacket surrounding the column.

for sample volumes up to 15 mL. Larger dimensions may be chosen if lowest detection limits are required. A T-fitting, connected to the helium supply line and closed off with a Teflon-coated silicone septum, serves as an injection port for liquid DMS standards. The sample drain line extends through the side arm to the bottom of the purge vessel. The helium outlet is connected to a Duran (Pyrex) glass drying tube (20 mm o.d., 10 cm length) fitted with a glass frit (Schott, pore size G0) to hold the desiccant (analytical grade K_2CO_3). Inlets and outlets of the purge vessel and the drying tube are made of 6.35 mm (1/4 in) o.d. glass tubing. The outlet of the drying tube is attached to a U-shaped Duran (Pyrex) glass tube (6.35 mm o.d., 30 cm unfold length) which serves both as a cryogenic trap and a separation column. The chromatographic packing (15 % OV3 on Chromosorb W-AW-DMCS 60–80 mesh, Supelco Inc.), held by two plugs of silanised glass wool, fills the last 20 cm towards the column outlet; the first 10 cm are left unfilled. About 1 m of heating wire (10 Ω total resistance) is wound around the U-tube and connected to the controller interface (Fig. 24-1) to allow controlled heating of the column at two adjustable power settings. To avoid sulphur gas adsorption onto metal surfaces, the transfer line extending from the column end into the FPD burner consists of glass-lined stainless-steel tubing (1.6 mm o.d.). All connections along the sample flow path are made of Teflon tubing (3.2 mm o.d.) and compression fittings with Nylon ferrules (Swagelok). All internal glass surfaces are silanised to minimize adsorptive losses of DMS during sample extraction. Preceding silylation, the glassware is cleaned with detergent solution, dilute hydrochloric acid and deionized water. The dried glassware is then deactivated by rinsing with a solution of 5 % dimethyldichlorosilane in toluene (Aldrich) at room temperature. Unreacted silylation agent is removed by successive rinsing with toluene, methanol and acetone; then, the glassware is dried at 200 °C. This treatment improves recovery and minimizes memory effects from reversible DMS adsorption onto active sites.

24.3.2 Cryogenic unit

Cooling of the chromatographic column is achieved by a basic cryogenic unit, schematically represented in Fig. 24-1. Cold nitrogen vapour is produced by controlled heating of a coil of heating wire (10 Ω total resistance), enclosed in a ceramic insulation, and submersed in liquid nitrogen at the bottom of a stainless-steel Dewar vessel (L'Air Liquide, Model TR 25, capacity 25 L). The nitrogen vapour is flushed through a short length of 6.35 mm o.d. Teflon tubing extending from the Perspex (Plexiglass) lid at the Dewar mouth into the left-hand side (*i.e.*, column inlet side) of a Teflon-coated aluminium jacket surrounding the column (Fig. 24-2). A second opening in the Perspex (Plexiglass) lid allows refilling of the Dewar and is closed off with a rubber stopper during analysis. As for the column heating, the controller interface allows time-programmed cooling of the column at two adjustable power settings. The cryogenic unit has also been described by *Andreae et al.* (1994).

24.3.3 Flame photometric detector and gas supply

A specially designed single-flame FPD (Max Planck Institute for Chemistry), equipped with an interference filter (Hewlett-Packard, peak transmission at 394 nm) and a Hamamatsu Model R268 photomultiplier was used by the author. However, most commercially available flame photometric detectors provide the required detection limit of better than 2 pmol of DMS. Constant gas flow rates should be maintained to ensure stable flame conditions and sulphur response characteristics. For this purpose, mass flow controllers have proved useful (*e.g.*, Tylan, FC 260 Model series), which also allow continuous monitoring of the gas flow rates. Alternatively, dual-stage regulators in combination with high precision flow controllers (*e.g.*, Vici Condyne, Model 202) may be used. Regulators with neoprene/rubber diaphragms must be avoided, because these sealing materials emit significant amounts of sulphur gases. Standard quality gases of 99.999 % purity are usually adequate, but still contain traces of sulphur gas impurities at the ppb level. To remove these impurities from the carrier gas stream, a re-usable stainless-steel scrubber cartridge (250 mL internal volume) packed with activated charcoal/5 Å molecular sieve (Merck, 1–2 mm pellets) is used. Prior to use, the cartridge is connected to an inert gas supply (*e.g.*, 30 mL min^{-1} nitrogen) and conditioned in an oven at 200 °C overnight. This procedure should be repeated, if a carrier gas blank is observed.

24.3.4   Sampling manifold

The purge and trap system is interfaced to a sampling manifold, essentially consisting of a 6-port sample injection valve (Vici Valco, Cheminert series, Model C12), a 3-way Teflon valve (Fluoroware, Galtek Model 203-3414) and a peristaltic pump with adjustable speed. A Teflon sample loop (3.2 mm o.d.) of gravimetrically calibrated internal volume is attached to the injection valve. Different sample loops of 1–15 mL may be used, depending on the expected DMS concentration. For semi-continuous operation, the injection valve is attached to a 10-port selection valve (Vici Valco, Cheminert series Model C15) which is connected to a clean seawater supply *via* an array of sample intakes (alternatively, discrete samples may be loaded *via* a selected port). Each intake is fitted with a 25 mm diameter polycarbonate filter housing holding a glass fibre filter (Whatman GF/F).

24.3.5   Control unit

A chromatography system (E-Lab, OMS Tech) installed in an IBM compatible PC controls the timing of column cooling/heating and valve operation, and processes the detector signal output. The controller interface fulfils the basic functions of valve triggering and power supply to the liquid nitrogen and column heatings.

24.4   Procedure

The following sequence of steps is repeated during sample analysis: first, the peristaltic pump is switched on, and the sample loop is flushed with filtered seawater from the selected intake, while the cryogenic unit is operated to cool the column. Then, the 6-port valve is actuated into the sample injection position and the trapped seawater volume is forced into the purge vessel by the carrier gas stream. The sample is degassed, and volatiles are cryoconcentrated at the column inlet. Following the degassing period, power is applied to the column heating to start the chromatographic separation, and the purge vessel is drained through 3-way valve and pump. Finally, the 10-port valve is actuated to select the next intake, and after 20 min the system is ready to load the next sample. A new filter must be selected for every sample in order to avoid DMS release from retained biological residues (see Section 24.8). Therefore, unattended operation is limited to a maximum number of 10 runs (*i.e.,* 10 h at a sampling interval of 60 min). The K_2CO_3 tube must be replaced at regular time intervals (usually every 12 h), preferably in parallel to the replacement of the glass fibre filters.

24.5   System optimisation and test procedures

Experimental variables such as sample volume, stripping time and gas flow rates are interrelated, and depend on the dimensions of the purge vessel and column as well as on the design of the FPD. Thus, the performance of each analytical system should be assessed individually. Using the above specified equipment and 3–12 mL samples, the author obtained optimum system performance with the following settings, which may serve as a guideline for system optimisation.

Gas flow rates (at standard pressure, 20 °C): purge/carrier gas, 90 mL/min He; FPD gas supply, 127 mL/min H_2, 120 mL/min air.

Time programme: pre-cooling of column, 3 min; degassing time, 12 min; column heating time, 5 min. The DMS retention time was 1.1 min at the set column heating rate.

The following steps should be carried out during initial system optimisation.

1. Pre-cool the column and obtain an FPD signal from a working standard injection corresponding to $\approx$ 20 pmol DMS, using an initial degassing time of 20 min (see Section 24.6). Repeat this step to find the required settings for liquid nitrogen heating and pre-cooling time. Adjust the column heating power to obtain a DMS retention time of 1–1.3 min.
2. Vary both air and hydrogen gas flow rates to find the maximum FPD response. Usually, an oxygen/hydrogen ratio of 0.2–0.3 is a good starting point. First, change the air flow rate in steps of 10 mL/min, since the FPD response varies more with oxygen input. It is recommended to measure and record accurately the optimum flow rates at the detector outlet for later trouble-shooting. Further information may be obtained from *Cardwell and Marriott* (1982).
3. Test for blanks by sampling the purge gas stream for 20 min with and without the purge and trap unit in line. The blank should be in the order of the detection limit or less (< 1 pmol), if a properly conditioned sulphur scrubber is installed, and the glassware has been cleaned and silanised as described.
4. Perform a calibration and record *n*-value and proportionality factor *k* of the FPD response function (Eq. 24-2) for later trouble-shooting. The *n*-values of different calibration curves should agree to within ± 10 %. Significant deviations are an indication of detector contamination or non-optimum flame conditions.
5. Stepwise, reduce the initial degassing period (20 min) to find the required degassing time for the specific conditions. The required degassing period must be redetermined, if

experimental variables (sample volume and concentration, purge vessel dimensions, He flow rate) have been changed significantly.

Tests for the system blank should be performed routinely. In between calibrations, the detector sensitivity should be monitored by injecting two working standards with DMS contents differing by a factor of ≈ 5. This procedure allows estimating the n-value of the response function. A full calibration should be performed, if the estimated n-value differs by more than 10 % from that of the last calibration.

24.6   Calibration

For calibration, microlitre amounts of a gravimetrically prepared standard solution of DMS in ethanediol are injected *via* the T-fitting into the helium line (see Fig. 24-1), as a sample of degassed (DMS-free) seawater is loaded into the purge vessel. This procedure provides a matrix-matched calibration and minimizes systematic errors by automatically correcting for degassing efficiency and potential DMS losses within the purge and trap unit. To avoid subsequent DMS release from particulates, the DMS-free seawater should be prepared from a filtered sample, ideally from deep water of low DMS concentration.

The working standard is prepared by sequential dilution of a primary standard. Fill three 20 mm crimp top vials (*ca.* 12 mL internal volume, with a Teflon-coated stirrer bar placed inside) to the top with a known mass of analytical reagent grade ethanediol (Merck, density 1.11 g/L). Seal off the vials for the primary and secondary standard with aluminium crimp caps fitted with silicone liners (Supelco, Part No. 2-7235); use a Mininert Teflon valve (Supelco, Part No. 3-3305) to seal off the working standard vial. To obtain a primary standard, inject a mass of approximately 0.1 g of DMS (Aldrich, > 99 % purity, packed under nitrogen, Cat. No. 27,438-0) through the silicone stopper of the first vial directly into the ethanediol. Determine the mass of DMS added using an analytical balance. Place the vial onto a magnetic stirrer and stir vigorously for about 15 min to ensure complete mixing (shaking is inefficient in the absence of a headspace). Proceed in this way to prepare a working standard in the range of 0.5–1 mg/g of DMS in the standard solution (*i.e.*, 7–15 pmol/mL of DMS in the standard solution) by two sequential dilutions of the primary standard. Analytical-reagent grade (a.g.) ethanediol is usually free of DMS, but should be tested for contamination prior to use, and cleaned by degassing with DMS-free helium if necessary. The accuracy of the working standard is better than 1.5 %, if an analytical balance of ±0.01 mg precision is used. Stored at 4 °C in a refrigerator, it is stable for several weeks. Further information on standard preparation and alternative calibration methods may be obtained from *Andreae and Barnard* (1983) and *Turner et al.* (1990).

In the absence of a detectable blank, the relationship of the logarithm of peak area to the logarithm of the mass of injected DMS should be linear over at least two orders of magnitude. However, blanks may occur as a consequence of decreasing system performance after prolonged use (*e.g.*, decreased sulphur scrubber efficiency, decreased surface silylation) or because of DMS traces in the seawater samples used for calibration. In this case, significant deviations from linearity are observed at low analyte levels. Therefore, the blank contribution, m_b, to the total amount of injected DMS must be taken into account:

$$\log A = \log k + n \log (m + m_b) \qquad (24\text{-}3)$$

Given the unknown blank, k and n can only be determined by an iterative approach. For high analyte levels ($m \geq 3\,m_b$), the response function can be approximated by

$$\log A = \log k' + n' \log m \qquad (24\text{-}4)$$

where k' and n' are estimates of the proportionality factor k and the linearity factor n, respectively. The blank can then be estimated from Eq. 24-4.

Substituting $m_b \gg (A_b/k')^{1/n'}$ for m_b in Eq. 24-3, an improved estimate of the response function can now be found. By successively repeating the last two steps, a good approximation of the response function is obtained. This iterative procedure can be implemented easily with common worksheet packages. As illustrated in Fig. 24-3 the use of uncorrected calibration curves may result in severe systematic errors at DMS levels below $3\,m_b$.

24.7   Precision

The precision of the method depends on the DMS concentration, the FPD used and experimental variables affecting DMS peak shape and detector response. For the manually operated version of the method, a relative standard deviation of ±6.2 % has been reported (*Andreae and Barnard*, 1983). For the automated system, the relative standard deviation has

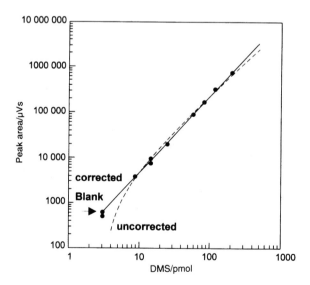

Fig. 24-3. Influence of blank correction on the FPD calibration curve. The fit to the blank-corrected calibration is represented by the solid line. For DMS levels below three times the blank level, the uncorrected calibration curve (dashed line) shows significant deviations from the true detector response. Both calibration curves are scaled to blank-corrected DMS (*i.e.*, blank plus injected DMS).

been estimated to ±3 % by nine replicate determinations of DMS in a seawater sample from the Northeast Atlantic (average DMS concentration, 4.38 nmol/L; amount of injected DMS, 16.8 pmol; sample volume, 3.83 mL). At DMS levels between 0.1 and 0.5 nmol/L, the relative standard deviation typically is $< \pm 8$ % (*G. Uher*, unpublished work).

24.8   Sampling and storage

To avoid sample degassing, subsampling should be performed in the same way as described for dissolved oxygen determinations. The seawater may be collected from standard water samplers as well as from continuous seawater supplies, as both sampling methods usually agree within experimental error (*Andreae et al.*, 1994; *G. Uher*, unpublished work). However, comparisons between samples from hydrocasts and continuous seawater supplies should be carried out routinely, as the quality of the pumping systems may vary from ship to ship.

Severe systematic errors may be caused by the formation of DMS from both particulate and dissolved DMSP during storage, especially in seawater samples containing high densities of phytoplankton. This DMS production can increase the DMS concentration of the sample by a factor of 1.5–4 within a few hours, (*Kieber et al.*, 1996; *Turner et al.*, 1988; *G. Schebeske*, personal communication). In contrast, previous work had indicated that refrigeration of samples, or storage in the dark at seawater temperature, maintains the sample integrity for periods up to 48 h (*Andreae and Barnard*, 1983). These conflicting results may be explained by differences in sample composition, since the DMS release is closely related to various biological factors such as species composition and physiological condition of the phytoplankton cells. Consequently, samples for DMS determination should be analysed immediately, unless the feasibility of a particular storage protocol has been verified for the respective sample composition. Given the sample-specific storage behaviour, sample preservation with $HgCl_2$ (*McTaggart and Burton*, 1992) should also be viewed with certain caution, and should be tested for the type of seawater under study.

Sample filtration as described in Section 24.4 is strongly recommended, as particulates in the purge and trap unit may release significant amounts of DMS, causing a persistent system blank (*Andreae and Barnard*, 1983). In addition, the presence of particulates in the purge and trap unit can artificially increase the DMS concentration by up to a factor of two (*Andreae et al.*, 1994; *Barnard et al.*, 1982; *Turner et al.*, 1990). Further, it has been speculated that sample filtration might stimulate DMS release from retained phytoplankton cells (*Turner et al.*, 1990). However, presently available data from various oceanic regions indicate that the potential DMS increase due to filtration is negligible compared with the DMS production from particulates during sample extraction.

References to Chapter 24

Andreae, M.O. (1990), *Mar. Chem.*, 30, 1.

Andreae, M.O., Barnard, W.R. (1983), *Anal. Chem.*, 55, 608.

Andreae, T., Andreae, M.O., Schebeske, G. (1994), *J. Geophys. Res.*, 99, 22, 819.

Barnard, W.R., Andreae, M.O., Watkins, W.E., Bingemer, H., Georgii, H.-W. (1982), *J. Geophys. Res.*, 87, 8787.

Benner, R.L., Stedman, D.H. (1989), *Anal. Chem.*, 61, 1268.

Brody, S.S., Chaney, J.E. (1966), *J. Gas Chromatogr.*, 4, 42.

Cardwell, T.J., Marriott, P.J. (1982), *J. Chromatogr. Sci.*, 20, 83.

Charlson, R.J., Lovelock, J.E., Andreae, M.O., Warren, S.G. (1987), *Nature* (London), 326, 655.

Dacey, J.W.H., Blough, N.V. (1987), *Geophys. Res. Lett.*, 14, 1246.

Dacey, J.W.H., Wakeham, S.G. (1986), *Science*, 233, 1315.

Farwell, S.O., Barinaga, C.J. (1986), *J. Chromatogr. Sci.*, 24, 483.

Gregory, G.L., Warren, L.S., Davis, D.D., Andreae, M.O., Bandy, A.R., Ferek, R.J., Johnson, J.E., Saltzman, E.S., Cooper, D.J. (1993), *J. Geophys. Res.*, 98, 23, 373.

Iverson, R.L., Nearhoof, F.L., Andreae, M.O. (1989), *Limnol. Oceanogr.*, 34, 53.

Keller, M.D., Bellows, W.K., Guillard, R.R.L. (1989), in: *Biogenic Sulfur in the Environment*: Saltzman, E.S., Cooper, D.J. (Eds.). Washington: American Chemical Society, 1989; pp. 169–179.

Kieber, D.J., Jiao, J.F., Kiene, R.P., Bates, T.S. (1996), *J. Geophys. Res.*, 101, 3715.

Kiene, R.P. (1992), *Mar. Chem.*, 37, 29.

Kirst, G.O., Thiel, C., Wolff, H., Nothnagel, J., Wanzek, M., Ulmke, R. (1991), *Mar. Chem.*, 35, 381.

Leck, C., Bågander, L.E. (1988), *Anal. Chem.*, 60, 1680.

Leckrone, K.J., Hayes, J.M. (1997), *Anal. Chem.*, 69, 911.

Ledyard, K.M., Dacey, J.W.H. (1994), *Mar. Ecol. Prog. Ser.*, 110, 95.

Liss, P.S., Malin, G., Turner, S.M., Holligan, P.M. (1994), *J. Mar. Systems*, 5, 41.

Lovelock, J.E., Maggs, R.J., Rasmussen, R.A. (1972), *Nature* (London), 237, 452.

McTaggart, A.R., Burton, H. (1992), *J. Geophys. Res.*, 97, 14, 407.

Nguyen, B.C., Belviso, S., Mihalopoulos, N., Gostan, J., Nival, P. (1988), *Mar. Chem.*, 24, 133.

Patterson, P.L. (1978), *Anal. Chem.*, 50, 345.

Shearer, R.L. (1992), *Anal. Chem.*, 64, 2192.

Turner, S.M., Liss, P.S. (1985), *J. Atmos. Chem.*, 2, 223.

Turner, S.M., Malin, G., Liss, P.S., Harbour, D.S., Holligan, P.M. (1988), *Limnol. Oceanogr.*, 33, 364.

Turner, S.M., Malin, G., Bågander, L.E., Leck, C. (1990), *Mar. Chem.*, 29, 47.

Uher, G., Schebeske, G., Rapsomanikis, S., Andreae, M.O. (1996), *Ann. Geophys.*, *Supplement II to Vol. 14*, C590.

Vairavamurthy, A., Andreae, M.O., Iverson, R.L. (1985), *Limnol. Oceanogr.*, 30, 59.

25 Determination of marine humic material

G. Liebezeit

25.1 Introduction

Marine humic material may be characterized as a highly complex mixture of dissolved and/or colloidal organic substances in seawater which cannot be separated into single compounds, even by the most powerful analytical techniques such as capillary gas chromatography or high performance liquid chromatography. *Kalle* (1937) was the first worker to publish scientific investigations of marine humic material, and who then coined the term 'Gelbstoff', the German word for 'yellow material', to denote this highly complex mixture of organic compounds. In near-shore sea areas under the influence of land run-off it has a more polyphenolic character, whereas aliphatic structures dominate in the open-ocean (*e.g.*, *Duursma*, 1965; 1974; *Stuermer and Payne*, 1976). This inseparable assemblage of organic compounds fluoresces, and therefore is also called 'dissolved fluorescent substances' (*e.g.*, *Laane*, 1981). Another name is 'marine humic substances' (*e.g.*, *Chen and Bada*, 1989; *Gagosian and Stuermer*, 1977) which is a somewhat unfortunate label as it implies structural similarity with largely lignin derived terrestrial humic material. Lignin, however, is a rare substance in the sea except near river mouths and in land-locked sea areas such as the Baltic Sea, and is not a major source material for indegenous marine humics.

In addition to introduction from terrestrial sources in coastal areas, *in situ* formation in open oceanic realms is considered to be the main source of dissolved humic substances (DHS) in marine waters (*Duursma*, 1965; *Harvey* et al., 1984). *Chen and Bada* (1989) showed that DHS may also diffuse from marine sediments into the overlying water column. DHS in marine porewaters may also be formed by condensation of mono- and oligomers previously liberated from polymeric organic material by bacterial action. *Poutanen and Morris* (1983) observed that a decaying diatom population was the source of high relative molecular mass compounds which may be components of DHS. With an ingeneous experiment *Harvey et al.* (1984) showed that exposure to sunlight of unsaturated triglycerides dissolved in seawater leads to the formation of a complex mixture of compounds with spectrometric properties closely resembling those of marine DHS.

Humic compounds in soils and sediments are subdivided operationally into two groups of compounds: humic and fulvic acids. Humic acids are soluble in aqueous alkali, while fulvic acids are soluble in both aqueous acids and bases. This distinction also applies to marine DHS.

Although humic compounds evade structural definition, their determination is an essential prerequisite for complete characterization of the marine organic carbon pool. This fraction, so far uncharacterized at the single compound level, makes up the major portion of organic carbon in seawater (*Degens and Ittekkot*, 1983). Any method capable of providing at least semiquantitative data therefore is a useful addition to the analytical arsenal of the marine chemist.

Various workers have attempted to relate DHS measurements to dissolved organic carbon concentrations (*e.g., Wheeler*, 1977; *Laane and Koole*, 1982; *Liebezeit*, 1988; *Ferrari et al.*, 1996). The results, however, were far from conclusive. After its introduction by *Kalle* (1949, 1963), fluorescence spectrometry has become the most widely used technique to determine DHS. Owing to their largely conservative nature (*e.g., Laane*, 1981), the fluorescence properties of DHS can be exploited for estuarine mixing studies (*Zimmerman and Rommets*, 1974) or in remote sensing applications (*e.g., Karabashev et al.*, 1993; *Reuter et al.*, 1993). Standardization of the measurements, however, still presents a major problem as the only seawater standard available is that from the International Humic Substances Society.

25.2 Instrumentation

Despite some disadvantages of conventional fluorimeters (discussed by *Chen and Bada*, 1990), their use is described in this chapter, because the more suitable laser induced fluorescence (LIF) instruments are not as readily available. Accounts of LIF applications have been published by *Chen and Bada* (1990) and *Donard et al.* (1989).

Generally, filter fluorimeters are more sensitive than those with monochromators. However, they lack scanning options and hence cannot be used for investigations related to molecular structure. Nevertheless, for a number of routine analyses filter instruments are quite satisfactory. If a filter fluorimeter is used, the excitation filter should be a narrow band filter (± 5 nm) centred at 325 nm. On the emission side, a wide band pass filter (± 10–20 nm) centred at 420 nm should be used. Scanning fluorimeters should have excitation and emission monochromators with slit widths of 5 nm or smaller.

In the original work *Kalle* (1949) used a combination of 365 nm excitation and 420 nm emission wavelengths which has been adopted by a number of other workers. However, with this combination Raman scattering may interfere in DHS determinations (*Morel*, 1974; *Gienapp*, 1979). The combination of the 325 nm excitation wavelength and emission at >420 nm avoids this problem as the Raman scattering peak then occurs at about 370 nm (Fig. 25-1). This combination was used by, *e.g., Hayase et al.* (1987), *Liebezeit* (1988) and *Chen and Bada* (1989).

Modern instruments usually support data recording in ASCII format for use with spread sheet, statistics and graphics programmes. Particularly with synchronous excitation fluorimetry (see following sections) digital data processing becomes a necessity.

25.3 Dissolved humic substances (DHS)

25.3.1 Sampling

Special precautions are unnecessary in DHS determinations, although extensive exposure to light should be avoided as UV radiation leads to the partial destruction of humic compounds (*Kieber et al.*, 1990). There is no need to filter open-ocean water samples. However, in turbid waters it might become necessary to remove particulate material by filtration. Particles do not influence the measurement itself, but their presence produces a noisy signal. Extended storage should be avoided, although detailed investigations on possible storage effects are still lacking.

25.3.2 Fluorimetry

Transfer the sample into the fluorimeter cell. Make sure the optical surfaces are clean. Set the excitation wavelength to 320 nm and scan the emission from 400 to 600 nm. The excitation spectrum is obtained by scanning the excitation wavelength from 200 to 370 nm at the fixed emission wavelength of 420 nm. Typical emission and excitation spectra are shown in Fig. 25-1.

25.4 Particulate humic compounds

Particulate humic compounds (PHC) can be determined as described above for DHS after extraction with alkaline agents.

25.4.1 Chemicals

0.1 mol/L NaOH–0.1 mol/L $Na_4P_2O_7$: Dissolve 44.606 g of sodium diphosphate decahydrate ($Na_4P_2O_7 \cdot 10H_2O$) in 500 mL of doubly distilled water. Dissolve 4 g of NaOH in 400 mL of doubly distilled water. After cooling add to the diphosphate solution and make up to 1 L. The extraction solvent is stable.

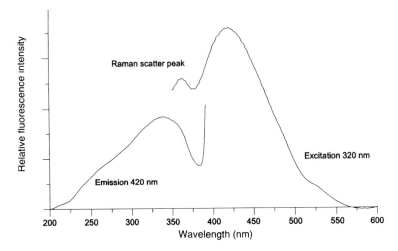

Fig. 25-1. Excitation/emission spectrum of dissolved humic substances from the coastal North Sea (Kontron Instruments SFM 25, 5 nm slit width, 500 V photomultiplier voltage).

25.4.2   Extraction

Place the wet filter containing the particulate material into a 20 mL scintillation vial, add 10 mL of 0.1 mol/L NaOH–0.1 mol/L Na$_4$P$_2$O$_7$ solution, close the vial and shake for 24 h. Transfer the extract into a centrifugation tube, centrifuge (6000 rpm are usually sufficient), and transfer the supernatant into a fluorimeter cell. Proceed as described under Section 25.3.2. It may be necessary to dilute the extract. With some experience the colour of the extract will tell the analyst whether this is necessary. Usually, a light brown discolouration indicates that the measurement will still be in the linear range (see Section 25.5.3) although in this case checking by dilution is recommended.

This extraction technique will give a value for total PHC including the lipid-associated fraction. If a distinction is desired, this fraction may be removed as described under Section 25.4.3 and analysed separately.

25.4.3   Lipid-associated humic compounds (LHC)

Air-dry the filter containing the particulate material and add to 20 mL of dichloro-methane-methanol 9+1 (v/v) in a scintillation vial. Shake for 24 h and centrifuge if necessary. Proceed as described earlier.

25.5   Calibration

Although time consuming, laboratory-prepared standards should be used when a specific sample type is analysed regularly. Alternatively, the marine standards available from the International Humic Substances Society (IHSS; web site: http://www.gatech.edu/ihss/) may be used. However, these materials have been obtained from open-ocean waters, and hence might differ in chemical composition and/or fluorescence characteristics from humic compounds present in estuarine and coastal waters.

25.5.1   Standardization with quinine sulphate

Kalle (1963) introduced quinine sulphate as a standard to measure relative fluorescence intensities. It was later used by, among others, *Laane* (1981) and *Chen and Bada* (1989). *Kalle* equated the fluorescence intensity of a solution of 0.1 mg/L of quinine sulphate in 0.1 mol/L sulphuric acid to 73 millifluorescence units (mFl). Values for mFl exceeding 100 were avoided by diluting the samples when necessary. *Laane* (1981), on the other hand, used a ten-fold higher concentration equating the fluorescence intensity of 1 mg/L to 700 mFl. In both cases no indications were given of the concentration range in which the fluorescence intensity was a linear function of the concentration. One may assume, however, that modern instruments as used by *Laane* have a wider linear range than the Pulfrich photometer used by *Kalle*. Nevertheless, the linear range of the instrument response should be checked by consecutive dilutions of a 1 mg/L standard (see Section 25.5.3).

25.5.2 DHS and PHC standards

As with DHS, the IHSS open-ocean standard may be used. If a reference material typical for the area under investigation is desired, it may be prepared following standard techniques (see Section 25.5.4). However, for PHC a considerable amount of particulate material is necessary, *i.e.*, at least several grams. Several protocols can be followed, *e.g.*, with or without removal of lipid material. The one given by *Kononova* (1966) is outlined in Section 25.5.4 with some modifications.

The humic acids offered commercially by some manufacturers are not recommended as reference materials. These substances are usually obtained from terrestrial sources such as peat or brown coal. They lack strict quality control, *i.e.*, the composition differs from batch to batch. Furthermore, these compounds are poorly characterized, even in terms of elemental composition and ash content. Further information on this problem has been provided by *Malcolm and McCarthy* (1986).

25.5.3 Linear range of instrument response

Self-quenching effects are observed when concentrated solutions of humic compounds are analysed fluorimetrically. They arise from absorption of emitted fluorescence radiation by higher concentrations of molecules of the same or different type present in solution. If this occurs, the function of fluorescence intensity *versus* concentration of humic material is linear only over a narrow concentration range. The self-quenching effect is detected most easily by measuring a series of consecutive dilutions of the sample with the extraction agent. If quenching does not occur, the fluorescence response is a linear function of the concentrations as shown schematically in Fig. 25-2. If this is not the case then dilution of the samples or extracts becomes necessary until linearity is attained. In the example given in Fig. 25-2, this is after the second dilution.

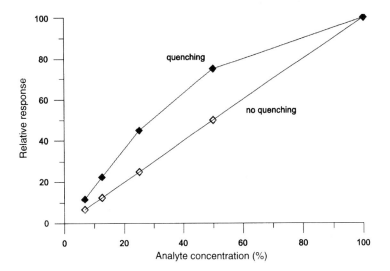

Fig. 25-2. Schematic illustration of quenching effects.

Such problems, however, will usually be encountered only for porewater samples from greater sediment depths, *i.e.*, usually several metres. Occasionally, estuarine waters may exhibit the same phenomenon. Open-ocean waters contain too little material to cause self-qenching.

25.5.4   Preparation of PHC standards

To avoid filtration of large volumes of water, particulate material (PM) is preferably concentrated by continuous centrifugation. All operations are carried out at room temperature and in polyethylene (PE) flasks. Alkaline agents liberate silicate from glass, which may become incorporated in the humic acid to be isolated and hence increase the ash content.

To remove carbonates, PM is treated with 1 mol/L HCl until no more CO_2 is liberated. Lipids are then removed by successive treatment with dichloromethane-methanol 9+1 (v/v) and n-hexane. The remaining PM is air dried and extracted with 0.1 mol/L NaOH –0.1 mol/L $Na_2P_4O_7$ for 24 h in the dark with continuous shaking. After centrifugation at 6000 rpm the supernatant is decanted and the extraction repeated twice. The supernatants are combined and acidified to pH 1–1.5 with concentrated HCl. After standing refrigerated overnight the humic acid precipitate is removed by centrifugation. Repeated redissolution in 5 % Na_2CO_3 solution and precipitation with concentrated HCl serves to reduce the ash content of the isolated humic acids. Prior to the last precipitation the isolate is dialysed against tap water (1 week) and distilled water (1 week). A dialysis membrane with 24 Å pore size has proved satisfactory. The water is replaced daily.

If desired, fulvic acids can be obtained from the humic acid free supernatants by adsorption on hydrophobic resins such as XAD-2 or XAD-8 (see *Malcolm*, 1990 for details).

A detailed account on the isolation of DHS in larger amounts has been given by *Malcolm* (1990). However, to obtain an amount useful for, *e.g.*, detailed spectrometric investigations, several thousand litres of seawater have to be processed. For example, *Stuermer and Harvey* (1977) reported on the recovery of 153 mg of fulvic acid from 1150 L of Sargasso Sea water. These workers also found that fulvic acids constituted the majority of DHS in open-ocean waters.

25.6   Synchronous excitation fluorimetry

Fluorescence spectrometry not only provides quantitative or semiquantitative data on concentrations, it may also be used to obtain information on the degree of condensation of humic compounds. Although only the fluorophor and the immediately adjacent parts of the molecule are involved, some useful information may be retrieved. Synchronous fluorescence spectrometry has also been used for remote sensing purposes (*Vodacek*, 1989).

This approach is based on the observation that higher degrees of condensation cause red shifts in both the excitation and the emission spectra. In synchronous excitation fluorescence spectrometry (SFS) both the excitation and the emission wavelengths are changed simultaneously at a constant wavelength difference. A $\Delta\lambda$ value of 18 nm has been shown to be most suitable for this purpose (*Miano et al.*, 1988). The resulting spectra do not provide quantitative information, but can be used to distinguish DHS and PHC of, *e.g.*, different origin or age. Two examples are shown in Fig. 25-3. While coastal North Sea water has its main peak at 350 nm the Skagerrak porewater has additional peaks at higher wavelengths indicating the presence of more highly condensed constituents.

Using peak deconvolution techniques (Fig. 25-4) it is seen that synchronous spectra of dissolved humic substances exhibit a highly complex structure. In this case the overall spectrum results from a combination of 14 components.

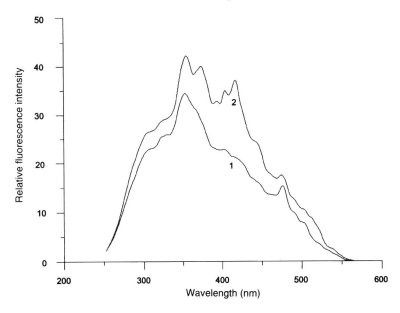

Fig. 25-3. Synchronous excitation spectra of DHS from 1) coastal North Sea water and 2) from Skagerrak sediment porewater ($\Delta\lambda$=18 nm).

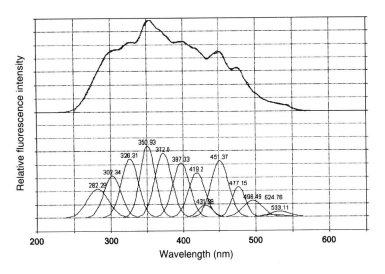

Fig. 25-4. Peak deconvolution of a synchronous excitation spectrum of DHS from coastal North Sea water. Component peaks were assumed to be Gaussian.

References to Chapter 25

Chen, R.F., Bada, J.L.(1989), *Geophys. Res. Lett.*, 16, 687.

Chen, R.F., Bada, J.L. (1990), *Mar. Chem.*, 31, 219.

Degens, E.T., Ittekkot, V. (1983), *Neth. Inst. Sea Res. Publ. Ser.*, 10, 179.

Donard, O.F.X., Lamotte, M., Belin, C., Ewald, M. (1989), *Mar. Chem.*, 27, 117.

Duursma, E.K. (1965), in: *Chemical Oceanography*: Riley, J.P., Skirrow, G. (Eds.). London: Academic Press, 1965; Vol. 1; pp. 433–475.

Duursma, E.K. (1974), in: *Optical Aspects of Oceanography*: Jerlov, N.G., Steemann Nielsen, E. (Eds.). London, New York: Academic Press, 1974; pp 237–256.

Ferrari, G.M., Dowell, M.M., Grossi, S., Targa, C. (1996), *Mar. Chem.*, 55, 299.

Gagosian, R.B., Stuermer, D.H. (1977), *Mar. Chem.*, 5, 605.

Gienapp, A. (1979), *Dtsch. Hydrogr. Z.*, 32, 204.

Harvey, G.R., Boran, D.A., Piotrowicz, S.R., Weisel, C.P. (1984), *Nature* (London), 309, 244.

Hayase, K., Yamamoto, M., Nakazawa, I. Tsubota, H. (1987), *Mar. Chem.*, 20, 265.

Kalle, K. (1937), *Ann. Hydrogr. Mar. Meteorol.*, 65, 276.

Kalle, K. (1949), *Dtsch. Hydrogr. Z.*, 2, 117.

Kalle, K. (1963), *Dtsch. Hydrogr. Z.*, 16, 153.

Karabashev, G.S., Khanaev, S.A., Kuleshov, A.F. (1993), *Oceanol. Acta*, 16, 115.

Kieber, R.J., Zhou, X., Mopper, K. (1990), *Limnol. Oceanogr.*, 35, 1503.

Kononova, M.M. (1966), *Soil Organic Matter*. Oxford: Pergamon Press, 1966.

Laane, R.W.P.M. (1981), *Netherl. J. Sea Res.*, 15, 89.

Laane, R.W.P.M., Koole, L. (1982), *Neth. J. Sea Res.*, 15, 217.

Liebezeit, G. (1988), *Mitt. Geol.-Paläontol. Inst. Univ. Hamburg*, 65, 153.

Malcolm, R.L. (1990), in: *Proceedings of Linköping, Sweden, Meeting on Humic Substances in the Environment*. Linköping University, 1990; pp. 390–417.

Malcolm, R.L., MacCarthy, P. (1986), *Environ. Sci. Technol.*, 20, 904.

Miano, T.M., Sposito, G., Martin, J.P. (1988), *Soil Sci. Am. J.*, 52, 1016.

Morel, A. (1974), in: *Optical Aspects of Oceanography*: Jerlov, N.G., Steeman Nielsen, E. (Eds.). London: Academic Press, 1974; pp. 1–24.

Poutanen, E.-L., Morris, R.J. (1983), *Estuar. Coastal Shelf Sci.*, 17, 139.

Reuter, R., Diebel, D., Hengstermann, T. (1993), *Int. J. Remote Sens.*, 14, 823.

Stuermer, D.H., Harvey, G.R. (1977), *Deep-Sea Res.*, 24, 303.

Stuermer, D.H., Payne, J.R. (1976), *Geochim. Cosmochim. Acta*, 40, 1109.

Vodacek, A. (1989), *Remote Sens. Environ.*, 30, 239.

Wheeler, J.C. (1977), *Limnol. Oceanogr.*, 22, 573.

Zimmerman, J.T.F., Rommets, J.W. (1974), *Neth. J. Sea Res.*, 8, 117.

26 Determination of amino acids and carbohydrates

G. Liebezeit and B. Behrends

26.1 Introduction

Amino acids and carbohydrates make up the majority of the compounds characterized in seawater. These substances, as primary products of photosynthesis, serve numerous functions, *e.g.*, as organic nutrients for heterotrophic organisms, as intermediate storage products in the form of high energy compounds, as catalysts in metabolic processes or as condensed molecules serving architectural functions. Besides monomeric amino acids and sugars a wide range of oligo- and polymeric species is found in dissolved seawater. These products may resemble primary products in the case of excretion or lysis products, but may also be further modified by secondary biological or chemical reactions.

Two approaches have been adopted for the study of natural organic compounds in seawater: compound class reactions based on the chemical reactivity of specific functional groups, *i.e.*, here the amine or aldehyde functionalities, and detailed investigations of individual compounds after chromatographic separation.

The techniques presented below are restricted to determinations of carbohydrates and amino acids by methods that have been found to be reproducible and suitable for use in field work. While the less specific techniques generally do not require sophisticated equipment and hence are adapted easily to shipboard work, chromatographic techniques are, from an instrumentation point of view, more complex. However, modern instruments are usually rugged enough to withstand conditions at sea, at least for some time.

26.2 General remarks

Natural organic compounds occur in seawater at levels in the pico- to low nanomolar range exceeded by several orders of magnitude by the concentrations of inorganic salts. To determine these compounds with a minimum amount of pre-treatment should be the ultimate aim of the marine analytical chemist. This in turn requires highly sensitive techniques employing small amounts of seawater.

However, high detector sensitivity offers few advantages without reducing the levels of contamination by minimizing manipulations between sampling and analysis. Since it is common knowledge that the levels of amino acids on human skin far exceed the amounts found in 1 L of seawater, extreme care is necessary to ensure absolute cleanliness of the sampling device and the equipment employed for subsequent analysis. All glassware should therefore be cleaned, preferably by combustion overnight at 450 °C. Washing with dilute acid followed by copious rinses with doubly distilled water serves to decontaminate more delicate apparatus. Chemically clean and mirobiologically sterile do not necessarily mean the same thing!

Filtration not only may contribute to sample contamination, but may also lead to other more or less systematic and quantifiable errors such as cell rupture or adsorption on filter materials. If possible, filtration should therefore be avoided. Filtration is usually unnecessary, when methods are employed which require small volumes of seawater (< 1 mL). In other cases, simple gravity fltration through a pre-combusted glass fibre filter appears to be the method of choice.

Samples of seawater intended for determination of their monomeric constituents may be subject to alteration during storage even when deep-frozen (*Webb and Wood*, 1966) although this might not be noticeable in total dissolved organic carbon contents (*Tupas et al.*, 1994). Fixation of samples with a variety of organic or inorganic agents to eliminate metabolic activity may lead to further drastic changes in the composition of unfiltered samples. Thus, if analysis directly in the field is not feasible, the analyst finds him or herself in a dilemma as to whether or not to accept the risks.

The methods described here advocate the determination of free amino acids in unfiltered samples whenever possible and soon after sampling. *Jørgensen et al.* (1981) suggested that deep-frozen (– 20 °C) the amino acid composition remains unaltered for short periods of time.

All reagents used in the subsequent analytical steps should be of the highest analytical purity; the water used for their preparation should be doubly distilled from a quartz distillation unit.

26.3   Dissolved free amino acids (DFAA)

26.3.1   Compound class reaction

The compound class reaction for total DFAA actually measures primary amine content and thus includes contributions of, *e.g.*, amino sugars, oligopeptides and ammonia to varying extents. The method is based on the reaction of primary amines with *o*-phthalaldehyde-mercaptoethanol (*Roth*, 1971) yielding highly fluorescent isoindole derivatives (Fig. 26-1).

The compounds detected by the method could, therefore, better be described as *o*-phthalaldehyde reactive substances (ORS) expressed as glycine equivalents. The automated techniques presented below are based on the procedures described by *Josefsson et al.* (1977) and *Delmas et al.* (1990) with modifications. Manual determinations are possible as well provided a fluorimeter for static measurements is available.

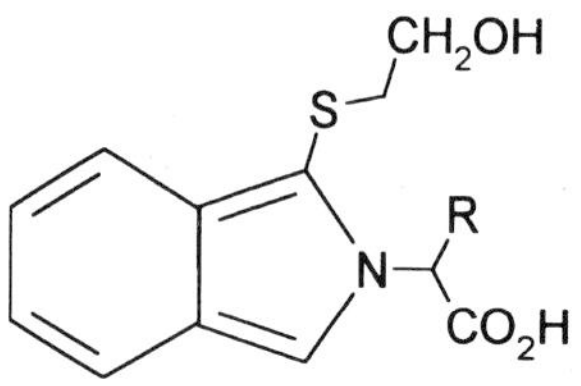

Fig. 26-1. Structure of fluorescent isoindole derivatives of amino acids. For R designations see Table 26-1.

26.3.2   Reagents

Sodium borate buffer: Dissolve 25 g of boric acid analytical reagent grade (a.g.) in 900 mL of distilled water, adjust to pH 10.5 with 8 mol/L sodium hydroxide solution and make up to 1 L with distilled water. Store in a polyethylene bottle. The solution is stable.

o-Phthalaldehyde stock solution: Dissolve 1 g of *o*-phthalaldehyde (for fluorescence purposes, Merck, Darmstadt, Germany) in 100 mL of chromatographic grade methanol. Store refrigerated.

2-Mercaptoethanol: The reagent should be stored refrigerated under nitrogen and be of the highest purity available. The reagent should be added with a syringe which is filled after piercing the the bottle septum cap.

26.3.3   Manual procedure

A sensitive filter fluorimeter with a 340 nm excitation and a 420 nm emission filter is required for detection and quantification of the fluorescent products. Standard fluorimeter cells may be used. Determinations are performed by mixing equal portions of sample (or standard solution) with the mixed reagent. The fluorescence signal is read after exactly 10 min against distilled water as the blank value. The reagent blank is calculated from the fluorescence signal of the reagent alone divided by 2. The native fluorescence of seawater at the wavelengths employed normally lies around 0.01 μmol of glycine equivalents per litre and may be ignored.

Ammonia interferes in the determination. However, its equivalent signal is a mere 5 % of the glycine response over the entire linear range of the method. If necessary, a correction may be applied after parallel determinations of the ammonia content (see Chapter 10). Samples with primary amine concentrations exceeding the linear range, *e.g.*, from anoxic pore waters, must be diluted with distilled water.

Mixed reagent: It should be made up daily as follows. To 100 mL of sodium borate buffer add 1 mL of *o*-phthalaldehyde stock solution followed by 0.05 mL of mercaptoethanol. After standing for a few minutes the reagent is ready for use.

26.3.4   Automatic procedure

The manual method may be automated according to the principles described in Chapter 10.

The manifold required (Fig 26-2) is simple, since only one reagent is to be added. The mixed reagent and the wash water should be protected from the atmosphere with a concentrated sulphuric acid trap. Air used for segmentation, likewise, should be scrubbed free of ammonia before entering the system. The cups used for the automatic sampler should be rinsed three to four times, either with the sample, if sufficient amounts are available, or with distilled water. A sample-wash interval of 2 min is sufficient for baseline resolution.

The fluorimeter is set to 340 nm excitation and 420 nm emission wavelengths for monochromator instruments. For filter instruments the corresponding filters are used. In this case the excitation filter should be a narrow band pass filter (± 10 nm), while on the emission side a cut-off filter (> 420 nm) may be used. The fluorimeter output is fed to a linear mV recorder, a data logger, or any other signal processing system. The peak heights are proportional

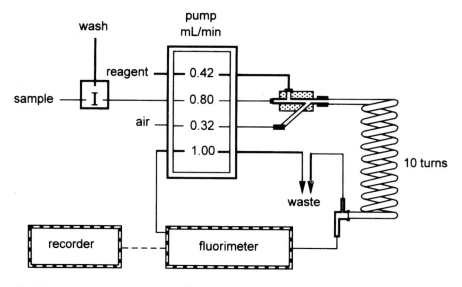

Fig. 26-2. Manifold for the determination of *o*-phthalaldehyde reactive substances.

to the concentrations of standard and sample. Blanks in this case result from contaminations of both wash water and reagent.

Mixed reagent: It should be made up daily as follows. To 1000 mL of sodium borate buffer add 5 mL of *o*-phthalaldehyde stock solution followed by 1 mL of mercaptoethanol. After allowing to stand for a few minutes the reagent is ready for use.

Calibrations may be performed as described under Section 26.3.6.

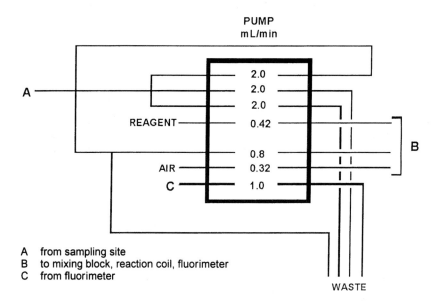

A   from sampling site
B   to mixing block, reaction coil, fluorimeter
C   from fluorimeter

Fig. 26-3. Manifold for continuous determination of *o*-phthalaldehyde reactive substances.

Measurements of discrete water samples suffer the disadvantage that the measured peak heights depend upon the baseline level of the wash water. At low natural concentrations the purity of the wash water thus becomes a considerable problem. A continuous technique has been developed, therefore, which avoids all possible sources of contamination during sampling and transfer (see Fig. 26-3). This modification again is calibrated with dilutions of the glycine stock standard that are pumped instead of the sample.

The technique, however, has some limitations. The inner diameter (i.d.) of the feed tube from the sampling site must be >1 mm, because tubes with smaller i.d. have increased flow resistances. Longer residence times of water in the feed tube should be avoided to prevent microbial growth and sample alteration. Thus, if the manifold of Fig. 26-3 is employed, the length of the feed tube should not exceed 40 m with an i.d. of 1 mm. The residence time will then be 5.23 min, which is well below the known microbial turn-over times of amino acids.

26.3.5 Flow injection analysis (FIA)

FIA is generally much faster than autoanalyser techniques and thus allows processing of about 100 samples per hour on average. However, it requires more sophisticated (and costly) equipment. An isocratic high-performance liquid chromatography (HPLC) system with the column removed may be used to advantage. Even a single piston pump can be employed, if care is taken to install a pulse damper such as a 10–20 m length of capillary tubing (0.1–0.15 mm i.d.) before the injector. The mixed reagent is delivered at a flow rate of 1 mL/min. Samples are injected manually using the full loop injection technique. A 20 μL loop is usually sufficient for ORS determination. A 2 m PTFE capillary of 0.3 mm i.d. ensures a reaction time of at least 2 min. The detector signal of the fluorimeter is fed to a mV recorder, a data logger or another data processing device.

Mixed reagent: It should be made up daily as follows. To 1000 mL of sodium borate buffer add 10 mL of *o*-phthalaldehyde stock solution followed by 2 mL of mercaptoethanol. After allowing to stand for a few minutes the reagent is ready for use.

26.3.6 Calibration

Dissolve 7.2 mg of glycine in 100 mL of 30 % (v/v) methanol-distilled water. This stock solution contains 1 mmol/L glycine. Exact dilutions of this stock standard with distilled water should be made up before use for a calibration range of 1–10 μmol/L glycine. The relative fluorescence signals are directly proportional to concentrations. The fluorescence response is linear from 0.01–100 μmol/L glycine. The technique does not show any salt dependence.

26.4 Dissolved combined amino acids (DCAA)

After suitable hydrolysis the techniques outlined above may be used for DCAA determination. As yields of vapour phase hydrolysis have been shown to be consistently higher than those of the classical 6 mol/L HCl/110 °C/24 h method (*Keil and Kirchman*, 1991); the first technique is described in the following (*Tsugita et al.*, 1987).

Samples (0.1 mL) are dried with a stream of nitrogen in 1 mL test tubes. These are inserted into wide-mouth 10–15 mL sealable ampoules to which 0.175 mL of a mixture of 7 mol/L HCl–10 % trifluoroacetic acid–0.1 % phenol is added. After flushing for about 5 min with either high-purity nitrogen or argon the ampoules are sealed under vacuum (> 600 mm Hg). Vapour phase hydrolysis is carried out at 156 °C for 23 min. After cooling, the ampoules are opened and the test tubes removed. Traces of acid which might be present are removed by nitrogen purging until the sample is completely dry again. If high concentrations of nitrate are present, amino acids might be oxidized. In this case ascorbic acid should be added to reduce the nitrate (*Robertson et al.*, 1987).

26.5   Dissolved carbohydrates

26.5.1   Compound class reactions

The relatively specific method outlined here combines a number of well–known reactions in carbohydrate chemistry; it is based on a procedure described by *Johnson and Sieburth* (1977) with minor modifications. Since formaldehyde is the only product actually detected, the responses of various sugars are fairly uniform. The method offers several advantages over earlier techniques especially as it avoids the use of concentrated mineral acids.

A number of steps are involved in the method and hence it is relatively time-consuming. On the other hand, a large number of samples may be processed at once, as only 1 mL of seawater is required. The technique is almost free of interferences and appears to detect monosaccharides only (MCHO). After a simple hydrolysis procedure the method may be employed to estimate concentrations of polysaccharides in seawater or in marine particulates. Since (in addition to glassware and chemicals) a photometer is the only equipment required, few problems are encountered when using the method in the field.

Briefly, the method comprises reduction of the monosaccharides to the respective sugar alcohols, periodate oxidation to formaldehyde and detection of this compound as a coloured complex with 3-methyl-2-benzothiazolinone hydrazone at 635 nm. After subtraction of a blank value and comparison with a standard curve the results are reported in glucose equivalents.

However, in addition to monosaccharides the technique will determine all compounds containing a 1,2-diol moiety such as alditols, amino sugars and uronic acids. Non-reducing disaccharides remain undetected by this procedure; reducing disaccharides show a somewhat lower response. Of the organic compounds tested so far only serine interferes at concentration levels higher than those expected in seawater.

26.5.2   Reagents

Sodium borohydride: Make sure that the crystalline sodium borohydride is stored in a desiccator; use fresh chemicals only. Immediately before use dissolve 70 mg of sodium borohydride in 5 mL of distilled water (5 °C).

0.36 mol/L HCl, 2 mol/L HCl: Dilute 10 mol/L HCl with distilled water to give the desired molarities.

0.025 mol/L Periodic acid: Dissolve 570 mg of periodic acid (a.g.) in 100 mL of distilled water. Store in a dark bottle at room temperature. Replace weekly.

0.25 mol/L Sodium arsenite: Dissolve 3.247 g of sodium arsenite (a.g.) in 100 mL of distilled water. The reagent is stable for at least 1 month when stored in a dark bottle at ambient temperature.

3-Methyl-2-benzothiazolinone hydrazone (MBTH): Dissolve 276 mg of MBTH hydrochloride in 10 mL of 0.1 mol/L HCl while heating slightly. Store in a dark bottle and replace weekly or when discoloured.

5 % Iron(III)chloride: Dissolve 5 g of iron(III)chloride hexahydrate (a.g.) in 100 mL of distilled water. Filter through a glass fibre filter and store at 5 °C.

Acetone. The acetone should be a.g.

26.5.3   Procedure

Determinations are best carried out in 5 mL test tubes with PTFE-lined screw caps. The test tubes should be cleaned with 8 mol/L aqueous sodium hydroxide and rinsed copiously with distilled water prior to use. Reagents are best added with automatic pipettes (*e.g.*, Eppendorf Multipette). Transfer 1 mL of standard solution or filtered seawater sample into a test tube. Add 0.05 mL of freshly prepared sodium borohydride solution to reduce the sugars to the corresponding alcohols. Seal tightly and incubate at ambient temperature for at least 6 h in the dark. It is advisable, however, to let the samples stand overnight. Add 0.05 mL of 0.36 mol/L HCl to destroy excess borohydride, allowing 10 min to de-gas in the dark. This will lead to complete destruction of excess borohydride and ensures that the pH for the following reaction is acidic enough. Add 0.1 mL of periodic acid solution to cleave the C–C bonds in the 1,2-diol groups. Allow to stand for 10 min in the dark at room temperature. Destroy excess periodic acid by adding 0.1 mL of sodium arsenite solution; wait at least 10 min before adding 0.2 mL of 2 mol/L HCl. The amber colour developed disappears rapidly. After addition of 0.2 mL of MBTH reagent the tubes are capped tightly and heated in a boiling water bath for 3 min. They are then cooled in a water bath to room temperature, and 0.2 mL of iron(III) chloride solution is added. Allow the colour to develop in the dark. The time required depends upon the ambient temperature and is roughly 30 min at 18 °C and 20 min at 25 °C. After colour development exactly 1 mL of acetone is added, mixed and the absorbance read as soon as possible at 635 nm against acetone + water, 1 + 1.9.

All samples should be analysed in duplicate. Blanks are treated in the same way except that the addition of periodic acid and sodium arsenite is omitted. Instead, a 1 + 1 mixture of both reagents is added, which has been allowed to react for at least 10 min. A correction can thus be made for aldehyde and serine contaminations in the reagents. The blanks are further treated as described above.

26.5.4   Calibration

Dissolve 18.0 mg of glucose (a.g.) in 100 mL of distilled water. Conserve by adding a few drops of 2 % $HgCl_2$ solution. This stock standard contains 1 mmol/L of glucose. For calibration, dilute with distilled water to give a range of 1–10 μmol/L of glucose immediately before analysis. After subtraction of the blank value, E^{635nm} in a 1 cm cuvette should read 0.035–0.038 for a 1 μmol/L glucose solution.

26.5.5   Analytical range

The amount of borohydride employed is sufficient to reduce monosaccharides up to a concentration of around 10 μmol glucose equivalents per mL. Samples containing higher amounts of monomeric sugars should be diluted to fall within this range. The concentration of periodic acid is sufficient for oxidation of around 8 μmol per mL of hexose. Opean-ocean samples are expected to contain 0.3–2.0 μmol/L glucose equivalents (*Burney et al.*, 1979), whereas coastal waters often show significantly higher levels (up to 80 μmol/L; *Liebezeit*, unpublished results).

26.5.6   Estimation of polysaccharide content (PCHO)

After hydrolysis the above technique may be employed to estimate polymeric carbohydrates both in the dissolved and particulate fractions. As for the combined amino acids, various hydrolysing agents have been suggested such as 2 mol/L *p*-toluene sulphonic acid, 2 mol/L HCl, 1 mol/L H_2SO_4 or 72 % H_2SO_4, followed by 3 % H_2SO_4, at elevated temperatures. However, these acids have been shown to cause some degree of destruction of labile monomers such as fructose. Also, in the presence of higher concentrations of amino acid containing compounds, browning reactions may occur in concentrated acids. Therefore, the method of *Burney and Sieburth* (1977) is recommended, since it has been tested successfully with a variety of marine polysaccharides. *Borch and Kirchman* (1997) also reported that 24 h hydrolysis with a dilute acid (0.85 mol/L H_2SO_4) gave higher results for natural waters than short-term treatment with more concentrated acids.

26.5.6.1   Procedure

Add 0.1 mL of 1 mol/L HCl to each mL of filtered sample in a Pyrex test tube with a PTFE-lined screw-cap and purge with high-purity nitrogen or argon. Hydrolyse for 20 h at 100 °C in a thermostated oven. Cool and neutralize by adding 0.1 mL of 1 mol/L NaOH solution for each mL of sample. Samples of marine particles resting on glass fibre filters may be hydrolysed with 0.2 mol/L HCl. Sufficient acid should be added to cover the filters; after hydrolysis at 100 °C for 20 h the supernatant is neutralized with an equal amount of 0.2 mol/L NaOH solution. (It may be advisable at this stage to centrifuge). The neutralized extract should be diluted to fall within the analytical range. The dilution factor depends upon the particle load and the amount of water filtered. Blank tests should be performed on unused filters.

After hydrolysis, the procedure is the same as described for the monomeric constituents; 1 mL portions are transferred from neutralized samples for further treatment. After blank correction and correction for the dilution arising from the additions of acid and alkali, the PCHO content is calculated for seawater samples by subtraction of the MCHO value from the total carbohydrate content obtained.

26.5.7   Mono- and polysaccharides after aqueous extraction of particulate matter

To estimate the content of easily extractable sugars from particulate matter, the particles collected from 250–2000 mL of seawater (depending on seston content) on pre-combusted (450 °C overnight) glass fibre filters are heated in 5 mL of distilled water at 80 °C for 4 h in a sealed tube. Periodic ultrasonification will accelerate the extraction. Other extraction techniques employing organic solvents (*e.g.*, toluene–methanol; *cf. Lee and Cronin*, 1982, for amino acids) are not recommended, because the organic solvents may contain contaminants that would interfere in the subsequent analysis.

The hot water procedure will extract free monosaccharides and soluble polysaccharides, such as starches and reserve cell material. Hydrolysis of more labile α-1,4 linked polysaccharides may occur during the extraction procedure at near-neutral pH and add to the pool of monosaccharides. Estimations of the monosaccharide contents of aqueous extracts follow the method earlier described after suitable dilution to attain the analytical range, which depends on the amount of extracted particulate matter. In regions of high productivity dilution up to 10-fold with distilled water may be necessary. Oligo- and polysaccharide contents of aqueous extracts may be determined after hydrolysis as outlined earlier.

Particulate matter may thus be characterized by three different sugar fractions: extractable monosaccharides, extractable polysaccharides (*i.e.*, starch-like substances) and total sugars after hydrolysis (including monosaccharides, starches and structural polysaccharides). An analogous approach may also be adopted to characterize three different amino acid fractions of particulate matter by employing the common analytical techniques of Section 26.3.

26.6   Chromatographic separation of amino acids

26.6.1   General remarks

High-performance liquid chromatography (HPLC) has become the method of choice for the determination of non-volatile organic compounds in a number of matrices including natural waters. Various techniques have been published for the HPLC determinations of amino acids (*Liebezeit*, 1985). Most of them deal with the separation of pre-column derivatives, either UV absorbing or fluorescent, although applications for underivatised compounds have also appeared in the literature (*das Neves and Morais*, 1997). Although a number of reagents have been tested for use with seawater samples (*Liebezeit*, 1985) only *o*-phthalaldehyde-mercaptoethanol has gained wider popularity. This presumably is due to the easy formation of the fluorescent isoindole derivatives (Fig. 26-1) making this reagent particularly suitable for use at sea (*Evens et al.*, 1982).

Regular reviews on HPLC methodology and applications including amino acid determination appear in the Fundamental Reviews (even years) and Application Reviews (odd years) series, respectively, of *Analytical Chemistry* in the first June issue of each year.

26.6.2   Equipment

HPLC systems suitable for routine separation of dissolved free amino acids should have the following characteristics:
- capability of working with a binary or ternary gradient system;
- on-line degassing to eliminate the need for continuous helium purging;
- cooled auto-injector capable of sample pre-treatment for unattended operation;
- column oven for separation at constant temperatures; and
- filter- or spectrofluorimeter.

26.6.3   HPLC columns

The most common stationary phase used in HPLC is chemically modified silica. Generally, octadecyl (C_{18}) phases are used in the determination of DFAA. Spherical 5 μm resins are commonly employed with standard column dimensions of 250 × 4 mm. Columns of smaller dimensions may be used as well, but the resin particle diameter should then be 3 μm. This configuration minimizes solvent consumption. A guard column should be used to protect the analytical column from small particles as present in unfiltered samples and polymeric material and thus extends the lifetime of the analytical column.

26.6.4   Detection

When excited at 340 nm, the isoindole derivatives of amine compounds fluoresce strongly at around 420 nm. The sensitivity is high, as only the products fluoresce and the mobile phase exhibits little or no native fluorescence at these wavelengths. Thus, with laser induced fluorescence, detection limits reach the lower femto- to attomolar range. Owing to the delicate nature of the optical equipment this technique is not suitable for shipboard use, but even with standard filter or spectrofluorimeters amino acid concentrations in the lower picomolar range can be determined (*Evens et al.*, 1982).

26.6.5   Analysis

Various eluent systems have been proposed for HPLC determination of DFAA. In most cases they are combinations of methanol with different buffers such as phosphate, citrate or acetate. *Mopper and Dawson* (1986) discussed problems associated with phosphate buffers which often cause clogging of check valves. In our experience acetate buffers present no such problems in routine analysis.

26.6.5.1   Mobile phases

Methanol: HPLC grade methanol should be employed.
0.05 mol/L Acetate buffer: Dissolve 4.102 g of sodium acetate (CH_3COONa) or 6.804 g of sodium acetate trihydrate ($CH_3COONa \cdot 3H_2O$) in 1 L of distilled water. This solution will

have a pH of 6.3. If required, the pH can be adjusted with either 8 mol/L NaOH solution or 50 % phosphoric acid.

Occasionally, addition of small amounts of modifiers such as tetrahydrofuran will improve critical separations.

Mobile phases should be free of particles. An inlet filter before the HPLC pump is a desirable precaution measure.

If no in-line degasser is available, mobile phases may be degassed by ultrasonication under aspirator vacuum (caution: non-spherical glass flasks may implode). Single-component mobile phases should not be used, as it has been found that even after the most careful degassing bubbles are formed upon mixing of methanol and aqueous buffers.

26.6.5.2   Gradient conditions

The actual gradient conditions depend on a number of factors including column age. The example given in Fig. 26-4 depicts a typical DFAA spectrum of a coastal water with high concentrations of serine, glycine and alanine. For peak assignments see Table 26-1. The gradient system used is given in Table 26-2.

Table 26-1. Amino acid structures and abbreviations.

Amino acid	R in Fig. 26-1	3-Letter code	1-Letter code
Alanine	$-CH_3$	ALA	A
Arginine	$-(CH_2)_3-NH-C(NH_2)_2^{+}$ [a]	ARG	R
Aspartic acid	$-CH_2-COO^-$ [a]	ASP	D
Asparagine	$-CH_2-CO(NH_2)$	ASN	N
Cysteine	$-CH_2-SH$	CYS	C
Glutamic acid	$-CH_2-CH_2-COO^-$ [a]	GLU	E
Glutamine	$-CH_2-CH_2-CONH_2$	GLN	Q
Glycine	$-H$	GLY	G
Histidine	$-CH_2-(C_3N_2H_4)^{+}$ [a]	HIS	H
Isoleucine	$-CH(CH_3)-CH_2-CH_3$	ILE	I
Leucine	$-CH_2-CH(CH_3)_2$	LEU	L
Lysine	$-(CH_2)_4-NH_3^{+}$ [a]	LYS	K
Methionine	$-CH_2-S-CH_2-CH_3$	MET	M
Phenylalanine	$-CH_2C_6H_5$	PHE	F
Serine	$-CH_2OH$	SER	S
Threonine	$-CH(OH)-CH_3$	THR	T
Tyrosine	$-CH_2-C_6H_4-p-OH$	TYR	Y
Valine	$-CH(CH_3)_2$	VAL	V

a At pH 6.0

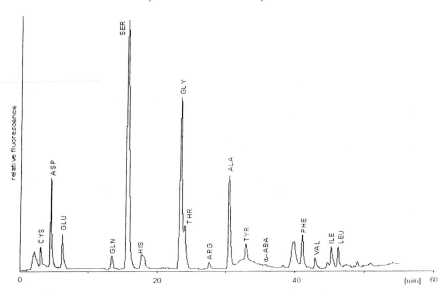

Fig. 26-4. Example of DFAA determination of coastal North Sea water. Running conditions as described in Table 26-2. For abbreviations see Table 26-1.

Table 26-2. Example of a gradient system for DFAA-isoindole determination.

Time (min)	% A (10 % sodium acetate, pH 6.30, 89 % methanol, 1 % tetrahydrofuran)	% B (100 % methanol)
0	100	0
10	90	10
50	10	90
55	10	90
60	100	0

This gradient system has been adapted for the analysis of coastal and interstitial waters where compounds derived from amino acid degradation such as β-alanine, taurine or amino butyric acids may occur in addition to the standard amino acids given in Table 26-1. For less complex samples such as, *e.g.*, hydrolysates the gradient run time may be abbreviated and a linear gradient employed. It should, however, be noted that under these conditions glycine and threonine are usually not separated.

26.6.6 Pre-column derivatisation

26.6.6.1 Reagents

o-Phthalaldehyde-mercaptoethanol: Dissolve 100 mg of *o*-phthalaldehyde in 0.5 mL of methanol (for fluorescence purposes) in a 1 mL vial with a PTFE-lined septum screw-cap. Add 0.05 mL of mercaptoethanol (see Section 26.6.3.2). The reagent is ready to use after standing for 10 min at ambient temperature. The reagent is stable for approximately 1 d; 0.05 mL of reagent contains enough *o*-phthalaldehyde to derivatise around 10 nmol of amino acids in total.

Borate buffer: Dissolve 610 mg of boric acid in 90 mL of distilled water. Adjust the pH to 13.5 with 8 mol/L NaOH solution, make up to 100 mL with distilled water and store in a polyethylene bottle.

26.6.6.2 Procedure

Add 0.1 mL of borate buffer and 0.01 mL of reagent to 1 mL of unfiltered seawater sample in a pre-cleaned conical reaction vial (capacity *ca.* 2 mL). Mix thoroughly and allow the mixture to react for exactly 10 min. *Liebezeit and Dawson* (1981) have shown that the ideal reaction pH for seawater is around 11.5. The reaction time of 10 min ensures more reproducible results and increases sensitivity.

Inject 0.1–0.2 mL of the reaction mixture corresponding to 0.09–0.18 mL of the original sample (rounded to the second decimal place). Factors to convert to pmol/L are then 1.11×10^4 to 5.55×10^3.

26.6.6.3 Use of internal standard

If it is not clear whether the reaction pH is correctly adjusted to around 11.5 by addition of the borate buffer, the use of an internal standard is recommended. This especially holds true if hydrolysates are to be analysed. Although small amounts may be present in particulate matter, non-protein amino acids such as α-amino butyric acid or norleucine should be employed, the former having the advantage of being well separated in the HPLC system under discussion (Fig. 26-4). The internal standard should be added before the reagent. The amount added depends upon the type of analysis to be carried out and may vary from 25 to 500 pmol per injection volume. Variations in the response of the internal standard compared with a calibration run allow correction for differences in reaction pH, time and temperature. However, if these parameters have been kept constant for both calibration and sample analyses, the response should be reproducible to within about 2 %.

26.6.7 Calibration

Commercially available calibration mixtures contain various combinations of individual amino acids, usually 2.5 μmol/mL. For most DFAA determinations a standard corresponding to a protein hydrolysate is suffcient. Amino acids such as glutamine or asparagine may be added, if necessary. This conserved stock standard should be diluted before use to give a final concentration of 1 nmol/mL. A few drops of 2 % HgCl$_2$ solution are added as preservative. Standards should be run with each batch of freshly prepared reagent or mobile phase. A reagent blank should be established by injecting organic-free seawater (or distilled water) instead of the calibration mixture. For determinations of dissolved free amino acids, amounts of 25–50 pmol normally are sufficient. For determinations of dissolved combined amino acids or particulate proteins amounts of 200–500 pmol can be used.

Peaks are identified by comparison of elution times in samples with those established for standards; in some cases comparison of elution patterns will further confirm peak identity.

26.7   Chromatographic separation of carbohydrates

Unlike dissolved amino acids, carbohydrates have not seen the development of sensitive techniques that are capable of routine chromatographic analysis at natural concentrations. Although there are procedures that in principle should allow determinations of dissolved saccharides at natural concentrations, these have so far not been tested in routine analysis. *Mopper et al.* (1992) used pulsed amperometric detection after separation of underivatised monosaccharides with an anion-exchange column. Thus, techniques which require sample pre-concentration/desalting and post-column derivatisation are still the ones that have been shown to be reliable for determinations of dissolved free monosaccharides in seawater (*Dawson and Liebezeit*, 1983).

Carbohydrate determination by HPLC has been treated by *Ben-Bassat and Grushka* (1990), while *Lee* (1990) reviewed applications of anion-exchange chromatography. Gas chromatographic techniques have been described particularly for the determination of particulate carbohydrates after hydrolysis (*e.g.*, *Leskovsek et al.*, 1994).

References to Chapter 26

Ben-Bassat, A.A., Grushka, E. (1991), *J. Liq. Chromatogr.*, 14, 1051.
Borch, N.H., Kirchmann, D.L. (1997), *Mar. Chem.*, 57, 85.
Burney, C.M., Sieburth, J.M. (1977), *Mar. Chem.*, 5, 15.
Burney, C.M., Johnson, K.M.. Lavoie, D.M., Sieburth, J.M. (1979), *Deep-Sea Res.* 26, 1247.
das Neves, H.J.C., Morais, Z.B. (1997), *HRC J. High Res. Chromatogr.*, 20, 115.
Dawson, R., Liebezeit, G. (1983), in: *Methods of Seawater Analysis: 2nd edn.*, Grasshoff, K., Ehrhardt, M., Kremling, K. (Eds.). Weinheim: Verlag Chemie, 1983; pp. 319–340.
Delmas, D., Frikha, M.G., Linley, E.A.S. (1990), *Mar. Chem.*, 29, 145.
Evens, R., Braven, J., Brown, L., Butler, I. (1982), *Chem. Ecol.*, 1, 99.
Johnson, K.M., Sieburth, J.M. (1977), *Mar. Chem.*, 5, 1.
Jørgensen, N.O.G., Lindroth, P., Mopper, K. (1981), *Oceanol. Acta*, 4, 465.
Josefsson, B., Lindroth, P., Östling, G. (1977), *Anal. Chim. Acta*, 89, 21.
Keil, R.G., Kirchman, D.L. (1991), *Mar. Chem.*, 33, 243.
Lee, C., Cronin, C. (1982), *J. Mar. Res.*, 40, 227.

Lee, Y.C. (1990), *Anal. Biochem.*, 189, 151.

Leskovsek, H., Perko, S., Zigon, D., Faganeli, J. (1994), *Analyst*, 119, 1125.

Liebezeit, G. (1985), *Oceanus*, 11, 503.

Liebezeit, G., Dawson, R. (1981), *J. High Res. Chrom. Chrom. Comm.*, 4, 354.

Mopper, K., Dawson, R. (1986), *Sci. Total Environ.*, 49, 115.

Mopper, K., Schultz, C.A., Chevolot, L., Germain, C., Revuelta, R., Dawson, R. (1992), *Environ. Sci. Technol.*, 26, 133.

Robertson, K.M., Williams, P.M., Bada, J.L. (1987), *Limnol. Oceanogr.*, 32, 996.

Roth, M. (1971), *Anal. Chem.*, 43, 880.

Tsugita, A., Uchida, T., Mewes, H.W., Atake, T. (1987), *J. Biochem.*, 102, 1593.

Tupas, L.M., Popp, B.N., Karl, D.M. (1994), *Mar. Chem.*, 45, 207.

Webb, K.L., Wood, L. (1966), in: *Automation in Analytical Chemistry*, Technicon Symposium 1966. New York: Mediad, Vol. 1; pp. 440–444.

27 Determination of photosynthetic pigments

P. Wallerstein and G. Liebezeit

27.1 Introduction

Marine phytoplankton and macrophytes use chlorophylls and carotenoids as primary and secondary light receptors in photosynthesis. These complex molecules serve to transfer light energy to the photoreaction centre, where chlorophylls are the main agents in converting this energy into a chemically oxidizing and reducing potential.

Chlorophylls are cyclic tetrapyrroles containing a central magnesium atom (Fig. 27-1). Numerous degradation products are known, beginning with phaeophytins which lack the central magnesium atom, and phaeophorbides, acidic compounds lacking the phytol side chain as well. A large number of further degradation products occur in marine sediments. With the exception of pyrophaeophorbide (loss of the carboxymethyl group at ring E) most of them have so far not been characterized in marine particulate material.

In addition to the ubiquitous chlorophyll a, chlorophyll b is found in terrestrial plants and green algae, while chlorophylls c1 and c2 are constituents of phaeophyceae and diatoms. Bacteriochlorophylls, besides showing structural variation in the tetrapyrrole system, are characterized by isoprenoid alcohol substituents other than phytol, such as farnesol or geraniol.

Carotenoids are isoprenoid polyene pigments (Fig. 27-2) with a high structural diversity especially in the marine environment (*Liaaen-Jensen*, 1976).

Recent reviews on the occurrence of these compounds in the marine environment have been provided by *Liaaen-Jensen* (1990; 1991). A full treatment of the subject of chapter 27 may be found in *Jeffrey et al.*, 1997. Spectrophotometry or fluorimetry are the classical methods to determine chlorophyll pigments. *Jacobsen* (1982) compared these techniques with

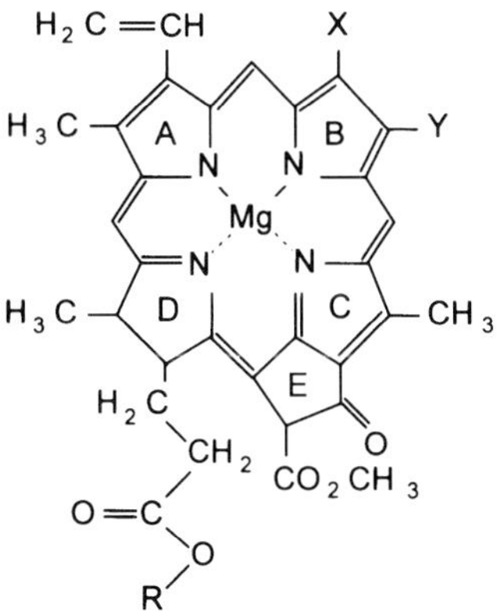

Fig. 27-1. Structure of chlorophylls.

Chlorophyll	X	R	Y
a	methyl	phytyl	ethyl
b	CHO	phytyl	ethyl
c1	methyl	H	vinyl

Note that chlorophylls c1 to c3 are fully aromatised ring D and an acrylic acid instead of the propionic acid side chain of chlorophylls a and b.

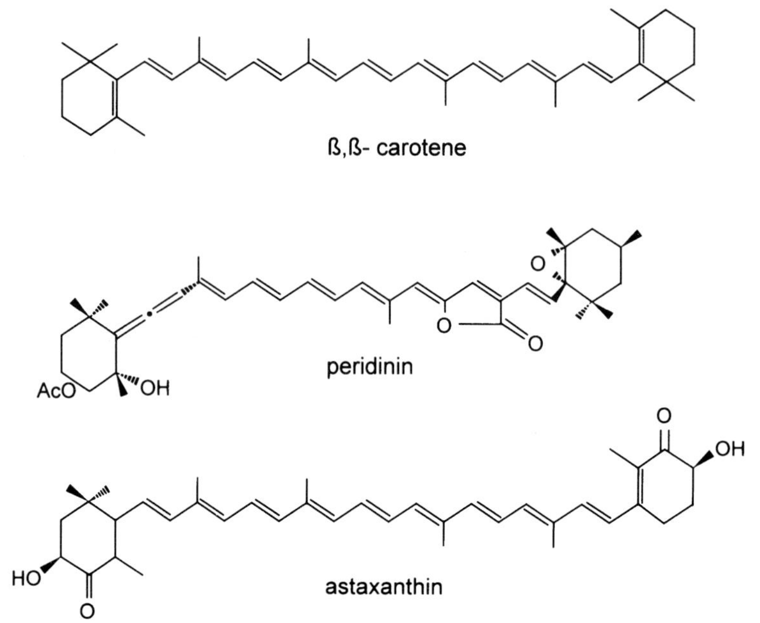

ß,ß- carotene

peridinin

AcO   ''OH

astaxanthin

HO
O

Fig. 27-2. Examples of marine carotenoids. While β,β-carotene occurs ubiquitously, peridinin is found in dinoflagellates and astaxanthin in eggs and shells of lobsters.

high-performance liquid chromatographic (HPLC) analysis and found significant differences that were attributed to the inability of bulk measurement techniques to determine the chlorophyll a content of natural samples accurately. This is due to the highly complex pigment matrices of marine samples. Similar absorbance maxima and only slightly lower molar absorptivity of, *e.g.*, chlorophyllide a and chlorophyll a or phaeophorbide a and phaeophytin a may account for the observed discrepancies between HPLC and total determination of chlorophyll a. Thus, HPLC has become the method of choice for the accurate quantitative determination not only of chlorophylls but also of carotenoids.

Pigment composition has been used to assess the contributions of various algal species to phytoplankton communities (*e.g., Barlow et al.*, 1993; *Letelier et al.*, 1993). The high structural variability of carotenoids proves particularly helpful in this regard. The diagnostic value of chlorophyll pigment composition is usually limited, but it may be used to determine the physiological state of the autotrophic assemblage. However, the presence of complex mixtures of steryl chlorines in sediment trap material and surface sediments has been inferred as an indicator of zooplankton grazing (*King and Repeta*, 1991). *Eckardt et al.* (1991), on the other hand, suggested phytoplankton senescence to be the major source for this class of compounds in which the phytol side chain has been replaced by sterols.

27.2 Sampling

Particulate material intended for pigment analysis can be obtained by classical water bottle techniques followed by batch filtration or by pumping and continuous filtration. Individual samples should be strained through 250 or 300 μm plankton gauze to remove larger zooplankton and then be filtered over pre-extracted glass fibre (*e.g.*, Whatman GF/F), silver or polycarbonate filters. To avoid pigment destruction, filtration should be carried out in subdued light whenever possible. Filters should be stored deep-frozen and analysed as soon as possible, although storage for up to five months does not appreciably alter pigment composition.

Variability between samples may be due to phytoplankton patchiness which may occur at scales of < 1 m (*Wangersky*, 1973). *Heileman and Mohammed* (1991) have described subsampling errors associated with the settling of particles in water samplers (*Gardner*, 1977). Artefacts may also arise from sample transport and storage (*Herve and Heinonen*, 1984). The recommendations given in Chapter 1 should be followed.

27.3 Extraction

A wide range of solvents used for extraction of pigments have been proposed (Table 27-1). It is the classical 90 % aqueous acetone (*Parsons and Strickland*, 1963) which is used most widely in total analysis with spectrophotometric or fluorimetric methods or prior to HPLC separation.

Table 27-1. Solvents used for the extraction of pigments from marine particles.

Conditions	Cell disintegration	Reference
90% Aqueous acetone	+	*Parsons and Strickland*, 1963
Acetone–methanol–water, 80 + 15 + 5	+	*Daley et al.*, 1973
Dimethylsulphoxide (DMSO)	–	*Hiscox and Israelstam*, 1979
DMSO–90 % acetone	–	*Shoaf and Lium*, 1976
N,N-dimethylformamide (DMF)	–	*Speziale et al.*, 1984
Dichloromethane–methanol, 9 + 1	+	*Bathmann und Liebezeit*, 1983

Various investigations have dealt with comparison of extraction efficiencies especially for freshwater phytoplankton. *Riemann and Ernst* (1982) found that for chlorophyceae and some cyanobacteria, which are notoriously difficult to extract, 100 % methanol or 96 % ethanol at extraction times >6 h work better than 90 % acetone. On the other hand, *Volk and Bishop* (1968) found that pigments from the cyanobacterium *Cyanidinium* sp. could not be extracted quantitatively with pure methanol. Similarly, *Speziale et al.* (1984) reported that dimethylformamide (DMF) and mixtures of dimethyl sulphoxide (DMSO) and acetone gave better results than 90 % aqueous acetone. Although the use of DMF or DMSO–acetone eliminates the need for grinding the cells, the use of the latter solvent system can cause problems in the final analysis due to phase separation and partial evaporation. Although chlorophyceae are not a common constituent of marine phytoplankton communities, cyanobacteria occasionally occur in bloom form, and these effects therefore have to be considered.

Wood (1985) suggested that chloroform might be a more effective extractant. However, chloroform has the disadvantage of forming hydrochloric acid upon storage, and hence its use will lead to the production of phaeophytins and phaeophorbides during the extraction procedure. While this may not seriously affect total analyses, it will surely be noted in HPLC analysis. Thus, if chloroform is to be employed as the extractant, frequent solvent clean-up becomes necessary.

However, no detailed investigation on the effects of different solvent systems has so far been carried out employing HPLC techniques. For most marine samples extraction with 90 % acetone appears to give satisfactory results.

27.4   Work-up artefacts

In methanol, allomerization of chlorophyll to the corresponding 10-hydroxy- and 10-methoxylactone compounds has been reported to occur instantaneously (*Schaber et al.*, 1984). This is the main reason why methanol should be avoided as the extraction agent. Lyophilisation of filtered material results in pigment losses (*Riaux-Gobin et al.*, 1987).

Storage of natural seston at low temperatures ($-20\,^{\circ}$C) for up to five months has no effect on seston samples; stored diatom culture material, however, showed considerable compositional changes after only one month. This has been attributed to ongoing chlorophyllase activity. *Jeffrey and Hallegreef* (1987) also reported on potential problems associated with enzymatic chlorophyll degradation. Thus, samples rich in phytoplankton should be extracted as soon as possible after collection; the extracts can then be stored deep-frozen for up to five months with only minor changes.

Extraction should be carried out in subdued light at low temperatures. If cell disintegration is deemed necessary, either ultrasonication or cell grinding (manual or automated) may be used. In our experience, 3 min of ultrasonication followed by low temperature centrifugation ($4\,^{\circ}$C) for 30 min at 6000 rpm is sufficient to extract well over 90 % of the pigments in particulate material and to remove particulate debris. The supernatant may be analysed by HPLC without any further clean-up. In fact, additional handling should be avoided as some carotenoids are extremely sensitive to oxidation .

So far there have been no reports on oxidation artefacts arising from the presence of dissolved oxygen in the extractants. This is, however, a possibility not to be dismissed.

27.5   HPLC analysis

A review on HPLC methods for pigment separations has been given by *Roy* (1987).

To avoid frustrations, the novice in HPLC should familiarize him or herself with the principles and basic trouble shooting techniques before starting actual analyses. For this purpose recommendable text books are available, *e.g.*, *Meyer* (1992).

27.5.1   Separation

In most published HPLC techniques reversed-phase columns have been employed, *i.e.,* silica esterified with a long-chain alcohol, usually octadecanol. *Hajibrabim et al.* (1978), among others, employed normal phase resins successfully for the separation of porphyrins, chlorines and carotenoids from ancient sediments. Stationary phases with 5 or 3 μm grain size and standard columns of 250 × 4.6 mm have been used successfully. Separation with capillary columns greatly reduces solvent consumption without affecting separation efficiency.

Methanol–water mixtures are the usual eluents, the actual composition being dependent on the specific separation problem. In some cases acetonitrile, ethyl acetate or acetone are added as third solvents in ternary mixtures. Modifiers such as tetrahydrofuran or *tert*-butyl-ammonium salts (*Mantoura and Llewellyn*, 1983) may also be added to solve particular separation problems. A detailed comparison of various columns and eluents has been given by *Schmid and Stich* (1995).

The complex pigment mixtures encountered in marine samples have so far not been resolved completely. Problems associated with co-elution of zeaxanthin and lutein may be overcome by using the method of *Wright et al.* (1991); this technique, however, does not allow the separation of monovinyl from divinyl chlorophyll a (*Goericke and Repeta*, 1993).

Table 27-2. Examples of gradient programmes for HPLC pigment separation. Percentages are v : v.

Time (min)	75 % Methanol 25 % Sodium acetate (0.05 mol/L, pH 7)	80 % Methanol 20 % Acetone	60 % Methanol 40 % Acetone
0	55	45	0
30	0	100	0
50	0	0	100
55	55	45	0
	Methanol	Acetone	Ammonium acetate
			1 mol/L [a]
0	80	0	20
0.5	80	0	20
15.5	60	30	10
20.5	60	30	10
30.5	30	60	10
50.5	5	90	5
75.5	5	90	5
78.5	30	60	10
83.5	60	30	10
85.5	80	0	20
95.5	80	0	20

[a] Ammonium acetate is used to improve the separation of acidic compounds such as phaeophorbides (*Lim and Peters*, 1984).

Thus, no generally applicable solvent composition or gradient programme can be given. The gradients given in Table 27-2 may, however, be used as guidelines for the development of suitable elution programmes. Both these solvent programmes have been employed not only in the analysis of marine particulate material, but also for more complex pigment mixtures derived from sediments or peat moss (*Wallerstein*, 1996; *Möhring*, 1997). An example for the HPLC analysis of pigments in coastal waters is given in Fig. 27-3.

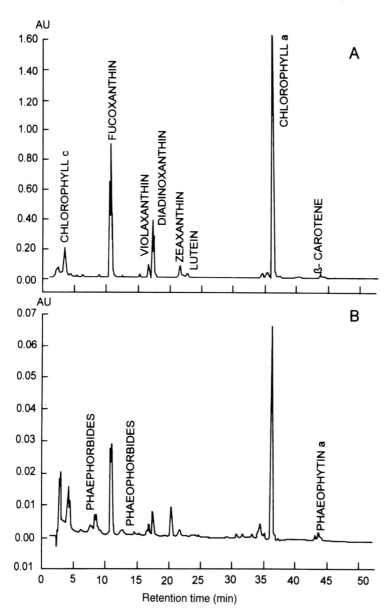

Fig. 27-3. HPLC separation of a 90 % aqueous acetone extract of a) *Cyclotella sp.* and b) Wadden Sea seston.

It should be noted that it is often difficult to reproduce published chromatograms. This may be due to reasons outlined in the following. Effects of injection conditions on HPLC separation efficiencies have been described by *Zapata and Garrido* (1991).

27.5.2 Matrix effects

Depending on the type of sample and extraction agent a number of other lipophilic compounds will be co-extracted. As sample clean-up usually is avoided due to the sensitive nature of the pigments to be determined, interaction between these other compounds and those of interest may lead to retention time shifts. This effect may be offset to some extent by using two or three internal standards, preferably one acidic standard eluting early and one non-polar compound eluting late. Care should be taken to resolve these compounds from the pigments of interest. For seston samples porphyrin derivatives such as octaethyl-porphyrin or porphyrin acids might be used to advantage.

27.5.3 Temperature effects

If HPLC analyses are carried out at ambient temperatures, even slight temperature variations will lead to retention time shifts making automated peak detection difficult if not impossible. The use of column ovens therefore is highly recommended.

27.5.4 Column ageing

Column ageing usually leads to increased retention times and reduced separation efficiency. It is thus advisable only to use columns that have been sufficiently conditioned.

If no guard column is used, lipophilic polymer material that has been co-extracted will be trapped at the front end of the column. This will change the retention characteristics of the column. Although previous conditions may be restored by replacing the top millimetres of the resin bed at regular intervals, usually every 25–30 analyses, this is undesirable as it involves frequent opening of the column with the associated risk of damaging capillary connections and threads. The use of a guard column is therefore recommended. There is some controversy as to whether the resin in the guard column should be of the same grain size or larger than that in the analytical column. If the same grain size is used, the guard column will contribute to the separation efficiency and hence any deterioration of the guard column resin will influence the retention behaviour of the pigments. We therefore advocate the use of resins coarser than that of the analytical column, *e.g.*, 10 or even 20 μm if a 5 μm resin is used for the actual separation.

27.5.5 Detection

The eluted compounds can be detected either by photometric or fluorimetric detectors. In the former case, chlorophylls and carotenoids can be distinguished by a band in the region > 600 nm wavelength, where carotenoids do not absorb. However, the a band absorption coefficient is lower than that of the major band, the Soret band, which means a loss of sensitivity. If both compound classes have to be determined simultaneously, the 400–450 nm range should be used. As the Soret band of the primary chlorophyll degradation products has its maximum at lower wavelengths than the parent compounds, a good compromise is to use broad band detection at 420 nm. Most carotenoids also have their absorption maxima in this region.

A more recent development in photometric detection is the photodiode array (PDA) detector which allows on-line recording of absorption spectra (*e.g.*, *Sims*, 1993; *Liang et al.*, 1993). As only a few standards are available, the identification of unknown pigments thus becomes much easier. It should, however, be noted that slight shifts in the absorption maxima may occur, as published spectra are usually obtained in pure solvents. *Mantoura and Llewellyn* (1983) and *Latasa et al.* (1996) have provided extensive lists of molar absorptivities for a wide range of pigments.

Fluorescence techniques so far have been used only for the specific detection of chlorophylls (*Liebezeit*, 1980), although fluorescence properties of carotenoids have been described (*e.g.*, *Gruszecki et al.*, 1990).

27.5.6   Standardization

Chlorophylls a and b are commercially available. It should be noted that chlorophyll a from spinach may contain substantial amounts of chlorophyll b. Hence preparations from phytoplankton algae should be preferred. The standard is dissolved in the extraction solvent and stored deep-frozen and protected from light. The stock standard concentration should be 1 mg per 100 mL.

Phaeophytin a can be prepared easily from the parent compound by acidification with a drop of 0.1 mol/L HCl per 100 mL of standard solution. Phaeophorbide a may be prepared following *Daood et al.* (1989), although *Möhring* (1997) recently reported on the formation of artefacts when this technique is used.

The preparation of chlorophylls c1 and c2 has been described by *Jeffrey* (1972). However, according to *Latasa et al.* (1996) it was not possible to produce stable standards for these two pigments.

Degradation products other than phaeophytin a may be isolated as described by, *e.g.*, *Fuhrhop and Smith* (1975). It is, however, easier to use known molar absorptivites for quantification purposes. Commercially available carotenoids include α- and β-carotene, lutein, peridinin and zeaxanthin.

27.5.7   HPLC systems

HPLC systems suitable for routine monitoring of algal pigments should have the following characteristics:
– capability of working with a ternary gradient system;
– on-line degassing to eliminate the need for continuous helium purging;
– cooled auto-injector for unattended operation;
– column oven for constant separation temperatures and
– photodiode array detector 400–800 nm, and fluorescence detector.

Most modern instrumentation will meet these requirements. During the last 20 years or so the hardware of most manufacturers has become comparable in reliability, performance and durability. The ease with which chromatograms (and spectra in PDA detection) can be evaluated, therefore, should be the major criterion for selecting a new HPLC system. The ease of software handling and its capabilities appear to be the most important issues of concern.

References to Chapter 27

Barlow, R.G., Mantoura, R.F.C., Gough, M.A., Fileman, T.W. (1993), *Deep-Sea Res. Pt. II-Top. St. Oceanogr.*, 40, 459.

Bathmann, U., Liebezeit, G. (1986), *P.S.Z.N. J. Mar. Ecol.*, 7, 59.

Daley, R.J., Gray, C.B.J., Brown, S.R. (1973), *J. Fish. Res. Board Can.*, 30, 345.

Daood, H.G., Czintokai, B., Hoschke, A., Biacs, P. (1989), *J. Chromatogr.*, 4723, 296.

Eckardt, C.B., Keely, B.J., Maxwell, J.R. (1991), *J. Chromatogr.*, 557, 271.

Fuhrhop, J.-H., Smith, K.M. (1975), in: *Porphyrins and Metalloporphyrins*: Smith, K.M. (Ed.). Amsterdam: Elsevier, 1975; pp. 757-859.

Gardner, W.D. (1977), *Limnol. Oceanogr.*, 22, 764.

Goericke, R., Repeta, D. (1993), *Mar. Ecol. Prog. Ser.*, 101, 307.

Gruszecki, W.I., Zelent, B., Leblanc, R.M. (1990), *Chem. Phys. Lett.*, 171, 563.

Hajibrahim, S.K., Tibbitts, P.J.C., Watts, C.D., Maxwell, J.R., Eglinton, G. (1978), *Anal. Chem.*, 50, 549.

Heileman, L.I., Mohammed, A. (1991), *Mar. Chem.*, 33, 353.

Herve, S., Heinonen, P. (1984), *Ann. Bot. Fenn.*, 21, 17.

Hiscox, J.D., Israelstam, G.F. (1979), *Can. J. Bot.*, 57, 1332.

Jacobsen, T.R. (1982), *Arch. Hydrobiol. Beih. Ergebn. Limnol.*, 16, 35.

Jeffrey, S.W. (1972), *Biochim. Biophys. Acta*, 279, 15.

Jeffrey, S.W., Hallegraeff, G.M. (1987), *Mar. Ecol. Prog. Ser.*, 35, 293.

Jeffrey, S.W., Mantoura, R.C.F., Wright, S.W. (1997) *Phytoplankton Pigments in Oceanography*. Paris: UNESCO.

King, L.L., Repeta, D.J. (1991), *Geochim. Cosmochim. Acta*, 55, 2067.

Latasa, M., Bidigare, R.R., Ondrusek, M.E., Kennicutt, M.C. (1996), *Mar. Chem.*, 51, 315.

Letelier, R.M., Bidigare, R.R., Hebel, D.V., Ondrusek, M., Winn, C.D., Karl, D.M. (1993), *Limnol. Oceanogr.*, 38, 1420.

Liaaen-Jensen, S. (1976), in: *Marine Natural Products Chemistry*: Faulkner, D.J., Fenical, W.H. (Eds.). New York : Plenum Press, 1976; pp. 239–259.

Liaaen-Jensen, S. (1990), *New. J. Chem.*, 14, 747A.

Liaaen-Jensen, S. (1991), *Pure Appl. Chem.*, 63, 1.

Liang, Y.Z., Brereton, R.G., Kvalheim, O.M., Rahmani, A. (1993), *Analyst*, 118, 779.

Liebezeit, G. (1980), *J. High Res. Chromatogr. Chromatogr. Commun.*, 3, 531.

Lim, C.K., Peters, T.F. (1984), *J. Chromatogr.*, 316, 397.

Mantoura, R.F.C., Llewellyn, C.A. (1983), *Anal. Chim. Acta*, 151, 297.

Meyer, V. (1992), *Praxis der Hochleistungsflüssigkeitschromatographie*. Frankfurt: Verlag Saller und Sauerländer.

Möhring, T. (1997), Diplomarbeit Universität Oldenburg.

Parsons, T.R., Strickland, D.H. (1963), *J. Mar. Res.*, 21, 155 .

Riaux-Gobin, C., Llewellyn, C.A., Klein, B. (1987), *Mar. Ecol. Prog. Ser.*, 40, 275.

Riemann, B., Ernst, D. (1982), *Freshwater Biol.*, 12, 217.

Roy, S. (1987), *J. Chromatogr.*, 391, 19.

Schaber, P.M., Hunt, J.E., Fries, R., Katz, J.J. (1984), *J. Chromatogr.*, 316, 25.

Schmid, H., Stich, H.B. (1995), *J. Appl. Phycol.*, 7, 487.

Shoaf, W.T., Lium, B.W. (1976), *Limnol. Oceanogr.*, 21, 926.

Sims, A. (1993), *Am. Lab.*, 25, 20.

Speziale, B.J., Schreiner, S.P., Giammatteo, P.A., Schindler, J.E. (1984), *Can. J. Fish. Aquat. Sci.*, 41, 1519.

Volk, S.L., Bishop, N.I. (1968), *Photochem. Photobiol.*, 8, 213.

Wallerstein, P. (1996), Diplomarbeit, Universität Oldenburg.

Wangersky, P.J. (1973), *Limnol. Oceanogr.*, 19, 980.

Wood, L.W. (1985), *Can. J. Fish. Aquat. Sci.*, 42, 38.

Wright, S.W., Mantoura, R.F.C., Llewellyn, C.A., Bjornland, T., Repeta, D., Welschmayer, N. (1991), *Mar. Ecol. Prog. Ser.*, 77, 183.

Zapata, M., Garrido, J.L. (1991), *Chromatographia*, 31, 589.

Appendix

Tables

1. Artificial seawater constituents after *Kester et al.* (1967).
2. Simplified artificial seawater after *Dickson* (1990) and *DOE* (1994).
3. Solubility of oxygen in water as a function of temperature and salinity (*UNESCO*, 1973).
4. Density of seawater (after *Millero and Poisson*, 1981).
5. Solubility of carbon dioxide in water (K_0) as a function of temperature and salinity (*Weiss*, 1974).
6. First dissociation constant of carbonic acid (K_1) in seawater as a function of temperature and salinity (after *Roy et al.*, 1993, 1996; *DOE*, 1994).
7. Second dissociation constant of carbonic acid (K_2) in seawater as a function of temperature and salinity (after *Roy et al.*, 1993, 1996; *DOE*, 1994).

General remarks to the tables

Tables 1 and 2

In many chemical and biological investigations seawater of a known composition is needed. Traditionally, the composition is restricted to the major constituents of seawater which normally represent more than 99.9 % of the total dissolved solids (see Chapter 11, Table 11-1). Although the concept of 'constant composition of seawater' is considered as being not strictly accurate (see Chapter 11) it is a useful device especially in simplifying the study of physicochemical properties of seawater.

The constituents in Table 1 was originally given by *Lyman and Fleming* (1940) and updated by *Kester* et al. (1967). The anhydrous salts must be dried and weighed very precisely and the chlorides containing water of hydration have to be standardized volumetrically.

The preparation can be made easier by direct weighing of analytical-reagent grade (a.g.) salts as follows:

1. Dissolve 23.9 g NaCl, 4.0 g Na_2SO_4, 0.7 g KCl, 0.2 g $NaHCO_3$, 0.1 g KBr, 30 mg H_3BO_3 and 3 mg NaF in 500 mL of distilled water.
2. The following chlorides are dissolved in 455 mL of distilled water: 10.8 g $MgCl_2 \cdot 6H_2O$, 1.5 g $CaCl \cdot 2H_2O$ and 25 mg $SrCl_2 \cdot 6H_2O$. (All reagents should be of a.g. standard).

While stirring add the chloride solution to the first solution and determine the exact salinity by measuring the conductivity (see Chapter 3).

In Table 2A a simplified artifical seawater recipe is offered (*Dickson*, 1990). Here bromide, fluoride, and total alkalinity are replaced by chloride, and strontium by calcium. The resulting species composition is given in Table 2B (*DOE*, 1994).

The various salts used (Table 2A) should be a.g. and must be purified further by recrystallization from water (twice for NaCl). The NaCl, Na_2SO_4 and KCl are dried in an oven at 110 °C before weighing. For $MgCl_2$ and $CaCl_2$ stock solutions of $\approx 1 \, mol \, kg^{-1}$ and $\approx 0.7 \, mol \, kg^{-1}$, respectively, are prepared.

These solutions should be analysed either by titration with a calibrated $AgNO_3$ solution (*e.g.*, using K_2CrO_4 as indicator; see also Chapter 11, 'Chlorinity' Section) or by gravimetric precipitation as AgCl.

The description of the preparation for artificial seawaters that are used by biologists in culture experiments are numerous and far beyond the scope of the 'Appendix'. These synthetic seawaters are normally enriched with fertilizers that facilitate the cell growth of organisms. The investigators, however, should be aware of the fact that analytical-reagent grade chemicals always contain traces of impurities (*e.g.*, organic compounds, trace metals or major nutrients) which, in an artificial seawater, can easiliy exceed the natural concentrations or nominal concentrations in marine culture media. Organic compounds can be reduced to acceptable levels by filtration through activated charcoal. Impurities of trace metals are removed with ion-exchange columns or by methods outlined in Chapter 12. It should, however, be mentioned that trace metal impurities introduced by 'dirty' equipment (flasks, tubes), inadequate sample handling or treatment very often exceed the levels introduced by stock solutions. Therefore, 'clean' equipment must be used as described in detail in Chapter 12.

Table 1. Artificial seawater constituents for a salinity of 35 after *Kester et al.* (1967).

Salt	g/(kg-soln)	Salt	g/(kg-soln)
NaCl	23.939	KBr	0.098
MgCl$_2$	5.079	H$_3$BO$_3$	0.027
Na$_2$SO$_4$	3.994	SrCl$_2$	0.024
CaCl$_2$	1.123	NaF	0.003
KCl	0.667	NaHCO$_3$	0.196

Table 2. Simplified artificial seawater of salinity 35.

A) Composition (after *Dickson*, 1990)

Salt	Molality (mol/(kg-H$_2$O)[a]
NaCl	0.42764
Na$_2$SO$_4$	0.02927
KCl	0.01058
MgCl$_2$	0.05474
CaCl$_2$	0.01075

B) Species concentrations (after *DOE*, 1994).[b]

Species	mol/(kg-soln)	g/(kg-soln)	mol/(kg-H$_2$O)	g/(kg-H$_2$O)
Na$^+$	0.46911	10.7848	0.48616	11.1768
Mg^{2+}	0.05283	1.2840	0.05475	1.3307
Ca^{2+}	0.01036	0.4152	0.01074	0.4304
K$^+$	0.01021	0.3992	0.01058	0.4137
Cl$^-$	0.54922	19.4715	0.56918	20.1791
SO$_4^{2-}$	0.02824	2.7128	0.02927	2.8117
Sum of column	1.11997	35.0675	1.16068	36.3424
Ionic strength	0.69713		0.72248	

[a] The molality of the components with salinity (other than 35) can be calculated from $m_s = m_{35} \times \frac{S}{35}$, and the ionic strength (I_S) from $I_s = \frac{0.019919\,S}{1-0.001002\,S}$

[b] The composition here is very slightly different from that used by *Dickson* (1990); see Table 2A.

Table 3 The solubility of oxygen in water (at equilibrium with air of 100 % humidity and 20.95 % oxygen at 1013.25 hPa) as a function of temperature and salinity is given by the expression (*UNESCO*, 1973):

$$\ln C(\mathrm{mL/L}) = -173.4292 + 249.6339 \cdot \left(\frac{100}{T}\right) + 143.3483 \cdot \ln\left(\frac{T}{100}\right) - 21.8492 \cdot \left(\frac{T}{100}\right)$$

$$+ S\left(-0.033096 + 0.014259 \cdot \left(\frac{T}{100}\right) - 0.0017000 \cdot \left(\frac{T}{100}\right)^2\right)$$

where

C	=	oxygen concentration in mL/L
T	=	absolute temperature in K
S	=	salinity

Check value:

At $S = 35$ and $t = 25\,°C$, $C = 4.7266$ mL/L

The conversion factor from oxygen concentration in units of mL/L to units of μmol/L is 44.615. Values in Table 3 are reported in units of μmol/L.

Table 4 The density of water in the temperature range 0–40 °C and the salinity range 0–42 is given by the expression (*Millero and Poisson*, 1981):

$$\rho_{\mathrm{SW}} = \rho_0 + A \cdot S + B \cdot S^{1.5} + C \cdot S^2$$

where

$$\rho_0 = 999.842594 + 6.793952 \cdot 10^{-2} \cdot t - 9.095290 \cdot 10^{-3} \cdot t^2 + 1.001685 \cdot 10^{-4} \cdot t^3$$
$$- 1.120083 \cdot 10^{-6} \cdot t^4 + 6.536336 \cdot 10^{-9} \cdot t^5$$
$$A = 8.24493 \cdot 10^{-1} - 4.0899 \cdot 10^{-3} \cdot t + 7.6438 \cdot 10^{-5} \cdot t^2 - 8.2467 \cdot 10^{-7} \cdot t^3$$
$$+ 5.3875 \cdot 10^{-9} \cdot t^4$$
$$B = -5.72466 \cdot 10^{-3} + 1.0227 \cdot 10^{-4} \cdot t - 1.6546 \cdot 10^{-6} \cdot t^2$$

where

ρ_{SW}	=	density of seawater in kg/m^3
ρ_0	=	density of pure water in kg/m^3 (*Bigg*, 1967)
S	=	salinity
t	=	temperature in °C

Check value:

At $S = 35$ and $t = 25\,°C$, $\rho_{\mathrm{SW}} = 1023.343$ kg/m^3

Values in Table 4 are reported as $\sigma = \rho - 1000$ (in kg/m^3).

Table 5 The solubility of carbon dioxide in water $[K_0 = (CO_2^*)/fCO_2]$, in mmol/(kg-soln) · atm of pure CO_2(gas) as a function of temperature and salinity is given by the expression (*Weiss*, 1974):

$$\ln\left(\frac{K_0}{k^0}\right) = 93.4517 \cdot \left(\frac{100}{T}\right) - 60.2409 + 23.3585 \cdot \ln\left(\frac{T}{100}\right)$$
$$+ S\left(0.023517 - 0.023656 \cdot \left(\frac{T}{100}\right) + 0.0047036 \cdot \left(\frac{T}{100}\right)^2\right)$$

where

k^0 = 1 mol/(kg-soln)
T = absolute temperature in K
S = salinity

Check value:

At $S = 35$ and $t = 25$ °C, $\ln\left(\frac{K_0}{k^0}\right) = -3.5617$

Values in Table 5 are reported as K_0 (in mmol/(kg-soln) · atm).

Table 6 The first dissociation constant of carbonic acid $[CO_2(aq) + H_2CO_3]$ in seawater as a function of temperature and salinity is given by the expression (after *Roy et al.*, 1993, 1996; modified after *DOE*, 1994):

$$\ln\left(\frac{K_1}{k^0}\right) = \frac{-2307.1266}{T} + 2.83655 - 1.5529413 \cdot \ln T + \left(\frac{-4.0484}{T} - 0.20760841\right) \cdot S^{0.5}$$
$$+ 0.08468345 \cdot S - 0.00654208 \cdot S^{1.5} + \ln(1 - 0.001005 \cdot S)$$

where

k^0 = 1 mol/(kg-soln)
T = absolute temperature in K
S = salinity

Check value:

At $S = 35$ and $t = 25$ °C, $\ln\left(\frac{K_1}{k^0}\right) = -13.4847$

Table 7 The second dissociation constant of carbonic acid $[CO_2(aq) + H_2CO_3]$ in seawater as a function of temperature and salinity is given by the expression (after *Roy et al.*, 1993, 1996; modified after *DOE*, 1994):

$$\ln\left(\frac{K_2}{k^0}\right) = \frac{-3351.6106}{T} - 9.226508 - 0.2005743 \cdot \ln T + \left(\frac{-23.9722}{T} - 0.106901773\right) \cdot S^{0.5}$$
$$+ 0.1130822 \cdot S - 0.00846934 \cdot S^{1.5} + \ln(1 - 0.001005 \cdot S)$$

where

k^0 = 1 mol/(kg-soln)
T = absolute temperature in K
S = salinity

Check value:

At $S = 35$ and $t = 25$ °C, $\ln\left(\dfrac{K_2}{k^0}\right) = -20.5504$

References of Appendix

Bigg, P.H. (1967), *Brit. J. Appl. Phys.*, 18, 521
Dickson, A.G. (1990), *Deep-Sea Res.*, 37, 755.
DOE (1994), *Handbook of Methods for the Analysis of the Various Parameters of the Carbon Dioxide System in Sea Water;* version 2, Dickson, A.G., Goyet, C. (Eds.). ORNL/CDJAC-74.
Kester, D.R., Duedall, I.W., Connore, D.N., Pytkowicz R.M. (1967), *Limnol. Oceanogr.*, 12, 176.
Lyman, J., Fleming, R.H. (1940), *J. Marine Res.*, 3, 134.
Millero, F.J., Poisson, A. (1981), *Deep-Sea Res.*, 28, 625.
Roy, R.N., Roy, L.N., Vogel, K.M., Porter-Moore, C., Pearson, T., Good, C.E., Millero, F.J., Campbell, D.M. (1993), *Mar. Chem.*, 44, 249.
Roy, R.N., Roy, L.N., Vogel, K.M., Porter-Moore, C., Pearson, T., Good, C.E., Millero, F.J., Campbell, D.M. (1996), *Mar. Chem.*, 52, 183.
UNESCO (1973), *Int. Oceanogr. Tabl.*, Vol. 2.
Weiss, R.F. (1974), *Mar. Chem.*, 2, 203

Table 3. Solubility of oxygen in water (in μmol/L, in equilibrium with air at 100 % humidity and 20.95 % oxygen at 1013.25 hPa) as a function of temperature and salinity (after *UNESCO*, 1973).

Temp. (°C)	Salinity																				
	0	2	4	6	8	10	12	14	16	18	20	22	24	26	28	30	32	34	36	38	40
0	456	450	444	438	432	426	420	414	409	403	398	392	387	382	377	371	366	361	356	352	347
1	443	437	431	426	420	414	409	403	398	392	387	382	377	372	367	362	357	352	347	343	338
2	431	425	420	414	409	403	398	392	387	382	377	372	367	362	357	352	348	343	338	334	329
3	420	414	409	403	398	393	387	382	377	372	367	362	358	353	348	344	339	334	330	326	321
4	409	403	398	393	388	383	378	373	368	363	358	353	349	344	339	335	331	326	322	318	314
5	398	393	388	383	378	373	368	363	358	354	349	345	340	336	331	327	323	318	314	310	306
6	388	383	378	373	368	364	359	354	350	345	341	336	332	328	323	319	315	311	307	303	299
7	379	374	369	364	359	355	350	346	341	337	333	328	324	320	316	312	308	304	300	296	292
8	369	365	360	355	351	346	342	338	333	329	325	321	316	312	308	304	301	297	293	289	285
9	361	356	351	347	343	338	334	330	325	321	317	313	309	305	301	298	294	290	286	283	279
10	352	348	343	339	335	330	326	322	318	314	310	306	302	298	295	291	287	284	280	277	273
11	344	340	335	331	327	323	319	315	311	307	303	299	296	292	288	285	281	278	274	271	267
12	336	332	328	324	320	316	312	308	304	300	297	293	289	286	282	279	275	272	268	265	262
13	329	325	321	317	313	309	305	301	298	294	290	287	283	280	276	273	269	266	263	259	256
14	321	317	314	310	306	302	298	295	291	288	284	281	277	274	270	267	264	261	257	254	251
15	314	311	307	303	299	296	292	289	285	282	278	275	271	268	265	262	258	255	252	249	246
16	308	304	300	297	293	290	286	283	279	276	273	269	266	263	260	256	253	250	247	244	241
17	301	298	294	291	287	284	280	277	274	270	267	264	261	258	254	251	248	245	242	240	237
18	295	292	288	285	281	278	275	271	268	265	262	259	256	253	250	247	244	241	238	235	232
19	289	286	282	279	276	272	269	266	263	260	257	254	251	248	245	242	239	236	233	231	228
20	283	280	277	274	270	267	264	261	258	255	252	249	246	243	240	237	235	232	229	226	224
21	278	275	271	268	265	262	259	256	253	250	247	244	241	239	236	233	230	228	225	222	220
22	272	269	266	263	260	257	254	251	248	245	243	240	237	234	231	229	226	224	221	218	216
23	267	264	261	258	255	252	249	246	244	241	238	235	233	230	227	225	222	220	217	215	212
24	262	259	256	253	250	248	245	242	239	237	234	231	229	226	223	221	218	216	213	211	208
25	257	254	252	249	246	243	240	238	235	232	230	227	225	222	219	217	215	212	210	207	205
26	253	250	247	244	242	239	236	233	231	228	226	223	221	218	216	213	211	209	206	204	202
27	248	245	243	240	237	235	232	229	227	224	222	219	217	214	212	210	207	205	203	200	198
28	244	241	238	236	233	231	228	225	223	220	218	216	213	211	209	206	204	202	199	197	195
29	240	237	234	232	229	227	224	222	219	217	214	212	210	207	205	203	201	198	196	194	192
30	235	233	230	228	225	223	220	218	216	213	211	209	206	204	202	200	197	195	193	191	189
31	231	229	226	224	221	219	217	214	212	210	207	205	203	201	199	196	194	192	190	188	186
32	227	225	223	220	218	215	213	211	209	206	204	202	200	198	195	193	191	189	187	185	183
33	224	221	219	217	214	212	210	208	205	203	201	199	197	195	192	190	188	186	184	182	180
34	220	218	215	213	211	209	206	204	202	200	198	196	194	192	190	188	186	184	182	180	178
35	217	214	212	210	208	205	203	201	199	197	195	193	191	189	187	185	183	181	179	177	175
36	213	211	209	206	204	202	200	197	196	193	192	190	189	186	184	182	180	178	176	174	173
37	210	208	205	203	201	199	197	194	193	191	189	187	185	183	181	179	177	176	174	172	170
38	206	204	202	200	198	196	194	192	190	188	186	184	182	180	179	177	175	173	171	169	168
39	203	201	199	197	195	193	191	189	187	185	183	181	180	178	176	174	172	171	169	167	165
40	200	198	196	194	192	190	188	186	184	183	181	179	177	175	173	172	170	168	166	165	163

Table 4. Density ρ of water (in kg/m³) as a function of temperature and salinity (after *Millero and Poisson*, 1981), here reported as σ = ρ − 1000.

Temp. (°C)	Salinity																				
	0	2	4	6	8	10	12	14	16	18	20	22	24	26	28	30	32	34	36	38	40
−1	−0.16										16.01	17.63	19.25	20.86	22.48	24.10	25.72	27.34	28.96	30.59	32.21
0	−0.10	1.48	3.10	4.72	6.34	7.95	9.57	11.18	12.79	14.40	16.01	17.62	19.24	20.85	22.46	24.07	25.68	27.30	28.91	30.53	32.15
1	−0.06	1.53	3.15	4.76	6.37	7.98	9.58	11.19	12.79	14.40	16.00	17.61	19.21	20.81	22.42	24.03	25.63	27.24	28.85	30.46	32.07
2	−0.03	1.56	3.17	4.78	6.38	7.98	9.58	11.18	12.78	14.38	15.97	17.57	19.17	20.77	22.37	23.97	25.57	27.17	28.77	30.38	31.98
3	−0.03	1.58	3.18	4.78	6.38	7.97	9.57	11.16	12.75	14.34	15.93	17.52	19.12	20.71	22.30	23.90	25.49	27.09	28.68	30.28	31.88
4	−0.03	1.58	3.18	4.77	6.36	7.95	9.53	11.12	12.71	14.29	15.88	17.46	19.05	20.64	22.22	23.81	25.40	26.99	28.58	30.17	31.77
5	−0.03	1.57	3.16	4.74	6.33	7.91	9.49	11.07	12.65	14.23	15.81	17.39	18.97	20.55	22.13	23.71	25.30	26.88	28.47	30.06	31.64
6	−0.06	1.54	3.12	4.70	6.28	7.85	9.43	11.00	12.58	14.15	15.72	17.30	18.87	20.45	22.03	23.60	25.18	26.76	28.34	29.93	31.51
7	−0.10	1.49	3.07	4.64	6.21	7.78	9.35	10.92	12.49	14.06	15.63	17.20	18.77	20.34	21.91	23.48	25.06	26.63	28.21	29.78	31.36
8	−0.15	1.43	3.00	4.57	6.14	7.70	9.27	10.83	12.39	13.96	15.52	17.09	18.65	20.22	21.78	23.35	24.92	26.49	28.06	29.63	31.21
9	−0.22	1.36	2.92	4.49	6.05	7.61	9.17	10.73	12.28	13.84	15.40	16.96	18.52	20.08	21.64	23.21	24.77	26.34	27.90	29.47	31.04
10	−0.30	1.27	2.83	4.39	5.95	7.50	9.05	10.61	12.16	13.72	15.27	16.82	18.38	19.94	21.49	23.05	24.61	26.17	27.73	29.30	30.86
11	−0.39	1.17	2.73	4.28	5.83	7.38	8.93	10.48	12.03	13.58	15.13	16.68	18.23	19.78	21.33	22.89	24.44	26.00	27.56	29.11	30.68
12	−0.50	1.06	2.61	4.16	5.70	7.25	8.79	10.34	11.88	13.43	14.97	16.52	18.06	19.61	21.16	22.71	24.26	25.81	27.37	28.92	30.48
13	−0.62	0.93	2.48	4.02	5.57	7.11	8.65	10.19	11.73	13.27	14.81	16.35	17.89	19.43	20.98	22.52	24.07	25.62	27.17	28.72	30.27
14	−0.75	0.80	2.34	3.88	5.41	6.95	8.49	10.02	11.56	13.09	14.63	16.17	17.71	19.25	20.79	22.33	23.87	25.42	26.96	28.51	30.06
15	−0.90	0.65	2.18	3.72	5.25	6.78	8.32	9.85	11.38	12.91	14.44	15.98	17.51	19.05	20.58	22.12	23.66	25.20	26.74	28.29	29.83
16	−1.06	0.49	2.02	3.55	5.08	6.61	8.13	9.66	11.19	12.72	14.25	15.78	17.31	18.84	20.37	21.91	23.44	24.98	26.52	28.06	29.60
17	−1.22	0.31	1.84	3.37	4.89	6.42	7.94	9.47	10.99	12.51	14.04	15.57	17.09	18.62	20.15	21.68	23.21	24.75	26.28	27.82	29.36
18	−1.40	0.13	1.66	3.18	4.70	6.22	7.74	9.26	10.78	12.30	13.82	15.35	16.87	18.39	19.92	21.45	22.98	24.51	26.04	27.57	29.11
19	−1.59	−0.06	1.46	2.98	4.49	6.01	7.53	9.04	10.56	12.08	13.60	15.12	16.64	18.16	19.68	21.21	22.73	24.26	25.79	27.32	28.85
20	−1.79	−0.27	1.25	2.77	4.28	5.79	7.31	8.82	10.33	11.85	13.36	14.88	16.40	17.91	19.43	20.95	22.48	24.00	25.53	27.05	28.58
21	−2.01	−0.48	1.03	2.54	4.05	5.56	7.07	8.58	10.09	11.61	13.12	14.63	16.14	17.66	19.18	20.69	22.21	23.73	25.26	26.78	28.31
22	−2.23	−0.71	0.80	2.31	3.82	5.33	6.83	8.34	9.85	11.35	12.86	14.37	15.88	17.40	18.91	20.43	21.94	23.46	24.98	26.50	28.02
23	−2.46	−0.94	0.56	2.07	3.57	5.08	6.58	8.09	9.59	11.10	12.60	14.11	15.62	17.13	18.64	20.15	21.66	23.18	24.69	26.21	27.73
24	−2.70	−1.19	0.32	1.82	3.32	4.82	6.32	7.82	9.32	10.83	12.33	13.83	15.34	16.85	18.35	19.86	21.37	22.89	24.40	25.92	27.43
25	−2.95	−1.44	0.06	1.56	3.06	4.56	6.05	7.55	9.05	10.55	12.05	13.55	15.05	16.56	18.06	19.57	21.08	22.59	24.10	25.61	27.13
26	−3.21	−1.71	−0.21	1.29	2.79	4.28	5.78	7.27	8.77	10.26	11.76	13.26	14.76	16.26	17.76	19.27	20.77	22.28	23.79	25.30	26.81
27	−3.48	−1.98	−0.48	1.01	2.50	4.00	5.49	6.98	8.48	9.97	11.47	12.96	14.46	15.96	17.46	18.96	20.46	21.97	23.47	24.98	26.49
28	−3.76	−2.26	−0.77	0.72	2.21	3.71	5.20	6.69	8.18	9.67	11.16	12.65	14.15	15.65	17.14	18.64	20.14	21.65	23.15	24.65	26.16
29	−4.05	−2.55	−1.06	0.43	1.92	3.40	4.89	6.38	7.87	9.36	10.85	12.34	13.83	15.33	16.82	18.32	19.82	21.32	22.82	24.32	25.83
30	−4.35	−2.85	−1.36	0.12	1.61	3.10	4.58	6.07	7.55	9.04	10.53	12.02	13.51	15.00	16.49	17.99	19.48	20.98	22.48	23.98	25.48
31	−4.66	−3.16	−1.67	−0.19	1.29	2.78	4.26	5.74	7.23	8.71	10.20	11.69	13.17	14.66	16.15	17.65	19.14	20.64	22.13	23.63	25.13
32	−4.97	−3.48	−1.99	−0.51	0.97	2.45	3.93	5.41	6.90	8.38	9.86	11.35	12.83	14.32	15.81	17.30	18.79	20.28	21.78	23.28	24.77
33	−5.29	−3.80	−2.32	−0.84	0.64	2.12	3.60	5.08	6.56	8.04	9.52	11.00	12.49	13.97	15.46	16.95	18.43	19.93	21.42	22.91	24.41
34	−5.62	−4.14	−2.66	−1.18	0.30	1.78	3.26	4.73	6.21	7.69	9.17	10.65	12.13	13.61	15.10	16.58	18.07	19.56	21.05	22.54	24.04
35	−5.96	−4.48	−3.00	−1.52	−0.05	1.43	2.90	4.38	5.86	7.33	8.81	10.29	11.77	13.25	14.73	16.22	17.70	19.19	20.68	22.17	23.66
36	−6.31	−4.83	−3.35	−1.87	−0.40	1.07	2.55	4.02	5.49	6.97	8.44	9.92	11.40	12.88	14.36	15.84	17.33	18.81	20.30	21.79	23.28
37	−6.67	−5.18	−3.71	−2.23	−0.76	0.71	2.18	3.65	5.13	6.60	8.07	9.55	11.02	12.50	13.98	15.46	16.94	18.43	19.91	21.40	22.89
38	−7.03	−5.55	−4.07	−2.60	−1.13	0.34	1.81	3.28	4.75	6.22	7.69	9.17	10.64	12.12	13.59	15.07	16.55	18.03	19.52	21.00	22.49
39	−7.40	−5.92	−4.45	−2.98	−1.51	−0.04	1.43	2.90	4.37	5.84	7.31	8.78	10.25	11.73	13.20	14.68	16.16	17.64	19.12	20.60	22.09
40	−7.78	−6.30	−4.83	−3.36	−1.89	−0.42	1.04	2.51	3.98	5.45	6.91	8.39	9.86	11.33	12.80	14.28	15.75	17.23	18.71	20.20	21.68

Table 5. Solubility of carbon dioxide (K_0, in mmol/(kg-soln)·atm of pure $CO_{2,gas}$) in water as a function of temperature and salinity (after *Weiss*, 1974).

Temp. (°C)	\multicolumn{21}{c}{Salinity}																			
	0	2	4	6	8	10	12	14	16	18	20	22	24	26	28	30	32	34	36	38	40
0	77.58	76.65	75.73	74.83	73.94	73.05	72.18	71.32	70.47	69.63	68.80	67.98	67.16	66.36	65.57	64.79	64.01	63.25	62.49	61.75	61.01
1	74.58	73.69	72.81	71.95	71.09	70.24	69.41	68.58	67.77	66.96	66.16	65.38	64.60	63.83	63.07	62.32	61.58	60.85	60.12	59.41	58.70
2	71.74	70.88	70.04	69.21	68.39	67.58	66.78	65.99	65.21	64.43	63.67	62.92	62.17	61.43	60.71	59.99	59.27	58.57	57.88	57.19	56.51
3	69.04	68.23	67.42	66.63	65.84	65.06	64.29	63.53	62.78	62.04	61.31	60.59	59.87	59.16	58.46	57.77	57.09	56.42	55.75	55.09	54.44
4	66.49	65.71	64.94	64.17	63.42	62.67	61.93	61.21	60.49	59.77	59.07	58.38	57.69	57.01	56.34	55.68	55.02	54.38	53.74	53.10	52.48
5	64.07	63.32	62.58	61.84	61.12	60.40	59.70	59.00	58.31	57.62	56.95	56.28	55.62	54.97	54.33	53.69	53.06	52.44	51.82	51.22	50.62
6	61.77	61.05	60.34	59.63	58.94	58.25	57.57	56.90	56.24	55.58	54.93	54.29	53.66	53.03	52.41	51.80	51.20	50.60	50.01	49.43	48.85
7	59.59	58.90	58.21	57.54	56.87	56.21	55.56	54.91	54.27	53.64	53.02	52.40	51.80	51.19	50.60	50.01	49.43	48.86	48.29	47.73	47.18
8	57.52	56.85	56.19	55.54	54.90	54.27	53.64	53.02	52.41	51.80	51.20	50.61	50.03	49.45	48.88	48.31	47.75	47.20	46.66	46.12	45.58
9	55.54	54.91	54.27	53.65	53.03	52.42	51.82	51.22	50.63	50.05	49.48	48.91	48.35	47.79	47.24	46.70	46.16	45.63	45.10	44.59	44.07
10	53.67	53.06	52.45	51.85	51.25	50.67	50.09	49.52	48.95	48.39	47.84	47.29	46.75	46.21	45.68	45.16	44.64	44.13	43.63	43.13	42.63
11	51.89	51.30	50.71	50.13	49.56	49.00	48.44	47.89	47.34	46.81	46.27	45.75	45.23	44.71	44.20	43.70	43.20	42.71	42.22	41.74	41.27
12	50.19	49.62	49.06	48.50	47.95	47.41	46.87	46.34	45.82	45.30	44.79	44.28	43.78	43.28	42.79	42.31	41.83	41.36	40.89	40.42	39.97
13	48.57	48.02	47.48	46.95	46.42	45.90	45.38	44.87	44.36	43.87	43.37	42.88	42.40	41.92	41.45	40.98	40.52	40.07	39.62	39.17	38.73
14	47.03	46.50	45.98	45.47	44.96	44.46	43.96	43.47	42.98	42.50	42.02	41.55	41.09	40.63	40.17	39.72	39.28	38.84	38.40	37.97	37.55
15	45.56	45.05	44.55	44.06	43.57	43.08	42.60	42.13	41.66	41.20	40.74	40.28	39.84	39.39	38.96	38.52	38.09	37.67	37.25	36.84	36.43
16	44.16	43.67	43.19	42.71	42.24	41.77	41.31	40.85	40.40	39.95	39.51	39.08	38.64	38.22	37.79	37.38	36.96	36.55	36.15	35.75	35.36
17	42.82	42.35	41.89	41.43	40.97	40.52	40.08	39.64	39.20	38.77	38.34	37.92	37.51	37.09	36.69	36.28	35.88	35.49	35.10	34.72	34.33
18	41.55	41.09	40.64	40.20	39.76	39.33	38.90	38.47	38.05	37.64	37.23	36.82	36.42	36.02	35.63	35.24	34.86	34.48	34.10	33.73	33.36
19	40.33	39.89	39.46	39.03	38.61	38.19	37.77	37.37	36.96	36.56	36.16	35.77	35.38	35.00	34.62	34.25	33.87	33.51	33.14	32.79	32.43
20	39.16	38.74	38.32	37.91	37.50	37.10	36.70	36.31	35.92	35.53	35.15	34.77	34.39	34.02	33.66	33.30	32.94	32.58	32.23	31.89	31.54
21	38.05	37.64	37.24	36.84	36.45	36.06	35.67	35.29	34.92	34.54	34.17	33.81	33.45	33.09	32.74	32.39	32.04	31.70	31.36	31.03	30.69
22	36.99	36.59	36.21	35.82	35.44	35.07	34.69	34.33	33.96	33.60	33.25	32.89	32.54	32.20	31.86	31.52	31.19	30.86	30.53	30.21	29.89
23	35.97	35.59	35.21	34.84	34.48	34.12	33.76	33.40	33.05	32.70	32.36	32.02	31.68	31.35	31.02	30.69	30.37	30.05	29.73	29.42	29.11
24	34.99	34.63	34.27	33.91	33.56	33.21	32.86	32.52	32.18	31.84	31.51	31.18	30.85	30.53	30.21	29.90	29.59	29.28	28.97	28.67	28.37
25	34.06	33.71	33.36	33.01	32.67	32.33	32.00	31.67	31.34	31.02	30.70	30.38	30.06	29.75	29.44	29.14	28.84	28.54	28.24	27.95	27.66
26	33.17	32.83	32.49	32.16	31.83	31.50	31.18	30.86	30.54	30.23	29.92	29.61	29.31	29.01	28.71	28.41	28.12	27.83	27.55	27.27	26.99
27	32.31	31.98	31.66	31.34	31.02	30.70	30.39	30.08	29.78	29.47	29.17	28.88	28.58	28.29	28.00	27.72	27.44	27.16	26.88	26.61	26.34
28	31.49	31.18	30.86	30.55	30.24	29.94	29.64	29.34	29.04	28.75	28.46	28.17	27.89	27.61	27.33	27.05	26.78	26.51	26.24	25.98	25.72
29	30.71	30.40	30.10	29.80	29.50	29.20	28.91	28.62	28.34	28.06	27.78	27.50	27.22	26.95	26.68	26.42	26.15	25.89	25.63	25.38	25.12
30	29.95	29.66	29.36	29.07	28.79	28.50	28.22	27.94	27.66	27.39	27.12	26.85	26.59	26.32	26.06	25.80	25.55	25.30	25.05	24.80	24.55
31	29.23	28.94	28.66	28.38	28.10	27.83	27.55	27.29	27.02	26.75	26.49	26.23	25.97	25.72	25.47	25.22	24.97	24.73	24.49	24.25	24.01
32	28.54	28.26	27.99	27.72	27.45	27.18	26.92	26.66	26.40	26.14	25.89	25.64	25.39	25.14	24.90	24.66	24.42	24.18	23.95	23.72	23.49
33	27.87	27.60	27.34	27.08	26.82	26.56	26.31	26.05	25.80	25.56	25.31	25.07	24.83	24.59	24.35	24.12	23.89	23.66	23.43	23.21	22.98
34	27.23	26.97	26.72	26.46	26.21	25.96	25.72	25.47	25.23	24.99	24.76	24.52	24.29	24.06	23.83	23.60	23.38	23.16	22.94	22.72	22.50
35	26.62	26.37	26.12	25.88	25.63	25.39	25.15	24.92	24.68	24.45	24.22	24.00	23.77	23.55	23.33	23.11	22.89	22.68	22.46	22.25	22.04
36	26.03	25.79	25.55	25.31	25.08	24.84	24.61	24.38	24.16	23.93	23.71	23.49	23.27	23.06	22.84	22.63	22.42	22.21	22.01	21.80	21.60
37	25.46	25.23	25.00	24.77	24.54	24.31	24.09	23.87	23.65	23.43	23.22	23.01	22.80	22.59	22.38	22.17	21.97	21.77	21.57	21.37	21.18
38	24.92	24.69	24.47	24.24	24.02	23.81	23.59	23.38	23.17	22.96	22.75	22.54	22.34	22.14	21.93	21.74	21.54	21.34	21.15	20.96	20.77
39	24.39	24.17	23.96	23.74	23.53	23.32	23.11	22.90	22.70	22.50	22.29	22.10	21.90	21.70	21.51	21.31	21.12	20.93	20.75	20.56	20.38
40	23.89	23.68	23.47	23.26	23.05	22.85	22.65	22.45	22.25	22.05	21.86	21.67	21.47	21.28	21.10	20.91	20.73	20.54	20.36	20.18	20.00

Table 6. First dissociation constant [ln (K_1/k^0)] of carbonic acid [CO_2(aq) + H_2CO_3] in seawater as a function of temperature and salinity (after *Roy et al.*, 1993, 1996; modified after *DOE*, 1994).

Temp. (°C)	Salinity 5	10	15	20	25	30	31	32	33	34	35	36	37	38	39	40
0	−14.47	−14.40	−14.31	−14.23	−14.16	−14.11	−14.10	−14.09	−14.08	−14.07	−14.06	−14.06	−14.05	−14.05	−14.04	−14.04
1	−14.45	−14.37	−14.28	−14.20	−14.13	−14.08	−14.07	−14.06	−14.05	−14.05	−14.04	−14.03	−14.03	−14.02	−14.02	−14.01
2	−14.42	−14.34	−14.26	−14.18	−14.11	−14.05	−14.05	−14.04	−14.03	−14.02	−14.01	−14.01	−14.00	−14.00	−13.99	−13.99
3	−14.40	−14.32	−14.23	−14.15	−14.08	−14.03	−14.02	−14.01	−14.00	−14.00	−13.99	−13.98	−13.98	−13.97	−13.97	−13.96
4	−14.37	−14.30	−14.21	−14.13	−14.06	−14.00	−14.00	−13.99	−13.98	−13.97	−13.96	−13.96	−13.95	−13.95	−13.94	−13.94
5	−14.35	−14.27	−14.18	−14.10	−14.04	−13.98	−13.97	−13.96	−13.95	−13.95	−13.94	−13.93	−13.93	−13.92	−13.92	−13.91
6	−14.33	−14.25	−14.16	−14.08	−14.01	−13.96	−13.95	−13.94	−13.93	−13.92	−13.91	−13.91	−13.90	−13.90	−13.89	−13.89
7	−14.30	−14.22	−14.14	−14.05	−13.99	−13.93	−13.92	−13.91	−13.91	−13.90	−13.89	−13.88	−13.88	−13.87	−13.87	−13.86
8	−14.28	−14.20	−14.11	−14.03	−13.96	−13.91	−13.90	−13.89	−13.88	−13.87	−13.87	−13.86	−13.85	−13.85	−13.84	−13.84
9	−14.25	−14.17	−14.09	−14.01	−13.94	−13.88	−13.87	−13.87	−13.86	−13.85	−13.84	−13.84	−13.83	−13.82	−13.82	−13.82
10	−14.23	−14.15	−14.06	−13.98	−13.92	−13.86	−13.85	−13.84	−13.83	−13.83	−13.82	−13.81	−13.81	−13.80	−13.80	−13.79
11	−14.21	−14.13	−14.04	−13.96	−13.89	−13.84	−13.83	−13.82	−13.81	−13.80	−13.80	−13.79	−13.78	−13.78	−13.77	−13.77
12	−14.18	−14.10	−14.02	−13.94	−13.87	−13.81	−13.80	−13.79	−13.79	−13.78	−13.77	−13.77	−13.76	−13.75	−13.75	−13.74
13	−14.16	−14.08	−13.99	−13.91	−13.85	−13.79	−13.78	−13.77	−13.76	−13.76	−13.75	−13.74	−13.74	−13.73	−13.73	−13.72
14	−14.14	−14.06	−13.97	−13.89	−13.82	−13.77	−13.76	−13.75	−13.74	−13.73	−13.73	−13.72	−13.71	−13.71	−13.70	−13.70
15	−14.12	−14.04	−13.95	−13.87	−13.80	−13.74	−13.73	−13.73	−13.72	−13.71	−13.70	−13.70	−13.69	−13.69	−13.68	−13.68
16	−14.09	−14.01	−13.93	−13.85	−13.78	−13.72	−13.71	−13.70	−13.70	−13.69	−13.68	−13.67	−13.67	−13.66	−13.66	−13.65
17	−14.07	−13.99	−13.90	−13.82	−13.75	−13.70	−13.69	−13.68	−13.67	−13.67	−13.66	−13.65	−13.65	−13.64	−13.64	−13.63
18	−14.05	−13.97	−13.88	−13.80	−13.73	−13.68	−13.67	−13.66	−13.65	−13.65	−13.64	−13.63	−13.62	−13.62	−13.61	−13.61
19	−14.03	−13.95	−13.86	−13.78	−13.71	−13.66	−13.65	−13.64	−13.63	−13.62	−13.61	−13.61	−13.60	−13.60	−13.59	−13.59
20	−14.01	−13.93	−13.84	−13.76	−13.69	−13.63	−13.62	−13.61	−13.61	−13.60	−13.59	−13.59	−13.58	−13.57	−13.57	−13.56
21	−13.98	−13.90	−13.82	−13.74	−13.67	−13.61	−13.60	−13.59	−13.58	−13.58	−13.57	−13.56	−13.56	−13.55	−13.55	−13.54
22	−13.96	−13.88	−13.79	−13.71	−13.65	−13.59	−13.58	−13.57	−13.56	−13.56	−13.55	−13.54	−13.54	−13.53	−13.53	−13.52
23	−13.94	−13.86	−13.77	−13.69	−13.62	−13.57	−13.56	−13.55	−13.54	−13.53	−13.53	−13.52	−13.51	−13.51	−13.50	−13.50
24	−13.92	−13.84	−13.75	−13.67	−13.60	−13.55	−13.54	−13.53	−13.52	−13.51	−13.51	−13.50	−13.49	−13.49	−13.48	−13.48
25	−13.90	−13.82	−13.73	−13.65	−13.58	−13.53	−13.52	−13.51	−13.50	−13.49	−13.48	−13.48	−13.47	−13.47	−13.46	−13.46
26	−13.88	−13.80	−13.71	−13.63	−13.56	−13.51	−13.50	−13.49	−13.48	−13.47	−13.46	−13.46	−13.45	−13.45	−13.44	−13.44
27	−13.86	−13.78	−13.69	−13.61	−13.54	−13.48	−13.48	−13.47	−13.46	−13.45	−13.44	−13.44	−13.43	−13.42	−13.42	−13.42
28	−13.84	−13.76	−13.67	−13.59	−13.52	−13.46	−13.45	−13.45	−13.44	−13.43	−13.42	−13.42	−13.41	−13.40	−13.40	−13.39
29	−13.82	−13.74	−13.65	−13.57	−13.50	−13.44	−13.43	−13.43	−13.42	−13.41	−13.40	−13.40	−13.39	−13.38	−13.38	−13.37
30	−13.80	−13.72	−13.63	−13.55	−13.48	−13.42	−13.41	−13.40	−13.40	−13.39	−13.38	−13.37	−13.37	−13.36	−13.36	−13.35
31	−13.78	−13.70	−13.61	−13.53	−13.46	−13.40	−13.39	−13.38	−13.37	−13.37	−13.36	−13.35	−13.35	−13.34	−13.34	−13.33
32	−13.76	−13.68	−13.59	−13.51	−13.44	−13.38	−13.37	−13.36	−13.36	−13.35	−13.34	−13.33	−13.33	−13.32	−13.32	−13.31
33	−13.74	−13.66	−13.57	−13.49	−13.42	−13.36	−13.35	−13.34	−13.34	−13.33	−13.32	−13.31	−13.31	−13.30	−13.30	−13.29
34	−13.72	−13.64	−13.55	−13.47	−13.40	−13.34	−13.33	−13.33	−13.32	−13.31	−13.30	−13.30	−13.29	−13.28	−13.28	−13.27
35	−13.70	−13.62	−13.53	−13.45	−13.38	−13.32	−13.31	−13.31	−13.30	−13.29	−13.28	−13.28	−13.27	−13.26	−13.26	−13.25
36	−13.68	−13.60	−13.51	−13.43	−13.36	−13.30	−13.30	−13.29	−13.28	−13.27	−13.26	−13.26	−13.25	−13.24	−13.24	−13.24
37	−13.66	−13.58	−13.49	−13.41	−13.34	−13.29	−13.28	−13.27	−13.26	−13.25	−13.24	−13.24	−13.23	−13.23	−13.22	−13.22
38	−13.64	−13.56	−13.47	−13.39	−13.32	−13.27	−13.26	−13.25	−13.24	−13.23	−13.22	−13.22	−13.21	−13.21	−13.20	−13.20
39	−13.62	−13.54	−13.45	−13.37	−13.30	−13.25	−13.24	−13.23	−13.22	−13.21	−13.21	−13.20	−13.19	−13.19	−13.18	−13.18
40	−13.60	−13.52	−13.43	−13.35	−13.28	−13.23	−13.22	−13.21	−13.20	−13.19	−13.19	−13.18	−13.17	−13.17	−13.16	−13.16

Table 7. Second dissociation constant [ln (K_2/k^0)] of carbonic acid [$CO_2(aq) + H_2CO_3$] in seawater as a function of temperature and salinity (after *Roy et al.*, 1993, 1996; modified after *DOE*, 1994).

Temp. (°C)	5	10	15	20	25	30	31	32	33	34	35	36	37	38	39	40
									Salinity							
0	-22.59	-22.38	-22.19	-22.01	-21.85	-21.72	-21.69	-21.67	-21.65	-21.63	-21.61	-21.59	-21.57	-21.55	-21.53	-21.51
1	-22.55	-22.34	-22.14	-21.96	-21.81	-21.67	-21.65	-21.62	-21.60	-21.58	-21.56	-21.54	-21.52	-21.50	-21.48	-21.47
2	-22.50	-22.29	-22.10	-21.92	-21.76	-21.63	-21.60	-21.58	-21.56	-21.53	-21.51	-21.49	-21.47	-21.46	-21.44	-21.42
3	-22.46	-22.25	-22.05	-21.87	-21.72	-21.58	-21.56	-21.53	-21.51	-21.49	-21.47	-21.45	-21.43	-21.41	-21.39	-21.38
4	-22.41	-22.21	-22.01	-21.83	-21.67	-21.54	-21.51	-21.49	-21.47	-21.44	-21.42	-21.40	-21.38	-21.37	-21.35	-21.33
5	-22.37	-22.16	-21.96	-21.78	-21.63	-21.49	-21.47	-21.44	-21.42	-21.40	-21.38	-21.36	-21.34	-21.32	-21.30	-21.29
6	-22.33	-22.12	-21.92	-21.74	-21.58	-21.45	-21.42	-21.40	-21.38	-21.36	-21.33	-21.31	-21.30	-21.28	-21.26	-21.24
7	-22.29	-22.08	-21.88	-21.70	-21.54	-21.40	-21.38	-21.36	-21.33	-21.31	-21.29	-21.27	-21.25	-21.23	-21.21	-21.20
8	-22.24	-22.03	-21.83	-21.65	-21.50	-21.36	-21.34	-21.31	-21.29	-21.27	-21.25	-21.23	-21.21	-21.19	-21.17	-21.15
9	-22.20	-21.99	-21.79	-21.61	-21.45	-21.32	-21.29	-21.27	-21.25	-21.22	-21.20	-21.18	-21.16	-21.15	-21.13	-21.11
10	-22.16	-21.95	-21.75	-21.57	-21.41	-21.27	-21.25	-21.23	-21.20	-21.18	-21.16	-21.14	-21.12	-21.10	-21.08	-21.07
11	-22.12	-21.91	-21.71	-21.53	-21.37	-21.23	-21.21	-21.18	-21.16	-21.14	-21.12	-21.10	-21.08	-21.06	-21.04	-21.02
12	-22.08	-21.87	-21.66	-21.48	-21.33	-21.19	-21.17	-21.14	-21.12	-21.10	-21.08	-21.06	-21.04	-21.02	-21.00	-20.98
13	-22.03	-21.82	-21.62	-21.44	-21.28	-21.15	-21.12	-21.10	-21.08	-21.05	-21.03	-21.01	-20.99	-20.97	-20.96	-20.94
14	-21.99	-21.78	-21.58	-21.40	-21.24	-21.11	-21.08	-21.06	-21.03	-21.01	-20.99	-20.97	-20.95	-20.93	-20.92	-20.90
15	-21.95	-21.74	-21.54	-21.36	-21.20	-21.06	-21.04	-21.02	-20.99	-20.97	-20.95	-20.93	-20.91	-20.89	-20.87	-20.86
16	-21.91	-21.70	-21.50	-21.32	-21.16	-21.02	-21.00	-20.98	-20.95	-20.93	-20.91	-20.89	-20.87	-20.85	-20.83	-20.82
17	-21.87	-21.66	-21.46	-21.28	-21.12	-20.98	-20.96	-20.93	-20.91	-20.89	-20.87	-20.85	-20.83	-20.81	-20.79	-20.77
18	-21.83	-21.62	-21.42	-21.24	-21.08	-20.94	-20.92	-20.89	-20.87	-20.85	-20.83	-20.81	-20.79	-20.77	-20.75	-20.73
19	-21.79	-21.58	-21.38	-21.20	-21.04	-20.90	-20.88	-20.85	-20.83	-20.81	-20.79	-20.77	-20.75	-20.73	-20.71	-20.69
20	-21.76	-21.54	-21.34	-21.16	-21.00	-20.86	-20.84	-20.81	-20.79	-20.77	-20.75	-20.73	-20.71	-20.69	-20.67	-20.65
21	-21.72	-21.50	-21.30	-21.12	-20.96	-20.82	-20.80	-20.77	-20.75	-20.73	-20.71	-20.69	-20.67	-20.65	-20.63	-20.61
22	-21.68	-21.46	-21.26	-21.08	-20.92	-20.78	-20.76	-20.73	-20.71	-20.69	-20.67	-20.65	-20.63	-20.61	-20.59	-20.57
23	-21.64	-21.43	-21.22	-21.04	-20.88	-20.74	-20.72	-20.70	-20.67	-20.65	-20.63	-20.61	-20.59	-20.57	-20.55	-20.53
24	-21.60	-21.39	-21.19	-21.00	-20.84	-20.71	-20.68	-20.66	-20.63	-20.61	-20.59	-20.57	-20.55	-20.53	-20.51	-20.49
25	-21.56	-21.35	-21.15	-20.96	-20.80	-20.67	-20.64	-20.62	-20.59	-20.57	-20.55	-20.53	-20.51	-20.49	-20.47	-20.46
26	-21.53	-21.31	-21.11	-20.93	-20.77	-20.63	-20.60	-20.58	-20.56	-20.53	-20.51	-20.49	-20.47	-20.45	-20.43	-20.42
27	-21.49	-21.27	-21.07	-20.89	-20.73	-20.59	-20.56	-20.54	-20.52	-20.50	-20.47	-20.45	-20.43	-20.41	-20.40	-20.38
28	-21.45	-21.24	-21.03	-20.85	-20.69	-20.55	-20.53	-20.50	-20.48	-20.46	-20.44	-20.41	-20.40	-20.38	-20.36	-20.34
29	-21.42	-21.20	-21.00	-20.81	-20.65	-20.51	-20.49	-20.47	-20.44	-20.42	-20.40	-20.38	-20.36	-20.34	-20.32	-20.30
30	-21.38	-21.16	-20.96	-20.78	-20.62	-20.48	-20.45	-20.43	-20.40	-20.38	-20.36	-20.34	-20.32	-20.30	-20.28	-20.27
31	-21.34	-21.13	-20.92	-20.74	-20.58	-20.44	-20.41	-20.39	-20.37	-20.34	-20.32	-20.30	-20.28	-20.26	-20.25	-20.23
32	-21.31	-21.09	-20.89	-20.70	-20.54	-20.40	-20.38	-20.35	-20.33	-20.31	-20.29	-20.27	-20.25	-20.23	-20.21	-20.19
33	-21.27	-21.05	-20.85	-20.67	-20.51	-20.37	-20.34	-20.32	-20.29	-20.27	-20.25	-20.23	-20.21	-20.19	-20.17	-20.15
34	-21.24	-21.02	-20.81	-20.63	-20.47	-20.33	-20.30	-20.28	-20.26	-20.23	-20.21	-20.19	-20.17	-20.15	-20.13	-20.12
35	-21.20	-20.98	-20.78	-20.59	-20.43	-20.29	-20.27	-20.24	-20.22	-20.20	-20.18	-20.16	-20.14	-20.12	-20.10	-20.08
36	-21.16	-20.95	-20.74	-20.56	-20.40	-20.26	-20.23	-20.21	-20.19	-20.16	-20.14	-20.12	-20.10	-20.08	-20.06	-20.04
37	-21.13	-20.91	-20.71	-20.52	-20.36	-20.22	-20.20	-20.17	-20.15	-20.13	-20.10	-20.08	-20.06	-20.04	-20.03	-20.01
38	-21.10	-20.88	-20.67	-20.49	-20.33	-20.19	-20.16	-20.14	-20.11	-20.09	-20.07	-20.05	-20.03	-20.01	-19.99	-19.97
39	-21.06	-20.84	-20.64	-20.45	-20.29	-20.15	-20.13	-20.10	-20.08	-20.06	-20.03	-20.01	-19.99	-19.97	-19.96	-19.94
40	-21.03	-20.81	-20.60	-20.42	-20.26	-20.12	-20.09	-20.07	-20.04	-20.02	-20.00	-19.98	-19.96	-19.94	-19.92	-19.90

Index